About Astronomia Nova

Astronomia Nova is a work of astonishing originality, and it stands, with Copernicus's *De Revolutionibus* and Newton's *Principia,* as one of the founding texts of the scientific revolution. When Kepler boldly chose the words *Astronomia Nova* for his title, he must have had some sense of how apt they were. Although the book is not lacking in historical antecedents (as Kepler himself was keenly aware), it is in most respects quite unlike anything that had appeared before. Astronomy is no longer seen as a deductive science, founded mainly upon geometry, whose aim is to construct an ideal system that matches the appearances of the celestial motions as closely as possible. Instead, it is an adventure in which the human being uses all the means at his disposal to explore the creation in which God has placed him as His image. Error is no longer equated with failure, but is seen as an indication of the way to the truth. Thus the rules of the game are no longer fixed, but are to be discovered in the playing. And so the hypotheses of the ancients, the uniform circular motions previously thought to be indispensable to astronomy, were tested, found wanting, and rejected. The question was no longer "How can the appearances be accounted for?" but "How does God make things move?"

In undertaking to subvert more than two thousand years of astronomical theory and tradition, Kepler knew he had set himself quite a challenge. Had he just presented his principles and theories in magisterial style, in the manner of Copernicus and Ptolemy, he would have appeared to be just another innovator, in the same cast as Bruno and Patrizi, whom he despised. Instead, he chose the profoundly rhetorical approach of taking us through a step-by-step account (heavily edited and dramatized, to be sure) of his "battle with Mars," aimed at convincing us that no conclusion other than his was possible. In the course of his tale, he sprinkles in images from literature and daily life: meat being squeezed in a sausage casing, a chaste maiden walking the streets behind a prostitute, pretzels, battles being fought on many fronts, ferryboats on the Danube, amusement park rides, biblical scenes, and lines from Horace and Virgil. Clearly, this is no ordinary astronomical treatise!

In the course of this adventure, Kepler shows us that the old geometrical principle of uniformity has to go: it was, he says, a geometrical approximation of a physical principle, according to which planetary speeds are governed by their distance from the sun, the source of driving power. Here we see the origin of what we now know as "Kepler's Second Law," which Kepler soon converts into the more familiar area/time formulation for the sake of easier computation. He then uses this physical principle to explore the shape of Mars's orbit: the planet takes more time over the parts where the distances and areas are greater. This leads to the discovery that the orbit must be squeezed in at the sides: it has to be an oval. But what sort of oval? There was no reason to suppose that it would be some neat curve, such as an ellipse; rather, Kepler expected it to be a complex shape resulting from the independent efforts of the solar whirling force and the planet's own striving towards and away from the sun. In the end, he shows, using both mathematics and physical causes, that the resulting curve could be nothing else but the ellipse itself, a form that had been ready at hand all along.

Kepler is remembered today chiefly for the mathematical laws of planetary motion that bear his name. His idiosyncratic physics has not stood the test of time. Nevertheless, his fundamental idea—one might call it his faith—that planetary astronomy must be based upon physics led ultimately to Newton's great synthesis of terrestrial and celestial motions.

Astronomia Nova

Johannes Kepler

Astronomia Nova

New Revised Edition

translated by William H. Donahue

Green Lion Press

Santa Fe, New Mexico

Manufactured in the United States of America.

Published by Green Lion Press.

www.greenlion.com

Green Lion Press books are printed on acid-free paper. Both softbound and clothbound editions have sewn bindings designed to lie flat and allow heavy use by students and researchers. Clothbound editions meet the guidelines for permanence and durability of the Committee on Production Guidelines for Book Longevity of the Council on Library Resources.

Second printing with corrections and minor revisions, 2020.

Printed and bound by McNaughton & Gunn, Saline, Michigan.

Cover design by Dana Densmore and William H. Donahue

Cataloging-in-Publication Data:

title of this book: Astronomia Nova

author: Johannes Kepler

translation from Latin of: Astronomia Nova / by Johannes Kepler

translated by: William H. Donahue

ISBN 978-1-888009-47-7 (sewn softcover binding)

ISBN 978-1-888009-48-4 (sewn hardcover binding)

1. History of Science. 2. Astronomy.

I. Kepler, Johannes (1571–1630) II. Donahue, William H. (1943–) III. Title.

Library of Congress Control Number 2015912481

CONTENTS

ASTRONOMIA NOVA

PART 1: ON THE COMPARISON OF HYPOTHESES

PART 2: ON THE FIRST INEQUALITY OF THE STAR MARS,
IN IMITATION OF THE ANCIENTS

PART 3: INVESTIGATION OF THE SECOND INEQUALITY;
THAT IS, OF THE MOTIONS OF THE SUN OR EARTH;
OR THE KEY TO A DEEPER ASTRONOMY,
WHEREIN THERE IS MUCH
ON THE PHYSICAL CAUSES OF THE MOTIONS.

PART 4: INVESTIGATION OF THE TRUE MEASURE
OF THE FIRST INEQUALITY
FROM PHYSICAL CAUSES
AND THE AUTHOR'S OWN IDEAS

PART 5: ON THE LATITUDE

Foreword

Owen Gingerich

Kepler's *Astronomia nova*, together with Copernicus's *De revolutionibus* and Newton's *Principia*, belongs in the select group of the most important books of the Scientific Revolution. Kepler's formidable treatise contains the first statement of elliptical orbits—a radical departure from the previous exclusive pre-eminence of the circle in astronomical hypotheses. The *Astronomia nova* also contains a powerful, but flawed, statement of the law of areas. But perhaps more important, it is the first published account wherein a scientist documents how he has coped with the multiplicity of imperfect data to forge a theory of surpassing accuracy.

For centuries, Kepler and his extraordinary creative genius have been overshadowed by his contemporary, Galileo. Like the *Astronomia nova*, both the *Sidereus nuncius* and the *Dialogo sopra i due massimi sistemi del mondo* played major roles in bringing about the acceptance of the heliocentric world view. Galileo's works are eminently readable and have long been accessible in English translation. Not so for Kepler's pioneering study.

Unlike Galileo's *Dialogo*, which was written in the vernacular and aimed at a general intellectual audience that extended far beyond academia, Kepler's book is a technical treatise written for the specialist in celestial mechanics. Nevertheless, like Galileo's *Dialogo*, the *Astronomia nova* is a polemical work; it is crafted to convince Kepler's readers that his revolutionary solution to the ancient problem of planetary motions is the only viable alternative. He apologizes at the outset for being too prolix, but its expansive presentation serves his purpose. In fact, 80% of the book—including the introduction, the title page, and dedicatory poem—was drafted before he had even arrived at the elliptical shape of the Martian orbit. Kepler touched up the introduction with a marginal reference to the ellipse, and tried, with only partial success, to make the new final chapters seamless with what had gone before. If he stumbled and left out an essential paragraph here or there, as D. T. Whiteside has assured me is so, no one seemed to notice—the rhetoric, planned or unwitting, had the desired effect. When Kepler plaintively sought the reader's pity because one wearying iterative procedure was carried out more than 70 times, Delambre remarked that the complaint was its own reward.

Kepler's *Astronomia nova* is more a book to be mined than to be read. Nevertheless it abounds with Kepler's sly humor. "I am going to give you a clown show," he says as he explains his fumbling attempt to make an observation at a geometrical configuration of Mars that his mentor, Tycho Brahe, had neglected. Elsewhere he writes, "Who would believe it! The hypothesis...goes up in smoke." Chapter 7, entitled "How I First Came to Work on Mars," contains one of Kepler's most important biographical statements, recounting the events that led to his working in Prague for Tycho.

The introduction to Kepler's treatise, salvaged and reworked from his earlier *Mysterium cosmographicum* where it had been censored out by the Tübingen University Senate, is one of the most interesting defenses of the Copernican theory in the entire seventeenth century. In fact, for over three centuries, the introduction to the *Astronomia nova* was the only significant piece of Keplerian writing turned into English—it was partially translated by Thomas Salusbury in his *Mathematical Collections and Translations* of 1661. In those decades when the heliocentric arrangement was still problematic, Kepler's discussion of the relation of Holy Scripture to the Copernican doctrine, found in this introduction, was an influential statement. In its Latin original, it was reprinted over and over with Galileo's *Dialogo*. Undoubtedly Galileo profited from it, though it would have been suicidal for the Italian Catholic to have acknowledged such a Lutheran source.

As long as Latin remained a working language for astronomers, it was unnecessary to translate the *Astronomia nova*. On the other hand, after the seventeenth century, astronomers had little reason to read Kepler's original text except as a historical curiosity. Its style was not easy to grasp—Kepler in a hurry is not a notably clear writer—and his allusions and puns and occasionally idiosyncratic punctuation contribute to the problem. Add to this erroneous diagrams—such as the one in Chapter 16 that Donahue is here the first to notice in print—and the difficulty becomes even more apparent.

The fine Latin edition prepared by Max Caspar in the *Johannes Kepler Gesammelte Werke* series in 1937 provided a good start for those determined to penetrate Kepler's prose. In the late 1960s we achieved a rough English translation with the aid of several Harvard students, most notably Ann Wegner, and the entry in the *Dictionary of Scientific Biography* indicates that a proper translation was on its way. After some years William Donahue challenged us to finish it, saying that otherwise he would undertake the task himself. As events unfolded, it became much the more sensible idea to let him profit at least in a minor way from our rough translation, and indeed, it provided Donahue with an occasional felicitous phrase or insight.

Today, students who are less interested in the specific results of Kepler's "Warfare on Mars" and far more intrigued by the process of scientific discovery can find an archeological treasure-trove in the strata of Kepler's treatise. We see a visionary driven by a search for physical causes, unsatisfied by the purely geometrical "hypotheses" used in the time-honored tool kit of astronomical procedures. As a thorough-going Copernican, Kepler believed the sun simply had to be near the center of the planetary system, for only the sun could serve as a source of the driving power for the planetary motions. Copernicus had already noticed that the nearer a planet was to the sun, the shorter its period—in fact, this must have been one of the primary reasons for considering the sun-centered arrangement. He pointed to the sun "seated upon its throne as a royal governor" and wrote that "only in this arrangement do we find a sure connection between the motion and the size of the orbit." For Kepler, the sun was the seat of a mysterious force, related to magnetism, that not only propelled the planets but which interacted with the planets' own magnetism to guide each one in its path of approach and regression to the hearth of the universe. With this translation a far wider audience can relive Kepler's excitement when he found that an ellipse seemed to answer to such a magnetic configuration. Appropriately, the sun fell at the focus—Kepler's word, its usage derived from its ancient meaning, "hearth."

Kepler, with his enthusiasm and his burning desire to publish, had already begun writing his treatise even before he had discovered the ellipse. Now, in 1605, he hastily finished it. Printing, however, was delayed by dissension with Brahe's heirs. The eccentric Danish observer had died in 1601 after Kepler had worked with him for only ten months. Emperor Rudolph had agreed to buy the observation books for Kepler's use, but had not made good on his promise. Finally, after the heirs were given an opportunity to include their own fulsome introduction, the tall, handsome volume was published in Heidelberg in 1609.

The Warfare on Mars was the anvil on which Brahe's observations were forged into a radical new celestial mechanics. But in those years between 1600 and 1605, Kepler himself passed through the refiner's fire. The youthful speculations of his *Mysterium cosmographicum* were now behind him, and, having acquired the finest treasury of observations that had ever existed, he could no longer be satisfied with the rather imprecise fit between model and data that he had put forward in 1596. His on-going task was to hammer the planetary theory into practical tables, and from the tables to make calculations of daily planetary positions—both a challenge and a drudgery, against which he exclaimed, "Don't sentence me completely to the treadmill of mathematical calculations. Leave me time for philosophical speculations, my sole delight!" And in that framework of cosmological speculation tempered by a newly won respect for the data itself, he completed his *Harmonice mundi*, an extension and refinement of his earlier ideas. In the course of this work he discovered, in 1618, the numerical relationship between the distance of a planet from the sun and its period, today called Kepler's third law. The *Harmonice* was followed by the *Epitome astronomiae Copernicanae*, Kepler's longest work and the clarifying summary of all his discoveries in celestial mechanics. And finally, the *Tabulae Rudolphinae* (1627) and the *Ephemerides* (1617, 1630) appeared.

Meanwhile Kepler had been busy with many other topics. His *Astronomiae pars optica* (1604) had prepared the way for a prompt theoretical treatment of the just-invented telescope, the *Dioptrice* (1611). His thin pamphlet on the six-cornered snowflake (*Strena*, 1611) is considered a foundation work in mineralogy, and his *Stereometria doliorum vinariorum* (1615) is a significant antecedent to the integral calculus. Having heard about logarithms, he independently figured out how to calculate them and published his own table in 1625. Nevertheless, as we survey this prodigious output, the *Astronomia nova* stands as the high-water mark, the achievement where he not only established his own professional credentials, but where he made his most lasting contribution to astronomy. His results, which swept away nearly two millennia of imperfect assumptions about planetary motions, were truly the new astronomy.

Translator's Acknowledgments

During the many years of work on two editions of this translation, I have been helped and encouraged by more people than I could possibly thank here. Preparing this second edition has required much additional work, revising the translation and reformatting, with help from many more people, all of whom merit heartfelt thanks. However, there are a few individuals and organizations that deserve special mention.

Foremost among these is Professor Owen Gingerich, of the Harvard-Smithsonian Center for Astrophysics. Professor Gingerich had hoped that he would be the one to produce the first English translation of this book, and was well on his way to doing so. Yet when I appeared with what was then a rival endeavor, he not only gave over his own hopes very graciously, but also has assisted me over the years in a great many ways. These range from giving me a copy of Ann Wegner's draft translation, through assistance in obtaining financial support and access to research materials, to help in finding a publisher. Indeed, although the translation does not bear his name, it is, in a sense, still very much his project.

Hardly less important, especially in the revision of the second edition, is long time friend and colleague Bruce M. Perry, Tutor at St. John's College, Santa Fe, who is a classicist and thoroughly familiar with Kepler's Latin. When Bruce asked me to compile the notes to his translation of Ptolemy's *Almagest* (now published by Green Lion Press), he offered, in exchange, to review the entire translation, sentence by sentence, a proposal I eagerly accepted. Perry has been an ideal collaborator, both expert and deferential.

A. E. L. Davis also deserves heartfelt thanks for coming to Santa Fe for a week of intense study and discussion of the translation. Dr. Davis has worked on the mathematics of *Astronomia Nova* for a great many years, and brings a mathematician's precision to the study. Although I have not incorporated all of the adjustments that Davis would have liked to see, the translation has benefited substantially from those sessions and subsequent discussions. In particular, Appendix C is adapted from Davis's work.

Yaakov Zik of Haifa University very generously brought his skill with astronomical software to bear on Kepler's treatment of Tycho Brahe's observations of Mars oppositions and Kepler's table of oppositions at the end of Chapter 15. Not only did he give modern values for the Tychonic and Keplerian naked-eye observations, but he also computed all the exact opposition times and positions and wrote a wonderfully illuminating account of how Kepler carried out the observation of the 1604 opposition. This work now constitutes Appendix B of the translation.

I am also deeply grateful to the faculty, students, and staff of St. John's

College in Santa Fe, New Mexico. The college provided the initial impetus for the translation by giving me the opportunity to lead a discussion class on *Astronomia nova*. Subsequently, in preparation of the first editon, the library staff, and Tracey Kimball in particular, were most helpful, both in arranging inter-library loans and in great forbearance regarding overdue books. In the final stages of work on the present edition, the Library Director, Jennifer Sprague, kindly allowed me to borrow the library's facsimile of the 1609 edition to scan a selection of diagrams.

Crucial funding for the early stages of the translation was provided by the United States National Science Foundation. This grant came at a time when work on the project had nearly stopped because of other responsibilities, and allowed me to concentrate my attention once again upon completing the work.

Several other individuals who were particularly helpful deserve special mention. Ann Wegner, in assocation with Professor Gingerich, wrote a draft translation of all but a few chapters of *Astronomia nova*, which proved most useful in spotting omissions and in suggesting particularly apt translations of several awkward terms. Curtis Wilson of St. John's College in Annapolis de-voted considerable time and effort to a number of technical questions I asked him. His replies were always both interesting and helpful. And Simon Mit-ton, of the Cambridge University Press, was singularly helpful and patient in seeing the first edition of the book through to publication. I would also like to thank an anonymous reviewer for the Cambridge University Press whose comments on part of the translation were of great assistance in the final revi-sion of the entire work.

To my wife and publishing partner, Dana Densmore, go very special thanks. She has unfailingly supported me in this apparently endless project, and has also been an inspiration in the high standards of excellence she sets for her own work.

This translation is dedicated to Mary Corinne Rosebrook of the Sidwell Friends School in Washington D. C., who never gave up on a most unpromis-ing Latin student.

Translator's Introduction

The title and its significance

When Kepler boldly chose the words "*Astronomia nova*" to head the title of this book, he must have had some sense of how apt they were. Although *Astronomia nova* is not lacking in historical antecedents (as Kepler himself was keenly aware), it is an astonishing book, utterly unlike anything that had appeared before. Astronomy is no longer seen as a deductive science, founded mainly upon geometry, whose aim is to construct an ideal system that matches the appearances of the celestial motions as closely as possible. Instead, it is an adventure in which the human being uses all the means at his disposal to explore the creation in which God has placed him as His image. Error is no longer equated with failure, but is seen as an indication of the way to the truth. Thus the rules of the game are no longer fixed, but are to be discovered in the playing. And so the hypotheses of the ancients, the uniform circular motions previously thought to be indispensable to astronomy, were tested, found wanting, and rejected. The question was no longer "How can the appearances be accounted for?" but "How does God make things move?"

It is nevertheless not certain that *Astronomia nova* is the title by which Kepler intended the work to be known. On the title page of the first edition, the largest typeface is reserved for the name of Kepler's patron, the Emperor Rudolph II, and that of Kepler's subject, the planet Mars. This would suggest that the intended short title was *De motibus stellae Martis* (that is, *On the motions of the star Mars*). And it would indeed have been somewhat awkward to refer to the work as *New Astronomy*, as I have discovered in telling others about the translation. Kepler himself usually referred to it modestly as *Commentaries on Mars*. Nevertheless, the book is now commonly called "*Astronomia nova,*" a title that both matches the original short title and serves to distinguish title from description. Therefore, although in the first edition of this translation I chose to use the English version of the title, I have now bowed to common usage and reverted to the Latin original.

On the composition of *Astronomia Nova*

About the history of the work, little will be said here: there is already much of value in other sources which it would be otiose to duplicate. Especially noteworthy is James R. Voelkel's book, *The Composition of Kepler's Astronomia Nova* (Princeton University Press, 2001), which shows how the book was shaped by litigation with Brahe's heirs, and by Kepler's rhetorical strategy in building his case for a physical astronomy. Also of great importance are Owen Gingerich's articles and papers, particularly his article on Kepler in the *Dictionary of Scientific Biography*, C. C. Gillispie, editor (New York, 1973), Vol. 7 pp. 289–312. Many other useful sources are mentioned in the biblиography to this article, among them two excellent articles by Curtis Wilson. Also highly recommended is D. T. Whiteside's "Keplerian Planetary Eggs, Laid and Unlaid, 1600–1605," *Journal for the History of Astronomy* 5, 1974, pp. 1–21 (too recent to appear in the Gingerich bibliography). More recently, A. E. L. Davis has published a number of articles on the geometry of the Mars orbit: see especially "Kepler's Resolution of Individual Planetary Motion" and the accompanying "modules" in *Centaurus* 35 (1992) pp. 97–191.

On the other hand, studies of the relevant manuscripts by Gingerich[1] and my own reading of the work itself and of the manuscript "Kepler Notebook"[2] suggest that there is much yet to be learned about the stages of composition. In this regard, an especially valuable resource is the recent publication of a transcription of the "Mars Notebook" in *Johannes Kepler Gesammelte Werke* (Vol. 20.2). So it seems best to include, instead of the usual historical sketch, only the few remarks on the the history of composition of the work that will assist the reader's understanding, and that may help guide further scholarly enquiry. These concern a certain obligation Kepler felt (and in the end was legally obligated to acknowledge) towards Tycho Brahe, as well as the ample evidence of frustration and even despair that frequently arose in the original investigations.

To treat the second point first, Kepler often points out errors that waylaid him in his campaign: the whole of Chapter 58 is a good example. But there is a great deal that, out of consideration for the reader, Kepler did not report. Sometimes (as in Chapter 16) he makes this clear, but not always. In particular, he never hints that he had mostly completed at least one draft even before discovering the elliptical shape of the orbit. The project had, indeed, progressed so far that Kepler had actually sent the manuscript to the Emperor as a preliminary to publication.[3] We know of the existence and contents of this "Proto-*Astronomia nova*" only from manuscript sources, especially a letter to his teacher Michael Maestlin (of March 5, 1605)[4] and a (much earlier) draft table of contents.[5] This work was not simply abandoned; rather, it was revised, the polished form of the elliptical orbit being superimposed

---

[1] "Kepler's Treatment of Redundant Observations; or, the Computer Versus Kepler Revisited," in F. Kraft, K. Meyer, and B. Sticker, editors, *Internationales Kepler-Symposium Weil der Stadt 1971*, (Hildesheim, 1973), pp. 307–314; also "Johannes Kepler and the New Astronomy," *Quarterly Journal of the Royal Astronomical Society* (1972) **13**, 346–373, especially p. 352.

[2] See for example the notes to chapter 53 of this work, as well as "Kepler's Approach to the Oval of 1602, from the Mars Notebook," *Journal for the History of Astronomy* **27** (1996) 281–295 and other articles.

[3] Letter to Herwart von Hohenburg, 10 February 1605, no. 325 in *KGW* 15 pp. 145–147. Kepler remarks, however, that "certain chapters are still missing."

[4] Letter number 335, in *KGW* 15 pp. 170–176.

[5] In *KGW* 3 pp. 457–460. For the date, see Gingerich, "Kepler's Treatment of Redundant Operations," in F. Kraft, K. Meyer, and B. Sticker, editors, *Internationales Kepler-Symposium Weil der Stadt 1971* (Hildesheim, 1973), p. 313.

upon earlier attempts, leaving nonetheless a few vestigial rough edges and mismatched seams.[6] This tendency of Kepler's to gloss over many of the difficulties he confronted has two main consequences for the reader. First, the computations presented were often intended as examples, to give an idea of the procedure used. This is most conspicuously so in Chapter 16, which would have at least doubled the size of the book had the entire iterative process been presented. Less obvious, but equally revealing, are the sample computations in Chapter 53, in comparison with the entirely different results displayed in the table at the end of that chapter. Second, the selection of data and arguments, and sometimes the data themselves, were determined by the conclusions. That is, although Kepler often seems to have been chronicling his researches, the *Astronomia nova* is actually a carefully constructed argument that skillfully interweaves elements of history and (it should be added) of fiction.[7] Taken as history, it is often demonstrably false, but Kepler never intended it as history. His introduction to the "Summaries of the Individual Chapters" makes his intentions abundantly clear. *Caveat lector*!

There is another feature of the work that may seem puzzling, that may at least partly be explained by Kepler's sense of obligation to Brahe. It will be seen that in the earlier parts (through Chapter 26), wherever a geometrical proof is carried out on the planetary orbit, the demonstration is presented in all three forms of hypotheses (Ptolemaic, Copernican, and Tychonic). In Kepler's words:[8]

> I, in the demonstrations that follow, shall link together all three authors' forms. For Tycho, too, whenever I suggested this, answered that he was about to do this on this own initiative even if I had kept silent (and he would have done it had he survived), and on his death bed asked me, whom he knew to be of the Copernican persuasion, that I demonstrate everything in his hypotheses.

There is nevertheless more to this cumbersome triple demonstration than the fulfillment of a solemn promise. Kepler's intention, as is shown in Chapter 6, was to establish the perfect geometrical equivalence of the three forms of hypotheses in order to show that geometry alone cannot decide which is correct. This prepares the reader for the climactic Part IV, in which the "first inequality" (the inequality in the heliocentric longitudes) is treated "from physical causes and the author's own ideas."[9] It would of course have been possible to treat the elliptical orbit generated in that part as just another geometrical hypothesis, and to show how it could have been accommodated to the Ptolemaic and Tychonic forms. That Kepler did not present such a demonstration is a consequence of his belief that to separate the geometry from the physics

---

[6]For example, at the beginning of Chapter 54, Kepler refers to a 15′ correction of the mean longitudes supposedly made in Chapter 53. No such correction appears in that chapter as it presently stands, and it seems clear that Kepler later revised Chapter 53 without correcting the cross reference.

[7]The entire table at the end of Chapter 53, for example, is based upon computed longitudes presented as derived from observations.

[8]Chapter 6, p. 106

[9]Part IV, title.

upon which it is based turns it into nonsense. His promise to Tycho was a promise about hypotheses, and is no longer valid in an "astronomy without hypotheses."[10]

Historical background and salient features.

As for the work itself, it is my intention to allow Kepler to speak for himself, providing notes and comments where his sense might not be clear to the modern reader, or where a reference or allusion might have been obscured by the passing of time. However, the opportunity to make a few remarks on the work as a whole is hard to resist. Also, one who has made a careful study of the book may have some insights that will prove useful. So I beg the reader's indulgence as I make a brief excursion into issues Kepler raises in this book and how he approaches them.

Historical context

The central question that guides Kepler's investigations throughout the *Astronomia nova* is, "What is the actual path of the planet through space?" This question occurs to us so naturally and seems so fundamental to planetary astronomy that its revolutionary implications come as something of a surprise. But at the time, for an astronomer to pose this question as insistently as Kepler did was unprecedented. Why was this so?

To find the answer, we need to look at the aims and methods of pre-Keplerian astronomy.[11] There was by no means a general consensus as to what these were, but certain characteristics are conspicuous. Planetary astronomy was the application of geometry to motion. On the relationship between the geometry and the motion, there were two main viewpoints. Some believed that the geometry did not represent the reality of the motion, but only served to predict it, while others held that an adequate geometrical model must somehow reflect physical reality. But despite the latter group's concern for the reality of planetary motions, the reality they saw was an expression of the model, and not the motion of a body through space. An early printed example is Peurbach's *Theoricae novae planetarum* (Nuremberg, 1473), in which the Ptolemaic epicycles were contained in eccentric channels in the planetary regions. Peurbach was not the originator of this idea, which is found in Islamic astronomy along with many other quasi-mechanical arrangements of spheres. But his work was widely read and reprinted, and so this model became well known and was associated with his name. Thus Kepler says, in Chapter 2 of this book,

---

[10]See the passage from Ramus on the verso of the title page, and Kepler's reply.

[11]The following account is based upon my research into the physics of celestial motion in the sixteenth and seventeenth centuries. For detailed references, I would direct the reader to my article, "The Solid Planetary Spheres in Post-Copernican Natural Philosophy," in Robert S. Westman, editor, *The Copernican Achievement* (University of California, 1975), pp. 244–275, or to my more detailed doctoral dissertation, published as *The Dissolution of the Celestial Spheres, 1595–1650* (New York: Arno Press, 1981). The former is clearer and more concise, while the latter is much more thorough, although the general analysis is confused in places.

Ptolemy has described these circles to us in their bare form, as geometry applied to the observations shows them. Peurbach set up a way for them to move around which follows Aristotle [i. e., makes sense in the context of Aristotelian physics].

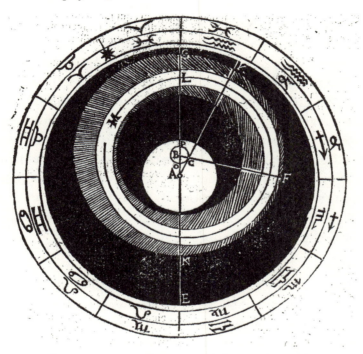

Contemporary diagram of the sun's motion

Looking at this diagram, a modern reader immediately sees a mechanism, a solid transparent device perhaps made of some crystalline substance.[12] A contemporary of Peurbach, however, would see something quite different. For medieval natural philosophy had surpassed even Aristotle in separating the heavens from earth. The most widely accepted view was that even the material of which the heavens are made is wholly different from elementary matter (Aristotle had never explicitly distinguished them materially, although he did believe in a "quintessence" that was different from the elements in its form). This material was described as intermediate between the eternal and the temporal, and was compared to the human intellect. In more extreme views, it has "no potential for existence, but only for location." And although mathematical quantities are the same everywhere, real quantities of the celestial material are utterly different in kind from quantities of elements. In other words, spatial relations in the heavens are not like those we experience on earth. Dante gives a vivid and (to the modern mind) startling account of such views in Canto 2 of the *Paradiso*.

In the sixteenth century, however, neoclassical humanism, Reformation theology, and a renascence of several non-Aristotelian natural philosophies,

---

[12]It should be remarked that the property of crystallinity was not applied to the planetary spheres until it was no longer usual to take them seriously, and was mostly attributed to them by those who did not believe in real spheres. In the heyday of real orbs, the *sphaera crystallina* or*coelum crystallinum* was invariably the starless ninth sphere, beyond the sphere of the fixed stars, which was given that name following Ezekiel 1. 22: "And the likeness of the firmament upon the heads of the living creatures was as the color of the terrible crystal, stretched forth over their heads above."

combined to bring the heavens closer to earth. As a result of this, the previously ideal celestial spheres acquired a decidedly un-Aristotelian solidity: the properties of non-interpenetrability, rigidity, and hardness were creeping into the ethereal regions. By the 1570's, when Tycho Brahe (philosophically a Paracelsian) applied his formidable instruments to the measurement of celestial parallax, the spheres had congealed enough to be highly vulnerable. So when his observations suggested that the comet of 1577 had passed through a region that should have been filled with solid spheres, instead of recalling Dante's verses describing dimensions interpenetrating each other, he threw out the whole apparatus. By the time Kepler was writing *Astronomia Nova*, the notion that the heavens are filled with very pure air was gaining respect, particularly among educated men outside the universities. To Kepler, the fluidity of this region, which he called "aethereal air," was a perfectly obvious consequence of Tycho's parallax observations.[13]

How the context shaped Kepler's views.

Therefore, when he approached the question of the physical nature of the planets' motions, Kepler saw a world dramatically different from Peurbach's. Epicycles no longer made sense, because they were no longer supported by any substance. In Kepler's words, they would require that the planet's mover "imagine for itself the center of its orb and its distance from it, or be assisted by some other distinguishing property of a circle in order to lay out its own circle."[14] The motion of the center of the epicycle upon the deferent is even more improbable:

> For it is also incredible in itself that an immaterial power reside in a non-body, move in space and time, but have no subject, moving itself (as I said) from place to place.[15]

What Kepler saw instead is depicted in the second diagram in Chapter 1: a series of interlocking spirals, each slightly different from the others, that never quite repeats itself. The appearance of this diagram is a dramatic moment in the history of thought. Nothing like it had ever been published before. Epicyclic motion had become the motion of a body, now freed from its spheres, and space had become a uniform medium in which this motion is performed. The astronomer's task was no longer to find a geometrical model to represent these spirals: he had to separate illusion from reality, and find the paths that the planets really traverse in that uniform medium. But to find what is illusory and what is real, one must go beyond the arcs and loops to find out what really moves the planets. And thus the "new astronomy" was born: the astronomy built upon physics.

---

[13]See, for example, Chapter 2 p. 84.

[14]Chapter 2 p. 85.

[15]Chapter 2 p. 86.

Kepler's response to the challenge.

The task Kepler had set himself was far more demanding than anything an astronomer had previously undertaken. No physical theory of the celestial motions had ever been required to yield accurate predictions,[16] nor had any predictive apparatus been required to satisfy the requirements of physics without the use of real spheres or orbs. Indeed, it was not even clear how to begin. Between the qualitative gropings of physics and the all-but-inscrutable records of observations lay a void. Something was needed to express the pattern hidden in the observations so as to make it accessible to physical explanation. In another stroke of genius, Kepler realized that this role could be played by geometry. For even if it was no longer thought to represent physical reality, geometry could simultaneously express the position of the path in space, and the exact cyclical nature of the path in time. Accordingly, throughout the *Astronomia Nova*, one sees a parallel development of theory on three levels: the physical theory, the geometrical model, and the predictive apparatus.

Like the dancing Graces in Renaissance art, these three distinct theoretical levels interact as the argument proceeds, coming in turns to the foreground while keeping the others ever within reach. Thus the "vicarious hypothesis" developed in Chapter 16, accurate for the longitudes though demonstrably false, guides the formation of the geometrical oval hypothesis of Chapters 46–50, which in turn was developed from physical principles in Chapter 45. And thus the physical conjecture that the sun is the source of power spurs an enquiry on the observational level, in Chapters 22-28, whose aim is the construction of a new geometrical model of the earth's motion. And the new geometrical model in turn confirms the original conjecture, which is then expanded in subsequent chapters. Kepler's goal, which he very nearly attained, was to intertwine the three into a single comprehensive theory: the interaction of physical forces and powers produces the elliptical orbit, which, together with the "Second Law" (areas swept out are proportional to the time), predicts the planet's position with perfect accuracy. It must be perfect, because it is true.

Technical aspects of Kepler's astronomy and mathematics.

Those bees having been liberated from the translator's bonnet, a return can now be made to the business of preparing the reader for the fray. A campaign against the war-god himself is not to be undertaken lightly, and as Kepler says, "the difficulties and thorns of *my* discoveries infest the very reading."[17] This is

---

[16] Although Copernicus certainly intended his construction to represent reality, it is what I would call a geometrically based system rather than a physically based one. Contemporary physically based systems were developed by Telesio, Patrizi, Campanella, Aslachus, Lydiat, and others, but they could only represent the motions qualitatively, if at all.

[17] "Summaries of the Individual Chapters," introduction.

perhaps even more so for the modern reader, for the mathematics and astronomy are unfamiliar, and many terms are used that do not have exact modern equivalents. Most of the unfamiliar or problematical terms are discussed in the Glossary. The few remaining points that need a more general explanation will be taken up here. These include the meaning of "anomaly" and the different kinds of anomaly, Kepler's trigonometry and arithmetic, and the use of tables.

In the traditional geometrical astronomy, the term **"anomaly"** denotes a planet's angular position about some point, measured from a line through apogee or aphelion. Anomalies are distinguished according to the points about which they are measured: the mean anomaly about the equant, the eccentric anomaly about the center of the eccentric, and the true or equated anomaly about the earth (for Ptolemy) or the center of the earth's orbit (for Copernicus). When Kepler operated with traditional planetary models, he used the conventional definitions. But when he departed from tradition, some changes in the definitions were required.

In Kepler's terminology, **Mean anomaly** is the measure of the time elapsed since the planet was at aphelion, expressed as an angle where the periodic time is 360°. Kepler uses a variety of means of expressing the mean anomaly geometrically. In its final formulation, however, the mean anomaly is measured by the area QAC in the adjacent figure. This is the sum of the circular sector QBC and the **"triangle of the equation"** AQB. (It is remarkable that in *Astronomia nova* Kepler never uses PAC, the actual area swept out, as a measure of time.)

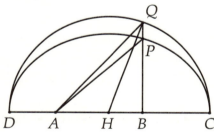

Eccentric anomaly is the angle about the center of the ellipse measured from aphelion to the point the planet would have occupied had it remained on the circle. In other words, if a line HP be erected perpendicular to the apsides CD and extended to intersect the eccentric circle (whose diameter CD coincides with the ellipse's major axis) at Q, the eccentric anomaly is the angle QBC.

Equated anomaly is the angle about the sun between the planet and the apsides: the angle PAC. It is usual to call this the "true anomaly." However, the Latin equivalent of "true anomaly" would be *"anomalia vera,"* or something similar, which is not the term used by Kepler. His term is *"anomalia coaequata,"* which evidently means "anomaly with the equation added or subtracted," or, more simply, "equated anomaly."

Kepler's trigonometry is very much like modern trigonometry, with two exceptions: there are no cofunctions, and no decimals. Where we would use a cosine, Kepler used either the sine of the complement[18] or what he calls the "secant," which is our exsecant (the secant with the radius subtracted). The

---

[18]The Latin for "sine of the complement" is *complementi sinus*, which became contracted into "cosine." So, in a sense, Kepler did use the cosine. See Glen van Brummelen, *The Mathematics of the Heavens and the Earth: the Early History of Trigonometry* (Princeton University Press, 2009).

lack of decimals is compensated by making the radius equal to 100,000, the standard procedure in those days, which produces five-place tables. Once one grows accustomed to allowing for the presence of a radius that is not unity, the computations become familiar and comfortable.

It may be useful to remark that Kepler used the tables of Philip van Lansbergen (Lanbergius) (*Triangulorum geometriae libri IV*, Leiden 1591) for at least some of the computations in this book. See the testimonial at the beginning of Chapter 15.

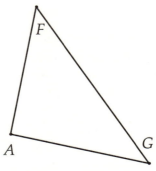

Kepler's arithmetic is also not difficult once the short cuts and conventions are understood. Let us take an example from Chapter 16 (which is mostly computations).

In the triangle *FAG*, the sides *FA* and *AG* and the included angle *FAG* are known. According to the law of tangents,

$$\frac{AF - AG}{AF + AG} \tan \frac{180 - \angle FAG}{2} = \tan \frac{\angle G - \angle F}{2}$$

The sides *FA* and *AG* are 59433 and 50703, respectively, and the angle *FAG* is 90° 56′ 23″, whose half is 45° 28′ 12″, complement 44° 31′ 48″. Here you see the computation as Kepler presents it.

		FAG 44° 31′ 48″ 98373		*GAD* 52° 23′ 4″ 129093		*DAE* 57° 23′ 4″ 156271		*EAF* 25° 50′ 41″* 48438	
Halves									
Tangents									
	AF	59433	*AG*	50703	*AD*	48052	*AE*	52302	
	AG	50703	*AD*	48052	*AE*	52302	*AF*	59433	
Differences		8730		2651		4250		7131	
Sums		110136		98755		100354		111735	
		770952	7	197510	2	401416	4	670410	6
		102048		67590		23584		42690	
		99123	9	59253	6	20071	2	33520	3
		2925		8337		3513		9170	
		2203	2	7900	8	3016	3	8938	8
		722	6	437	4	497	5	232	2
Quotients		7926		2684		4235		6382	
Tangents		98373		129093		156271		48438	
		6886	11	2581	86	6250	84	2906	86
		885	33	774	54	312	54	145	34
		19	66	103	20	46	86	38	72
		5	88	5	16	7	81		96
Tangents		7797		3465		6618		3142	
Differences	*F* 4° 27′ 30″		*D* 1° 59′ 4″		*D* 3° 47′ 10″		*F* 1° 47′ 59″		

First he finds the tangent (line 2). The third and fourth lines are the two sides, and the fifth and sixth their difference and sum, respectively.

Lines 5–12 show Kepler's version of **long division**. Note that the dividend (line 5) is treated as if it had been multiplied by the radius: this will be divided back out later. Line 7 has two numbers: the first digit of the quotient on the right, and on the left the product of this digit and the divisor (line 6). This left number on line 7 is subtracted from line 5: the difference appears on line 8. On the right side of line 9 is the next digit of the quotient, and on the left, the product of that digit and the divisor (with the last digit struck). This product is subtracted from line 8, and the process is repeated on lines 10-12. So the quotient appears vertically on the right side of the computation, instead of horizontally at the top, as it would appear today (in those few places where the arcane rite of Long Division is still practiced).

Lines 13-19 are an example of **multiplication**. Line 13 is the quotient of the previous division, and line 14 the tangent, brought down from line 2. Note that line 3 is treated as the multiplier, and line 14 as the multiplicand, the inverse of our customary procedure. That is, line 15 is the product of line 14 and the first digit of line 13, line 16 is the product of line 14 and the second digit of line 13, and so on. In each successive partial product a digit is dropped from the multiplicand (line 14); thus, line 16 is the product of 9837 and 9, line 17 the product of 983 and 2, and so on. The vertical line passing through the midst of all the numbers in lines 15-18 serves much the same purpose as a decimal point: note that it has been so placed that the product (line 19) has in effect been divided by the radius (which, in line 5 above, had tacitly been multiplied in). This product is the tangent of the required angle (line 20), which is half the difference of the angles at *F* and *G*.

Kepler's **parallax computation** is carried out with the help of the parallactic table in the *Astronomiae pars optica* (Frankfurt, 1604).[19] Since he explained the use of the table in that work, he felt justified in referring the reader to that explanation, for the sake of brevity. But although the *Astronomiae pars optica* has now been translated into English, complete with enormous table, readers may find it inconvenient or impossible to take his advice. Here, for their sake, is how Kepler reckons parallax.

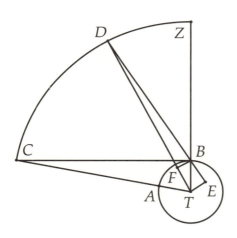

The first step is to find the **nonagesimal**. This is the point on the ecliptic 90° from the ecliptic's intersection with the horizon. It is also necessary to find the altitude of the nonagesimal, or its complement, which is the arc between the nonagesimal and the zenith. This is the same as the angle between the ecliptic and the horizon, or its complement. The importance of the nonagesimal is that a great circle drawn from it to the zenith is always perpendicular to the ecliptic. Therefore, when the planet is at the nonagesimal, the parallax affects only the latitude. Furthermore, the latitudinal part of the parallax is always the same as the parallax at the nonagesimal, regardless of where the planet is. The longitudinal part of the parallax, in turn, is a function of the planet's distance from the nonagesimal.

Once the nonagesimal and its zenith distance are known, the rest follows directly. Suppose *AB* is the earth's surface, *T* its center, with an observer at *B*, *Z* the zenith, *D* the nonagesimal, and *C* the point where the great circle

---

[19] English translation: *Optics*, trans. W. H. Donahue (Green Lion Press, 2000).

through Z and D intersects the horizon. And let TE be perpendicular to DE extended. The angle at C is the **horizontal parallax**, found by observation or by assuming (or calculating) a magnitude for the distance CT. The sine of this angle is $\frac{BT}{CT}$, or $\frac{BT}{DT}$. The sine of the angle at D, the latitudinal parallax, is $\frac{ET}{DT}$. Therefore,

$$\frac{\sin D}{\sin C} = \frac{ET}{BT} = \sin \angle DBZ,$$

or the sine of the latitudinal parallax is the product of the sines of the horizontal parallax and the zenith distance of the nonagesimal. Or, since the parallax angles are very small, the angles themselves can be used instead of the sines, and thus the latitudinal parallax is the product of the horizontal parallax and the sine of the zenith distance.

The **parallactic table** is designed to perform this computation. Horizontal parallaxes are set out across the top, and zenith distances at the side, and one can find the corresponding latitudinal parallax by finding the intersection of the appropriate row and column. If the horizontal parallax is not an integral number of minutes, one finds the number of seconds at the top and finds the entry opposite the zenith angle, which will give the extra seconds and "thirds" of latitudinal parallax.

Longitudinal parallax may be a little harder to visualize: it is the part of the parallax that occurs in the plane of the ecliptic, which is perpendicular to the plane of the diagram in the above figure, intersecting it at the line DT. So a new figure is required, showing that plane. But first, in the figure above, let BF be drawn perpendicular to DT, intersecting it at F. Thus F is the point in the plane of the ecliptic corresponding to the position of the observer B, and $FT = BE$. Now let the new figure be drawn.

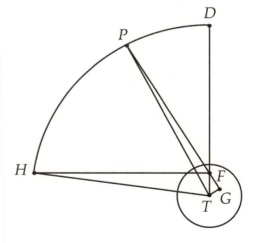

Here, DH is the ecliptic, with D the nonagesimal as before, and P the planet. The sine of the horizontal longitudinal parallax (the angle at H) is $\frac{FT}{TH}$ or (in the previous figure) $\frac{BE}{TC}$, which may also be expressed as the product of $\frac{BE}{BT}$ and $\frac{BT}{CT}$. Therefore, the sine of the horizontal longitudinal parallax is the product of the sines of the altitude of the nonagesimal and the total horizontal parallax. This may also be computed from the table, the total parallax being found at the top, and the horizontal longitudinal parallax being read opposite the altitude of the nonagesimal (note that this is the complement of the zenith distance).

Finally, the actual longitudinal parallax (the angle at P) is found. The sine of this angle is equal to the product of the sines of the angle at H and the angle PFD, which is the planet's elongation from the nonagesimal. This is exactly analogous to the latitudinal parallax, and the table is used in the same way. That is, after reading the horizontal longitudinal parallax as in the preceding paragraph, one finds that number at the top and reads the actual longitudinal parallax opposite the planet's elongation from the nonagesimal. Odd seconds may be accounted for as before, if desired.

As for the **calendar**, Kepler held to the old Julian dates. The Gregorian calendar, it must be remembered, was a Popish innovation and was not very widely used, particularly in Lutheran countries. Sometimes, when he is using a date obtained from another astronomer, he gives both the Julian and the Gregorian dates. But when a single date is given, it is invariably in the old calendar. The Gregorian date is obtained by adding 10 days to Kepler's dates.

Time is measured from noon. Although Kepler usually reports morning observations by giving the number of hours from midnight, in his calculations he refers the time to noon on the previous day. This can be a source of confusion, since "5 am on June 21" becomes "17^h on June 20." To avoid this confusion, Kepler often reports morning observations by saying (using the same example), "on the morning following June 20", or more succinctly, "June 20/21." If you are checking computations, and find a discrepancy of about 31′ in the mean longitudes, it is probably the result of using the wrong date for a morning observation.

Observations and computations performed upon them. In Chapter 11 Kepler presents a series of his own observations of Mars, which he calls "a clown show," contrasting his attempt with the fine observations made by Tycho Brahe and his assistants. The series provides a fascinating view of Kepler's treatment of observational error, and is also a good illustration of Brahe's procedures, which Kepler followed. For this reason, I include here an account of the first of these observations, with diagrams and explanatory notes.

In outline, the procedure is to measure the angular distance between Mars and one or more fixed stars, and to obtain Mars's declination by observing its meridian altitude and subtracting the altitude of the equator. The distance combined with the declination are used to calculate the right ascension and declination of Mars, which are then converted to latitude and longitude.

Here is Kepler's description:

> On the night between Thursday and Friday, which was February 17/27 [1604], while Corvus was on the meridian, there was 9° 44′ between Mars and Spica [*MS*, in the diagram], and between it and the Northern Pan [β Librae], 17° 41′ [*ML*]; and between Mars and Arcturus, 29° 13′ [*MA*].

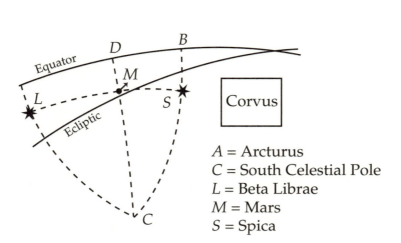

A = Arcturus
C = South Celestial Pole
L = Beta Librae
M = Mars
S = Spica

It is evident from the diagram that β Librae and Spica are suitable reference stars, since they have nearly the same declination as Mars. It is convenient to use right ascensions and declinations of the stars, since the distances between the stars are mostly longitudinal. Arcturus was chosen as the third reference star because it was nearly at the same right ascension as Mars, and therefore provides a good check on the declination.

Kepler continues:

> Also, to test the sextant, we measured the interval between Arcturus and Spica as 32° 57′, which should have been 33° 1′ 45″, as is clear if you calculate it using the right ascensions and declinations, or latitudes and longitudes, which Tycho assigned to these stars in Book I of the *Progymnasmata*.

There are two published star catalogs, one from the *Progymnasmata* of 1602, *TBOO* 2 pp. 258–280, and the other published separately in 1598, which is in *TBOO* 3 pp. 344–377. The former gives only longitude and latitude, while the latter has a separate table (pp. 375–377) giving right ascensions and declinations of prominent stars, especially those used as reference stars. According to this catalog, the right ascensions and declinations of these three stars, together with corrections for precession, are as follows.

	Rt. Asc.	Dec.	100 yr. change in R. A.	100 yr. change in Dec.	In 1604 Rt. Asc.	Dec.
Spica	196° 4 ′	9° 1 ′ S.	+1° 19$\frac{1}{2}$′	32$\frac{1}{2}$′ S.	196° 7$\frac{1}{4}$′	9° 2$\frac{1}{4}$′ S.
β Librae	223° 54$\frac{1}{2}$′	7° 50 ′ S.	+1° 21$\frac{1}{2}$′	24 ′ S.	223° 57$\frac{3}{4}$′	7° 51 ′ S.
Arcturus	209° 23$\frac{1}{2}$′	21° 18$\frac{1}{2}$′ N.	+1° 11 ′	29$\frac{1}{2}$′ S.	209° 26$\frac{1}{4}$′	21° 17$\frac{1}{4}$′ N.

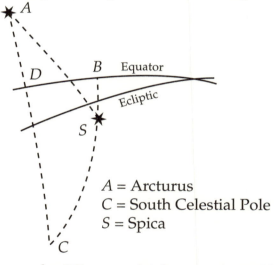

A = Arcturus
C = South Celestial Pole
S = Spica

In the diagram, the difference of right ascensions of Spica and Arcturus is the arc *BD*, which is numerically equal to the angle at *C*. Since the arcs *BC* and *CD* are both 90°, the arc *SC* is found by subtracting the declination of Spica (*BS*) from 90°, and the arc *AC* is found by adding the declination of Arcturus (*AD*) to 90°.

Therefore, in spherical triangle *ACS*,

$$AC = 111° \ 17\frac{1}{4}' \quad SC = 80° \ 57\frac{3}{4}' \quad \angle C = 13° \ 19'.$$

This can be solved by procedures set forth in contemporary trigonometry books. A cumbersome procedure is derived by Regiomontanus, *De triangulis omnimodis* IV.28; Brahe developed a more succinct algorithm in his *Triangulorum praxis arithmetica*, Dogma 6 (*TBOO* 1 pp. 289–290). This procedure is equivalent to the modern formula:

$$\cos AS = \cos AC \cos SC + \sin AC \sin SC \cos C.$$

Applying this formula to the numbers given above, one obtains a distance of 33° 1′ 30″, which is nearly the same as Kepler's result.

Kepler continues:

> Therefore, my distances are smaller than the true distances by $4\frac{3}{4}$ minutes. I applied this correction to the distances of Mars from the fixed stars, so as to make it 9° 48′ 45″ from Spica, 17° 45′ 45″ from the Pan [β Librae], and 29° 17′ 45″ from Arcturus.
>
> I then used the quadrant to obtain the meridian altitude of Mars, 32° 4′, and of Spica, 30° 50′. Since the latter has a declination of 9° 2′, Mars is left with a declination of 7° 48′.

The meridian altitude is the angular distance of Mars above the horizon as it crosses the meridian circle, which runs from the horizon due south of the observer up through the zenith and on to the point on the horizon due north. From these transit measurements, one sees that Mars was 1° 14′ farther north than Spica; therefore, subtracting this amount from Spica's declination of 9° 2′ S., one obtains for Mars a declination of 7° 48′ S. Kepler notes in passing that the measured transit altitude for Spica should have been higher; this does not affect the calculated declination for Mars. He then continues:

> Now, from the declination of Mars and its distance from the fixed star, its right ascension comes out:

	°	′	″
From Spica	205	57	36
From the Pan	206	3	17
Difference	0	5	41
Resultant mean	206	0	26

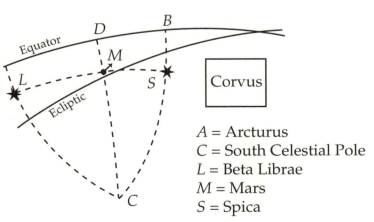

A = Arcturus
C = South Celestial Pole
L = Beta Librae
M = Mars
S = Spica

The procedure here is the inverse of the one given above, where the distance between stars is calculated from given coordinates. Here, we are given the distance, and we must calculate Mars's position. Subtracting the declinations of Mars (*MD*) and Spica (*SB*) from the 90° arcs *DC* and *BC*, we find the sides MC and SC of spherical triangle MSC, and side MS is also known, since it is the measured distance between Mars and Spica. The sides of the triangle are:

$$MC = 82° \; 12' \quad SC = 80° \; 57\frac{3}{4}' \quad MS = 9° \; 48\frac{3}{4}' \, .$$

Again, procedures for solving this problem are presented in trigonometry books: in Brahe's *Triangulorum praxis arithmetica*, it is Dogma 9 (*TBOO* 1 pp. 291–292). Brahe's algorithm is equivalent to a rearrangement of the terms of the equation given above:

$$\cos C = \frac{\cos MS - \cos MC \cos SC}{\sin MC \sin SC} \, .$$

Angle C is numerically equal to the arc BD; therefore, this angle is added to the right ascension of Spica to obtain the right ascension of Mars. Computation shows the angle to be 9° 50′ 26″, which, added to Spica's right ascension, gives a right ascension for Mars that is within a few arc seconds of Kepler's figure.

The next step is to convert the coordinates to longitude/latitude. Although we think of this as the algebraic problem of transforming coordinates, Kepler and his contemporaries treated it as a problem in spherical triangles. Alert for ways of using tables to avoid lengthy computation, and with an eye to showing the usefulness of his enormous "*tabula parallactica*" in the *Optics*, he next shows how to do the transformation using existing tables.

> From the right ascension and declination of a star, to find its longitude and latitude without calculation, with the help of tables. From this right ascension [206° 0′ 26″], I first select the degree of the right sphere that was rising at the same time, 28° 1′ 0″ Libra, from Tycho's table of right ascensions.

The table is in *TBOO* 2 p. 74. The phrase "degree of the right sphere that was rising at the same time" requires some explanation. The "right sphere" is the celestial sphere as seen from the earth's equator: thus the celestial equator is at right angles to the horizon, and all stars rise perpendicularly (see, for example, *Almagest* I.16 and Toomer's note on p. 18 of his translation). What Kepler needs here is the point on the ecliptic that is intersected by the horizon when Mars is exactly on the horizon. In the table, the right ascensions corresponding to whole degrees on the ecliptic are in the body of the table, and opposite 28° Libra one finds 205° 59′ 25″, with one degree on the ecliptic corresponding to 57′ 5″ at this place. Our right ascension is 1′ 1″ greater, which adds about a minute to the longitude: the corresponding point on the ecliptic is therefore 28° 1′ 0″.

> Its declination, from another of that author's tables, is 10° 48′ 30″, and Mars's is 7° 48′.

The table of declinations is in *TBOO* 2 pp. 65–71. For 28° 0′ the declination is 10° 48′ 2″, and a 10′ increase will add 3′ 35″ to the declination, or about 22″ for 1′. Kepler rounds up by a few seconds to the nearest half minute. The declination of Mars is given above. The configuration is shown in the adjacent diagram. Here, A is the autumnal equinox point (0° Libra),

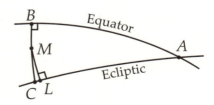

AB is the arc on the equator corresponding to the right ascension of Mars (26° 0′ 26″),
M is Mars,
BM the declination of Mars (7° 48′),
MC is the arc that Kepler calls "the base of the latitude,"
AC is the arc on the ecliptic (28° 1′ 0″),
BC the declination of the ecliptic at this place (10° 48′ 30″), and
C is the point on the ecliptic that rises with Mars in the right sphere.

To find the latitude and longitude, Kepler finds the sides of the triangle *MCL*: *ML* is the latitude, and the longitude *AL* is found by subtracting the side *CL* from the known arc *AC*.

> Therefore, [Mars] is distant from the ecliptic, the oblique path, by 3° 0′ 30″ on the circle of declination [*MC*, = *BC* − *BM*]. But the angle which the circle of declination makes with the ecliptic [∠*C*] is 68° 59′ from the appropriate table. Its complement is 21° 1′.

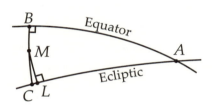

Kepler does not specify a table; however, this would be a table giving the angle between the ecliptic and a meridian. Brahe seems not to have published such a table, though there is one in Copernicus, *De revolutionibus* II.4, and they were also available elsewhere, though the obliquity of the ecliptic would probably differ from Brahe's value of 23° 31′ 30″ (Copernicus's table, for example, would give 69° 2′). The angle can be computed using the law of sines for right spherical triangles:

$$\sin C = \frac{\sin AB}{\sin AC}.$$

The final two operations use the "Tabula Parallactica" that Kepler included in his *Optics* (in the English translation, it is in a pocket inside the back cover). This table allows one to find the angle whose sine is the product of the sines of two other angles, provided that one of the angles is no more than a few degrees. Although the table only goes up to 66′ for the second angle, it can be extended using the approximation $n \sin x = \sin nx$, which holds for small angles. Kepler does this in the present example.

The final steps involve the right spherical triangle *MCL*. The right angle is at *L*, and the side *MC* and the angle *C* are known. Therefore, by the law of sines for right spherical triangles,

$$\sin ML = \sin C \sin MC, \quad \text{and} \quad \sin CL = \sin M \sin MC.$$

Since side *MC* is greater than 66′, Kepler divides it into three arcs of 1° each and one arc of 30″, finds the partial sine for each one, and adds them up:

> And in my table of parallax, under the heading of 60′, I find, opposite 68° 59′, the entry 56′ 1″. Under [the heading of] 30″, however, I find 28″.

Note that rather than interpolating, one can treat the column head as the number of seconds (rather than minutes, as in the initial entry) and then add the number in the body of the table, now likewise treated as seconds, to the number obtained in the first entry.

But since I have thrice 60′ in this distance [*MC*] of Mars from the ecliptic (which I call the base of the latitude), I multiply what I extracted under 60′ by 3. This gives me a latitude of 2° 48′ 31″.

That is,

Opposite 69°	beneath 60′:			56′	1″
Multiply by 3:		2°	48′	3″
Opposite 69°	beneath 30″:				28″
Sum:		2°	48′	31″

The longitude (arc *AL*) is found by subtracting *CL* from *AC*, which was found above. Since triangle *MCL* is small, it can be treated as if it were a plane triangle, which makes things simpler: ∠*M* is the complement of ∠*C*, and the law of sines is again applied, using the table:

The same operation opposite 21° 1′ shows me what has to be subtracted from the place rising at the same time, namely, 1° 5′ 4″. Accordingly, Mars's position will be 26° 56′ Libra.

Here is the computation, in tabular form:

Opposite 21°	beneath 60′:			21′	30″
Multiply by 3:		1°	4′	30″
Opposite 21°	beneath 30″:				11″
Sum:		1°	4′	41″
Subtract from *AC*,		28°	1′	0″
		26°	56′	19″

To test the latitude of Mars, I also consulted the distance from Arcturus, through the star's latitude and longitude provided from Tycho, and Mars's longitudinal position just found, and it replied to me that Mars was at a latitude of 2° 47′ 48″. Before, it was at 2° 48′ 31″.

This is a different kind of computation, and so is worth presenting. Here, *AM* is given, together with the difference in longitudes *DE*. The arc *AP* is the complement of Arcturus's latitude. Since *P* is the pole of the ecliptic, ∠*P* = $\overset{\frown}{DE}$. Therefore, in spherical triangle *MAP*, two sides and one angle are given, and the side *PM* is sought.

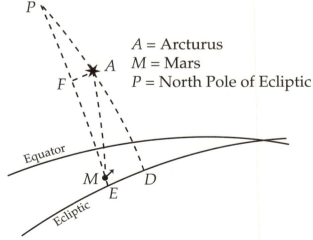

A = Arcturus
M = Mars
P = North Pole of Ecliptic

The procedure in Brahe's *Triangulorum praxis arithmetica*[20] is to drop a perpendicular from *A* to side *PM*, and to treat the right spherical triangles *PFA*

---

[20]Dogma 8, in *TBOO* 1 p. 291.

and *MFA* separately. First, *AF* is found, and through it angle *M* is found. *PF* and *FM* may then be computed, each in its own triangle, and the sum is subtracted from 90° to obtain the latitude, arc *EM*.

On the translation

General remarks

Translation is always a more or less criminal act against both author and readers; and, as in other cases, it is incumbent upon the guilty party to excuse his crimes as much as possible by imputing them either to the author or to unavoidable circumstances. Let me then say at the outset that Kepler's Latin is not simple. He was given a thorough classical education of the kind that had evolved during the sixteenth century,[21] and sometimes appears to have valued style more than clarity. Often he will choose an unusual word or turn of phrase for no apparent reason, leaving the translator wondering whether to represent it with something awkward or to try to force it to make sense. I am unabashedly on the side of sense. I allow Kepler the courtesy of supposing that what he writes can be understood, and try to put my understanding into reasonably good English. Sometimes, if the Latin is irremediably awkward or odd, the translation will reflect this, but I always give him the benefit of doubt. This approach admittedly involves the risk of a mistranslation based upon a misunderstanding. To this I can only say that any attempt at understanding involves the risk of misunderstanding, and it is a risk I willingly take. It is my fervent hope that the light my efforts shed upon this difficult work will outweigh any errors in translation that remain.

The task of proofreading and searching for blunders was originally made much easier by the use of the nearly complete draft translation by Ann Wegner which Owen Gingerich kindly gave me. I have compared my entire work with hers, sentence by sentence, and many errors have thus been caught. In passages where the meaning is doubtful, I have also referred to Caspar's German translation.[22]

As I was preparing this second edition, my St. John's colleague Bruce Perry very kindly offered to read the translation against the Latin and offer corrections and suggestions. Although an additional review would undoubtedly delay publication, Dr. Perry's offer was too attractive to reject. He is a classicist familiar with both Kepler's Latin (he had assisted in my translation of the *Optics*) and the ancient astronomical tradition (as translator of *The Almagest*). The translation has benefited enormously from his unstinting work.

Mathematical arguments present special problems for the translator. In some respects, mathematical passages require less actual translation than the

---

[21] For educational trends in the sixteenth century, see R. R. Bolgar, "Education and Learning," *New Cambridge Modern History* (Cambridge, 1968), vol. III ch. 14. For Kepler's familiarity with classical rhetorical conventions, see N. Jardine, *The Birth of History and Philosophy of Science* (Cambridge, 1983), pp. 74–79.

[22] Johannes Kepler, *Neue Astronomie*, übersetzt und eingeleitet von Max Caspar (Munich and Berlin, 1929; new edition 1990). Caspar's introduction to the translation contains much mathematical material that is not in the notes to the Latin text, including series expansions of many orbital equations.

other parts (symbols and letters, for example, remain unchanged). But it is my experience that no argument can be well translated unless it is well understood. Often the correct translation of a particular preposition cannot be known unless the accompanying operation is known, and the only way to be certain that the operation is known is to repeat it. Therefore, to test my comprehension, I have repeated every computation in the book. Although this may seem an extreme measure, I have time and again found this precaution justified, both in fitting the translation to the argument and in gaining a perspective on the whole that could be acquired in no other way.[23]

There is another difficulty in Kepler's language that is not really his fault. As a highly inflected language, Latin possesses an inherent syntactic clarity that far surpasses that of English. As a result, it is possible to write a very long Latin sentence which is perfectly clear but which is very cumbersome at best when turned into a single English sentence. Therefore, I have nearly everywhere dissected Kepler's rambling sentences mercilessly in the interest of clarity. The sole exception is the Letter of Dedication, a delightful rhetorical bagatelle, where I have kept the sentence divisions intact. This may give some idea of the flavor of Kepler's language, and will also perhaps inspire in the reader's bosom effusions of gratitude that the rest of the book was treated differently.

As was the custom, the book is prefaced by an assortment of poetry, much of it in praise of Kepler and his work. However, Kepler has also included a substantial poem by Tycho Brahe urging the youth of the day to assist in the reformation of astronomy. He has responded with a poem of his own, depicting his work as arising from Tycho's and answering his call. It is an interesting and informative interchange, and the poetry is not without a degree of elegance. I have accordingly taken the trouble to retain the dactylic hexameter of the original in the translation, in an attempt to capture some of the feeling of the verse.

The text

The text used is the excellent modern edition by Max Caspar, volume III of *Johannes Kepler Gesammelte Werke* (Munich, 1937). Since no manuscript is known (other than a few early drafts of some chapters), Caspar based his edition entirely upon Kepler's printed edition, correcting typographical errors and noting the computational errors he found. Caspar's emendations have been followed in the translation, usually without comment. Errors that remain in the Caspar edition have been checked in the facsimile of the first edition, where appropriate, and the ensuing correction has been documented in a footnote.

In preparing his edition, Caspar added many notes, some of which are most helpful, and some, in my opinion, merely distracting. He was at great pains to translate much of Kepler's mathematics into modern form, presumably so that a modern reader could understand it. This approach seems to me

---

[23]Recomputation has also resulted in the discovery of many hitherto undetected errors. Where these are significant (that is, where the discrepancy is greater than about $2'$ of arc, or about 0.1% in linear magnitudes), corrected figures are given in the notes. Some, but not all, of the data have also been checked.

misguided: what the original geometry shows is very different from what a polar equation tells. Both are valuable, but we learn much more about Kepler by trying to see things as he saw them. Moreover, Caspar's analysis has been surpassed, most notably in the articles by D. T. Whiteside and A. E. L. Davis cited above. I have therefore kept only what seems useful in Caspar's notes, incorporating it into my own footnotes, giving credit where I have been aided by his insight or am relying upon his information.

Among Caspar's notes are to be found two remarkable long selections from the surviving manuscripts, one a draft table of contents (apparently from 1601–1602,[24] and the other a document of several pages entitled "*Axiomata physica de motu stellarum.*" These are, of course, of great interest. However, to include them in this translation would unduly focus attention upon them at the expense of other important manuscripts and printed letters. I have therefore decided not to include translations of any of the manuscript material, other than a few brief passages in footnotes.

Format

In the interest of fidelity to the author's intentions, and of ease of reference to the original, Kepler's layout and typography have been largely retained. Paragraph divisions have mostly been kept even when they seemed odd or illogical (an exception is the setting of Kepler's numbered topics in the "Summaries" as numbered lists). Kepler's marginalia are in the margin. Kepler's idiosyncratic use of italics to distinguish mathematical passages, which reveals much about the distinction then drawn between mathematics and physics, has been followed.

For this second edition, the page size has been substantially enlarged. Kepler's original was a grand volume, more than 35 cm tall, and although practical concerns gainsaid an attempt to match the original, there were good arguments for a larger format. There would be ample room for the displayed computations and the many tables. Fewer diagrams would have to be repeated in order to satisfy the publisher's requirement that no page should need to be turned to follow a reference in the text to a figure. Adjusting page breaks would be less challenging. There would be more space for Kepler's larger and more complex diagrams. The magnificent Synoptic Table could appear in a single view, as Kepler had intended, instead of being distributed over eight pages, as in the 1992 edition. A taller volume is, of course, something of a nuisance for librarians, but most libraries that include the translation will also own the *Gesammelte Werke*, which is also oversized, and with which the translation will accordingly be shelved.

The diagrams deserve special mention. In the past, it has been usual for diagrams to be drawn (or cut!) by graphic artists following the author's or editor's instructions. The advent of computer drawing software has made it much easier for an author to produce print-quality diagrams directly. This avoids the kind of error that can creep into the production of geometrical fig-

---

[24]Gingerich, "Kepler's Treatment of Redundant Observations," in F. Kraft, K. Meyer, and B. Sticker, editors, *Internationales Kepler-Symposium Weil der Stadt 1971*, (Hildesheim, 1973), p. 313.

ures. An example is the diagram to Chapter 16: apparently the artist who did the original woodcut neglected to make it a mirror image of Kepler's drawing, and the result is that the lettering on the printed diagram has been reversed in all editions and translations up to the present one. All the diagrams in this edition have been constructed in accordance with instructions in the text, so as to function properly. This has occasionally resulted in a correction or modification of Kepler's original.

Nevertheless, the original woodcuts are in general impressively accurate (if sometimes difficult to read); and therefore, where possible, the re-drawn versions have followed the pattern of the originals fairly closely. Further, in certain especially significant diagrams, the woodcuts have been enhanced with ornamentation which can be viewed as possessing symbolic importance. Therefore, although the lines and curves have been re-drawn, some of the more significant ornaments have been scanned from the originals and included on a separate page (p. 437).

Notation and abbreviations

For the sake of clarity and economy of expression, the following notations and abbreviations have been uniformly used in the translation:

Numerals (with a very few exceptions) are Arabic. Although Kepler often used Roman numerals, especially for dates and times, I can see no reason to follow his usage, and much to be gained by abandoning it.

Dates and times are expressed by stating the year, month, and day, in that order. Times are indicated by superior characters (superscripts); thus, where Kepler writes "H. VII M. XXXVI" the translation reads "$7^h 36^m$" Where a time interval is greater than one day, the number of days is indicated by a superscript "d."

Angles are expressed in degrees, minutes, and seconds, denoted by the conventional symbols (°, ′, ″). These are never used to express times. Longitudinal positions and elongations are often expressed in signs, degrees, minutes, and seconds (a sign is 30°). The number of signs is denoted by a superscript "s." It should be noted that Kepler himself most often used these symbols, but sometimes used the words they stand for (or their abbreviations), and sometimes even the symbols and the words together. No attempt has been made to follow that usage in the translation.

Abbreviations of frequently cited titles are:

KGW—W. von Dyck, M. Caspar, and F. Hammer, eds., *Johannes Kepler Gesammelte Werke* (Munich, 1937–).

TBOO—J. L. E. Dreyer, ed., *Tychonis Brahe Opera Omnia* (Copenhagen, 1913–1929).

Numbers in the margins are page numbers of the 1609 edition. Since these numbers are also in the margin of *KGW* vol. 3, the translation is conveniently keyed to both of the presently available editions. Where pages in the 1609 edition are not numbered (as in all the prefatory and introductory matter), pages are identified by the signature and folium number.

Other material in the margins is Kepler's.

Footnotes are the translator's, unless otherwise identified.

Glossary

This glossary serves three purposes: it defines a number of technical or otherwise unusual English words, it discusses the meaning of a number of Latin words which have no exact English equivalents, and it explains a number of instances where context suggests an unusual translation.

Anomaly: any of several angular measures of a planet's motion. For an explanation and illustration of the different kinds of anomaly, see "Technical aspects of Kepler's astronomy and mathematics" in the translator's introduction.

Anomaly of commutation: the angular measure of the relative motion of the earth and a planet about the sun or about the center of the earth's orbit. The former is called the "true" or "equated" anomaly of commutation, and the latter, the "mean" anomaly of commutation.

Artifex: "Practitioner." An *artifex* is generally the master of some art, but in astronomy it has a sense that is closer to the English "artificer," in that it denotes one who creates artifices. One might say that this is a theorist, but "artifex" also covers the art of observation; hence, following Duncan and Jardine, I have used "practitioner."

Aspect: The apparent angle between two celestial bodies. In astrology, aspects are based on equal divisions of the full circle. Thus, the **trine** aspect is $\frac{1}{3}$ of a circle, **quadrature** is $\frac{1}{4}$ of a circle, and so on. Aspects and their angles (from Kepler's *Harmonice Mundi* IV.9, trans. Aiton et al. p. 340) are:

Conjunction	0°
Opposition	180°
Trine	120°
Quadrature or Quartile	90°
Octile or Semiquartile	45°
Sesquiquadrate	135°
Quintile	72°
Biquintile	144°
Half-quintile or Decile	36°
Tridecile	108°
Sextile	60°
Semisextile	30°
Quincunx	150°

Aura aetherea: "aethereal air." The term Kepler uses to describe the interplanetary space.

Base of the Latitude: The difference between the declination of a celestial body and the declination of the ecliptic at the same right ascension. See Chapter 11 and the Translator's Introduction under "Observations and computations."

Biquintile, *see* aspects.

Calendars: In 1582, Catholic countries adopted the **Gregorian Calendar**, which inserted a small adjustment to the prevalent **Julian Calendar**. At the same time, the Pope moved the date ten days forward, to bring the calendar into agreement with the seasons and with liturgical dates, such as Easter. Lutherans did not accept the change; hence, all of Kepler's dates (unless otherwise specified) are Julian dates. To express Kepler's dates in modern terms, subtract ten days.

Complementum: Used to represent both "complement" and "supplement." Usually one must tell which is meant by the context; sometimes, however, Kepler distinguishes them by calling the former "*complementum ad rectum,*" and the latter "complementum ad semicirculum" or "ad duos rectos." For the "full circle complement," see below.

Doctrina: This word denotes a coherent body of learning, and performs much the same function as the English suffixes "-ology" or its relatives "-ography" and "-ometry." Therefore, some phrases in which it appears, such as "*doctrina triangulorum,*" have been Englished with the appropriate suffix (in this instance, "trigonometry'). Its use also has close affinity with the use of the word "theory" in such phrases as "theory of relativity," and so, when no suitable compound word was found, "doctrina" has been translated "theory." For example, "*doctrina de gravitate*" becomes "theory of gravity."

Duplicate ratio: "When three magnitudes are proportional, the first is said to have to the third the **duplicate ratio** of that which it has to the second." (Euclid, *Elements*, Book V def. 9). When the magnitudes are expressed in numbers, the ratio of the third to the first is the square of the ratio of the second to the first. As Kepler uses the term, it is practically equivalent to "square."

Eccentric equation: The amount that must be added to or subtracted from the planet's mean longitude (q. v.) to give its eccentric position (q. v.). For circular orbits, Kepler also calls this the "physical part of the equation."

Eccentric position: the planet's position on the eccentric, in zodiacal coordinates, measured about the center of the eccentric.

Elongation: *see* middle elongation.

Equation: the angular quantity that must be added to or subtracted from the measure of a planet's position about one center to give its measure about a different center. Or, in terms of Kepler's final theory, the angular difference between any two measures of a single planetary position.

Full circle complement: the difference between a given angle and the full circle (360°).

Hypothesis: "We, however, call 'a hypothesis' generically whatever is set out as certain and demonstrated for the purpose of any demonstration whatsoever. [...] Specifically, however, when we speak in the plural of 'astronomical hypotheses,' we do so in the manner of present-day learned discourses. We thereby designate a certain totality of the views of some notable theorist [*artifex*], from which totality he demonstrates the entire basis of the heavenly motions." (Kepler, *Apologia pro Tychone*, fol. 266r, in N. Jardine, *The Birth of History and Philosophy of Science* (Cambridge 1984) pp. 138–9.

Lex: "law." The preposterous idea that inanimate objects "obey" the "laws of motion" much as rational beings obey laws laid down by their ruler has a fascinating history, but this is not the place to go into it. Suffice it to say that the notion of "laws of motion" was ubiquitous in sixteenth century astronomical works, and that it was plausible in this context because the motions of the planets were thought to be governed by angels. Kepler is thus apparently only following current usage when he mentions *leges motuum*. For more sources, see N. Jardine, *The Birth of History and Philosophy of Science* (Cambridge 1984), p. 240.

Libration: *see* reciprocation.

Mean longitude (*longitudo media*): A measure of elapsed time, in degrees, where 360 degrees is set equal to the planet's tropical period. Since it is a measure of time and not of position, it is always given as an angle: so many "signs" of 30° each (abbreviation "s"), plus so many degrees, minutes, and seconds. The mean longitude is so adjusted that when the planet's eccentric position (q. v.) is on the line of apsides, the mean longitude is the same as the true longitude. Thus a change in the position of the apsides requires a change in the mean longitudes, an adjustment to which Kepler gives considerable attention.

Kepler creates much confusion in using these same words, usually in the plural, to denote either the mean distance of the planet from the sun, or the region of the orbit that is at or near the mean distance from the sun. See *Middle elongation*.

Middle elongation (*longitudo media*): The word *longitudo* here does not have its usual astronomical meaning (see the previous entry). In Chapter 47, p. 350, Kepler says that when a planet is equally distant from the center of the eccentric and the center of vision (i. e., the sun), it is said to be at the *longitudo media*, that is, at the mean distance of the planet from the sun. He distinguishes this from what he characterizes as the incorrect modern usage, by which the *longitudo media* is the point on the circumference that possesses the mean distance. It is notable that neither sense refers to the planet's angular relation to aphelion and perihelion, as in the previous glossary entry. Nevertheless, an attempt has been made to reflect the ambiguity in the word *longitudo* in the translation: "elongation" seems to do that reasonably well, since it can refer either to an angle or a length. So wherever the English reads "middle elongation," it is a translation of *longitudo media*.

Mora: Literally, "delay." Kepler uses this word to denote the amount of time a planet takes to traverse a given orbital arc; hence, it has been translated "elapsed time," or simply "time," when it describes a finite arc with a fixed starting point; and "increment of time" when it denotes the time to traverse any of the small equal parts into which Kepler divides a finite arc. It is, I believe, of great significance that Kepler treated time as a *dependent* variable, as if more or less of it elapses as the planet traverses uniform arcs. It is at least arguable that he would not have formulated his "second law" (the law of areas) had he not held this concept of "time-per-unit-distance," which is the inverse of our Cartesian-Galilean-Newtonian concept of velocity.

Motus: Although this usually means "motion," Kepler often uses it where we would say "position'; for example, *"motus medius Solis est..."* is best translated, "the sun's mean position is...." The choice of translation is made on the basis of context: what makes sense here?

Mundus: "world." This usually (but not always) denotes the entire visible creation, including the fixed stars (see, for example, Chapter 6, p. 106).

Nonagesimal: the point on the ecliptic 90° west of its intersection with the eastern horizon. It is also called the "medium coeli," or M. C. for short. The nonagesimal is an important point in parallax computations, for which see the Translator's Introduction.

Orbis: "orb." Although this usually means "circle," the Latin has the same ambiguity as the English "orb': it may be a sphere.

Planum: "plane area." Kepler's usage is significantly different from the meaning of the English "plane:" Kepler's *planum* is finite, and usually denotes what we mean by "area."

Potentia: literally, "power." In mathematics, it usually means "second power," or "square," but can also mean "sum of the squares." Usually translated "square."

Potest: literally, "is able" or "has the power." This is the verb corresponding to the noun *"potentia"* (above), and in mathematical usage means "is equal in square."

Quadrature: *See* Aspects.

Reciprocation (*libratio*): Kepler evidently meant to suggest the image of a balance beam swaying to and fro. However, the in-and-out motion that this word denotes is closer to the English "reciprocation."

Sesquiquadrate, *see* **Aspects.**

Species: This word, related to the verb *"specio"* (see, observe) has an extraordinarily wide range of meaning. Its root meaning is "something presented to view," but it also can mean "appearance," "surface," "form," "semblance," "mental image," "sort," "nature," or "archetype," to mention only a selection of its most diverse senses. It is in fact the Latin equivalent of the Greek "εἶδος," which is Plato's word for his "forms" or "ideas." The Epicureans used it to denote a "surface film given off by physical objects" (*Oxford Latin Dictionary*; cf. Lucretius 4. 602 and 6. 993). Robert Grosseteste and Roger Bacon used what is apparently a Neoplatonized form of this technical meaning, in which the material "surface film" is transformed into an "immaterial form" (see A. C. Crombie, *Robert Grosseteste and the Origins of Experimental Science,* second ed. (Oxford, 1962) pp. 104–116 and 144–147). Accordingly, in his translation of Kepler's *Mysterium cosmographicum* (*The Secret of the Universe,* Abaris

Books, 1981), A. M. Duncan has translated *species* as "emanation." Ann Wegner, in her draft translation of *Astronomia nova,* independently chose the same word, and Gingerich used it in his published translation of Chapter 34 (*The Great Ideas Today, 1983,* Chicago, 1983, pp. 325–329).

C. G. Wallis, on the other hand, stayed closer to the root meaning and the Platonic nuances in translating *species* as "form" (Kepler's *Epitome,* in *Great Books of the Western World* 16 (Chicago, 1952) p. 897 (translating the first few words of Kepler's p. 517), *et passim*). I began by using "form," possibly influenced by Wallis's translation, but soon abandoned this as being too far from English usage. I also rejected the nearly suitable "emanation," for two reasons. First, Kepler also used various Latin cognates of that word which seem to me best translated by their English kin. Hence, if *species* were also translated "emanation" the reader would have no indication of what word Kepler was using. And second, there are places where "emanation" does not make good sense. Take, for example, this passage from Chapter 34, which, in Gingerich's translation (*op. cit.,* p. 326) reads:

> Therefore,…when any particle of the solar body moves toward some part of the world, the particle of the immaterial emanation which from the beginning of creation corresponded to that particle of the body also always moves toward the same part. *If this were not so, it would not be an emanation,* and would come down from the body in curved rather than straight lines. (Italics supplied)

The sense of the italicized phrase is much clearer if "form," or simply *"species,"* is substituted for "emanation." Kepler's point is that the *species* is a property of the solar body and not a self-subsisting entity, and hence does not behave like a body. This point is lost in the proposed translation, and a puzzled reader would have to consult the original to tell whether the Latin read *"species"* or something like *"emanatio."*

Having rejected "form" and "emanation," I was tempted to use the English word "species," since it has had a wide range of meanings many of which correspond to the various Latin senses, and since it has also been used to denote "A supposed emission or emanation from outward things,…" (*Oxford English Dictionary*). Unfortunately, all the relevant senses are obsolete, so its use would serve only to confuse matters. I have therefore thrown up my hands, admitted defeat, and declined to translate it at all. It appears as the Latin word *"species,"* in italics.

Speculatio: "Theory," "theorizing," "theoretical argument," "theoretical consideration." This is not a classical word. It is derived from the verb *"speculor,"* "spy into, examine," but was later used in a philosophical context by Boethius (*De consolatione philosophiae* 4. 1 and elsewhere) to mean something like the English "speculation." But the *speculatio* that Kepler has in mind is a reasoned argument, not a wild conjecture. For example, in the *Epitome,* on p. 654, where Kepler supports a mathematical argument "by the force of physical *speculatio,*" he obviously means that the physical theory requires a certain mathematical relation to hold. Therefore, even though "theory" is also used to translate other words, it seems the best translation for *"speculatio,"* a choice that is amply confirmed by Kepler's usage in *Astronomia Nova.*

Theoricae: "Theories" or "theories of the planets." This is a technical term used in several widely read medieval books on planetary theory, and denotes the geometrical configurations used to account for planetary motions. It is nearly equivalent to "hypotheses," although the latter term is not restricted to the geometrical models. However, "hypotheses" as a translation has the disadvantage of blurring the subtle distinction between the two. This problem is avoided by the use of "theories." Further, despite the use of "theory" as a translation of other words, it should not be difficult to tell from the context what the Latin is.

Theory: This is the English translation of three distinct Latin words: *doctrina, speculatio,* and *theoricae* (the last always plural). The Latin can be distinguished by the usage: if the English is "theory of...," the Latin is *"doctrina'*; if "physical theory," the Latin is *"speculatio'*; and if "theories of the planets," the Latin is *"theoricae."*

Trine: *see* aspects.

Versed sine:

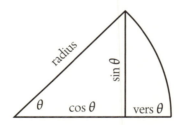

The difference between the radius and the cosine, as illustrated above.

Virtus: "power," and
Vis: "force." In classical usage, *vis* properly refers to physical action, while *virtus* (from *vir,* "man") denotes those qualities pertaining to manly excellence, both physical and moral. *Vis,* for example, is frequently used classically of violent action, while *virtus* is never used in this way. On the other hand, there appears to be no classical instance of *virtus* where it would not have made sense to use *vis,* and hence, in a physical context, *virtus* would be a particular and more abstract kind of *vis.* With "power" and "force," the opposite is true: we think of force as a particular and more concrete form of power. Nevertheless, the two words in most respects adequately represent the meanings of *virtus* and *vis* in classical Latin.

In Kepler's physics, these two words are supremely important, and so the question of their meaning merits serious study. Although in *Astronomia Nova* Kepler usually uses *"virtus"* to denote the abstract motive power that emanates from the sun as a *species,* and *"vis"* to denote the actual pushes and pulls on the planet, counterinstances can be found. Further, his use of the words in the *Epitome* is different, and it is not clear whether his later usage represents a refinement, a modification, or some combination of the two. In the present translation, no attempt has been made to answer these questions. Instead, it seems most sensible to choose a translation that expresses the classical usage, since it is that usage which Kepler's audience knew, and which therefore formed the context within which Kepler developed his ideas. Also, since the meanings are not perfectly clear, it is desirable to use a consistent translation so that the reader may easily infer the Latin original: hence, "*virtus*" is always "power" and "*vis*" is always "force," and these two English words are used only to translate these two Latin words.

Whole sine: The sine of a right angle, equal to the radius (100,000 units, in Kepler's trigonometry).

Astronomia Nova

ASTRONOMIA NOVA
ΑΙΤΙΟΛΟΓΗΤΟΣ,
SEV
PHYSICA COELESTIS,
tradita commentariis

DE MOTIBVS STELLÆ
MARTIS,
Ex observationibus G. V.
TYCHONIS BRAHE:

Juſſu & ſumptibus
RVDOLPHI II.
ROMANORVM
IMPERATORIS &c:

Plurium annorum pertinaci ſtudio
elaborata Pragæ,

A Sᵉ. Cᵉ. M.ᵗⁱˢ Sᵉ. Mathematico
JOANNE KEPLERO,

Cum ejusdem Cᵉ. M.ᵗⁱˢ privilegio ſpeciali
ANNO æræ Dionyſianæ ↀↃ IↃ C IX.

NEW ASTRONOMY
BASED UPON CAUSES
OR
CELESTIAL PHYSICS

Treated by means of Commentaries
ON THE MOTIONS OF THE STAR

MARS

From the Observations of
TYCHO BRAHE, GENT.

By Order and Munificence of

RUDOLPH II
EMPEROR OF THE ROMANS, &c.

Worked out at Prague in a tenacious study
lasting many years,

By His Holy Imperial Majesty's Mathematician
JOANNES KEPLER

With the same Imperial Majesty's special privilege
In the Year of the Dionysian era MDCIX

P. Ramus, *Scholae Mathematicae* Book II, p. 50

"Thus the contrivance of hypotheses is absurd; nevertheless, in Eudoxus, Aristotle, and Callippus,[1] the contrivance is simpler, as they supposed the hypotheses to be true—indeed, they have been venerated as if they were the gods of the starless[2] orbs. In later times, on the other hand, the tale is by far the most absurd, the demonstration of the truth of natural phenomena through false causes. For this reason, Logic above all, as well as the Mathematical elements of Arithmetic and Geometry, will provide the greatest assistance in establishing the purity and dignity of the most noble art.[3] Would that Copernicus had devoted more energy to this idea of establishing an astronomy without hypotheses! For it would have been far easier for him to describe an astronomy corresponding to the truth about its stars, than to move the earth, as if some gigantic task, so that we might observe the stars at rest in relation to the earth's motion. Why could there not rather arise someone from among the great number of celebrated schools of Germany, a philosopher as well as a mathematician, who would attain the prize of eternal praise that is publicly offered? And if any fruit of transitory usefulness can be offered to compare with a prize of such excellence, I will solemnly promise you the Regius Professorship at Paris as a prize for an astronomy constructed without hypotheses, and will fulfill this promise with the greatest pleasure, even by resigning our professorship."

The Author to Ramus.

Conveniently for you, Ramus, you have abandoned this surety by departing both life and professorship. Were you now to hold the latter, I would, by my rights, claim it for myself, inasmuch as, by this work, I shall succeed, even by the judgement of your own Logic. As you ask the assistance of Logic and Mathematics for the noblest art, I would only ask you not to exclude the support of Physics, which it can by no means forgo. And unless I am mistaken, you readily grant this, seeing that you surround your *Conformator* with Philosophy as well as Mathematical matters. Thus with the same ease, do you yourself also hearken to Philosophy when she is defending something commonly considered most absurd, not with a gigantic effort, but with the best arguments. For when it functions, it effects nothing new, nothing unaccustomed, but only fulfills the function for which it was invented.

It is a most absurd tale, I admit, to demonstrate natural phenomena through false causes, but this tale is not in Copernicus. For he too considered his hypotheses true, no less than those whom you mention considered their old ones true, but he did not just consider them true, but demonstrates them; as evidence of which I offer this work.

But would you like to know who originated this tale, at which you wax so wroth? "Andreas Osiander"[4] is written in my copy [sc. of Copernicus's *De revolutionibus*], in the hand of Hieronymus Schreiber of Nuremberg. This Andreas, when he was in charge of publishing Copernicus, thought this preface most prudent which you consider so absurd (as may be gathered from his letters to Copernicus), and placed it upon the frontispiece of the book, Copernicus himself being dead, or certainly unaware of this. Thus Copernicus does not mythologize,[5] but seriously presents paradoxes;[6] that is, he philosophizes.[7] Which is what you wished of the astronomer.

---

[1] Eudoxus, a pupil of Plato (as was Aristotle), attempted to account for planetary motion through a nest of concentric spheres each of which imparted its motion to the axis of the sphere it immediately contained. Callippus later tried to bring Eudoxus's rather unsatisfactory attempt more nearly into accord with the phenomena by adding more concentric spheres to the schema. Aristotle subsequently gave the Eudoxian-Callippian homocentric system his blessing, adding intelligent beings whose task was to move the spheres, most of which were starless—hence, Ramus's sarcastic remark. See Aristotle, *Metaphysics*, Book XII ch. 8.

[2] ἀνάστρων.

[3] Logic, Arithmetic, Geometry, and Astronomy are four of the seven "Liberal Arts," the other three being Grammar, Rhetoric, and Music. Astronomy was frequently regarded as the noblest of the arts because of the excellence of the objects of its study.

[4] For more on the theologian Osiander (1498–1552) and his notorious "instrumentalist" preface to Copernicus's work, see N. Jardine, *The Birth of History and Philosophy of Science* (Cambridge 1984) pp. 150–154.

[5] μυθολογεῖ.

[6] παραδοξολογεῖ.

[7] φιλοσοφεῖ.

TO RUDOLPH II
THE EVER AUGUST EMPEROR OF THE ROMANS
KING OF GERMANY, HUNGARY, BOHEMIA, &c.
ARCHDUKE OF AUSTRIA, &c.
MOST AUGUST EMPEROR

In order that Your Holy Imperial Majesty, as well as the entire House of Austria, might be happy and prosperous in most serene renown, I am now at last exhibiting for the view of the public a most Noble Captive, won some time ago through a difficult and strenuous war waged by me under the auspices of Your Majesty. I do not fear he will object to the name of Captive, since for some time he has been accustomed to dropping for a little while his vaulted shield and his arms and giving himself over freely and playfully to capture and bondage, whenever custody, prison, or chains are ordered.

I will not be able to make this spectacle any more brilliant by any means other than by writing a panegyric upon this most distinguished captive, and proclaiming it publicly.

However, one who would venture forth upon this battlefield encounters an astonishing brightness, and averts his squinting eyes, made accustomed to the feeble light of Night, and to scholastic shadows.

I therefore leave it to the writers of history books to describe the greatness of our Stranger, which he acquired in the art of war.

They would certainly say that he it is through whom all armies conquer, all military leaders triumph, and all kings rule, without whose aid no one ever honorably took a single captive. Let them now feast their eyes with looking upon him, captured through my martial efforts.

Those who admire Roman greatness would say that he is the begetter of the Kings Romulus and Remus, the preserver of the City, protector of the Citizens, Supporter of the Empire, by whose favor the Romans discovered military discipline, improved and perfected it, and subjugated the orb of the world. Let them therefore give thanks at his being penned up and at his being acquired as a happy omen for the House of Austria.

I, for my part, retreat hence to other ground better suited to my powers. Nor will I set foot upon that part of my profession in which strife arises between me and my fellow soldiers.

They, for their part, would surely rejoice with a different joy: he has been restrained by the bonds of Calculation, who, so often escaping their hands and eyes, was accustomed to deliver vain prophecies of the greatest moment, concerning War, Victory, Empire, Military Greatness, Civil Authority, Sport, and even the cutting off or calling forth of Life itself. Let them congratulate Your Majesty that the Master of the Horoscope[1] has been brought under control,

---

[1] It is undoubtedly the Emperor's horoscope that Kepler means, and the details mentioned (such as the rising sign and the moon sign) are probably those of the time of his birth. According to a horoscope attributed to Nostradamus (Cod. Guelf. 208 Extrav. in the Herzog August Bibliothek in Wolfenbüttel), Rudolph was born on 18 July 1552 at $6^h 45^m$ after noon. The planetary positions given by Kepler are consistent with those in Nostradamus's chart. The translator has been unable to find information on the *astragalis lusum trigonicum*, here translated "the triangular game with knucklebones." *Astragalus* is a Greek word, denoting (among other things) any of a number of different bones, and by derivation, a knucklebone used as a die in gaming.

contains the Heart of Heaven;[2] in Capricorn, which is rising, he is exalted; in Cancer, into which the moon was entering, he customarily plays the triangular game with knucklebones; in Leo, where the Sun plays host, he is recognized as being one of the family; and finally, he is the ruler of Aries, beneath which Germany is supposed to be, over which he rules in complete harmony with Your Holy Imperial Majesty.

Let them be occupied in this part of the triumph; I do not mind. I shall give them no cause for quarreling on such a festive day: let this impertinence pass as a soldiers' joke. I myself shall occupy myself with Astronomy, and, riding in the triumphal car,[3] will display the remaining glories of our captive that are particularly known to me, as well as all issues of the war, both in its waging and in its conclusion.

For he is not to be held without honor among us, whom the eternal Architect of this world, and the Father of Heavens and Humans in common, Jupiter, located in the front lines of the visible bodies, so that he might serve as a soldier for the glory of his Creator through his perennial course through the ethereal regions, and so that he might raise human minds, lulled to sleep by a deep somnolence, from the slanderous reproach of idleness and igno-

(**) 2 v   rance, train them by his own sorties, and provoke them forcefully to carry out investigations in the heavens for the praise of their Creator.

It is he who is the most potent conqueror of human inventions, who, ridiculing all the sallies of the Astronomers, smashing their machines, and striking down the hostile troops, kept safe the secret of his empire, well guarded throughout all ages past, and performed his rounds in perfect freedom with no restraints: hence, that Priest of Nature's Mysteries and most distinguished of the Latins, C. Pliny, registered his chief complaint that "Mars is the untrackable star".[4]

The rumor is that Georg Joachim Rheticus[5] (a disciple of Copernicus not lacking honor in the memory of our forebears, and who, as the first to dare to yearn for a reconstruction of Astronomy, thereupon strove for it through observations and discoveries that are not to be scorned), when he was brought up short in amazement by the motion of Mars, and did not disentangle himself, fled to the oracle of his familiar Genius, either intending (the gods willing) to explore that being's erudition, or driven by a headstrong desire for the truth, whereupon that stern patron, exasperated, alternately caught the importunate inquirer by the hair and dashed his head against the low-hanging paneled ceiling, and then threw him down, flattening him on the paved floor, adding the response: "This is the motion of Mars." A rumor—cause of evil:[6] there is noth-

---

[2] The Heart of Scorpius is the star Antares. "Ant-Ares", that is, "Counterpart of Mars", was so named because of its red color. Astrologers supposed it to have a particular affinity for Mars.

[3] In this connection, see the illustration of Urania's chariot included in the diagram for Protheorem XI in Chapter 59, p. 436.

[4] Pliny, *Natural History*, II. 17.

[5] For Rheticus see Dennis R. Danielson, *The First Copernican: Georg Joachim Rheticus and the Rise of the Copernican Revolution* (Walker & Co., 2006)

[5] *Fama malum:* presumably an allusion to Virgil's description of Rumor in *Aeneid* IV.174 ff.

ing else more injurious to good reputation, for it is as tenacious of deception and distortion as it is informative of the truth. It is nevertheless not unbelievable that Rheticus himself, when his speculations were not succeeding and his spirit was in turmoil, leapt up in fury and pounded his head against the wall. For what wonder would it be if the same thing happened to Rheticus, who provoked Mars, as once happened to C. Octavius Augustus Caesar when he lost five legions under the command of Quintilius Varus, surrounded by his enemy Arminius, protégé of our Germanic Mars?[7] (**) 3 r

Nevertheless, here too, as in other kingdoms, the ruling influence of our enemy was sustained and supported, more than any other thing, by the persuasion and confusion of the multitude of people, the defiance of which I have always considered the path to victory. Indeed, when I was but indifferently well versed in this theater of Nature, I formed the opinion, with practice [*usus*] as my teacher, that, just as one human being does not greatly differ from another, neither does one star differ much from another, nor one opponent from another, and hence, no account is to be received easily that recklessly spreads abroad something unusual about a single individual of the same kind.

In this place chief praise is to be given to the diligence of Tycho Brahe, the commander-in-chief in this war, who, under the auspices of Frederic II and Christian, Kings of Denmark, and finally of Your Holy Imperial Majesty as well, explored the habits of this enemy of ours nearly every night for twenty years, observed every aspect of the campaign, detected every stratagem, and left them fully described in books as he was dying.

I, instructed by those books as I succeeded Brahe in this charge, first of all ceased to fear [the enemy] whom I had to some extent come to know, and then, having diligently noted the moments of time at which he was accustomed (**) 3 v to arrive at his former positions, as if going to bed, I directed the Brahean machines thither, equipped with precise sights, as if aiming at a particular target, and besieged each position with my inquiry as the chariots of the great Mother Earth were driven around in their circuit.

The campaign did not, however, succeed without sweat, since it frequently happened that machines were lacking where they were most needed, or that they were transported over muddy roads by inexperienced charioteers at great expense of time and material, or that the launching of some of them, where I had not yet investigated the matter, occurred in other directions than I had had in mind. Often the brightness of the sun or of the moon, and often an overcast sky, cheated the commander's eyes; and more often the interposition of vaporous air deflected the globe, forcing it from the straight path. Also not infrequently, the walls, where they were presented most obliquely, received ineffectual blows, however numerous they might be. Add to this the enemy's enterprise in making sallies; his vigilance for ambuscades, while we were frequently asleep; and finally, his constancy in defense: whenever he was driven or fled from one castle, he repaired to another: neither was the procedure the same for conquering all the castles, nor was the journey from one to the others easy—either rivers lay in the way, or brambles impeded the attack, but

---

[7]See Suetonius, *Lives of the Caesars*, II.23; also Tacitus, *Annals*, Book I passim.

most of the time the route was unknown; each of which things is thoroughly described in its own place in this commentary.

Meanwhile, in my camp, is there any sort of carnage, any kind of disaster that he has not raged? The overthrow of the Most Distinguished Leader, rebellion, plague, pestilences, domestic matters both good and bad, destined in either case to take time; a new, unforeseen, and terrifying rear attack by the enemy, as I have recounted in the book *On the New Star*;[8] at another time, an

(**) 4 r enormous Dragon with a very long tail, vomiting fire and attacking my camp; desertion and poverty of the soldiers; the inexperience of novices; and, at the head of all, the extreme deficiency of provisions.

At last, when he saw that I held fast to my goal, while there was no place in the circuit of his kingdom where he was safe or secure, the enemy turned his attention to plans for peace: sending off his parent Nature, he offered to allow me the victory; and, having bargained for liberty within limits subject to arbitration, he shortly thereafter moved over most eagerly into my camp with Arithmetic and Geometry pressing closely at his sides.

However, from the time when, after surrendering, he abode by our house's fair laws of friendship, he, through hidden illusions (being unaccustomed to rest), did not cease to incite among us I know not what further fears of war, and if we happened to become terrified, we would give him much to laugh at. But, seeing us strong in spirit, he agreed to live with us in earnest, and, dropping the appearance of hostility, confirmed his faith with us.

This one thing he begs of Your Majesty: since his alliance in the ethereal regions is great (for indeed, his father is Jupiter, his grandfather Saturn, Venus is his sister as well as his mistress, and for some time now the chief alleviation of his chains, Mercury his brother and faithful herald); and since he is possessed by desire of them, and they of him, owing to their similar ways, he wishes that they too might live among humanity, becoming partakers as well of the honor with which he is bestowed; and that Your Majesty might restore them to him

(**) 4 v as soon as possible, since the remnants of this expedition have been severely diminished, who, now that he himself has surrendered, no longer pose any threat. To this end, I readily offer Your Majesty a work that is not without usefulness (it being trained in the most combative circumstances, and well acquainted with the terrain) and no less trustworthy than its predecessor.[9] I pray and beseech you for this one favor (seeing that throughout these nine years, conversation in this hall, packed with soldiers, centurions, and commanders, has supplied me with the word "beseech", as well as the rest of the oration): that Your Imperial Majesty command the chiefs of the treasury to take thought for the sinews of war and supply me with new funds to enlist the army. I pray thus, seeing that I both know that these things already are approved by Your Majesty, and consider that they aim at the glory of God and the immortality

---

[8] In the autumn of 1604 a nova appeared in Ophiuchus, about which Kepler wrote the pioneering work *De stella nova in pede Serpentarii* (Prague 1606). According to Caspar, the "Dragon" is the comet that appeared in the winter of 1607, about which Kepler wrote *Ausfuhrlicher Bericht von dem...Cometen...* (Hall/Sachsen 1608).

[9] A reference to the proposed *Tabulae Rudolphinae* (not published until 1627), which applied the conclusions of the present work to the other planets.

of the Name of Your August Majesty, to Whom I have devoted all my work for a long time; and to Him I now most humbly commend myself.

> March the 28th, in the year of the Dionysian[10] era 1609.
> Your Holy Imperial Majesty's
> Most Humble Mathematician
> Joannes Keppler[11]

---

[10]Here and on the title page, Kepler refers to the Roman Abbot Dionysius Exiguus, who established the commonly used Christian chronology in the early sixth century.

[11]Spelled as in the original.

(**) 5 r

EPIGRAMS[1]
ON THESE COMMENTARIES
ON THE MOTIONS OF MARS
URANIA TO KEPLER[2]

"Cease, O Keplerides, in a war against Mars to do battle:
    Mars submits to none, but it be he to himself.
Vainly therefore you strive to make him submit to your bondage
    Who has lived quite free over numerous ages."
Thus speaks the Muse. But in response he thus answers: "What then?
    Have the accounts of Pallas Minerva escaped your mind?
Pallas was able to prostrate horrific Mars with a boulder—
    True this is, if the songs sung by Homer are yours.
Why, therefore, with the help of mighty Minerva, could not
    Mars now, however fierce, also submit to the yoke?
Look at the book that we've published, under the Rudolphine aegis:
    'Surely,' you will say, 'Mars now too suffers great hardship.'"

ANOTHER

The Liparaean god[3] once captured Mars, ensnared in some netting:
    He was on his way, Venus, to your embrace.
Now once again the War god is chained in the same bonds and fetters,
    Nor is Venus to blame: it's your fault, O Pallas.
You, O Minerva, gave these nets to Tycho, and Tycho
    Gave them to Kepler, who threw them around Mars's ankles.
This is a wondrous thing: Vulcan's art, and that of another;
    Kepler yet surpasses both the one and the other.
Short was the time over which the Vulcanian chains endured.
    Against the ages, however, remain these Keplerian shackles.

Saxirupius wrote this at Prague
in the year 1609.

---

[1] An attempt has been made in the translations to retain approximately the same meter as in the originals, although the time values in Latin verse cannot be reproduced in English. That is, where the English as an accented syllable, the Latin has a temporally long syllable; thus, the rhythm is quite different.

[2] This poem and the two following are written in elegiac couplets. Each such couplet consists of two dactylic hexameters the second of which lacks an arsis in the third and sixth feet. It is a form of verse peculiarly suited to Greek and Latin, and is not easily rendered in English. Furthermore, the verses here are sometimes defective. So although the elegiac form is imitated in the translation, no attempt is made to adhere to it strictly.

[3] This is of course Vulcan, associated with the Aeolian island of Lipara; the story of his capture of Mars is brilliantly told in the *Odyssey*, VIII.302–410.

ANOTHER

Keplerius, an alumnus of earth, makes assault on the heavens:
Seek not for ladders: the earth itself takes flight.

J. Seussius wr. at Dresden.

A HORTATORY ODE[4]   (**) 5 v
of
THYCHO BRAHE
The Highest Astronomer, to Those Who Cultivate Astronomy,
Appended to the Restitution of the Fixed Stars,
in the *Progymnasmata* Vol. I page 295.[5]

Paved is the road now, that many past ages no one could travel,
Though it indeed required the greatest, most vigilant, labors.
Now it allows one to climb the peaks of the unapproached heavens,
And to go through to the highest of houses, Divinities' dwellings.
Or one may wish to describe heaven's fires, whether fixed or traversing
Various paths, and prove their celestial course and position.
Thus can the wonders of Jupiter, highest of gods, be established.

Come then: bestir yourselves, youths, possessing the keen high
Vigor of wit and favor of Talent, for whom, from her birth, famed
Goddess Urania has inspired a divine love of heaven,   10
Granting that they hold earth and the earthly second to heaven's
Goods. For you do not care for the thoughtless beliefs that are held by
Common louts, nor give heed to the gloomy talk of the lazy.
You dismiss those moles to dwell in their dingy dark caverns,
Sightless that they may remain, for that is their dearest desire.
Hither bring your glad spirits, and leave the many behind you;
Hither turn, and do not let Mind, an inspired part of heaven,
Lose the good of its homeland. Your studies as well as your labors
Hither bring with one spirit, to let help come to the weary
King Alfonso,[6] who, as successor to Atlas his neighbour,   20
Bore the weight in like manner, with forces not up to the effort;
Likewise, to let great Copernicus sense the help that is ready,

---

[4]Latin *Paraeneticum*, which is a transliteration of the Greek παραινετικός, meaning "hortatory". Tycho's name is given here as it is spelled in the text. There is a copy of the first ten lines of this poem in Kepler's hand in the Leningrad Kepler manuscripts, vol. XIV f. 372, in which Tycho's name (in the genitive case) is spelled "Trehenis Brahe'. Elsewhere, Kepler consistently has "Tycho," genitive "Tychonis." The poetry, which is heroic verse (dactylic hexameter), is much more elegant than the previous examples.

[5]*TBOO* 2 pp. 302–303.

[6]Alfonso X, "the Wise" (1221–1284), King of Castile and Leon, under whose direction the Alphonsine Tables were prepared. These were the standard planetary tables of the Middle Ages and the early Renaissance, and were still widely used in Kepler's time, although super-seded by Erasmus Reinhold's *Prutenicae tabulae coelestium motuum* (Tubingen, 1551). According to legend, Atlas was supposed to have been turned to stone in North Africa, becoming the Atlas Mountains; hence, he was Alfonso's "neighbor".

Lest, as he gives himself o'er to Herculean labor, approaching
Trustingly, he might succumb to a burden excessively heavy.
Thus are driven the poles, already starting to nod, and
Wanting the pillars of Atlas and Hercules, ruin to bring,
Monstrous, and with it removing the earthly globe from its station,*
Stirring up welcome of barbarousness (that's born of a stupid
Ignorance of matters celestial), and harrying in double downfall
All of humankind, with all the wild beasts and the cattle,
Mingling the courts of the world with blind darkness and primeval Chaos.
30   Do not accept this obscenity: battle such decadent ruin,
Come with me to ascend high Olympus with forces redoubled,
Let us now hasten to close the cracks that have lately been broken,
Firm up the panelled ceiling of heaven with sturdy new crossbeams:
(**) 6 r   Now is the time to act, ere the whole machine tumbles to pieces.

Is there then anyone here who would fly to acquire this crown,
Beautiful for its pure gold, and striking with ivory, gems, and
Red-glowing bronze, but stronger, enduring throughout all the ages,
40   Also wishing to mingle his soul with the souls of the highest?
Will there be any among the numberless earth-dwellers that the
Orb holds, who has accepted things so sublime in his heart?
Is there any who strives to acknowledge the Author of All, for
So many marvelous spectacles set in the measureless heavens?
Do you all indeed keep silent about such great questions?
What use is this silence? The hand must be set to the labor,
Bringing to light at last the heavens' recondite secrets.
Some may be stayed by ambition, by opulence, ignorance, lucre,
From such lofty attempts, driven down to basest endeavors:
50   At least let them spare the others, and not stay their utmost assistance.

I myself, if divine wills give me their favoring glances,
Letting me (as hitherto) surmount any hindrance whatever,
Alas   Lofty and constant in spirit—I will exert myself further,
Stretching each sinew to open great heaven's innermost secrets
Unto the natives of earth, revealing its roofed-in recesses.

Nod in wondrous way thine assent, O Olympus's sapient
Founder, and grant thine aid to one heeding thy deeds so astounding.

The Author of the Work replies:

O hero refulgent in lineage, brilliant in highness of birth, to
Whom the undoubted celestial source of your spirit has granted
Precedence over the rest in achievement, and giving new life to
Those who are dying, through singing and through exhortation:
Why are you scourging with wind and flame a soul, apprehensive,
Nurturing so long such mighty infernos, of things it had hoped for?
For although such great undertakings, surpassing my forces,
No other masters demand than those that your Muse bears; and
Nature according to law gave me in my generation

Margin note (left): * Meaning, that the poles go to ruin. For here he indicts the imperfection of astronomy, and ignorance of it, but not the hypothesis of Copernicus, which makes the earth mobile.

Less intellect than spirit, than intellect less strength of arm: 10
Nevertheless, the Ninth Sister "inspired a divine love of heaven."[7]

Ominous love, to what does it not drive mortal breasts?
Intellect it gave me, and vigorous arms it gave me,
Vivifying with hope disproportionate. Yet you and I are
Sundered by unfair Juno's prejudiced countenance into
Contrary aims: to you she gave the means of acquiring
Power; to me the harsh one denies it: her cunning recurs;
Shutting me off from places aethereal, also more closely
Watching the holy Fires, stung by Prometheus' thieving.
Thus she indulgently heaps you with riches, bedazzling eyes with   20
Lustre of metal, to make them more slothful in seeking Celestial
Lights, and to make them prefer to cleave to the purple-clad pomp that
Flattering whispered praise of the popular whim ever follow.   (**) 6 v
Fortune, when spurned, might threaten to bring on unspeakable sorrow.

Blessed be for the strength of your spirit, O victor of Gods and
Humans, and of your own character, you who approve of the things that
Are to be striven for by Reason's fine eye: having pursued this,
Ceaselessly daring, you could spurn patrimonial riches.
Summon no longer your comrades to utter these praises in private,
Write not your words on the waters. Virtue and treasure do not make   30
Friendly companions: the earth and the Pole are enormously far from
One another, and one is easily viewed from the other.

Scornful of me, the goddess potent begrudged me honor
Measureless. Binding me closely by means of her almighty wishes,
She has bestowed on me nothing that I could scornfully treat as
Second in worth to the Muses, or to block astronomical studies.
Loathing might have won out, might hinder feats of great daring;
Envious Nemesis, too, might have pressed to the ground a talent
Having the power to fly, to sail through heaven's high reaches,
But that the love of singing heaven's mysteries, in which   40
I had followed your footsteps, forestalled me at life's first crossroads.

Therefore, surveying in thought the worn-out orbs of the planets,
Portents immense, and the walls of the world bound for ruin because of
Great gaping spaces that never were braced by the placement of columns,
Causes meanwhile hid by Stygian night, while, sure that they're right, the
Crowd of sages slumbers, not knowing the Prussian Master:[8]
Trustingly I approach to take up such a great weight, and
"Firm up the panelled ceiling of heaven with sturdy new crossbeams:"[9]
Famous timbers from Samos, the five regular solids;
Euclid provided the measure, and Pallas renowned the mind.   50
Urania, redoubling the cheers from more than one learned

---

[7]Cf. Tycho's *Hortatory Ode*, line 10. The "Ninth Sister" is Urania, muse of astronomy.

[8]Copernicus

[9]Cf. *Hortatory Ode*, line 35.

Judge, sang out the festive triumph, pleased 'twas successful.

Having admired your daring, O Brahe, and your delightful labor,
Even though you preferred not to stray from received opinion,
Doubting much above earth, and many things in the heavens:
Me nonetheless you were pleased to number among your disciples,
Spreading your nights out before me, and secrets you've found in your searching
Over long years, and to shine a clear light on your great undertakings.

Would that you had lived on, that the Fates had never snatched from you
60   Prizes matching such deeds, and such well-merited triumphs.
You would have found no orbs, "the great heaven's innermost secrets,"
Spreading themselves out for sight and instruments most precise,
Other than those very ones that are buttressed by my new crossbeams.

Now, since the Goddesses have carried off our too-hasty Master,
And our Patron, abducted, has upset the festive days, and
Well-adorned celebrations, that ought to have given him gladness,
What can I do but in veneration call you again to
Life, O manifest Hero, in an artful image of mind.

(***) r   You will stand near, a mighty Priest in star-studded vestments.

The opinion of Aristarchus and Copernicus.

Here where the offerings heaped up to God the author in the
Temple cerulean rise, while the six curved paths move around in
70   Order; the same number moved by the lamp's most rapid gyration;
And, in the middle, the Hearth,[10] and flames of eternal Light.

Supplicant, I approach, and these my written exertions
In a scholarly book—'tis sweetest frankincense, sweated
From your own trees, O parent of things, and while you were waiting,
gathered by my care—to you with raised hands I give them.
Ah! Let the offering burn![11] Lo, pure I follow, with mighty
Cry, and join the chaste prayers: "May Olympus's sapient
80   Founder grant his kind aid to one heeding his deeds so astounding."[12]

---

[10] Latin, *Focus*. It is fascinating that, although Kepler had already applied this term to the ellipse in the *Optics* of 1604, this is the sole appearance of the word in *Astronomia Nova*. Even as he constructed the elliptical orbit (in Chapter 59) he was evidently unaware that the sun was at the point that had the focal property. Could it be that, before writing this poem (which was presumably written after the rest of the work), Kepler realized the connection? Or, still more fascinating, could the use of the word here in a poetical context have reminded him of the technical meaning he had given it earlier, leading him to discover the connection?

[11] It is difficult to make sense of this line (*Eia adole purus; sequor en…*) as it stands in all editions. Bruce Perry suggests moving the semicolon so as to fall before *purus*, which then clearly modifies the subject of *sequor* (i.e., Kepler himself) rather than standing mysteriously stranded.

[12] Cf. *Hortatory Ode* lines 56–7. The Latin, like the English, is slightly different in the two passages.

The same author's Elegy written in the friendship book[13]
next to the hand and Motto of Brahe:
In looking up I look down.

O noble one, give room, and do not disdain one who follows:
    All that I am is your gift, and all that I will be.
You who have marveled at all of humanity's empty worries,
    Only to you have I stopped my playing the Satyr.
All your monuments give me rest from my worries: without you
    Shadow I was; fathered by you, will be body.
Though for me the earth weaves starlike in airy gyrations,
    Let the same earth stand fixed for you, central in place:
I, however, ascribe these beliefs to the Ancient Masters,
    Not from myself: they displeased you while you were living.
Nevertheless, I, weak as I am, only set out these lights* to
    Mars's red-glowing fires by your nights' labors.
Not unless you, "looking up," had turned the dioptra correctly,
    Could I myself have "looked down" from there, upon the earth's courses;
Could have measured swift paces, Libra confronting the Goat, and,
    What proportion may give the path's center to you, O Phoebus,[14]
That by a similar step it might seek and flee from the sun, while
    Nonetheless not being spun with a spin like the Lord's, but
Gathering forces as it approaches[15] its source, and in turn,
    Languishing as it retires to more distant places;
Whence the seven Globes are borne by the sevenfold Minds in
    Sequence, and by the eighth soul, from Father Sun.
Nature thus is exempted from numberless twistings and turnings:
    Nine times five* gods thus depart from the flock.
Cheat against reason by tens, O Tycho; cheat it by minutes,
    Which none but you would number: the whole will collapse.
O human cares – the amount of vanity in our affairs! To
    Think that the stars can't be reached by a different road.

> As if in the diagram on p. 149 [of the first edition] of this work, an eye were depicted at the letter η (the star Mars).

> * Arist. *Metaph.* XII. 8.

The same author's epigram on the studies of Tycho Brahe   (***) v

Tycho has described the course of the sun and the fixed stars,
    To this he joined the Moon's circuit, and then died.
Phaeton grieves that he ascended his light-bearing four-horse
    Chariot. This artful care did you no harm, O Tycho:
Loving Diana, Endymion slept with a slumber eternal;
    Love of this Goddess gives you too a sleep everlasting.

---

[13]Latin, *Philothesio.* This puzzling word does not appear in classical Latin, nor (despite its obvious Greek origin) in any ancient Greek text. Caspar translates it *Freundschaftsbuch,* literally, "friendship book," a reading which is at least plausible, and which suggests that the book might be a counterpart to a modern guest book.

[14]The sun, identified with Phoebus Apollo.

[15]*Vires acquirat eundo,* alluding to Virgil, *Aeneid* IV.175.

TO THE READER[16]

GREETING

I had resolved to say many things to you (reader), but the heap of political business, by which I am detained more than usual these days, and the over-hasty departure of our Kepler, at this very moment about to go to Frankfurt, has hardly left me even this little scrap of an occasion to write. And so I thought I should give you just three words' warning, lest you be moved by Kepler's liberty in certain matters, but especially in his disagreeing with Brahe in physical argumentations, groundlessly complicating the work on the *Rudolphine Tables*, and habitual to all philosophers from the creation of the universe to the present. As for the rest, you will come to know from the work itself that he has constructed it upon Brahe's foundation, that is, upon his restitution of the fixed stars and the sun, and that all the materials (observations, that is) were amassed from the labor of Brahe. Meanwhile, the republic of letters is for now condescending, treating this important work of Kepler's, appearing among these tumults of rebellion and wars sprouting again from time to time, as if blessed by a precursor of the Tables, and after them the observations (seeing the light of day later); and also, more eager for the further progress of the work so fervently desired, condescending to pray with us for more felicitous times from God, the Best and Greatest.

Franz Gansneb Tengnagel von Campp.
His Imperial Majesty's Councillor

---

[16]The author of this preface was Tycho's son-in-law, who, through his position as Tycho's heir and his influence with the Emperor, delayed the publication of *Astronomia Nova* for several years. Although for some time he succeeded in preventing Kepler from publishing anything without his consent, he finally agreed to allow Kepler's work to appear on condition that he be allowed to write a preface stating what he saw as the correct Tychonic position. Clearly, he postponed the writing until the last possible moment—the incoherence and somewhat jumbled syntax, preserved in the translation, make this plain even without Tengnagel's lame excuses. The result is a comical ending to a profoundly aggravating situation. As Arthur Koestler put it, "…in Tengnagel's preface to the *New Astronomy*, we hear the braying of a pompous ass echoing down the centuries."

For a thorough account of Kepler's conflict with Brahe's heirs, and the effect that conflict had upon the structure of *Astronomia Nova*, see James R. Voelkel, *The Composition of Kepler's* Astronomia Nova (Princeton, 2001), especially Ch. 7.

Introduction to this Work

It is extremely hard these days to write mathematical books, especially astronomical ones. For unless one maintains the truly rigorous sequence of proposition, construction, demonstration, and conclusion,[1] the book will not be mathematical; but maintaining that sequence makes the reading most tiresome, especially in Latin, which lacks the articles and that gracefulness possessed by Greek when it is expressed in written symbols. Moreover, there are very few suitably prepared readers these days: the rest generally reject such works. How many mathematicians are there who put up with the trouble of working through the Conics of Apollonius of Perga? And yet that subject matter is the sort of thing that can be expressed much more easily in diagrams and lines than can astronomy.

I myself, who am known as a mathematician, find my mental forces wearying when, upon rereading my own work, I recall from the diagrams the sense of the proofs, which I myself had originally introduced from my own mind into the diagrams and the text. But then when I remedy the obscurity of the subject matter by inserting explanations, it seems to me that I commit the opposite fault, of waxing verbose in a mathematical context.

Furthermore, prolixity of phrases has its own obscurity, no less than terse brevity. The latter evades the mind's eye while the former distracts it; the one lacks light while the other overwhelms with superfluous glitter; the latter does not arouse the sight while the former quite dazzles it.

These considerations led me to the idea of including a kind of elucidating introduction to this work, to assist the reader's comprehension as much as possible.

I conceived this introduction as having two parts. In the first I present a synoptic table of all the chapters in the book. I think this is going to be useful, because the subject matter is unfamiliar to most people, and the various terms and various procedures used here are very much alike, and are closely related, both in general and in specific details. So when all the terms and all the procedures are juxtaposed and presented in a single display, they are mutually explanatory. For example, I discuss the natural causes that led the ancients, though ignorant of them, to suppose an equant circle or equalizing point. However, I do this in two places, namely, in Parts Three and Four. A reader who encounters this subject in Part Three might think I am dealing here with the first inequality, which is a property of the motions of each of the

---

[1] Proposition, construction, demonstration, and conclusion are the parts of a Euclidean theorem, as formalized by Proclus. See Proclus, *A Commentary on the First Book of Euclid's Elements*, Glenn R. Morrow, trans. (Princeton 1970, 1992) p. 159 (p. 203 of Friedelein's Greek text).

planets individually. And indeed, this is the case in Part Four. However, in the third part, as the summary indicates, I am discussing that equant which, under the name of the second inequality, varies the motion of all the planets in common, and primarily governs the theory of the sun. Thus the synoptic table will serve to make this distinction clear.

Nevertheless, the synopsis does not indeed assist all equally. There will be those to whom this table (which I present as a thread leading back out of the labyrinth of the work) will appear more tangled than the Gordian Knot. For their sake, therefore, there are many points that should be brought together here at the beginning which are presented bit by bit throughout the work, and are therefore not so easy to attend to in passing. Furthermore, I shall reveal, especially for the sake of those professors of the physical sciences who are irate with me, as well as with Copernicus and even with the remotest antiquity, on account of our having shaken the foundations of the sciences with the motion of the earth—I shall, I say, reveal faithfully the intent of the principal chapters which deal with this subject, and shall propose for inspection all the principles of the proofs upon which my conclusions, so repugnant to them, are based.

The introduction to this work is aimed at those who study the physical sciences.

For when they see that this is done faithfully, they will then have the free choice either of reading through and understanding the proofs themselves with greatest exertion, or of trusting me, a professional mathematician, concerning the sound and geometrical method presented. Meanwhile, they, for their part, will turn to the principles of the proofs thus gathered for their inspection, and will examine them thoroughly, knowing that unless they are refuted the proof erected upon them will not topple. I shall also do the same where, in the manner of physics professors, I mingle the probable with the necessary and draw a plausible conclusion from the mixture. For since I have mingled celestial physics with astronomy in this work, no one should be surprised at a certain amount of conjecture. This is the nature of physics, of medicine, and of all the sciences which make use of other axioms besides the most certain evidence of the eyes.

On the schools of astronomers.

The reader should be aware that there are two schools of thought among astronomers, one distinguished by its chief, Ptolemy, and by the assent of the large majority of the ancients, and the other attributed to more recent proponents, although it is the most ancient. The former treats the individual planets separately and assigns causes to the motions of each in its own orb, while the latter relates the planets to one another, and deduces from a single common cause those characteristics that are found to be common to their motions. The latter school is again subdivided. Copernicus, with Aristarchus of remotest antiquity, ascribes to the translational motion of our home the earth the cause of the planets' appearing stationary and retrograde, and I too subscribe to this opinion. Tycho Brahe, on the other hand, ascribes this cause to the sun, in whose vicinity he says the eccentric circles of all five planets are connected as if by a kind of knot (not physical, of course, but only quantitative). Further, he says that this knot, as it were, revolves about the motionless earth, along with the solar body.

(***)2v

For each of these three opinions concerning the world[2] there are several other peculiarities which themselves also serve to distinguish these schools, but these peculiarities can each be easily altered and amended in such a way that, so far as astronomy, or the celestial appearances, are concerned, the three opinions are for practical purposes equivalent to a hair's breadth, and produce the same results.

My aim in the present work is chiefly to reform astronomical theory (especially of the motion of Mars) in all three forms of hypotheses, so that what we compute from the tables may correspond to the celestial phenomena. Hitherto, it has not been possible to do this with sufficient certainty. In fact, in August of 1608, Mars was a little less than four degrees beyond the position given by calculation from the Prutenic tables.[3] In August and September of 1593 this error was a little less than five degrees, while in my new calculation the error is entirely removed.

The twofold aim of the work.

Meanwhile, although I place this goal first and pursue it cheerfully, I also make an excursion into Aristotle's *Metaphysics*, or rather, I inquire into celestial physics and the natural causes of the motions. The eventual result of this consideration is the formulation of very clear arguments showing that only Copernicus's opinion concerning the world (with a few small changes) is true, that the other two are proved false, and so on.

On the physical causes of the motions.

Indeed, all things are so interconnected, involved, and intertwined with one another that after trying many different ways by which I might attain to the reform of astronomical calculations, some well trodden by the ancients and others constructed in emulation of them and by their example, none other could succeed than the one founded upon the motions' physical causes themselves, which I establish in this work.

Now my first step in investigating the physical causes of the motions was to demonstrate that that coming together of the eccentrics occurs in no other place than the very center of the solar body (not some nearby point), contrary to what Copernicus and Brahe thought.

The first step towards those causes is taken: the planes of all six eccentrics intersect at a single point, namely, the center of the solar body.

If this correction of mine is carried over into the Ptolemaic theory, Ptolemy will have to investigate not the motion of the center of the epicycle, about which the epicycle proceeds uniformly, but the motion of some point whose distance from that center, in proportion to the diameter, is the same as the distance of the center of the solar orb from the earth for Ptolemy, which point is also on the same line, or on lines parallel to it.

Here the Braheans could raise the objection against me that I am a rash innovator, for they, while holding to the opinion received from the ancients and placing the intersection of the [planes of the] eccentrics not in the sun but near the sun, nevertheless constructed on this basis a calculation that corresponds to the heavens. And in translating the Brahean numbers into the Ptolemaic form, Ptolemy could say to me that as long as he upheld and expressed the observations, he would not consider any eccentric other than the one described by the center of the epicycle, about which the epicycle proceeds

---

[2]Latin, *Mundus*. This comprises the entire corporeal universe, including the fixed stars

[3]Erasmus Reinhold, *Prutenicae Tabulae Coelestium Motuum* (Tübingen, 1551).

uniformly; and thus that I ought to look again and again at what I would do, so that, using a new method, I should not do what they had already done by the old method.

So to counter this objection, I have demonstrated in the first part of the work [Chapters 1–6] that exactly the same things can result or be presented by this new method as are presented by the method of those ancients.

In the second part of the work [Chapters 7–21] I took up the main subject, and described the positions of Mars at apparent opposition to the sun, not worse, but indeed much better, with my method than they had expressed the positions of Mars at mean opposition to the sun with the old method.

Meanwhile, throughout the entire second part (as far as concerns geometrical demonstrations from the observations) I have left in suspense the question of whose procedure is better, theirs or mine, seeing that we both matched a great many observations (this is, indeed, a basic requirement for our theorizing). However, my method is in agreement with physical causes, and their old one is in disagreement, as I have partly shown in the first part, especially Chapter 6.

But finally in the fourth part of the work [Chapters 41–60], in Chapter 52, I have considered certain other observations, no less trustworthy than the previous ones, which their old method could not match, but which mine matched most beautifully. I have thereby demonstrated most soundly that Mars's eccentric is so situated that the center of the solar body lies upon its line of apsides, and not any nearby point, and hence, that all the [planes of the] eccentrics intersect in the sun itself.

This should, however, hold not just for the longitude, but for the latitude as well. Therefore, in the fifth part [Chapters 61–70] I have demonstrated the same from the observed latitudes, in Chapter 67.

(***)3r

The second step towards the physical causes of the motions is taken: there is also an equant in the theory of the sun or the earth, and thus the solar eccentricity must be bisected.

These things could not have been demonstrated earlier in the work, because the entry into these astronomical demonstrations is through an exact knowledge of the causes of the second inequality in the planets' motion, for which some other new thing had likewise to be discovered in the third part, unknown to our predecessors, and so on.

For I have demonstrated in the third part [Chapters 22–40] that whether the aforementioned old method, which depends upon the sun's mean motion, is valid, or my new one, which uses the apparent motion, nevertheless, in either case there is something from the causes of the first inequality that is mixed in with the second, which pertains to all planets in common. Thus for Ptolemy I have demonstrated that his epicycles do not have as centers those points about which their motion is uniform. Similarly for Copernicus I have demonstrated that the circle in which the earth is moved around the sun does not have as its center that point about which its motion is regular and uniform. Similarly for Tycho Brahe I have demonstrated that the circle on which the common point or knot of the eccentrics mentioned above moves does not have as its center that point about which its motion is regular and uniform. For if I concede to Brahe that the common point of the eccentrics may be different from the center of the sun, he must grant that the circuit of that common point, which in magnitude and period exactly equals the orbit of the sun, is eccentric and tends towards Capricorn, while the sun's eccentric circuit tends towards Cancer. The same thing befalls Ptolemy's epicycles.

However, if I place the common point or knot of the eccentrics in the center of the solar body, then the common circuit of both the knot and the sun is indeed eccentric with respect to the earth, and tends towards Cancer, but by only half the eccentricity shown by the point about which the sun's motion is regular and uniform.

And in Copernicus, the earth's eccentric still tends towards Capricorn, but by only half the eccentricity of the point about which the earth's motion is uniform, also in the direction of Capricorn.

Likewise, in Ptolemy, on those diameters of the epicycles that run from Capricorn to Cancer, there are three points, the outer two of which are at the same distance from the middle ones; and their distances from one another have the same ratio to the diameters as the whole eccentricity of the sun has to the diameter of its circuit. And of these three points, the ones that are middle in position are the centers of their epicycles, those that lie toward Cancer are the points about which the motions on the epicycles are uniform, and finally those that lie toward Capricorn are the ones whose eccentrics (described by them) we are tracing out *if instead of the sun's mean motion we follow the apparent motion,* just as if those were the points at which the epicycles were attached to the eccentric. The result of this is that each planetary epicycle contains the theory of the sun in its entirety, with all the properties of its motions and circles.

With these things thus demonstrated by an infallible method, the previous step towards the physical causes is now confirmed, and a new step is taken towards them, most clearly in the theories of Copernicus and Brahe, and more obscurely but at least plausibly in the Ptolemaic theory.

> The earth is moved and the sun stands still. Physico-astronomical arguments.

For whether it is the earth or the sun that is moved, it has certainly been demonstrated that the body that is moved is moved in a nonuniform manner, that is, slowly when it is farther from the body at rest, and more swiftly when it has approached this body.

Thus the physical difference of these three opinions is now immediately apparent—by way of conjecture, it is true, but yielding nothing in certainty to conjectures of doctors on physiology or to any other natural philosophers whatever.

Ptolemy is certainly hooted off the stage first. For who would believe that there are as many theories of the sun (so closely resembling one another that they are in fact equal) as there are planets, when he sees that for Brahe a single solar theory suffices for the same task, and it is the most widely accepted axiom in the natural sciences that Nature makes use of the fewest possible means?

That Copernicus is better able than Brahe* to deal with celestial physics is proven in many ways.

> * Of whom, in all fairness, most honest and grateful mention is made, and recognition given, since I, borrowing all the materials from him, am building this entire structure from the bottom up upon his work.

First, although Brahe did indeed take up those five solar theories from the theories of the planets, bringing them down to the centers of the eccentrics, hiding them there, and conflating them into one, he nevertheless left in the world the effects produced by those theories. For Brahe no less than for Ptolemy, besides that motion which is proper to it, each planet is still actually moved with the sun's motion, mixing the two into one, the result being a spiral. That it results from this that there are no solid orbs, Brahe has demonstrated most firmly. Copernicus, on the other hand, entirely removed the five planets from this extrinsic motion, the cause of the deception arising from the circumstances of observation. Thus the motions are still multiplied to no purpose by Brahe, as they were before by Ptolemy.

II.   2. Second, if there are no orbs, the conditions under which the intelligences and moving souls must operate will be made very difficult, since they have to attend to so many things to endow the planet with two intermingled motions. They will at least have to attend at one and the same time to the principles, centers, and periods of the two motions. But if the earth is moved, I show that most of this can be done with physical rather than animate faculties,[4] namely, magnetic ones. But these are more general points. There follow others arising specifically from demonstrations, upon which we now begin.

(***)3v   For if the earth is moved, it has been demonstrated that it receives the laws
III.   of speed and slowness from the measure of its approach towards the sun and of its receding from the same. And in fact the same happens with the rest of the planets: they are urged on or held back according to the approach toward or recession from the sun. So far, the demonstration is geometrical.

And now, from this very reliable demonstration, the conclusion is drawn, using a physical conjecture, that the source of the five planets' motion is in the sun itself. It is therefore very likely that the source of the earth's motion is in the same place as the source of the other five planets' motion, namely, in the sun as well. It is therefore likely that the earth is moved, since a likely cause of its motion is apparent.

IV.   That, on the other hand, the sun remains in place in the center of the world, is most probably shown by (among other things) its being the source of motion for at least five planets. For whether you follow Copernicus or Brahe, for both the source of motion of five of the planets is in the sun, and in Copernicus, of a sixth as well, namely, the earth. And it is more likely that the source of all motion should remain in place rather than move.

V.   But if we follow Brahe's theory and say that the sun moves, this first conclusion still remains valid, that the sun moves slowly when it is more distant from the earth and swiftly when it approaches, and this not only in appearance, but in fact. For this is the effect of the circle of the equant, which, by an inescapable demonstration, I have introduced into the theory of the sun.

Upon this most valid conclusion, making use of the physical conjecture introduced above, might be based the following theorem of natural philosophy: the sun, and with it the whole huge load (to speak coarsely) of the five eccentrics, is moved by the earth; or, the source of the motion of the sun and the five eccentrics attached to the sun is in the earth.

Now let us consider the bodies of the sun and the earth, and decide which is better suited to being the source of motion for the other body. Does the sun, which moves the rest of the planets, move the earth, or does the earth move the sun, which moves the rest, and which is so many times greater? Unless we are to be forced to admit the absurd conclusion that the sun is moved by the earth, we must allow the sun to be fixed and the earth to move.

---

[4] Latin, *facultas animalis*. This is the faculty of the soul that controls local motion in animals.

What shall I say of the motion's periodic time of 365 days, intermediate VI.
in quantity between the periodic time of Mars of 687 days and that of Venus
of 225 days? Does not the nature of things cry out with a great voice that the
circuit in which these 365 days are used up[5] also occupies a place intermediate
between those of Mars and Venus about the sun, and thus itself also encircles
the sun, and hence, that this circuit is a circuit of the earth about the sun, and
not of the sun about the earth? These points are, however, more appropriate
to my *Mysterium cosmographicum*, and arguments that are different from those
examined in this work need not be mentioned here.

Hence, for other metaphysical arguments that favor the sun's position in VII.
the center of the world, derived from its dignity or its illumination, see my
little book just mentioned, or look in Copernicus. There is also something in
Aristotle's *On the Heavens*, Book II, in the passage on the Pythagoreans, who
used the name "fire" to signify the sun.[6] I have touched upon a few points in
the *Astronomiae pars optica* Ch. 1 p. 7, and also Ch. 6, especially p. 225.[7]

But that it is reasonable for the earth to have a circular motion in some place VIII.
other than the center of the world, you will find a metaphysical argument in
Chapter 9, p. 322 of that book.[8]

I trust the reader's indulgence if I take this opportunity to point out certain *Objections to the earth's motion.*
remedies to a number of objections which capture people's minds, and in this
way use these arguments to shed darkness. For these replies are by no means
irrelevant to matters that I argue in the book regarding the physical causes of
the planets' motion, chiefly in parts three and four [Chapters 22–60].

The motion of heavy bodies prevents many from being able to believe that *I. On the motion of heavy bodies.*
the earth is moved by an animate motion, or better, by a magnetic one. They
should ponder the following propositions.

A mathematical point, whether or not it is the center of the world, can *The theory of gravity is in error.*
neither effect the motion of heavy bodies nor act as a object towards which
they tend. Let the physicists prove that this force is in a point which neither is

---

[5] Behind this odd phrase lies Kepler's peculiar treatment of time as a dependent variable: he
makes consistent use of the amounts of time "used up" in traversing equal distances, rather
than considering the distances traversed in equal times (as Galileo and his successors did). It is
quite likely that this different viewpoint was of importance in developing the "area law" which
later became known as Kepler's Second Law. See especially the beginning of Chapter 40.

[6] Aristotle, *On the Heavens*, II.13, 293a 20. Kepler wrote a German translation of this, with commentary. Although it was not published in Kepler's lifetime, it was included in Frisch's edition
of Kepler's works, and more recently, in *KGW* 20.1 pp. 150–167.

[7] Kepler, *Optics*, p. 20 in the English translation: "The sun is accordingly a particular body; in it is
this faculty of communicating itself to all things, which we call light, to which, on this account
at least, is due the middle place in the whole world, and the center, so that it might perpetually
pour itself forth equably into the whole orb."
 Optics, English trans. p. 241: "You will see in the sun a palpable image of the world; in the
world, that of God the Creator."

[8] Kepler, *Optics*, English trans. pp. 331–2: "And so it was evidently not fitting that the human
being, destined to be the inhabitant and watchman of this world, should reside in its middle, as
if in a closed cubicle, under which circumstance he would never have made his way through to
the contemplation of heavenly bodies that are so remote; but rather, by the annual translatory
motion of the earth, his domicile, he circumambulates and strolls around in this most ample
building, so as to be able more rightly to perceive and measure the individual members of the
house."

a body nor is grasped otherwise than through mere relation.

It is impossible that, in moving its body, the form of a stone seek out a mathematical point (in this instance, the center of the world), without respect to the body in which this point is located. Let the physicists prove that natural things have a sympathy for that which is nothing.

Nor again, do heavy bodies to tend towards the center of the world simply because they are seeking to avoid its spherical extremities. For, compared with their distance from the extremities of the world, the proportional part by which they are removed from the world's center is imperceptible and of no effect. Also, what would be the cause of such antipathy? With how much force and wisdom would heavy bodies have to be endowed in order to be able to flee so precisely an enemy surrounding them on all sides? Or what ingenuity would the extremities of the world have to possess in order to pursue their enemy with such exactitude?

Nor are heavy bodies driven in towards the middle by the rapid whirling of the primum mobile, as objects in whirlpools are. For that motion (if we suppose it to exist) does not carry all the way down to these lower regions. If it did, we would feel it, and would be caught up by it along with the very earth itself; indeed, we would be carried ahead, and the earth would follow. All these are absurdities for the opposing view, and it therefore is clear that the common theory of gravity is in error.

(***)4r

True theory of gravity

The true theory of gravity rests upon the following axioms.[9]

Every corporeal[10] substance, to the extent that it is corporeal, has been so made as to be suited to rest in every place in which it is put by itself, outside the orb of power of a kindred[11] body.

Gravity is a mutual corporeal disposition among kindred bodies to unite or join together; thus, the earth attracts a stone much more than the stone seeks the earth. (The magnetic faculty belongs to this order of things.)

Heavy bodies (most of all if we establish the earth in the center of the world) are not drawn towards the center of the world because it is the center of the world, but because it is the center of a kindred spherical body, namely, the earth. Consequently, wherever the earth be established, or whithersoever it be carried by its animate faculty, heavy bodies are drawn towards it.

---

[9] As Max Caspar notes in his edition of *Astronomia Nova*, the theory presented in these terse statements constitutes a rejection of the Aristotelian view of gravity and plays an important role in Kepler's physical thought. In later works Kepler refers back to them, especially in Book I part 4 of the *Epitome of Copernican Astronomy*, where he develops them further. Especially important in providing an insight into Kepler's early thoughts on the subject are his letters to David Fabricius of 11 October 1605 and 10 November 1608 (*KGW* 15 p. 240 and 16 p. 194). In the former, he likens gravity to magnetism, and says, "...not only does a stone approach the earth, but the earth also approaches the stone, and they divide the space between them in the inverse ratio of their weights." Also illuminating is Kepler's letter to Herwart von Hohenburg of January 1607 (*KGW* 15 p. 386).

[10] Latin, *corporea*. There is a close relation in Kepler's thought here between *corpus* (body) and *corporea* that is not made entirely clear in this translation. Other possible renderings would be "physical" or "bodily." The former would not adequately represent Kepler's meaning in the first axiom, while the latter is sufficiently at odds with correct usage to lead the translator to reject it, though with some regret.

[11] Latin, *cognata*, "of the same origin."

If the earth were not round, heavy bodies would not everywhere be drawn in straight lines towards the middle point of the earth, but would be drawn towards different points from different sides.

If two stones were set near one another in some place in the world outside the sphere of influence of a third kindred body, these stones, like two magnetic bodies, would come together in an intermediate place, each approaching the other by an interval proportional to the bulk [moles] of the other.

If the moon and the earth were not each held back in its own circuit by an animate force or something else equally potent, the earth would ascend towards the moon by one fifty-fourth part of the interval, and the moon would descend towards the earth about fifty-three parts of the interval, and there they would be joined together; provided, that is, that the substance of each is of one and the same density.[12]

If the earth should cease to attract its waters to itself, all the sea water would be lifted up, and would flow onto the body of the moon.

The orb of the attractive power in the moon is extended all the way to the earth, and calls the waters forth beneath the torrid zone, in that it calls them forth into its path wherever the path is directly above a place. This is imperceptible in enclosed seas, but noticeable where the beds of the ocean are widest and there is much free space for the waters' reciprocation. It thus happens that the shores of the temperate latitudes are laid bare, and to some extent even in the torrid regions the neighboring oceans diminish the size of the bays. And thus when the waters rise in the wider ocean beds, it can happen that in its narrower bays, if they are not too closely surrounded, the moon being present, the water might even seem to be fleeing the moon, though in fact they are subsiding because a quantity of water is being carried off elsewhere.

Reason for the ebb and flow of the sea.

But the moon passes the zenith swiftly, and the waters are unable to follow so swiftly. Therefore, a westward current of the ocean arises beneath the torrid zone, which, when it strikes upon the far shores, is thereby deflected. But when the moon departs, this congress of the waters, or army on the march towards the torrid zone, because it is abandoned by the traction that had called it forth, is dissolved. But since it has acquired impetus, it flows back (as in a water vessel) and assaults its own shore, inundating it. In the moon's absence, this impetus gives rise to another impetus until the moon returns and submits to the reins of this impetus, moderates it, and carries it around along with its own motion. So all shores that are equally exposed are flooded at the same time, while those more remote are flooded later, some in different ways because of the different approaches of the ocean.

I will point out in passing that the sand dunes of the Syrtes[13] are heaped

Effects of the sea's ebb and flow.

---

[12]The apparent angular size of the moon is about half a degree, so the apparent size of its radius is about $\frac{1}{4}$ degree. The moon's distance from earth is about 60 times the earth's radius (these numbers were well known to the ancient Greeks). By elementary trigonometry, 60 times the sine of $\frac{1}{4}$ degree will give the ratio of the radii of earth and moon, which comes out to a little less than 1:4. Cubing this gives the ratio of volumes, to which, Kepler supposes, the attractive power is proportional.

[13]Shoals on the coast of North Africa. Kepler is probably referring to Syrtis Minor, the Gulf of Gabès, in Tunisia. This is one of the few parts of the Mediterranean that has appreciable tides, and extensive sandy shoals are exposed at low tide.

up in this way; that thus are created or destroyed countless islands in bays full of eddies (such as the Gulf of Mexico); that it seems that the soft, fertile, and friable earth of the [East] Indies was thus at length broached and penetrated by this current, this perpetual inundation, with help from a certain all-pervading motion of the earth. For it is said that India was once continuous from the Golden Chersonese[14] towards the east and south, but now the ocean, which was once farther back between China and America, has flowed in, and the shores of the Moluccas and of other neighboring islands, which are now raised on high because of the subsidence of the surface of the sea, suppress the credibility of this matter.[15]

The Taprobane of the ancients is lost today.

Taprobane,[16] too, seems to have been submerged through this cause (as is consistent with the account of the Calcuttans that several localities there too were once submerged), when the China Sea burst in through breaches into the Indian Ocean, with the result that nowadays nothing of Taprobane remains but the peaks of the mountains, which take the form of the innumerable islands known as the Maldives. For it is easy to prove, from the geographers and Diodorus Siculus, that the site of Taprobane was once there, namely, to the south opposite the mouths of the Indus and the promontory of Corium.[17] Moreover, in ecclesiastical history one individual is said to have been bishop of Arabia and Taprobane together, and so the latter must surely have been nearby and not five hundred German miles to the east (indeed, more than a thousand, following the roundabout routes used in those days). The island of Sumatra, nowadays considered to be Taprobane, I think was once the Golden Chersonese, joined to the Indian isthmus at the city of Malacca. For Chersonesus,[18] which nowadays we believe to be the Golden, seems to have no more right than Italy to the name "Chersonese."

Although these things were appropriate to a different topic, I wanted to present them all in one context in order to make more credible the ocean tide and through it the moon's attractive power.

For it follows that if the moon's power of attraction extends to the earth, the earth's power of attraction all the more extends to the moon and far beyond, and accordingly, that nothing that consists to any extent whatever of terrestrial material, carried up on high, ever escapes the mightiest grasp of this power of attraction.

(***)4v
True theory of levity.

Nothing that consists of corporeal material is absolutely light. It is only comparatively lighter, because it is less dense, either by its own nature or through an influx of heat. By "less dense" I do not just mean that which is porous and divided into many cavities, but in general that which, while occu-

---

[14]"Chersonese" is a Greek-derived word equivalent to the Latin-derived "peninsula." The Golden Chersonese is usually identified with the Malay Peninsula

[15]Suppress: *opprimunt*. Possibly Kepler means that the existence of these coastal bluffs make it seem implausible that the land could once have been continuous.

[16]This name usually refers to the island now known as Sri Lanka, though Kepler has his doubts.

[17]A search in Zedler's massive *Grosses Universal-Lexicon* (1731–1754) suggests that Corium may be in the province of Iran now known as Kerman.

[18]The Thracian Chersonese (Gallipoli).

pying a place of the same magnitude as some heavier body, contains a lesser quantity of corporeal material.

The motion of light things also follows their definition. For it should not be thought that they flee all the way to the surface of the world when they are carried upwards, or that they are not attracted by the earth. Rather, they are less attracted than heavy bodies, and are thus displaced by heavy bodies, whereupon they come to rest and are kept in their place by the earth.

But even if the earth's power of attraction is extended very far upwards, as was said, nevertheless, if a stone were at a distance that was perceptible in relation to the earth's diameter, it is true that, the earth being moved, such a stone would not simply follow, but would mingle its forces of resistance with the earth's forces of attraction, and it would thus detach itself somewhat from the earth's grasp. In just the same way, violent[19] motion detaches projectiles somewhat from the earth's grasp, so that they either run on ahead if they are shot eastwards, or are left behind if shot westwards, thus leaving the place from which they are shot, under the compulsion of force. Nor can the earth's revolving effect impede this violent motion all at once, as long as the violent motion is at its full strength.

> To the objection that objects projected vertically fall back to their places.

But no projectile is separated from the surface of the earth by even a hundred thousandth part of the earth's diameter, and not even the clouds themselves, or smoke, which partake of earthy matter to the very least extent, achieve an altitude of a thousandth part of the semidiameter.[20] Therefore, the resistance of the clouds, smokes, or objects shot vertically upwards can do nothing, nor, I say, can the natural inclination to rest do anything to impede this grasp of the earth's, at least where this resistance is negligible in proportion to that grasp. Consequently, anything shot vertically upwards will fall back to its place, the motion of the earth notwithstanding. For the earth cannot be pulled out from under it, since the earth carries with it anything sailing through the air, linked to it by the magnetic force no less firmly than if those bodies were actually in contact with it.

When these propositions have been grasped by the understanding and pondered carefully, not only do the absurdity and falsely conceived physical impossibility of the earth's motion vanish, but it will also become clear how to reply to the physical objections, however they are framed.

Copernicus preferred to think that the earth and all terrestrial bodies (even those torn away from the earth) are informed by one and the same motive soul, which, while rotating its body the earth, also rotates those particles cast

> The opinion of Copernicus.

---

[19] A technical Aristotelian term, for which there is no satisfactory modern equivalent. Aristotle categorized all motions as being either "natural" or "violent," depending upon whether they are carried out in accordance with some inner principle or are caused by something external. Here, the "natural" motion is the coming together of all terrestrial bodies, while the "violent" motion is the separation of those bodies. Kepler is arguing that when a body is separated from kindred bodies, there is some faculty that brings into action a force tending to bring the bodies back together.

[20] In the *Optics*, Ch. 4 Prop. 11, English trans. pp. 141–2, Kepler argued on the basis of atmospheric refraction that the altitude of the air cannot be greater than half a German mile (about 3.7 km), and therefore concluded that the highest mountain peaks must be above the atmosphere. His chief error was to suppose that the air maintains its density with increasing altitude, and has a well-defined surface, as water does.

away from it. He thus held it to be this soul, spread throughout the particles, that acquires force through violent motions, while I hold that it is a corporeal faculty (which we call gravity, or the magnetic faculty), that acquires the force in the same way, namely, through violent motions.

Nevertheless, this corporeal faculty is sufficient for anything removed from the earth: the animate faculty is superfluous.

II. To objections concerning the swiftness of the earth's motion.

Although many people fear the worst for themselves and for all earth's creatures on account of the rapidity of this motion, they have no cause for alarm. On this point see my book, *On the New Star*, Chapters 15 and 16, pp. 82 and 84.[21]

III. To objections concerning the immensity of the heavens.

In the same place, you will find the full-sail voyage along the world's immense orbit, which, in objection to Copernicus, is usually held to be unnatural. There it is demonstrated to be well-proportioned, and that, on the contrary, the speed of the heavens would become ill-proportioned and unnatural were the earth ordered to remain quite motionless in its place.

IV. To objections concerning the dissent of holy scripture, and its authority.

There are, however, many more people who are moved by piety to withhold assent from Copernicus, fearing that falsehood might be charged against the Holy Spirit speaking in the scriptures if we say that the earth is moved and the sun stands still.[22]

But let them consider that since we acquire most of our information, both in quality and quantity, through the sense of sight, it is impossible for us to abstract our speech from this ocular sense. Thus, a great many things happen each day where we speak in accordance with the sense of sight, although we are quite certain that the truth of the matter is otherwise.

This verse of Virgil furnishes an example:

> We are carried from the port, and the land and cities recede.[23]

Thus, when we emerge from the narrow part of some valley, we say that a great plain is opening itself out before us.

Thus Christ said to Peter, "Lead forth on high,"[24] as if the sea were higher than the shores. It does seem so to the eyes, but the writers on optics show the cause of this fallacy. Christ was only making use of the common idiom, which nonetheless arose from this visual deception.

Thus, we call the rising and setting of the stars "ascent" and "descent," though at the same time that we say the sun ascends, others say it descends. See the *Optics* Ch. 10 p. 327.[25]

---

[21] In *KGW* 1. Here Kepler argues that the extreme motion of the heavens required by the supposed daily rotation is much more incredible than the comparatively moderate motion of the earth in the Copernican schema.

[22] The following arguments on the interpretation of scripture were to become the most widely read of Kepler's writings. They were often reprinted from the seventeenth century on, and translated into modern languages. Indeed, this part of the Introduction was the only work of Kepler's to appear in English before the 1870s.

[23] *Aeneid* III. 72. This line was also quoted by Copernicus, Book I Chapter 8 of *De revolutionibus*.

[24] Luke 5:4. The Latin *altum* can mean either "high" or "deep." However, Kepler cannot have been unaware that the original Greek verse unambiguously has the latter meaning, and hence he must be charged with making a rather silly distortion in order to prove a point.

[25] Kepler, *Optics*, English trans. pp. 337–8.

Thus, the Ptolemaic astronomers even now say that the planets are stationary when they are seen to stay near the same fixed stars for several days, even though they think the planets are then really moving downwards in a straight line, or upwards away from the earth.

Thus writers of all nations use the word "solstice," even though they in fact deny that the sun stands still.[26]

Thus there has not yet been anyone so doggedly Copernican as to avoid saying that the sun is entering Cancer or Leo, even though he wishes to signify that the earth is entering Capricorn or Aquarius. And there are other like examples.

Now the holy scriptures, too, when treating common things (concerning   (***)5r
which it is not their purpose to instruct humanity), speak with humans in the human manner, in order to be understood by humans. The scriptures make use of what is generally acknowledged among humans, in order to weave in other things more lofty and divine.

No wonder, then, if Scripture also speaks in accordance with human perception when the truth of things is at odds with the senses, whether or not humans are aware of this. Who is unaware that the allusion in Psalm 19 is poetical? Here, under the image of the sun, are sung the spreading of the Gospel and even the sojourn of Christ the Lord in this world on our behalf; and in the singing the sun is said to emerge from the tabernacle of the horizon like a bridegroom from his marriage bed, exuberant as a strong man for the race. Which Virgil imitates thus:

> Aurora leaving Tithonus's saffron-colored bed.[27]

(The Hebrew poetry was, of course, earlier.)

The psalmodist was aware that the sun does not go forth from the horizon as from a tabernacle (even though it may appear so to the eyes). On the other hand, he considered the sun to move for the precise reason that it appears so to the eyes. In either case, he expresses it so because in either case it appeared so to the eyes. He should not be judged to speak falsely in either case, for the perception of the eyes also has its truth, well suited to the psalmodist's more hidden aim, the adumbration of the Gospel and also of the Son of God. Likewise, Joshua [Joshua 10:12 ff.] makes mention of the valleys against which the sun and moon moved, because when he was at the Jordan it appeared so to him. Yet each writer was in perfect control of his meaning. David (and Syracides[28] with him) was describing the magnificence of God made manifest, which he expressed so as to exhibit them to the eyes, and possibly also for the sake of a mystical sense spelled out through these visible things. Joshua meant that the sun be held back in its place in the middle of the sky for an entire day with respect to the sense of his eyes, since for other people during the same interval of time it would remain beneath the earth.

---

[26]The Latin *solstitium* means literally "sun-standing."

[27]*Aeneid* IV. 585. In Kepler's time it was generally believed that Virgil was familiar with the Hebrew scriptures.

[28]Yehoshua ben Sirach, author of the apocryphal Biblical text *Ecclesiasticus*, or *Sirach*.

But thoughtless persons pay attention only to the verbal contradiction, "the sun stood still" versus "the earth stood still," not considering that this contradiction can only arise in an optical and astronomical context, and does not carry over into common usage. Nor are these thoughtless ones willing to see that Joshua was simply praying that the mountains not remove the sunlight from him, which prayer he expressed in words conforming to the sense of sight, as it would have been quite inappropriate to think, at that moment, of astronomy and of visual errors. For if someone had admonished him that the sun doesn't really move against the valley of Ajalon, but only appears to do so, wouldn't Joshua have exclaimed that he only asked for the day to be lengthened, however that might be done? He would therefore have replied in the same way if anyone had filed a lawsuit against him about the sun's perpetual rest and the earth's motion.

Now God easily understood from Joshua's words what he meant, and proved it by stopping the motion of the earth, so that the sun might appear to him to stop. For the gist of Joshua's petition comes to this, that it might appear so to him, whatever the reality might meanwhile be. Indeed, that this appearance should come about was not vain and purposeless, but quite conjoined with the desired effect.

But see Chapter 10 of the *Astronomiae pars optica*,[29] where you will find reasons why, to absolutely everyone, the sun appears to move and not the earth: it is because the sun appears small and the earth large, and also because, owing to its apparent slowness, the sun's motion is perceived, not by sight, but by reasoning alone, through its changed proximity to the mountains over a period of time. It is therefore impossible for a previously uninformed reason to imagine anything but that the earth, along with the arch of heaven set over it, is like a great house, immobile, in which the sun, so small in stature, travels from one side to the other like a bird flying in the air.

What all people indeed imagine, gave the first line of holy scripture. "In the beginning," says Moses, "God created the heaven and the earth," because it is these two parts that chiefly present themselves to the sense of sight. It is as though Moses were to say to man, "This whole worldly edifice that you see, light above and dark and widely spread out below, upon which you are standing and by which you are roofed over, has been created by God."

In another passage [Jeremiah 31:37.], Man is asked whether he has learned how to seek out the height of heaven above, or the depths of the earth below, because to the ordinary man both appear to extend through equally infinite spaces. Nevertheless, there is no one in his right mind who, upon hearing these words, would use them to limit astronomers' diligence either in showing the contemptible smallness of the earth in comparison with the heavens, or in investigating astronomical distances. For these words do not concern measurement arrived at by reasoning. Rather, they concern physical measurement, which is utterly impossible for the human body, fixed upon the land and drawing upon the free air. Read all of Chapter 38 of *Job,* and compare it with matters discussed in astronomy and in physics.

---

[29]Kepler, *Optics*, Ch. 10, English trans. pp. 335–346.

Suppose someone were to assert, from Psalm 24, that the earth is founded upon rivers, in order to support the novel and absurd philosophical conclusion that the earth floats upon rivers. Would it not be correct to say to him that he should regard the Holy Spirit as a divine messenger, and refrain from wantonly dragging Him into physics class? For in that passage the psalmodist intends nothing but what men already know and experience daily, namely, that the land, raised on high after the separation of the waters, has great rivers flowing through it and seas surrounding it. Not surprisingly, the same figure of speech is adopted in another passage, where the Israelites sing that they were seated upon the waters of Babylon [Psalm 137], that is, by the riverside, or on the banks of the Euphrates and Tigris.

If anyone easily accepts this, why can he not also accept that in other passages usually cited in opposition to the earth's motion we should likewise turn our eyes from physics to the aims of scripture?

A generation passes away (says Ecclesiastes [1:4]), and a generation comes, (***)5v but the earth stands forever. Does it seem here as if Solomon wanted to argue with the astronomers? No; rather, he wanted to warn people of their own mutability, while the earth, home of the human race, remains always the same, the motion of the sun perpetually returns to the same place, the wind blows in a circle and returns to its starting point, rivers flow from their sources into the sea, and from the sea return to their sources, and finally, as these people perish, others are born. Life's tale is ever the same; there is nothing new under the sun.

You do not hear any physical dogma here. The message[30] is a moral one, concerning something self-evident and seen by all eyes but little pondered. Solomon therefore urges us to ponder. Who is unaware that the earth is always the same? Who does not see the sun rise again daily from the east, rivers perennially flowing towards the sea, the winds returning in regular alternation, and men succeeding one another? But who really considers that the same drama of life is always being played, only with different characters, and that not a single thing in human affairs is new? So Solomon, by mentioning what is evident to all, warns of that which almost everyone wrongly neglects.

It is said, however, that Psalm 104, in its entirety, is a physical discussion, since the whole of it is concerned with physical matters. And in it, God is said to have "founded the earth upon its stability, that it not be laid low unto the ages of ages."[31] But in fact, the psalmodist is very far from speculating about physical causes. For he entirely finds comfort in the greatness of God, who made all these things: he has composed a hymn to God the creator, in which he treats the world in order, as it appears to the eyes.

If you consider carefully, you will see that it is a commentary upon the six days of creation in Genesis. For in the latter, the first three days are given to the separation of the regions: first, the region of light from the exterior darkness; second, the waters from the waters by the interposition of an extended region;

---

[30] Νουθεσία.

[31] The Latin of the Vulgate, quoted by Kepler, differs markedly from the Greek (and hence from most English translations) here.

and third, the land from the seas, where the earth is clothed with plants and shrubs. The last three days, on the other hand, are devoted to the filling of the regions so distinguished: the fourth, of the heavens; the fifth, of the seas and the air; and the sixth, of the land. And in this psalm there are likewise the same number of distinct parts, analogous to the works of the six days.

In the second verse, he enfolds the Creator with the vestment of light, first of created things, and the work of the first day.

The second part begins with the third verse, and concerns the waters above the heavens, the extended region of the heavens, and atmospheric phenomena that the psalmodist ascribes to the waters above the heavens, namely, clouds, winds, tornadoes, and lightning.

The third part begins with the sixth verse, and celebrates the earth as the foundation of the things being considered. The psalmodist relates everything to the earth and to the things that live on it, because, in the judgement of sight, the chief parts of the world are two: heaven and earth. He therefore considers that for so many ages now the earth has neither sunk nor cracked apart nor tumbled down, yet no one has certain knowledge of what it is founded upon.

He does not wish to teach things of which men are ignorant, but to re-call to mind something they neglect, namely, God's greatness and potency in a creation of such magnitude, so solid and stable. If an astronomer teaches that the earth is carried through the heavens, he is not overturning what the psalmodist says here, nor is he subverting human experience. For it is still true that the land, the work of God the architect, has not toppled as our build-ings usually do, consumed by age and rot; that it has not slumped to one side; that the dwelling places of living thing have not been set in disarray; that the mountains and coasts have stood firm, unmoved against the blast of wind and wave, as they were from the beginning. And then the psalmodist adds a beau-tiful sketch of the separation of the waters from the continents, and adorns his account by adding springs and the benefits that springs and crags provide for bird and beast. He also does not fail to mention the adorning of the earth's surface, included by Moses among the works of the third day, although the psalmodist derives it from its prior cause, namely, a humidification arising in the heavens, and embellishes his account by bringing to mind the benefits ac-cruing from that adornment for the nurture and pleasure of humans and for the lairs of the beasts.

The fourth part begins with verse 20, and celebrates the work of the fourth day, the sun and the moon, but chiefly the benefit that the division of times brings to humans and other living things. It is this benefit that is his subject matter: it is clear that he is not playing the part of an astronomer here.

If he were, he would not fail to mention the five planets, than whose mo-tion nothing is more admirable, nothing more beautiful, and nothing a better witness to the Creator's wisdom, for those who take note of it.

The fifth part, in verse 26, concerns the work of the fifth day, where he fills the sea with fish and ornaments it with sea voyages.

The sixth is added, rather obscurely, in verse 28, and concerns the animals living on land, created on the sixth day. At the end, in conclusion, he in general declares the goodness of God in sustaining all things and creating new things. So everything the psalmodist said of the world relates to living things. He tells nothing that is not generally acknowledged, because his purpose was to praise things that are known, not to seek out the unknown. It was his wish to invite men to consider the benefits accruing to them from these works of each of the days.

I, too, implore my reader, when he returns from the temple and enters astronomical studies, not to forget the divine goodness conferred upon men, to the consideration of which the psalmodist chiefly invites. I implore that, with me, he praise and celebrate the Creator's wisdom and greatness, which I unfold for him in the more perspicacious explanation of the world's form, the investigation of causes, and the detection of errors of vision. Let him thus not only extol the safety of the living things of all of Nature in the firmness and stability of the earth, as a gift of God, but also acknowledge His wisdom expressed in its motion, at once so well hidden and so admirable.

Advice for astronomers.
(***)6r

But whoever is too stupid to understand astronomical science, or too weak to believe Copernicus without affecting his faith, I advise him that, having dismissed astronomical studies and having damned whatever philosophical opinions he pleases, he mind his own business and betake himself home to scratch in his own dirt patch, abandoning this wandering about the world. He should raise his eyes (his only means of vision) to this visible heaven and with his whole heart burst forth in giving thanks and praising God the Creator. He can be sure that he worships God no less than the astronomer, to whom God has granted this, that with the mind's eye he see more penetratingly, and that he himself is able and willing to celebrate his God above whatever he discovers.

Advice for idiots.

At this point, a modest (though not too modest) commendation to the learned should be made on behalf of Brahe's opinion of the form of the world, since in a way it follows a middle path. On the one hand, it frees the astronomers as much as possible from the useless apparatus of so many epicycles and, with Copernicus, it embraces the causes of motion, unknown to Ptolemy, giving some place to physical theory in accepting the sun as the center of the planetary system. And on the other hand, it serves the mob of literalists and eliminates the motion of the earth, so hard to believe, although through it the theories of the planets in astronomical discussions and demonstrations are entangled in many difficulties, and the physics of the heavens is no less disturbed.

Commendation of the Brahean hypothesis.

So much for the authority of holy scripture. As for the opinions of the pious[32] on these matters of nature, I have just one thing to say: while in theology it is authority that carries the most weight,[33] in philosophy it is reason. Therefore, Lactantius is pious, who denied that the earth is round,[34] Augustine is

V. To objections concerning the authority of the pious.

---

[32]Latin, *Sancti*, literally, "the Saints," or "the Holy." Context shows, however, that in modern usage "pious" or "saintly" fits Kepler's meaning better, even though it does miss his verbal play on Sanctum Officium (see note 35, below).

[33]*KGW* has *poneranda* here, an error (rare for this edition). The correct reading is *ponderanda*, as in the 1609 original and in Frisch.

[34]In *De Revolutionibus*, in his dedicatory letter to Pope Paul III, Copernicus also mentions Lactantius as a revered theologian whose cosmological opinions are acknowledged to be false. See Lactantius, *Institut. Divin.*, III. 24, and Augustine, *The City of God*, XVI. 9.

pious, who, though admitting the roundness, denied the antipodes, and the Inquisition[35] nowadays is pious, which, though allowing the earth's small-ness, denies its motion. To me, however, the truth is more pious still, and (with all due respect for the Doctors of the Church) I prove philosophically that the earth is round, and also that it is inhabited all the way around at the antipodes, and also that it is contemptibly small, and finally that it is carried along among the stars.

But enough about the truth of the Copernican hypothesis. Let us return to the plan I proposed at the beginning of this introduction.

I had begun to say that in this work I treat all of astronomy by means of physical causes rather than fictitious hypotheses, and that I had taken two steps in my effort to reach this central goal: first, that I had discovered that the planetary eccentrics all intersect in the body of the sun, and second, that I had understood that in the theory of the earth there is an equant circle, and that its eccentricity is to be bisected.

The third step towards the physical explanation. The eccentricity of Mars's equant is to be precisely bisected.

Now we come to the third step, namely, that it has been demonstrated with certainty, by undertaking a comparison of Parts 2 and 4, that the eccentricity of Mars's equant is also to be precisely bisected, a fact long held in doubt by Brahe and Copernicus.[36]

Therefore, by induction extending to all the planets (carried out in Part 3 by way of anticipation) it has been demonstrated that, since there are (of course) no solid orbs, as Brahe demonstrated from the paths of comets, the body of the sun is therefore the source of the power that drives all the planets around. Moreover, I have specified the manner [in which this occurs] as follows: that the sun, although it stays in one place, rotates as if on a lathe,[37] and out of itself sends into the space of the world an immaterial *species*[38] of its body, analogous to the immaterial *species* of its light. This *species* itself, as a consequence of the rotation of the solar body, also rotates like a very rapid whirlpool throughout the whole breadth of the world, and carries the bodies of the planets along with itself in a gyre, its grasp stronger or weaker according to the greater density or rarity it acquires through the law governing its diffusion.

Once this common power was proposed, by which all the planets, each in its own circle, are driven around the sun, the next step in my argument was to give each of the planets its own mover, seated in the planet's globe (you will recall that, following Brahe's opinion, I had already rejected solid orbs). And this, too, I have accomplished in Part 3.

---

[35] Latin, *Officium*, the so-called "Holy Office," by which name the Inquisition was officially known. Kepler literally says, "the Office is Holy," referring to its name and not implying approval.

[36] For these terms, see the Glossary. Kepler means that the eccentricity of the equant point is twice as great as the eccentricity of the eccentric.

[37] It should be pointed out that this remarkable conjecture was written several years before the telescopic observation of the sun's rotation by Galileo and others.

[38] This word can mean form, image, kind, emanation, spectacle, outward appearance, apparition, or idea (to name a few possibilities). Since there is no English word that can embrace so many meanings, I have chosen to leave it untranslated. See the Glossary for more about this term.

By this train of argument, the existence of the movers was established. The amount of work they occasioned me in Part 4 is incredible to speak of, when, in producing the planet-sun distances and the eccentric equations[39] that were required, the results came out full of flaws and in disagreement with the observations. This is not because they were falsely introduced, but because I had bound them to the millstones (as it were) of circles, under the spell of common opinion. Restrained by such fetters, the movers could not do their work.

But my exhausting task was not complete until I had made a fourth step towards the physical hypotheses. By most laborious proofs and by computations on a very large number of observations, I discovered that the course of a planet in the heavens is not a circle, but an oval path, perfectly elliptical.

> Fourth step to the physical explanation. The planet describes an oval path.

Geometry stepped in, and taught that such a path results if we assign to the planet's own movers the task of making the planet's body reciprocate[40] along a straight line extended towards the sun. Not only this, but also the correct eccentric equations, agreeing with the observations, resulted from such a reciprocation.

> (***)6v

Finally, the pediment was added to the structure, and proven geometrically: that such a reciprocation is apt to be the result of a magnetic corporeal faculty. Consequently, these movers belonging to the planets individually are shown with great probability to be nothing but properties of the planetary bodies themselves, like the magnet's property of seeking the pole and catching up iron. As a result, every detail of the celestial motions is caused and regulated by faculties of a purely corporeal nature, that is, magnetic, with the sole exception of the whirling of the solar body as it remains fixed in its space. For this, a vital faculty[41] seems required.

Next, in Part 5, it was demonstrated that the physical hypotheses we just introduced also give a satisfactory account of the latitudes.

There are some, however, who are alarmed by a few extraneous and seemingly valid objections and do not wish to put such great trust in the nature of bodies. Therefore, in Parts 3 and 4, some room was left for Mind, so that the planet's proper mover could attach the faculty of Reason to the animate faculty[42] of moving its globe. Let such a one allow the mind to make use of the sun's apparent diameter as a measure of the reciprocation, and to be able to sense the angles that astronomers require.

Let this much be said for the sake of the physicists. The astronomers and geometers will find the rest in the following summaries of the individual chapters, each in its proper place. I intentionally made them rather detailed, partly so that it might serve as an index, and partly so that a reader who gets stuck

---

[39] The mathematical expressions giving the planet's angular positions around the sun with respect to the apsides at specified times. The distances and the eccentric equations together are sufficient to specify the planet's position in space. Kepler's chief difficulty was that he could not find a single theory that could give both of these correctly. The solution, as he implies here, was to abandon the circular orbit and to introduce an ellipse.

[40] This is explained in Chapter 57, below.

[41] Latin, *facultas vitalis*. In animal physiology, the faculty that facilitates bodily motions (other than motion of the body as a whole from place to place). See footnote 4, above.

[42] Latin, *facultas animalis*. See footnote 4 above.

The Synoptic Table. here and there in the synoptic table, whether because of the obscurity of the material or of the style, might seek some additional light from these summaries. If the reasons for the order and the coherence of topics lumped together in the same chapter turn out to be rather less clear in the text itself, the reader might perceive them more readily among the summaries, which are divided into paragraphs. I therefore ask the reader to be well content.

Let the Synoptic Table be introduced here.

Synopsis of the Entire Work

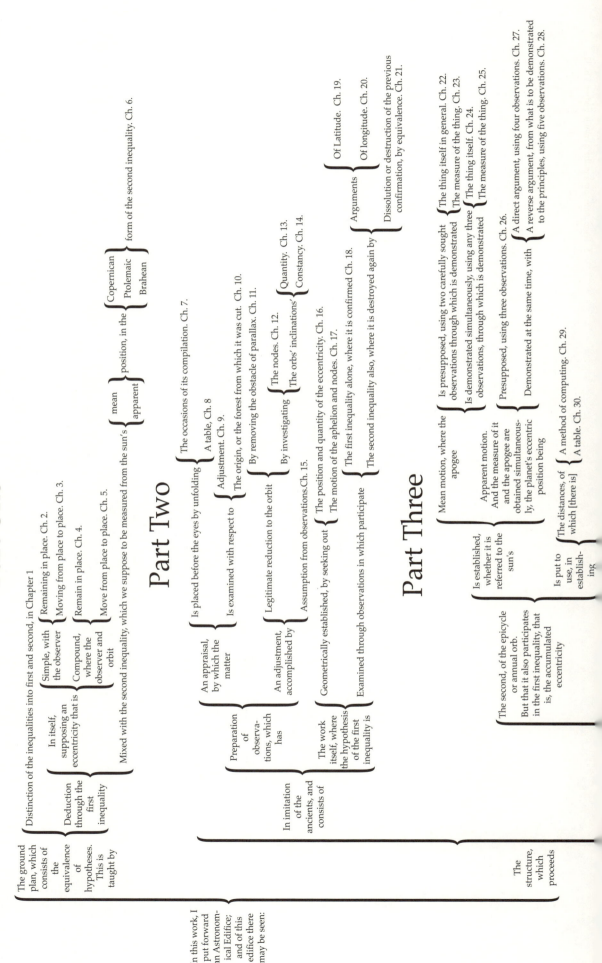

In this work, I put forward an Astronomical Edifice; and of this edifice there may be seen:

Part One

The ground plan, which consists of the equivalence of hypotheses. This is taught by

- Distinction of the inequalities into first and second, in Chapter 1
- Deduction through the first inequality
  - In itself, supposing an eccentricity that is
    - Simple, with the observer { Remaining in place. Ch. 2. / Moving from place to place. Ch. 3. }
    - Compound, where the observer and orbit { Remain in place. Ch. 4. / Move from place to place. Ch. 5. }
  - Mixed with the second inequality, which we suppose to be measured from the sun's { mean / apparent } position, in the { Copernican / Ptolemaic / Brahean } form of the second inequality. Ch. 6.

Part Two

In imitation of the ancients, and consists of

- Preparation of observations, which has
  - An appraisal, by which the matter
    - Is placed before the eyes by unfolding { The occasions of its compilation. Ch. 7. / A table, Ch. 8 }
    - Is examined with respect to Adjustment. Ch. 9.
  - An adjustment, accomplished by
    - The origin, or the forest from which it was cut. Ch. 10.
    - Legitimate reduction to the orbit
      - By removing the obstacle of parallax. Ch. 11.
      - By investigating { The nodes. Ch. 12. / The orbs' inclinations' { Quantity. Ch. 13. / Constancy. Ch. 14. } }
    - Assumption from observations. Ch. 15.
- The work itself, where the hypothesis of the first inequality is
  - Geometrically established, by seeking out { The position and quantity of the eccentricity. Ch. 16. / The motion of the aphelion and nodes. Ch. 17. }
  - Examined through observations in which participate
    - The first inequality alone, where it is confirmed Ch. 18. — Arguments { Of Latitude. Ch. 19. / Of longitude. Ch. 20. }
    - The second inequality also, where it is destroyed again by Dissolution or destruction of the previous confirmation, by equivalence. Ch. 21.

Part Three

The structure, which proceeds

- Mean motion, where the apogee Is established, whether it is referred to the sun's
  - Is presupposed, using two carefully sought observations through which is demonstrated { The thing itself in general. Ch. 22. / The measure of the thing. Ch. 23. }
  - Is demonstrated simultaneously, using any three observations, through which is demonstrated { The thing itself. Ch. 24. / The measure of the thing. Ch. 25. }
- Apparent motion. And the measure of it and the apogee are obtained simultaneously, the planet's eccentric position being
  - Presupposed, using three observations. Ch. 26.
  - Demonstrated at the same time, with { A direct argument, using four observations. Ch. 27. / A reverse argument, from what is to be demonstrated to the principles, using five observations. Ch. 28. }
- The second, of the epicycle or annual orb. But that it also participates in the first inequality, that is, the accumulated eccentricity Is put to use, in establishing The distances, of which [there is] { A method of computing. Ch. 29. / A table. Ch. 30. }

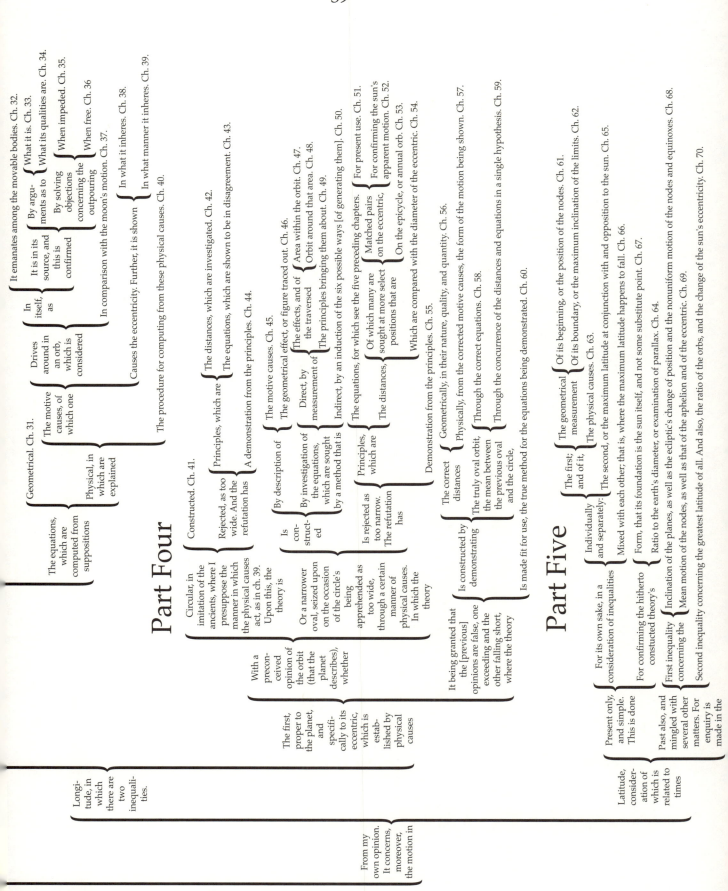

Longitude, in which there are two inequalities.

The equations, which are computed from suppositions:

- The motive causes, of which one
  - Geometrical. Ch. 31.
  - Drives around in an orb, which is considered — In itself, as
    - It emanates among the movable bodies. Ch. 32.
    - By arguments as to — It is in its source, and this is confirmed
      - What it is. Ch. 33.
      - What its qualities are. Ch. 34.
    - By solving objections concerning the outpouring
      - When impeded. Ch. 35.
      - When free. Ch. 36
  - In comparison with the moon's motion. Ch. 37.
- Physical, in which are explained — Causes the eccentricity. Further, it is shown
  - In what it inheres. Ch. 38.
  - In what manner it inheres. Ch. 39.

The procedure for computing from these physical causes. Ch. 40.

Part Four

The first, proper to the planet, and specifically to its eccentric, which is established by physical causes

With a preconceived opinion of the orbit (that the planet describes), whether

- Circular, in imitation of the ancients, where I presuppose the manner in which the physical causes act, as in ch. 39. Upon this, the theory is
  - Constructed. Ch. 41.
  - Rejected, as too wide. And the refutation has
    - Principles, which are
      - The distances, which are investigated. Ch. 42.
      - The equations, which are shown to be in disagreement. Ch. 43.
    - A demonstration from the principles. Ch. 44.
- Or a narrower oval, seized upon on the occasion of the circle's being apprehended as too wide, through a certain manner of physical causes. In which the theory
  - Is constructed
    - By description of
      - The motive causes. Ch. 45.
      - The geometrical effect, or figure traced out. Ch. 46.
    - By investigation of the equations, which are sought by a method that is
      - Direct, by measurement of
        - The effects, and of the traversed
          - Area within the orbit. Ch. 47.
          - Orbit around that area. Ch. 48.
        - The principles bringing them about. Ch. 49.
      - Indirect, by an induction of the six possible ways [of generating them]. Ch. 50.
  - Is rejected as too narrow. The refutation has
    - Principles, which are
      - The equations, for which see the five preceding chapters.
        - For present use. Ch. 51.
        - Matched pairs on the eccentric,
          - Of which many are sought at more select positions that are
            - For confirming the sun's apparent motion. Ch. 52.
            - On the epicycle, or annual orb. Ch. 53.
          - Which are compared with the diameter of the eccentric. Ch. 54.
      - The distances,
        - Geometrically, in their nature, quality, and quantity. Ch. 56.
        - Physically, from the corrected motive causes, the form of the motion being shown. Ch. 57.
    - Demonstration from the principles. Ch. 55.

It being granted that the [previous] opinions are false, one exceeding and the other falling short, where the theory

- Is constructed by demonstrating — The correct distances
  - Through the correct equations. Ch. 58.
  - The truly oval orbit, the mean between the previous oval and the circle, — Through the concurrence of the distances and equations in a single hypothesis. Ch. 59.
- Is made fit for use, the true method for the equations being demonstrated. Ch. 60.

From my own opinion. It concerns, moreover, the motion in

Part Five

Latitude, consideration of which is related to times

- For its own sake, in a consideration of inequalities (Present only, and simple. This is done)
  - Individually and separately:
    - The first; and of it,
      - The geometrical measurement
        - Of its beginning, or the position of the nodes. Ch. 61.
        - Of its boundary, or the maximum inclination of the limits. Ch. 62.
      - The physical causes. Ch. 63.
    - The second, or the maximum latitude at conjunction with and opposition to the sun. Ch. 65.
  - Mixed with each other; that is, where the maximum latitude happens to fall. Ch. 66.
- For confirming the hitherto constucted theory's (Past also, and mingled with several other matters. For enquiry is made in the)
  - Form, that its foundation is the sun itself, and not some substitute point. Ch. 67.
  - Ratio to the earth's diameter, or examination of parallax. Ch. 64.

First inequality concerning the

- Inclination of the planes, as well as the ecliptic's change of position and the nonuniform motion of the nodes and equinoxes. Ch. 68.
- Mean motion of the nodes, as well as that of the aphelion and of the eccentric. Ch. 69.

Second inequality concerning the greatest latitude of all. And also, the ratio of the orbs, and the change of the sun's eccentricity. Ch. 70.

Summaries of the Individual Chapters

Since there is one method which the nature of a subject teaches, and another which our understanding requires, and both are contrived by art, the reader should expect neither from me undiluted. The scope of this work is not chiefly to explain the celestial motions, for this is done in the books on Spherics and on the theories of the planets. Nor yet is it to teach the reader, to lead him from self-evident beginnings to conclusions, as Ptolemy did as much as he could. A third way enters in, which I hold in common with the orators, which, since I present many new things, I am constrained to make plain in order to deserve and obtain the reader's assent, and to dispel any suspicion of cultivating novelty.

No wonder, therefore, if along with the nobler methods I mingle the third, familiar to the orators; that is, an historical presentation of my discoveries. Here it is a question not only of leading the reader to an understanding of the subject matter in the easiest way, but also, above all, of the arguments, meanderings, or even chance occurrences by which I the author first came upon that understanding. Thus, in telling of Christopher Columbus, Magellan, and of the Portuguese, we do not simply treat indulgently the errors by which the first opened up America, the second, the China Sea, and the last, the coast of Africa; rather, we would indeed not wish them omitted, which would be to deprive ourselves of an enormous pleasure in reading. So likewise, I would not have it ascribed to me as a fault that with the same concern for the reader I have followed this same course in the present work.[1] For although we by no means share in the exploits of the Argonauts by reading, the difficulties and thorns of *my* discoveries infest the reading itself—a fate common to all mathematical books. Nevertheless, since we are human beings who take delight in different things, there will appear some who, having overcome the difficulties of perception, and having placed before their eyes all at once this entire sequence of discoveries, will be inundated with a very great sense of pleasure.

That the whole work is well ordered through this method will now become apparent through the summaries of the individual chapters.

I have, moreover, seen to it that whenever the text presents a geometrical

---

[1]The reader must beware of supposing that Kepler is presenting a historical account of his enquiry. Although he does lead us through much of his erroneous reasoning, he has also spared us many thorny byways. Nor does he claim to present such a history: he says only that he is mingling some history with the theoretical and didactic matter of the book. What is remarkable, however, is his sense that mistakes are important, that we approach truth by first being wrong.

demonstration, construction, or lemma, it is printed in "running letters",[2] as the print shops call them. If this is not so everywhere, you should attribute it to the material, which mingles physics with geometrical matters, or to the typesetters, who might not always have noticed my signs.

PART I

Chapter 1

Explains the manner in which astronomers perceived a difference between the first motion and the second or planetary motions, and also, the manner in which two inequalities, called the first and the second, were discovered in a planet's proper motion.

The occasion of this chapter, as well of the whole first part, is this. When I first had come to Brahe, I became aware that in company with Ptolemy and Copernicus he reckoned the second inequality of a planet in relation to the mean motion of the sun. Now four years previously, it seemed to me that for physical reasons it ought to be measured by the apparent motion of the sun, as is stated in the *Mysterium cosmographicum*. So when a dispute arose between us, Brahe said in opposition to me that when he used the mean sun he accounted for all the appearances of the first inequality. I replied that this would not prevent my accounting for the same observations of the first inequality using the sun's apparent motion, and thus it would be in the second inequality that it would be seen who was more nearly correct.

What I answered then is what has been set out to be proved in the first part of the book.

Chapter 2

Next, since what was proposed is the perplexing and difficult matter of the equivalence of hypotheses, I have begun with the first and simplest equivalence: when a concentric with an epicycle is changed to an eccentric.

And lest the geometrical distinction be thought trivial, I have discussed the causes, both physical and rational or mental, by which it would be reasonable to suppose both of the equivalent hypotheses to be administered and to carry out their motions. This is done in one way if solid orbs be granted, and in another if they be denied. (Brahe has in fact demonstrated from the trajectories of comets that there are no solid orbs).

Chapter 3

This simple eccentric, or the concentric with a single epicycle equivalent to it, being supposed, it is shown what difference it makes to the eye or in the natural causes of motions, if the mean motion of the sun is exchanged for the apparent; that is, if the observer, or rather if the source of power, be moved in imagination to another place.

---

[2]*Litera cursoria;* that is, italics.

Chapter 4

1. Transition is made from the unconditioned and simple eccentric to an eccentric with an equant having twice the eccentric's eccentricity, which Ptolemy had assigned to the first inequality of the five planets.[3]
2. This is shown to be absurd on the assumption of solid orbs, but elegant and physically supportable if such orbs be denied.
3. It is next shown how Copernicus changed this eccentric with an equant into a concentric with two epicycles.
4. This hypothesis of Copernicus's is shown to be rather poor physics on the assumption of solid orbs, but absurd if they be denied.
5. But it is also proved that the hypothesis is deficient in geometrical beauty in the planetary path.
6. Nor is it everywhere equivalent to the Ptolemaic eccentric. The discrepancy is small in the first inequality, but greater in the second.
7. In the same place there is a demonstration of a method for easily computing the equation in either form of hypothesis.
8. A way of eliminating the discrepancy between the two hypotheses.
9. Finally, another form of the Copernican hypothesis, using a concentric with an epicycle.

Chapter 5

Chapter 5 is to 4 as 3 is to 2. For the business at hand is more weighty, and concerns:

1. The changes in the hypothesis that result from moving the observer or seat of power from its original position to another, by substituting the motion of the apparent sun for the mean. This is all done in the form of Copernican hypothesis presented at the end of Chapter 4.
2. Which things in turn would be changed in the physical causes of the motions [resulting] from the same hypothesis.
3. This transposition is outlined and constructed in the Ptolemaic form of the first inequality.
4. It is demonstrated that, if one allows two lines of apsides (the ancient one and another arising from the transposition), and thus changes the form of hypothesis, it will follow that appearances of two kinds are given, though the planet's path in the heavens remains the same.   (****) v
5. But if a single line of apsides be set up, passing through the original center of the eccentric, it is demonstrated that the same planetary orbit does not give rise to the original required appearances, even though the path remains the same, nor is even the same form of hypothesis retained.
6. Finally, assuming a new line of apsides passing through the center of the equant, and keeping the same form of hypothesis, it is demonstrated that the orbit changes position in the heavens.

---

[3]In Kepler's text, the numbered lists of topics in many of the chapters were typeset as single paragraphs rather than itemized lists. Exceptions were Chapters 58, 59, and 60. In the interest of clarity, the translation presents them in list form. In the few places where Kepler included paragraph breaks in the enumerated text, extra space has been inserted between the items.

7. The position and size of the circle of greatest difference or aberration from the appearances caused by this transposition is demonstrated geometrically from what is supposed.

8. It is demonstrated that all these phenomena occur if, the center of vision remaining fixed, the center of the equant be moved an equal amount in the opposite direction.

9. All things said concerning an eccentric with an equant, which Ptolemy adopted, are applied to the Copernico-Brahean concentric with two epicycles (which, by Chapter 4, is equivalent).

Chapter 6

Here things demonstrated in Chapter 5, especially numbers 6, 7, and 8, are to some extent put into practice. And hitherto, respecting these hypotheses, those things in particular concerning the first inequality, different for different authors, have been in question. Now, however, those things that are attributed to the second inequality are also added, which, as the principal hypotheses (ahead of those discussed previously), are named for their authors Ptolemy, Copernicus, and Tycho Brahe. Usually when we say "Copernican Hypothesis" we mean that of the second inequality.

1. Therefore, to begin, I compare these.

2. In the Copernican, I show how the hypothesis of the first inequality was derived from the sun's mean motion, and how the eccentricity arises from the point taken in place of the sun.

3. I argue from physics that this is not done correctly, and that the eccentricity ought to be reckoned from the center of the sun's body itself.

4. If we judge the second inequality to take place according to the apparent motion of the sun, why physical arguments support this.

5. It is demonstrated that in this case positions in longitude in the first inequality do not vary much, but distances of the body of a planet from the body of the sun vary considerably.

6. The position on the earth's orbit[4] at which an observer would encounter the greatest difference in distances as well as the greatest error is demonstrated geometrically.

7. It is concluded by arithmetic operations that the value of the error can amount to one degree and about 20 minutes.

8. In the Ptolemaic hypothesis I show how the hypothesis of the first inequality was derived from the sun's mean motion.

9. In a general manner, on physical or metaphysical grounds, many objections are raised against this hypothesis as well as the sun's mean motion.

10. But in particular some objections are raised upon the same basis against the sun's mean motion specifically.

11. If we reckon the second inequality from the sun's apparent motion, the physical objections are satisfied.

---

[4]*Orbis magnus,* Copernicus's name for the earth's path around the sun.

12. The position, size, and shape of the new hypothesis is shown by transposing the point of uniform motion.

13. The discrepancy of the appearances of the first inequality, the place on the epicycle where the error in appearances of the second inequality is greatest, and the amount of this error, are adapted from the above considerations.

14. In the Brahean hypothesis I show how the hypothesis of the first inequality was derived from the sun's mean motion, and how the center of Mars's orbit was thus attached to the solar orbit, not at the center of the sun's body, but nearby.

15. A few general objections are raised against the Brahean hypothesis, but in particular, on physical grounds, I raise many objections against this type of attachment, maintaining (speaking to what has been supposed) that it ought to occur at the very center of the sun's body.

16. The position, size, and shape of the new hypothesis is shown by transposition of the point to which the planetary orbits are attached, and by applying the above considerations, the places on the eccentric as well as on the great orb bearing the eccentric (or concentric with epicycles) are found at which the error is greatest.

And that is the extent of Part I.

PART II

Chapter 7

I explain in greater detail the occasions both of my having taken up the theory of Mars, and of what induced me to follow the sun's apparent motion and to begin the work with the first part, just completed, as it stands here. You have a synopsis of this chapter in the summary of Chapter 1.

Chapter 8

Displays the hypothesis of Mars's first inequality, as constructed by Brahe, and shows it in a table, which has foundations (namely, acronychal observations), a result (namely, computed positions compared with those observed), and an examination of these, aimed at revealing whether this hypothesis might agree with the observations in minute detail.

Chapter 9

Concerns the corrected adoption of observed positions.

1. The need is shown of constructing the position on the ecliptic corresponding to the planet instead of its position in its own circle.

2. The notion, followed in the table, that the arc from a node to the planet's apparent position is equal to the arc from a node to the corresponding position on the ecliptic, is refuted.

3. The notion of equality is likewise refuted where one of the arcs ends at the true position on the orbit instead of the apparent position.
4. The method of reduction by means of the angle of apparent latitude is also refuted, and a method of reduction by means of the angle of inclination of planes is established.

Chapter 10

Pertains to the same thing, and concerning the table's adopted positions inquires whether they were correctly and aptly deduced as being at opposition to the mean sun from observations close to opposition. Also, gentle admonitions on other subtle matters, chiefly parallax, are added. And this concludes the examination of the table.

Chapter 11

For the sake of beginning my accommodation of the data to the sun's apparent [position] with a legitimate reduction and deduction, so as not to err in any way, I first investigate the diurnal parallaxes of Mars.

1. I tell what Brahe thought about it.
2. I prove from Brahe's observations, through hourly and daily motions, that it is almost imperceptible, less than what we consider the solar parallaxes to be.
3. For fun, I bring in my own observations, in the course of which I explain my special method of investigating the diurnal parallax by means of a stationary latitude.

Chapter 12

1. Brahe's own way of finding the nodes of Mars from nearby observations, and a criticism of it.
2. Another way, which presupposes knowledge of the eccentric equations from the *Prutenics*, Ptolemy, or Brahe. By which it is simultaneously demonstrated that the descending node (found by four observations) and the ascending node (by two) are at opposite positions on the ecliptic.

(****) 2 r

Chapter 13

1. The reckoning of the inclination of the planes is shown to be somewhat more intricate, in all three forms of hypotheses.
2. One way, presupposing known eccentric equations, when Mars, setting in the evening or rising in the morning, is at the limits by the first inequality, for then the apparent latitude is equal to the true inclination of the limit from the ecliptic.
3. The size of the arc of elongation from the sun for which this is true is shown, in the Copernican as well as the Ptolemaic hypotheses, and this is carried out using several observations about either limit.

4. A second way, requiring nothing besides select and rare observations, in which the sun is at the nodes and Mars is 90° from the sun. This is carried out by means of a few observations.

5. This is extended, so that Mars, other things being the same, can be at some angle other than 90° from the sun, and hence an inclination at a certain place, other than that of the limit but nonetheless fixed, may be found.

6. This way is also applied to the Ptolemaic hypothesis, involving some difficulty.

7. A third way is advanced, using latitudes observed at a position opposite the sun's, with the help of the previously known ratio of the orbs. This is treated in all three forms of hypotheses.

Chapter 14

Next, from what has been demonstrated in Chapter 13, the opinion of the ancients, that the planes of the eccentrics are subject to oscillation, is refuted. For it is demonstrated that the inclination remains constant, at least within the limits of one or two centuries.

Chapter 15

From nearby observations, the positions Mars has at the moments of opposition to the apparent motion of the sun are sought out arithmetically, and are corrected by precautions already considered. Finally, these are presented in a table, as a foundation for new operations.

Chapter 16

Now, in imitation of the ancients, physical causes aside, it is posited that the course of the planet is a circle, that within this circle there exists some point about which the planet traverses equal angles in equal times, and that between this point and the center of the sun lies the center of the planetary circle, at some unknown distance. With these assumptions, four acronychal observations are taken, with zodiacal positions and their time intervals; and through these are sought, by a most laborious method, the zodiacal positions of both centers, their distances from the center of the sun, and the ratios of the two eccentricities, both to one another and to the radius of the circle.

Chapter 17

By comparing the positions of the aphelion and nodes of Ptolemy's time with those which they are found to have in our time, their motion, required in the next chapter, is found.

Chapter 18

Finally it is shown that this hypothesis, constructed in the way described and depending upon the sun's apparent motion, accounts for every observed motion in longitude near opposition to the sun, and does so much more accurately than before, since the Brahean hypothesis was founded upon the sun's mean motion.

Chapter 19

1. Although so far the hypothesis that has been constructed has worked well for motion in longitude near opposition to the sun, it is nonetheless demonstrated that it does not work well for motions in latitude near opposition to the sun.
2. It is further demonstrated that the Brahean hypothesis is similarly deficient. In both instances, this is done in the Copernican form.
3. The same, in the Ptolemaic and Brahean form of hypothesis.
4. It is shown that the error in latitude arises from the eccentricity's not having been bisected.
5. But if the eccentricity be bisected, then the hypotheses are in error in motions of longitude. Which makes it clear why I was compelled to forsake the ancients and to search more diligently into these matters.

Chapter 20

Just as in the preceding chapter my hypothesis was convicted of error in motions of latitude near opposition to the sun, it is now so convicted in motions of longitude at other places.

2. So also the Brahean hypotheses, founded upon the sun's mean motion.
3. The demonstration is also applied to the Ptolemaic and Brahean form of the motions.
4. A finger is pointed at the sources of the errors and at the way to correct them.
5. A protheorem is inserted that states the sort of line in the plane of the ecliptic that may be substituted for the lines of the planet's distance from the sun in the plane of the planet's eccentric, when the planet has some latitude.

Chapter 21

Causes are sought from geometry that result in a false hypothesis yielding something true; and it is shown to what extent this can happen.

And this is the end of the second part, in which I have imitated the ancients.

PART III

Chapter 22

Now, following my method, I begin the whole inquiry anew, not with the first inequality but with the second. And

1. The circumstances are recounted that led me to suspect a governing equant circle in the theory of the sun.

2. I demonstrate in the three forms of hypotheses that on the supposition of an equant (which I have been favoring), the earth's orbit (or epicycle, for Ptolemy) appears to increase and decrease, as Brahe used to assert.
3. A method is presented for finding suitable observations from which this equant may be proved.
4. The proof is carried out, using two select observations, even supposing the Brahean rendition, which depends upon the sun's mean motion.

(****) 2 v

Chapter 23

Using the sun's distance from the earth at two zodiacal positions (found in the preceding chapter) together with the position of the sun's apogee (or earth's aphelion), a geometrical demonstration is used to find out the eccentricity of the sun's (or earth's) circle. It is presupposed that this is a perfect circle.

Chapter 24

The same as in Chapter 22 is demonstrated, but by means of four observations less discriminately selected, which nonetheless have Mars in the same eccentric position. That is, it is shown that some part of the sun's or the earth's eccentricity must be given to the equant circle. This is also shown in the three forms of hypotheses compared with one another, and likewise even on the supposition of the Brahean rendition of Mars's motion, which is founded upon the sun's mean motion.

Chapter 25

Using the distances of the zodiacal positions of the sun from the earth found in the preceding chapter, three by three, through a geometrical demonstration that presupposes nothing but a perfectly circular path, I discover not only the eccentricity of the sun's or earth's circle (as in Ch. 23), but also the position of the sun's apogee or the earth's aphelion opposite it, and this is nearly the same as that found by Brahe using observations specific to the sun, while here there are only observations of Mars.

Chapter 26

These four observations of Chapter 24 are transferred from the sun's mean motion to the true motion, from the Brahean rendition to mine, and the same thing is concluded as in Ch. 25. And the demonstration is presented in all three forms of hypotheses.

Chapter 27

By an even more daring method, assuming no rendition[5] of Mars whatever, and using no less than four observations of Mars, different from the preceding

---

[5]*Restitutio.* Normally, this word means "restoration, giving back," but in Kepler's time it had the special astronomical sense of reconstructing in theory the motions that are actually occurring in the heavens.

ones and compared among themselves as before, I demonstrate not only the eccentricity of the sun or earth, and at the same time the aphelion as before, as well as the ratio of the orbs at this eccentric position, but also the actual eccentric position of Mars beneath the fixed stars, which before was presupposed as known from the rendition.

Chapter 28

By nearly the same form of demonstration, but taking as given the eccentricity of the sun or earth, and the aphelion (now so many times established), and adding a number of observation, namely, the five that are here compared with one another as before, it is shown that one and the same eccentric position of Mars always results, nearly as in Chapter 27. You may recall, however, that in all preceding chapters of Part 3 I have presupposed that the earth's path is perfectly circular, as it indeed is to the senses. For because of the small eccentricity of its ellipse it can only barely depart from circularity.

Chapter 29

A perfectly circular eccentric with known eccentricity and an point of uniform motion with doubled eccentricity are posited. Then from these are sought by geometry, first, the distances at apogee and perigee, second, those at 90° of equated anomaly, and third, those elsewhere. In the same place a short cut for finding four distances in one operation is shown. Further, the point at which the circle is distant from the sun's center by the magnitude of the semidiameter, is demonstrated. And finally, the point on the circle, different from the other, at which one part of the equation is a maximum, is demonstrated.

Chapter 30

Distances of the sun from the earth are set out in a table, and a way of making selections is shown which, even though it is shown to overstep the bounds of the principles by making the star's orbit oval (and thus justifiably refers ahead to Chapters 31, 40, 44, and 55, where doubts about this are quelled), is nevertheless shown to depart imperceptibly from what has been previously demonstrated.

Chapter 31

Brahe feared that if I bisected the sun's eccentricity I would vitiate his equations of the sun's motion. Therefore, this fear is shown to be groundless by a demonstration that, whether there be a simple eccentricity or a bisected one or one formed by doubling the halved eccentricity, the equation of the sun's motion remains ever the same. Thus the doubt raised here in Chapter 31 is different from that raised in Chapter 30. There, the concern was about distances, while here it is about the Brahean equations; there the cause of concern was the shape of the orbit, while here it is the ratio of the eccentricity. In the one, the consideration was anticipated, while in the other it finds its true place.

Chapter 32

First comes an inductive argument that all planets, without exception, have an equant circle: one in which the eccentricity of the point of uniform motion bisected.

Upon this principle is constructed, by a geometrical demonstration, the following universal proposition: the increments of time[6] of a planet over equal arcs of the eccentric are proportional to the planet's distances from the center whence the eccentricity originates. Natural scientists, prick up your ears! For here is raised a deliberation involving an inroad to be made into your province.

Chapter 33

Now, from the conclusion of the preceding demonstration, with the help of a few generally accepted and purely physical axioms, it is evinced that the distances of a planet from the center whence the eccentricity is calculated are the administrative causes[7] of the planet's increments of time over equal arcs of the eccentric.

Second, it is argued that these administrative causes reside in the one terminus of the distances common to them all; that is, in the center of the planetary system.

Third, in relation to what was thus demonstrated, it is assumed—partly on the basis of Part I, as demonstrated probably, partly on the basis of Parts IV and V, as demonstrated necessarily and geometrically, and also partly made probable in this very place and in Part II—that the sun's body is itself at the center of the planetary system.

Fourth, it is shown to be consistent with the preceding that the motive power or the power that administers the time increments is in the body of the sun. Physical arguments are added.

Then, in passing, this also is inferred: that the sun is at rest in the center of   (****) 3 r
the world, and that the earth is moved around the center of the world. Here the natural scientist should observe that these physical theories are supported by the earth's motion, but are deduced from elsewhere, and are valid in the Brahean system as well as in the Copernican. In fact, on the contrary, it is upon these very theories that the earth's motion and the sun's rest are now being founded.

Fifth, it is demonstrated that the motive power is quantifiable exactly as light is, and is more tenuous in a greater circle, and more concentrated in a lesser.

Sixth, it is demonstrated as a consequence that that which moves the planets from place to place is an immaterial *species* of the power in the sun's body, similar to the immaterial *species* of light.

---

[6] *Moras*, literally, "delays". In his kinematics, Kepler usually kept the distances constant and considered the times required to traverse them, a procedure that resulted in cumbersome proofs. The arguments of this chapter are a good example of this. On Kepler's use of time as a dependent variable, see footnote 5 to the Introduction.

[7] *Causas dispensatrices.*

Chapter 34

The physical theory is brought to completion: it is demonstrated from the foregoing that the *species* of the power that conveys the planets moves around through the extent of the world like a river or whirlpool,[8] more swiftly than do the planets.

Second, it is shown as a consequence that the sun's body also rotates upon its axis. An enquiry is made into the probable periodic time of this rotation, and at the same time the question is raised what moves the earth and what the moon.

Third, the sun's body is proved to be a sort of magnet; and, by the example of the earth, it is shown that there are magnets in the heavens.

Chapter 35

A reply is made to the objection that if the motion of the heavenly bodies comes from the sun, it is impeded by the interposition of bodies, as light is. At the same time, many points from the preceding chapter are illustrated, namely, how the motive power and light are akin, and how each accompanies the other.

Chapter 36

Other objections are dispelled. The first, which actually has a geometrical foundation, argues from a point on the sun's body to a line, from this to its surface (to all appearances a plane), and so on to its volume, in order to establish that the ratio in which the density of light spreads differs from that which lets it be equivalent to the motive power. The reply to this is from the principles of optics: the argument cannot begin with a point or a line; it must start with the surface itself. Further, it is denied that the apparent magnitude of the sun's disc needs to be considered in the physical effect – this could have been shown on many grounds. For it could not possibly be an indication of this physical effect, since it varies in a different ratio. (It does, however, become an indication of something else, below.) And thus is attributed to Light a mode of spreading that is wholly commensurate with the administration of the planetary motions.

Another objection argues, to the contrary, that light is not fitted for association with motion, since light is also spread towards the poles. This is dispelled by means of the principles already adopted (that is, physical principles), but in a purely geometrical manner, in order that the solution show the natural causes of the zodiac, and why the planets are never outside it.

Chapter 37

From the physical principles laid down, the causes are sought for the anomaly in the moon that Brahe called the "variation', which makes the new and full moon swifter than it is otherwise. Here, two false opinions on the subject are

---

[3] *Vortex.*

rejected. Next, from the same principles, the causes are sought by which the moon's equation is made greater at the quadratures than at opposition and conjunction with the sun. Other remarks are added pertaining to the explanation of the particular power by which the moon is moved.

Chapter 38

Apart from the common motive force coming from the sun, the individual planets are shown to administer their own motions by other, individual motive causes. This is done through two arguments: one drawn from the motion in longitude, the other from the motion in latitude.

Chapter 39

To begin, there are adopted as premises six physical axioms necessary for investigating the power that is attributed solely to the individual planets.

At the same time, however, there are these two preconceived opinions that hold for this chapter as a whole: first, that the planet's orbit is arranged in a perfect circle; second, that this orbit is administered by a Mind. The question thus arises how this mind would make a circle of the planet's path. And it is first demonstrated that could happen if the planet's own power were to endeavor to move its body in a perfect epicycle whilst the body is also being carried by the solar power. In opposition to this way, five physical absurdities are raised. Second, it is demonstrated that this can happen if the planet observes a fixed point outside the sun from which it maintains a constant distance in its entire circuit about the sun. But this keeping track of a certain incorporeal point is also refuted in three ways.

Third, it is demonstrated that a perfect circle could result if the planet's own power were to make it reciprocate upon the diameter of the epicycle directed towards the sun, doing so by a prescribed law as if it were running around the epicycle's circumference. But at the same time it is shown that the exact reciprocations cannot be described by the planet if it moves on the diameter of the epicycle, nor do they correspond to eccentric arcs traversed, nor to the time, nor to the equated anomaly – on the assumption, that is, that a perfect circle ought to result from the planet's composite path.

Fourth, this too is denied: that by means of a mind the force proper to the planet somehow conceives an imaginary eccentric or epicycle, and by using it as a rule sets up the distances required for a perfectly circular orbit.

So, as long as we consider the planet's orbit to be a perfect circle, it remains in doubt by what standard the planet's own mind might measure out these reciprocations of its body.

Now that I have aired out the question of the standard of this reciprocation, I proceed further to consider the means by which the planet's mind might grasp this standard and the reciprocation defined by it. Whether the epicycle be taken as the standard, or its diameter, or the center of the eccentric, all must be rejected as ill suited for comprehension: they all need a commensurate means, suited to being comprehended, by which the reciprocations might be comprehended by the mind. Here it is added that the planet's mind observes the increase and decrease of the sun's diameter, and uses this to work out the distance of its body from the sun (this is shown by a probable argument drawn from the latitudes). A reply is also made to objections based upon the narrow angle under which the sun appears, and the planet's lack of senses. And at the end it is proposed that the opinion of a governing mind is nevertheless not entirely to be despised.[9]

Finally, a difficulty is aired about the local motion of the planet's body from an inherent animate force. And thus, since so many difficulties have been encountered on all points, there is but one thing achieved, namely, to call into question the preconceived notion of a perfectly circular planetary orbit (and also, to some extent, that of a governing mind of the reciprocation) on physical grounds. Shortly hereafter, in Chapter 44, this notion will be torn to shreds on geometrical grounds.

Chapter 40

(****) 3 v

1. A method by which the physical part of the equation, that is, the increment of time of a planet over any arc of the eccentric, may be found from the distances of the points of its arc from the sun.

2. Here there is a geometrical proof of how the distances from the sun of the infinity of points on an arc are almost exactly contained in the area enclosed by the arc and the lines drawn to the sun from the ends of the arc. Also, how the one triangle defined by the sun, the center of the eccentric, and the end of the arc, displays both parts of the equation: the optical part by the angle at the end of the arc; the physical by the area.

3. A demonstration that in the [theory of the] sun the optical and physical parts of the equation are equal to the senses.

4. As a preliminary, it is proved that triangles on equal bases are to one another as their altitudes.

5. By this theorem it is demonstrated that the area of the triangle of the equation[10] increases with the sine of the eccentric anomaly; whence there is a short cut for computing this area. At the same time it is shown by a numerical example that the parts of the equation do not perceptibly differ anywhere. This is done first at 90°, then at 45°.

6. There follows a minor exception, showing that the area is somewhat less than the sum of the distances at all degrees of the eccentric, and somewhat greater than the sum of the distances at all degrees of equated anomaly.

7. A geometrical construction of a quadrilateral conchoid equivalent to the distances from the sun at all degrees of the eccentric. Here, geometers are challenged to find the quadrature of this area.

8. The area between the two conchoids is shown not to be of the same width at places equidistant from the middle. (Of this, more in Chapter 43.)

---

[9] ἀναντίλεκτον.

[10] *Triangulum aequatorium.*

PART IV

Chapter 41

On the assumption that the planet's path is a perfect circle, and by taking the distances from the sun's body of three points on Mars's eccentric (demonstrated with great certainty in Part III), a geometrical proof is used to conjure up a false position for the apogee, a false eccentricity, and a false ratio of eccentricity.

Chapter 42

In a new approach, the distances of two points on the eccentric near aphelion are sought by means of five observations, and near perihelion, by three. Then, by halving the periodic time and the circle of the zodiac, the position of the aphelion is investigated very precisely, and is found to be the same as in Parts I and II. The mean longitude of Mars is thereby corrected. Then, by comparing the two distances, the true eccentricity is found, and the ratio of Mars's orb to the earth's. With the eccentricity of the eccentric found very precisely (if not delicately) by observations of the sun, it becomes clear at the same time that it is half the eccentricity of the equant, found elsewhere. Therefore the preliminary theories of Chapter 32 are also valid for Mars.

Chapter 43

A fundamental principle is laid down, which was demonstrated previously in Chapter 42: eccentricities are to one another in a double ratio. Second, it is supposed that the planet's orbit is arranged in a perfect circle. Third, it is supposed (as demonstrated in Chapter 33) that the increments of time of the planet over equal arcs of the orbit are proportional to the distances of those arcs from the sun. On these suppositions, faulty equations, disagreeing with experience, are conjured up. Then a cautionary mark is made as to where that falsehood does not lie hidden.

2. For this investigation, it is necessary to have a measure of the area between the two conchoids of Chapter 40, to the solution of which geometers' attention is drawn, as it involves many contrivances.[11]

Thus it is manifest that some one of the premises of this false conclusion is itself false.

Chapter 44

It is demonstrated by two arguments that the planet's orbit is not a circle, but an oval figure.

In the first place, those things demonstrated in Chapters 41 and 42 are presupposed. Now a perfect circle results in certain distances, and the diameter of this circle was found in Chapter 42. But the observations which were reintroduced in Chapter 41 require different distances, namely, ones that are shorter

---

[11] ἀτεχνίαν.

at the sides. But an oval figure admits such distances. Therefore, the orbit is an oval.

In the second argument, the same things are presupposed as in Chapter 43. The time increments shown to us by experience are not admitted by a circular figure, but are admitted by an oval. Therefore, the orbit of the planet is an oval.

Chapter 45

In what follows, the reader should overlook my credulity, since I am judging everything by my own wits. Indeed, the occasions by which people come to understand celestial things seem to me not much less marvelous than the nature of the celestial things itself. I therefore display these occasions scrupulously, with, no doubt, some attendant ennui for the reader. Nonetheless, that victory is sweeter that was born in danger, and the sun emerges from the clouds with redoubled splendor. Therefore, O reader, pay heed to the dangers of our army, and contemplate the clouds horrifying in their darkness. Contemplate, I say; for beyond these clouds the sun of truth truly lies hidden, and shortly will emerge. Therefore, an account is given of the occasions that enticed me to suppose what turned out false: that the planet, by its inherent force,[12] strives for a perfect epicycle, describing equal parts of it in equal times; but that the same planet is swept around by the extrinsic force[13] of the sun, through unequal arcs in equal times, as before. From this it is demonstrated that the orbit or path shaped by the two causes comes out to be an oval figure.

Chapter 46

1. First, this physical hypothesis, which is properly epicyclic, is transformed into an eccentric.
2. Next is taught one way of describing the line of the planet's motion in accord with this opinion.
3. Four obstacles to calculation[14] associated with this method are recounted. It is shown here that the mean of the sums of the terms is not the same as the mean of the terms themselves.
4. A second way of describing this curve is proposed, and the obstacle to calculation[15] of this method is likewise shown. Both methods are meanwhile useful in numerical operations.
5. A third way of describing the orbit of the planet is proposed, through the conjunction of the two hypotheses.
6. A fourth way which one might propound is rejected.
7. It is demonstrated that the line thus created is truly oval, not elliptical.

---

[12] *Vi insita.*

[13] *Vi extranea.*

[14] ἀμηχανίαν.

[15] ἀμηχανία.

Chapter 47

Now, on the assumption that the line of the planet's path is perfectly elliptical, it is demonstrated that the area of the ellipse is less than the area of the circle by the small area of the epicycle or of the circle described by the eccentricity of the eccentric, very nearly.

2. The area of this circle is sought, and so also the area of the oval-shaped plane.
3. It is shown also to be necessary to find a geometrical means of cutting this oval-shaped area in a given ratio – a problem to which geometers' attention is drawn.
4. The lunule by which the oval area differs from a circle, is stretched out straight, in a manner as nearly geometrical as possible.
5. It is proposed to geometers for study, whether this figure thus extended is double the true lunule.
6. Although there is no ready means of dividing an ellipse or oval in isolation, it is demonstrated that an ellipse can be easily divided by means of a circle.
7. Therefore, the ellipse being supposed, and being divided by means of the circle, a way is shown of computing both the distance and the equation.
8. The equation is computed at an anomaly of 90°, with the area expressed numerically in terms of the square on the diameter.
9. A way of correcting the eccentricity by means of the physical equation.   (****) 4 r
10. The equation is computed at the octants of anomaly, where the area of the triangle of the equation is expressed in numbers significant in seconds.
11. As these equations are also seen to be false, no less than those above in Ch. 43, the causes of the error are sought.

Chapter 48

I have tried to eliminate all the inconveniences and geometrical imperfections of Chapter 46 by avoiding areas and having recourse to numerically defined divisions of the ovoid's circumference.

1. 1. A way is taught of using geometry to find the corresponding portion of the oval path, using the method of distances found at equal intervals of time, from what has been demonstrated in Chapter 33, as well as the supposition that the total length of the oval is known.
2. A geometrical account is given for a contrivance[16] which substitutes the single distance of the midpoint of an arc for the two distances of the beginning and end of that arc.
3. By another contrivance,[17] which nonetheless proceeds by a geometrical path, the approach of the ends of the segments of the oval towards the center of the eccentric, and hence the angle that a segment of the oval subtends at the center, and thus finally the angle that the same segment of the oval subtends at the center of the sun, are demonstrated.

---

[16] Ἀτεχνία.

[17] Ἀτεχνία.

4. Another contrivance[18] for seeking the length of the oval path, which is, however, accompanied by other geometrical theories. For two circles are given, and two means between them, one arithmetical and the other geometrical, by the former of which a greater circle is constructed, while by the latter, a lesser. Then, by two arguments, the ellipse is shown to be equal to the arithmetic mean; the one, more general, depending upon a contraction of the extremes, and the other, purely geometrical, showing that the ellipse certainly exceeds the lesser mean and therefore probably equals the greater.

5. One procedure for seeking the equations, which ignores what has been said under numbers 3 and 4, exactly as if, as in the whole, so also in the parts, the two cancel each other.

6. It is demonstrated geometrically that the visual lengthening in the parts caused by the approach of no. 3, and the contrary shortening of the elliptical arcs of no. 4, are not equal.

7. A genuine process is described that is in agreement with all that is demonstrated in this chapter, and the equations found thereby are still convicted of error.

Chapter 49

1. The above method is shown to beg the question, and to offend against what it was proposed to do.

2. The areas of Chapter 46 and 47, as well as the oval circumferences of Chapter 48, are dismissed, and a return is made to the causes by which the oval is formed. And because hitherto the epicycle has been transformed into an eccentric, thus confounding the planet's own power with the power coming from the sun, the epicycle on a concentric is taken up again, and the physical causes treated in Ch. 45 are applied in order that by this means a proper foundation for seeking the equations may be obtained.

3. The actual method of constructing the equations is reviewed, and the equations are convicted, by experience, of the same error that occurred above in Ch. 47.

4. Therefore, the suspicions of error in the computations carried out above in Ch. 47 are dismissed, and it is concluded that the hypothesis in Ch. 45 is itself at fault.

Chapter 50

Has six attempts at finding the equation by the distances themselves; that is, at finding the time increment of the planet on a certain arc of the eccentric, the distances having been taken before I knew that their sum is contained in the plane surface. For it is most certain from Ch. 33 that the increments of time are to be obtained from the distances. But since there are three anomalies: one, which is the measure of the time; a second, of the arc of the eccentric; and a third, of the angle that this arc subtends around the sun; I have given individ-

---

[18]Ἀτεχνία.

ual distances to the 360 equal parts of all three anomalies. So by this token the consideration of distances has been made threefold. Now, from the same Ch. 33 it is clear that the diurnal path of a planet at aphelion to the diurnal path at perihelion, as if seen from the center of the sun, is inversely in the duplicate ratio of the aphelial and perihelial distances of the planet from the sun. I have therefore squared all the distances and divided by the standard 100,000, so that the result, compared to the standard 100,000, might represent that duplicate ratio that holds between the apparent diurnal motions as seen from the center of the sun. So to the three kinds of distances the same number of kinds of third proportionals are added. With these things diligently sought out, I hoped I had left out nothing pertaining to the effects of natural causes (which tell us to seek the eccentric position of the planet by means of the distances). Thus were the six methods developed.

In the first and second, which have the distances of the second anomaly (that is, the eccentric anomaly), there occurs a geometrical matter worthy of consideration. For the sum of the 360 third proportional lines has come out equal to the sum of the 360 radii, or first proportional lines. This is proposed to geometers for demonstration.

Aside from this, here is how the six methods compare. Two (the fourth and the fifth) result in absurdities, and double the errors of the equations. The remaining four coincide with the methods of the preceding chapters. Two (the second and the third) suppose the planet's path to be a circle, and two (the first and the sixth) transfer the distances and set up the oval path as in the theory in Chapter 45. And the latter err in defect by the same amount as the former in excess: they have the truth in the middle.

Chapter 51

Now that it is clear that the oval of Chapter 45 produces false equations, the question is raised whether it is also in error concerning the distances.

Consequently, in this chapter, first the observations and second the distances of the sun from the earth are adopted, such as were established with great certainty in Part III. Aside from these, nothing is assumed or admitted as a principle of demonstration. From these, in turn, are demonstrated the distances of Mars from the sun in many places on the eccentric throughout the entire orbit, and especially in those places so chosen that individual pairs on their respective ascending and descending semicircles, are equidistant from the position of the aphelion, as determined above by more than one route. Whence the position of the aphelion is corroborated, and at the same time the trustworthiness of the vicarious hypothesis is investigated.

Chapter 52

From what was demonstrated in the preceding chapter, it is demonstrated further, that parts equally removed from the established aphelion, and equidistant from the sun, are not equidistant from any other point not on the line joining the sun and the aphelion. Therefore, the line of Mars's apsides passes right through the body of the sun, since all other lines divide Mars's eccentric in an absurd manner, namely, into two unequal segments. It is added in anticipation of anyone who would construct that eccentric upon another point, such that it would be cut into two equal parts by some line other than that which passes through the sun, that he will be refuted by the observations. In the same way, it is demonstrated that since the sun is on the longer diameter of the oval eccentric, the point that Copernicus used in place of the sun in constructing the eccentric is not on that longer diameter. And it is by no means probable that the line of apsides of an oval eccentric be other than the longer diameter of the oval; hence, the line of apsides does not bypass the sun, and so the lines of apsides of all planets come together in the very center of the sun, and not in some point of the sun's mean position.

Chapter 53

(****) 4 v   A unique method of finding the distances of Mars from the sun near its opposition to the sun, and at the same time a demonstration of the point on the earth's orb[19] at which an error in the distance appears greatest. Here it is presupposed that the difference of two eccentric positions and the difference of their distances from the sun are known approximately. At the same time, by this means, the trustworthiness of the vicarious hypothesis is investigated, as in Ch. 51 above.

Chapter 54

By gathering together things that were demonstrated in various places, the ratio of the eccentricity and of the orbs is very carefully established and adjusted.

Chapter 55

Finally we return to the course from which we had departed in Chapter 45. For, by a complete induction,[20] it is demonstrated that, just as the circle in Chapter 44 was too broad at the sides, so the oval of Chapter 45 is too narrow. The arguments are two. One is drawn from the distances: those observed and presented in Ch. 51 and 53 are compared with the distances computed from the hypothesis, using the ratio of the orbs given in Chapter 54 and the form of the motions given in Chapters 45, 46, and 49. And it is shown that the observed distances are greater. The other argument is taken from the equations. For the equations computed from the circle in Ch. 43 were in error in one direction, while those of the oval of Chapter 45, computed in Chapters 46, 47, 48, 49, and 50, were in error by the same amount in the opposite direction.

---

[19] *Orbis magnus.*

[20] *Inductio omnium. Inductio,* in the signification given it by logicians of the sixteenth century, usually denoted enumeration of the next instance in a well-defined class or sequence; hence, *inductio omnium* means an enumeration of all instances, or a "complete induction." See N. Jardine, *The Birth of History and Philosophy of Science,* (Cambridge University Press, 1984) p. 77.

Chapter 56

It is now demonstrated from the preceding that the distances are not to be derived from the circumference of the epicycle, whether the planet moves uniformly upon it, as in Ch. 45, or retains the ratio of the eccentric motion, as in Ch. 41, but are to be taken from the diameter of the epicycle. The premises are the same as before.

Chapter 57

Since the physical arguments of Chapter 45 necessarily have some admixture of falsehood, on account of their false conclusion, while now the genuine conclusions are made clear, those physical arguments are reformed, and the theoretical argument of Chapter 39 is continued.

First it is shown that the reciprocation on the diameter of the epicycle (which supplies distances in agreement with the observations) follows the natural laws of bodies.

2. Since reciprocation is a translation from place to place, it is shown that this translation of the body of the planet is caused and carried through by the sun, no less than the motion of revolution in Part III, but in such a manner that the reins of the reciprocation are in the hands of the planet itself. This is made clear by two examples, one imperfect (using oars), the other more perfect (using a magnet).
3. In applying the magnetic example, I suppose two faculties in both the planet and the magnet, one of direction and the other of appetency. The magnet is directed towards the pole, and seeks out iron. Just so, the globe of the planet is directed with respect to the fixed stars, and seeks out the sun. Now the function of direction, upon which depend the motion and position of the aphelion, I at first leave in doubt, whether it belongs to Mind or Nature. The function of appetency, upon which depends the eccentricity, I ascribe to Nature, and I show approximately that the measure of the reciprocation obtained by observing is in agreement with the physical causes by parts.
4. Afterwards I treat this more accurately, initially considering the faculty of direction, and conceding that a certain amount of declination is removed from it arising from the appetency of the sun, just as a magnet directed towards the pole is nonetheless somewhat deflected by iron and nearby mountains. I then demonstrate that the position and the very slow eastward motion of the aphelion can be accounted for by a natural and corporeal faculty, with no assistance from a mind.
5. The measure of appetency I show to be analogous to the balance; or, more particularly, the sine of the equated anomaly measures the strength of appetency at any given point of time.
6. Concerning the distance traversed in the reciprocation at any given time, hear, reader, what I shall demonstrate. Its measure is clear from Chapter 56, namely the versed sine of the eccentric anomaly, not of the equated anomaly. This measure is founded upon the observations. I therefore also had to work out here a way to demonstrate this measure of the line

traversed in the reciprocation, namely, the versed sine of the eccentric anomaly, from the stated measure of strength at any given place (which, however, was the right sine of the equated anomaly). To obtain this, I had to show that, with a quadrant divided into any number of equal parts, the versed sine of any arc has an imperceptibly smaller ratio to the versed sine of the whole quadrant, than the sum of the sines on the arc has to the sum of the sines in the quadrant.

7. Here, there seemed to be two considerations tending to make the latter premise inconsistent with the former conclusion. First, the eccentric anomaly, which gives the measure of the reciprocation, was greater in the upper semicircle, and displayed a greater number of sines, than the equated anomaly, which gives the measure of the strength. But the answer is that this happens correctly, since in the equated anomaly the planet also takes more time, and consequently also pours out more forces.

8. The other obstacle: that the sines of the equated anomaly are less than the sines of the eccentric anomaly in the upper semicircle. It has therefore been shown that the versed sine as well is somewhat less than the sum of the sines of its arc, and thus is equivalent to the sum of the smaller sines.

9. Objections which may be raised against the example of the magnet are in part dispelled, and in part give occasion for calling Nature into question and passing over to Mind, in order to see whether and by what means Mind might cause the eccentricity through reciprocation.

10. And so, positing that which is demonstrated most certainly in Ch. 56, that the versed sine of the eccentric anomaly measures the reciprocation, it is now demonstrated that the versed sine of the equated anomaly measures the increment in the apparent diameter of the sun. That is, the sun's apparent diameter not only begins to increase as does the versed sine of the equated anomaly, and reaches a maximum when the latter does so, but also stands at a mean between the extremes when the versed sine of the equated anomaly is the semidiameter, the versed sine of the eccentric anomaly then being greater.

11. On the contrary, if the versed sine of the eccentric anomaly be the semidiameter, it is demonstrated that the apparent diameter of the sun is smaller yet, since it is a mean between extremes.

12. In order to show that this measure is appropriate and comprehensible to the planet's mind, a comparison is first of all made between the eccentric anomaly and the equated anomaly, and it is denied that the angle of the eccentric anomaly (if it had been proposed as a measure) could have been comprehended by the planet's mind.

13. But it turns out to be arguable that the angle of equated anomaly, whose versed sine is proportional to the increase of the sun's diameter, is comprehended by the planet's mind.

14. But since it is not this angle but its versed sine that measures the increment of the sun's diameter, it is shown by reasoning, by physical hypotheses, and by examples of natural phenomena, to be arguable that the planet's mind can comprehend the sine (the strength, in physical terms) of this angle.

15. A comparison is made of the two ways so far considered by which the

motions proper to the planetary bodies (that is, the reciprocations) may be carried out: the former placed under the care of Nature, and the other, of Mind. A conclusion is finally made in favor of Nature, repudiating Mind.

16. Among the arguments for this conclusion, the chief is the geometrical uncertainty admitted by that form of motion administered by Mind. An explanation of this is given.

17. It is shown that this uncertainty might provide occasion for the progressive motion of the aphelion. But since in Ch. 35 above another cause for the progression of the aphelion was suggested, the two are compared here, and it is shown that a single interposed body, if it is left with any efficacy, does not cause the progression of the aphelia, neither if Nature nor if Mind moves them.

18. And so the physical positions are given limits, lest an interposition affect anything else.

19. But in order that the progression of the aphelion might arise hence, it is shown that the peculiar mental function which was rejected in no. 17 as   (****) 5 r
absurd must be associated with any interposition. That we might be freed from this, a conclusion is made in favor of the opinion which, in no. 4, ascribed the motion of the aphelion to Nature.

Chapter 58

Now that the true proportion of the planetary reciprocation have been found, it is shown how, on that basis, an orbit can be made for the planet (by compounding the two motions of revolution and reciprocation) in a "puff-cheeked" form as well. It is also shown how, through an error that seemed like the truth, I happened upon this puff-cheeked orbit.

2. This orbit is convicted of error by the equations, though the distances are correct. This is contrary to what happened previously, when both the equations and the distances were simultaneously in error.

3. I show how, in ostensibly doing something else, and bringing the ellipse back in, I unwittingly corrected the error.

4. It is demonstrated that the orbit is made puff-cheeked by the erroneous hypothesis I used.

5. It is shown that since the elliptical orbit provided the correct equations, the reciprocation that was deformed into the puff-cheeked orbit was called into question.

Chapter 59

1. The geometry of the ellipse in 10 propositions, by which

2. it is demonstrated in proposition 11 that the distances constructed by the reciprocation and supported by the observations are contained in the perfect ellipse no less than in the puff-cheeked orbit introduced in Ch. 58 and convicted of falsity. And so, since the ellipse furnishes both the equations and the distances, the orbit of the planet is therefore elliptical.

3. On the same basis, it is demonstrated, in proposition 12, that the area of the ellipse is a most perfect measure of the distances of the unequal arcs

on an ellipse corresponding to equal arcs on a circle.

4. By solving the objection to using unequal arcs on the ellipse, it is shown, in proposition 13, that this ellipse is precisely in accord with the physical principles of Part III.

5. It is demonstrated, in proposition 14, that the arcs of the ellipse are to be bounded by lines drawn ordinatewise through the individual degrees of the circle. At the beginning and end of a quadrant, this is done with two perfect demonstrations, but for intermediate motion it is done less perfectly, though by an indication[21] that is clear enough – and to this problem geometers' attention is directed.

6. With these conclusions, and especially those appearing under no. 3, with the help of those under no. 1, it is demonstrated more fully by this, in proposition 15, that the area of the circle itself is also a very perfect measure of the distances that are assigned to unequal arcs of the ellipse (which arcs must be defined by lines drawn ordinatewise through equal arcs of the circle). This is also confirmed by numerical computation, by which the observations are satisfied in both of these methods.

Chapter 60

1. From what was demonstrated in Chapter 59, a method of finding the equations is established.

2. Demonstration of a precept telling how the mean anomaly and the equated anomaly may be found from a given eccentric anomaly.

3. One way of finding the eccentric anomaly from a given equated anomaly and eccentricity. This is founded upon a most beautiful and purely geometrical theoretical consideration of the short lines of the planet's incursion from the circumference of the circle towards the line of apsides. The consideration comprises five problems, and is carried out by means of the rectangle on the quadrant.

4. Another approach to this problem, using analytical rules.

5. A clumsy[22] method, using a kind of iteration, for finding the eccentric anomaly and the equated anomaly, given the mean anomaly or the time; and the reason why a geometrical method cannot be given.

PART V

Chapter 61

Now that the hypothesis of the longitudes has been found, the position of the two nodes is investigated more accurately, from the observations.

Chapter 62

Now that the distances have been found, the inclination of the planes is investigated more accurately, from acronychal observations in both semicircles.

---

[21] ἔνδειξιν.

[22] ἄτεχνος.

2. The ratio of the apparent latitude to the inclination at any point is demonstrated to be the inverse of the ratio of the distances of the sun and the earth from the planet.
3. A small table of observed latitudes at opposition to the sun, compared with the computations from our hypothesis.

Chapter 63

1. The physical cause of the latitudinal deviation is presented.
2. It is demonstrated geometrically that a plane is swept out in this deviation.
3. The question is raised, whether this is the work of corporeal Nature or of Mind, and a conclusion is made in favor of Nature.
4. The question is raised, whether the axis of the latitudes is the same as, or different from, the axis causing the eccentricity, and it is shown what form a body must have if its Nature alone is to do everything.
5. On the supposition of solid orbs, a simple and easy hypothesis for the latitudes is presented.

Chapter 64

The theory of latitudes has been finished; now, the diurnal parallax is examined more accurately, and is shown to be quite imperceptible by two arguments, one from the position of the nodes and the other from the inclination of the planes.

Chapter 65

The value of the maximum latitudes at both opposition and conjunction is determined, assuming knowledge of the cycles[23] of all the motions under all circumstances, and of the exact space of the ages.

2. The same value is determined for our time.

Chapter 66

The value of the maximum latitudes elsewhere than at syzygy is investigated, and their positions are determined.

2. The cause of a paradox concerning the latitude at opposition to the sun is revealed.
   (****) 5 v
3. An accurate method for computing the latitude elsewhere than at acronychal position.

Chapter 67

The same thing as in Chapter 52 is demonstrated: that the eccentricities have their common origin in the very middle of the sun, not in some other point substituting for the sun. This is shown by two arguments, the first from the position of the nodes, and the other from the inclination of the planes.

---

[23] ἐξελιγμῷ.

Chapter 68

1. A theory of the change in latitude of the fixed stars, which is proposed through physical causes and the mean ecliptic, or rather, through the introduction of the royal circle (as we call the *via regia*).
2. It is shown that the northern limit of the ecliptic is in $5\frac{1}{2}°$ Aries, and this lends plausibility to the notion that that mean or constant path passes through the positions of the planets' apsides.
3. The mean ecliptic (or rather, royal circle) is added, by changing the obliquity of the common or true ecliptic. In the margin is the theory of the precession of the equinoxes, done by means of a cylindrical annual translation of the earth's axis and poles, and a very slow inclination deflecting them into a cone.
4. Hence it is shown that the inclination of the planes of Mars and the ecliptic does not stay the same over the ages.
5. The same is concluded less clearly from a comparison of the Ptolemaic observations with our own.

Chapter 69

1. The observations the ancients made of Mars, and left in writing.
2. On the nonuniformity of the precession of the equinoxes, pro and con.
3. On the inconvenient number of spheres in recent theories.
4. Was the sun's eccentricity once greater? Or, on the length of summer and winter, at the time of Ptolemy.
5. The sun's apogee at the time of Hipparchus is uncertain; and the means he used to investigate it.
6. That the positions of the fixed stars at the time of Ptolemy are somewhat uncertain; and his means of investigating them.
7. Side effects of errors in the positions of the fixed stars upon the theory of Mars.
8. From three of Ptolemy's acronychal observations, adjusted to the modern equations, the correction of the motions for Ptolemy's time is found. This is done eight times, as one or another of Ptolemy's preconceptions, already discussed, is changed.
9. Next, in order to settle this confusion, it is shown that, when refraction and the defect of the solar eccentricity, which cancel each other, are neglected, the fixed stars retain the zodiacal positions assigned by Ptolemy.
10. Upon this foundation, the epoch of the mean motion of Mars is established for the times of Ptolemy and of Christ.
11. The epoch of the sun's mean sidereal motion, at the times of Ptolemy and of Christ, is added.

Chapter 70

Through two ancient and untrustworthy observations, the values in ancient times for the ratio of the orbs of Mars and the sun, the latitude of Mars, and the sun's eccentricity, are examined.

Index of a great many terms in the margins, as well as of authors of whom mention is made.[24]

A.

---

[24]Page numbers are from the first edition (included in the margins of the present translation). Entries are rearranged so as to be in alphabetical order in English. Names are as given by Kepler (Latin endings removed where appropriate), and entered where he chose to enter them (thus, David Fabricius is under D, while Philip Lansberg is under L). Where the placement does not square with modern usage, a cross reference in provided in square brackets.

In the Name of the Lord

Part One

of the
Commentaries

On the Motions of the Star Mars

The Comparison of Hypotheses

CHAPTER 1

On the distinction between the first motion and the second or proper motions; and in the proper motions, between the first and the second inequality.

The testimony of the ages confirms that the motions of the planets are orbicular. Reason, having borrowed from experience, immediately presumes this: that their gyrations are perfect circles. For among figures it is circles, and among bodies the heavens, that are considered the most perfect. However, when experience seems to teach something different to those who pay careful attention, namely, that the planets deviate from a simple circular path, it gives rise to a powerful sense of wonder, which at length drives people to look into causes.

It is just this from which astronomy arose among humans. Astronomy's aim is considered to be to show why the stars' motions appear to be irregular on earth, despite their being exceedingly well ordered in heaven, and to investigate the specific circles whereby the stars may be moved, so that by their aid the positions and appearances of those stars at any given time may be predicted.

Terms:
1. The first motion is that of the whole heaven and of all its stars from the east through the meridian to the west, and from the west through the lowest part of the heavens to the east, in the period of 24 hours; in the present diagram, *ABCD*.
2. The second motions are those of the individual planets from the west to the east, from *A* to *E*, from *F* to *G*, in longer periods.
3. The smaller circles are those which are nearer to one of the poles, as *HLK* which is closer to pole *Q* than to pole *R*.
4. The greatest circle of the sphere is that which is equidistant from each of its poles.

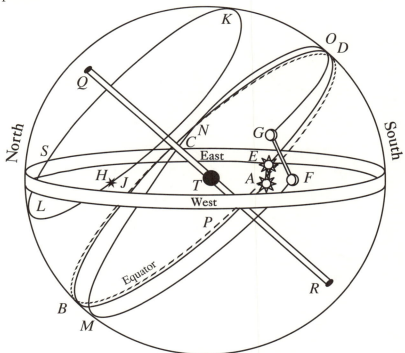

Before the distinction between the first motion[1] and the second motions[2] was established, people noted (in contemplating the sun, moon and stars) that their diurnal paths were visually very nearly equivalent to circles. These were, however, entwined one upon another like yarn on a ball, and the circles were for the most part smaller[3] circles of the sphere, rarely the greatest[4] (*such*

75

2   *as here ABCE, FMNG cutting the equator AB in CN*), part of them north and part south. They also saw that the stars have different speeds in this diurnal and apparent motion. The fixed stars are fastest of all, since those that are in conjunction with any of the planets on the preceding day (*such as H with A and F*) come to their setting first (*such as H, moving along LK back to I*). The sun (*on ABE*) is slower, as *on the following day it stands at E* and so its setting follows that of the fixed stars *at I* with which on the previous day it was conjoined on *HA*. Slower still than this, slowest of all the stars, is the moon, since after setting with the sun today (*at A, the moon being at F*), it lags by an appreciable interval (*EG*) tomorrow when the sun sets (*at E, the whole heaven and the moon along with it having made a circuit around the earth along FMNOG*). Hence the Pythagoreans, when they shared out musical tones among the stars, gave the lowest (the *hypate* among the strings of the lyre)[1] to the moon, because the motions of both were slowest. Hence have originated the words προηγούμενος and ὑπολειπτικός. [2] The former of these terms originally corresponded to a star which, on the next day, comes to its setting earlier (*the sun E is said to be προηγούμενος with respect to the moon G*). The latter term corresponded to a star that is slower in the first motion (*such as the moon here*), which is, as it were, abandoned and left behind (*at G*) by the swifter ones (*E and I*). For more on this subject see our *Optics*, Ch. 10. [3]

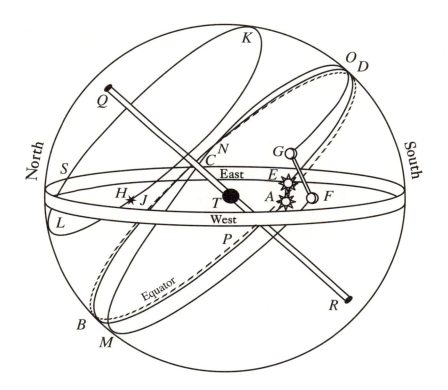

---

[1]The Greek word *hypate* actually means "highest"; however, the Greek convention regarding "high" and "low" in music is the opposite of our own.

[2]"Leading" and "left behind", respectively

[3]*Astronomiae pars optica*, Frankfurt, 1604, Ch. 10, esp. pp. 333-4, in *KGW* 2 pp. 286-7; English trans. pp. 343–344.

This first adumbration of astronomy consists, not of the unfolding of a cause, but solely of the experience of the eyes, extremely slowly acquired. It cannot be explained in figures or numbers, nor can it be extrapolated into the future, since it is always different from itself, to the extent that no spiral is equal to any other in increments of time, and none carries over into the next with a curvature of the same quantity. Nevertheless, there are some people today who, riding roughshod over two thousand years' work, care, erudition, and knowledge, are trying to revive this, obtruding admiration of themselves far and wide (an attempt which has not been fruitless among the ignorant). Those with more experience consider them with good reason to be incompetent, or (if, like that man Patricius,[4] they want to be known as philosophers) to act mad with reasoning.

For it was helpful to astronomers to understand that two simple motions, the first one and the second ones, the common and the proper, are intermingled, and that from this mingling that continuous sequence of conglomerated motions follows. And so, when that common and extrinsically derived diurnal revolving effect is removed, the fixed stars are suddenly no longer the swiftest and the moon slowest, but quite the opposite, the latter being the swiftest in itself *and in its proper motion FG* while the former are clearly very slow or immobile. When a planet (*such as the moon at G*) is "left behind" (*by the sun at E or the fixed stars at I*), it is carried in consequence* *through FG* more swiftly than the sun (*through AE*) or the fixed stars (*through HI*). If, however, it appears to be 'leading' among the fixed stars, it goes along with a retrograde motion. *For example, if the sun at A along with a fixed star at H had been released from the same starting line AH on the previous day, so as to traverse BCDE and arrive at P while the fixed star traversed HLK and arrived at I, the sun, in the space of one day, would have retrogressed through the interval AP.*

This turned out to be of great profit in astronomy in grasping the simplicity of the motions. Instead of unending spirals, *a new one always being added to the end of the earlier one at E or G,* there remained little but the solitary circles *FG* and *AE,* and a single common motion, either of all the planets and the whole world as well in a direction opposite to the proper motions, or (with the world standing still, according to Aristarchus) of the earth's globe *T* around the axis *QR* in the same direction as the proper motions.

Now that the first and diurnal motion had thus been set aside, and only those motions that are apprehended by comparison over a period of days, and that belong to the planets individually, were considered, there appeared in these motions a much more complicated mingling than before, when the diurnal and common motion was still mixed in with them. For although this residual mingling was there before, it was less observed—less striking to the eyes—because the diurnal motion was very swift. And so this residual motion was divided into minute parts and spread out over several days and several

Term:
* "In consequence" means "according to the sequence of signs" (Aries, Taurus, and so on), which sequence runs from the west through the meridian to eastern parts, then down towards the nadir and again the west: from *F* to *G*, from *A* to *E*.

3

---

[4]Francesco Patrizi (1529–1597), professor of philosophy at Ferrara and at the *Sapienza* in Rome. His chief work, *Nova de universis philosophia*, Venice 1593, delineates in an engaging manner an innovative Hermetic-Platonic cosmology. Patrizi made no secret of his dislike for astronomers.

diurnal spirals. But now, that minute division and distribution of the star's proper motions over so many days was removed (that is, by the removal of the diurnal motion), and so all the proper motions of the stars, as many as they are, and all the confusion of this multitude shone forth more obviously. First, it was apparent that the three superior planets, Saturn, Jupiter, and Mars, attune their motions to their proximity to the sun. For if the sun would approach them they moved forward and were swifter than usual; where the sun would come to the signs opposite the planets they retraced with crablike steps the road they had just covered; between these two times they became stationary; and these things always used to occur, no matter what the signs of the zodiac in which the planets might have been seen. At the same time, it was clear to the eye that the planets appeared large when retrograde, and small when anticipating the coming of the sun with a direct and swift motion. From this, the conclusion was easily reached that when the sun approaches they are raised up and recede from the lands, and when the sun departs towards the opposite signs they descend again towards the lands.[5] And finally, it was observed that these phenomena of retrogressions and increase of luminosity, just described, was moved through the signs of the zodiac in the order that tended from west through the meridian eastward, so that whatever has happened at one time in Pisces soon would come to pass similarly in Aries, then in Taurus, and so on in consequence.

If one were to bundle all this together, and were at the same time to believe that the sun really moves through the zodiac in the space of a year, as Ptolemy and Tycho Brahe believed, he would then have to grant that the circuits of the three superior planets through the aethereal space, composed as they are of several motions, are real spirals, not (as before) in the manner of balled up yarn, with spirals set side by side, but more like the shape of a pretzel,[6] as in the following diagram.

---

[5]Latin, *a terris, ad terras*. Use of the plural usually denotes the earth as the place of human habitation, or the various lands that the earth contains. Kepler may have intended to plant our feet "on the ground," as observers, rather than having us imagine a "view from earth," as a more conceptualized body.

[6]The Latin is *panis quadragesimalis*, that is, "bread of the forty [days]", or lenten bread. Pretzels were invented by monks of southern Germany, who made it a practice to give them to children as treats during Lent.

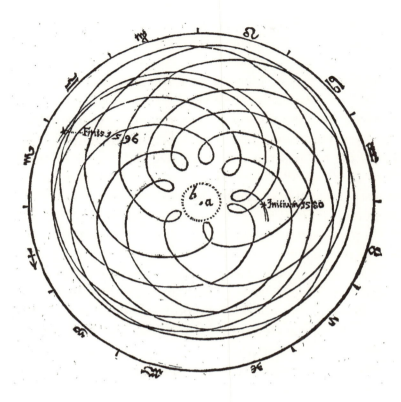

4

This is the accurate depiction of the motions of the star Mars, which it traversed through the aethereal air from the year 1580 until the year 1596, on the assumption that the earth stands still, as Ptolemy and Brahe would have it. These motions, continued farther, would become unintelligibly intricate, for the continuation is boundless, never returning to its previous path. Take note, too, that since the orb of Mars requires such a vast space, the spheres of the sun, Venus, Mercury, the moon, fire, air, water, and earth, have to be included in the tiny little circle around the earth *A*, and in its little area *B*. In addition, the greatest part even of this little space is given to Venus alone, much greater in proportion than is given to Mars here out of the whole area of the diagram. Moreover, we are forced to ascribe similar spirals to the remaining four planets, and much more complicated ones to Venus, if the earth stands still. Ptolemy and Brahe offer explanations of the causes, order, permanence, and regularity of these spirals, the former using individual epicycles carried around on the eccentrics of the individual planets, in imitation of the sun's motion, and the latter by having all the eccentrics carried around upon the single orb of the sun. Nevertheless, both leave the spirals themselves in the heavens. Copernicus, by attributing a single annual motion to the earth, entirely rids all the planets of these extremely intricate spirals, leading the individual planets into their respective orbits,[a] quite bare and very nearly circular. In the period of time shown in the diagram, Mars traverses one and the same orbit as many times as the 'garlands' you see looped towards the center, with one extra, making, say, nine times, while at the same time the earth repeats its circle sixteen times.

---

[a]Here and in the following sentence, Kepler strikingly reverses the classical sense of the word "orbit." It originally denoted the track made by a wagon wheel, and Pliny and others had used it to describe the wanderings of the planets. Here, in contrast, Kepler uses it to refer to a fixed, closed path in the heavens which a planet traverses repeatedly. For Kepler, as for astronomers who followed him, the task of astronomy has shifted radically, having become the discovery and description of orbits, rather than the construction of hypothetical spherical models.

Again, however, it was noticed that these loops in each planet's spirals are unequal in different signs of the zodiac, so that in some places the planet would retrogress through a longer arc of the zodiac, at others through a shorter, and now for a longer, now for a shorter time. Nor is the increment of brightness of a retrograde planet always the same. Because, if one were to compute the times and distances between the midpoints of the retrogressions, neither times nor arcs would be equal, nor would any of the times answer to its arc in the same proportion. Nevertheless, for each planet there was a certain sign of the zodiac from which, through the semicircle to the opposite sign in either direction, all those things successively increased.

From these observations it came to be understood that for any planet there are two inequalities mixed together into one, the first of which completes its cycle with the planet's return to the same sign of the zodiac, the other with the sun's return to the planet.

Now the causes and measures of these inequalities could not be investigated without separating the mixed inequalities and looking into each one by itself. They therefore thought they should begin with the first inequality, it being more nearly constant and simple, since they saw an example of it in the sun's motion, without the interference of the other inequality. But in order to separate the second inequality from this first one, they could not do anything but consider the planets on those nights at whose beginning they rise while the sun is setting, which thence were called ἀκρονυχίους [acronychal, or "night rising"]. For since the presence and conjunction of the sun makes them go faster than usual, and the opposition of the sun has the opposite effect, before and after these points they are surely much removed from the positions they were going to occupy through the action of the first inequality. Therefore, at the very moments of conjunction with and opposition to the sun they are traversing those very positions that are their own. But since they cannot be seen when in conjunction with the sun, only the opposition to the sun remains as suitable for this purpose.

But since the sun's mean and apparent motions* are two different things, for the sun, too, is subject to the first inequality, the question is raised which of these releases the planets from the second inequality, and whether the planets should be considered when at opposition to the sun's apparent position or its mean position. Ptolemy chose the mean motion, thinking that the difference (if any) between taking the mean sun and the apparent sun could not be perceived in the observations, but that the form of computation and of the proofs would become free from difficulty if the sun's mean motion were taken. Copernicus and Tycho followed Ptolemy, carrying over his assumptions. I, as you see in Ch. 15 of my *Mysterium cosmographicum*, instead establish the apparent position, the true body of the sun, as my reference point, and will vindicate that position with proofs in Parts 4 and 5 of this work.

But before that, I shall prove in this first part that one who substitutes the sun's apparent for its mean motion establishes a completely different orbit for the planet in the aether, whichever of the more celebrated opinions of the world he follows. Since this proof depends upon the equivalence of hypotheses, we shall begin with this equivalence.

Marginal notes:

The sun has only a single inequality, with respect to the time within which it is completed. But as for the causes of this in equality, the same two factors combine as much for the sun as for the other planets, as will be explained below.

5

Terms:
* The sun's apparent position is that which it is perceived to occupy through its inequality. The mean position is that which it would have occupied if it had not had its inequality.

CHAPTER 2

On the first and simple equivalence, that of the eccentric and the concentric with an epicycle, and their physical causes.

And now, to begin, I take up the equivalence of hypotheses adopted to save [the appearances of] the first inequality,[1] which were demonstrated by Ptolemy in Book III and by Copernicus in Book III Ch. 15.

There, an eccentric is shown to square accounts with an epicycle on a concentric, provided, that is, that the line of apsides in the eccentric and the line through the center of the epicycle and the planet on the concentric always remain parallel, and that the semidiameter of the epicycle in the latter is equal to the eccentricity in the former, while the semidiameters of eccentric and concentric are equal. And also provided that, in the former, the planet is moved uniformly on its eccentric, so as to traverse equal arcs in equal times.

6

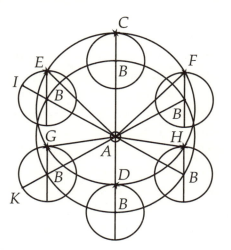 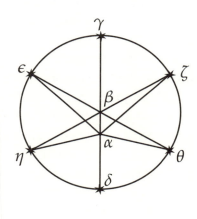

First, let A be the position of the observer and the center of the concentric BB on which is the epicycle BC, BE. Let the arcs between two B's, or the angles BAB, be equal, and the planet be first at C, then at E and G, with the lines BE, BG parallel to BC. Next, let β be the center of the eccentric γζ, with βγ, βε equal to AB, and let α be the point at which the observer is, and βα (the eccentricity) be equal to the semidiameter of the epicycle BC, BE, and parallel to them. Also, let the arcs γε, γζ, that is, the angles γβε γβζ, be equal both among themselves and to the former angles BAB. I say that the distances AC, αγ, are equal, and likewise AE to αε, AG to αη, AD to αδ, AH to αθ, and AF to αζ; and again, that the angles EAC, εαγ are equal; and that in each instance the planet, although its motion is uniform, will appear from A, α slow at C, γ, and swift at D, δ. As I said, Ptolemy demonstrated this in Book III,

---

[1]Kepler uses the late Latin word *salvare*, "save," which is a translation of the Greek σῴζειν, used by Proclus in *Hypotyposes* in the phrase σῴζειν τὰ φαινόμενα to denote the aim of planetary hypotheses. To "save an inequality" would accordingly mean to show how apparently irregular events can be presented as resulting from regular causes.

81

so there is no need for further discussion. To geometers, the diagram speaks for itself, and others may go to Ptolemy.

As for the physical accounts of these models, there is a greater difference between the two. That this may be very clear, it must be researched at somewhat greater depth, in one way from Peurbach[2] using Aristotle's principles, and in another way from Tycho.

Ptolemy has described these circles to us in their bare form, such as geometry applied to the observations shows. Peurbach set up a way for them to be traversed that follows Aristotle, who attempted this same thing upon the geometrical suppositions of Eudoxus and Calippus, by which they had treated their astronomy.

And while these authors used 25 orbs to demonstrate all the inequalities of the planets, Aristotle (since he believed the heavens to be filled with solid orbs)[3] thought that 24 other revolving orbs had to be interposed in order to free each lower orb from the revolving effect which, on account of the contiguity of surfaces, it was going to receive from the orb above it. So, having thus accumulated 49 orbs in all (or 53 or 55, following Calippus), he attributed to each its own mover. Any one of these would be responsible for the perfectly uniform motion of its own orb and all inferior ones which it encompassed. This motion would take place inside the closest surrounding superior orb, as if in a sort of place,[4] and from the mover would proceed a constant determination [*ratio*] of both the direction in which the motion was to occur, and the swiftness with which the orb was to return to its starting position. Moreover, since that

7   philosopher held the motion to be eternal, he also stipulated that the movers were eternal. Since they created motion for an infinite time, and since Aristotle knew that nothing material could receive the form of infinity, he maintained that they are also immaterial, and separable principles, and consequently immobile. Also, since he had constructed the world's eternity from the eternity of motion, and this duration of essence, the goodness and perfection of the whole world, was opposed to destruction, which was bad, he therefore attributed to those principles the highest perfection and the understanding thereof, and from good understanding the will to see it accomplished, lest the good not be done well. In this way, he introduced to us separate minds which, it turned out, were gods, as the perpetual administrators of the heavens' motions. They also bestowed a moving soul [*anima motrix*], more closely attached to the orbs and giving them form, so that the mind would only have to give assistance. This was either because it seemed necessary that the mover and the moving thing have some common ground, or because the potentiality, considered in

---

[2]George Peurbach (1423–1461), an astronomer remembered for his work in trigonometry, as well as for his textbook on planetary theory. He had as his disciple Johannes Müller of Königsberg, better known by the Latinized form of his home city, "Regiomontanus." In 1472, Regiomontanus published his mentor's chief astronomical work, entitled *Theoricae novae planetarum*, to which Kepler is referring.

[3]One must beware of supposing that "solid" implies "hard." The distinction being made here is that between mere circles, which (being two-dimensional, are not solid) and real spheres.

[4]...*Orbem suam proxime ambeunte, tanquam in loco quodam, ...* This is a paraphrase of the Aristotelian theory of place, to which Kepler adhered, and would be recognized as such by all contemporary readers. See Aristotle, *Physics*, Book IV Ch. 4.

relation to the distance to be traversed, should not be infinite, just as there exists no infinite motion, but only motion through a certain space in a certain time. They therefore transferred this potentiality for creating motion to a soul, for the meantime becoming subject to matter for the purpose of inhering in the orbs of the heavens.

Now this coupling of mind and soul is indeed quite in agreement with the detailed considerations of the astronomers, even though the philosophers' mode of argument is chiefly metaphysical. For it is the same in humans: the moving faculty is one thing; that which makes use of the moving faculty according to the indications of the senses—the Will—is another. The senses differ from the moving faculty both in the means they use and in the excellence of their structure, which in the organs of sense is more admirable than in the seats of the motive faculty. Similarly, if we should propose these Aristotelian orbs as objects of contemplation, two things will present themselves to us: 1) the motive force, sufficient for the round orb, from whose activity and constant strength the time of revolution arises; 2) the direction in which it acts. The former is more correctly ascribed to the animate faculty, and the latter to its intelligent or remembering nature. Now although through this solidity of orbs all motions or celestial appearances are so provided for that nothing is left to the providence of the presiding movers, while the whole variety of motions is a consequence of the number and disposition of the orbs; an although nothing else is required but that the moving souls receive and retain their activity and be set going from the first moment of creation in whatever direction is theirs, sent forth from their prisons, as it were, into space;[5] nevertheless, it must be kept in mind that there is need of the supreme mind to launch any of the planets in its own direction, as if into its fixed and proper province. Aristotle, who knew nothing of the world's beginning and did not believe in it, of necessity ascribed this function instead to the governors of the motions. The followers of Aristotle, and even Scaliger,[6] who professes to be a Christian, openly contend that this motion of the orbs is voluntary, and that the principle of volition for them is intellectual intuition and desire.

So, to return to Peurbach, certain others along with him (chiefly authors of books on the sphere), explain the first model by imagining for themselves one solid concentric orb of the thickness of the whole epicycle, with an epicycle in it, and in the epicycle a planet. Then they attributed two moving souls to these two orbs (if they should carry through with their physical considerations), both with the same amount of power, proportionally, so that they would complete their periods in the same time, although moving in opposite directions.

---

[5]Cf. Virgil, *Aeneid* I.52–54: Hic vasto rex Aeolus antro luctantes ventos tempestatesque sonoras imperio premit ac vinclis et carcere frenat."Here King Aeolus...restrains the winds ...with chains and prison."

[6]Julius Caesar Scaliger (1484–1558), a mercenary turned scholar who is remarkable chiefly for his introduction of siege tactics into scholarly debates. He attacked the views of the mathematician, physician, philosopher, and gambler Girolamo Cardano (1501–1576) in a great, rambling work entitled *Exercitationum exotericarum adversus Cardanum liber quintus decimus...*, Paris 1557. The title, incidentally, suggests (falsely) that this is merely the fifteenth volume in a series of such onslaughts. On the present subject, see especially *exercitatio* 61.5, and 359.8.

The other model requires two deferents (which remain motionless so long as we abide by the simplicity of the motions, mentally removing the progression of the apogees), and one orb with the thickness of the planetary body. In this orb is a soul which drives it around with a uniform effort in that direction in which it was projected in the beginning. Thus, if this solidity of the orbs and the other assumptions be granted, in the first model BC and BE will remain parallel, and in the second, the orb $\gamma\epsilon$ will go around the center β, even though the movers in the former pay no attention to AC, nor in the latter, to β. For they are governed by material necessity or by the arrangement and contiguity of the orbs.

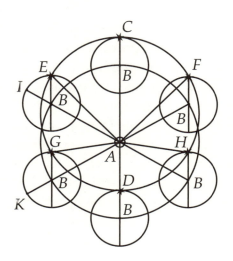

But, with arguments of the greatest certainty, Tycho Brahe has demolished the solidity of the orbs, which hitherto was able to serve these moving souls, blind as they were, as walking sticks for finding their appointed road; and hence the planets complete their courses in the pure aether, just like birds in the air. Therefore, we shall have to philosophize differently about these models.

Let it then be assumed among the first principles, that every force by which motions of this sort are administered dwells in the body of the planet itself, and is not to be sought outside it.

Now the planet must execute a perfect circle in the pure aether by its inherent force, an epicycle in the first model and an eccentric in the second. It is therefore clear that the mover will to have two jobs: first, it must have a faculty strong enough to move its body about, and second, it must have sufficient knowledge to find a circular boundary in the pure aethereal air, which in itself is not divided into such regions. This is the function of mind. Please don't tell me that the motive faculty itself, a scion of a simple and brute soul, has a native aptitude for circular motion, exactly like a stone's nature to descend in a straight line. For I deny that God has created any perpetual non-rectilinear motion that is not ruled by a mind. Even in the human body, all the muscles move according to the principles of rectilinear motions. They either swell by contracting into themselves, or stretch out, the ends moving apart: in the former case, the member approaches the muscle, while in the latter, the member recedes. This same thing takes place in its own way in the circular muscles that are set up as guards for orifices. When they are extended by the circular filaments, they relax and open the passage, and constrict it when the filaments return into the form of a smaller circle. There is no member whatever that rotates uniformly and without impediment. On the contrary, the bending of the head, feet, arms, and tongue is expressed in certain mechanical devices by many straight muscles carried across or stretched out from one place to another. In this way it is brought about that the motive faculty, which by its own nature tends in a straight line, swings its member in a gyre. Likewise, certain machines raise water to great heights, not because the nature of the body, which conveys the motion, tends to an exalted position, but because, by an arrangement of channels, it is brought about that the water necessarily gives way upwards when a greater weight tends downwards. And even if the motion of certain members were perfectly circular, they[7] nonetheless could not be perpetual. There should be no great wonder at this, since in the human body mind presides over the animate faculty. Surely, then, if there had been any way of so constructing some moving faculty that it might be able to rotate some body, it would not have been neglected in the human body.

Besides, it is quite impossible for any mind to manifest a circular path without recourse to the guidepost either of a center or of some body which might appear under a greater or smaller angle according to its approach or recession. For a circle is both defined and brought to perfection by the same criterion, namely, equality of distance from the middle. No matter how many of these motive faculties you set up, a circle, even for God, is nothing other than what was just said. Geometers do, of course, show how, given three points on a circumference, to form a continuous circle, but in this very proof it is presupposed that some portion of the circumference (that which passes through the three points) is already constructed. Who, then, will show the planet this starting place, in conformity with which it will make the rest of its path? This is possible in no other way than if the planet's mover (as in Avicenna's opinion) imagine for itself the center of its orb and its distance from it, or if it be assisted by providing some other property of a circle in order to lay out its own circle.

We will therefore now form the physical hypothesis of these two models in another way. In the latter, simpler one, if our supposition is valid that the mover driving the planet around the path $\gamma\epsilon\delta$ is in the planet itself, there will have to accrue to the planet's mover some sort of awareness of the apparent magnitude of the body at α seen (or as if seen) from γ, ϵ, η, δ. The planet will therefore have to strive both to move forward uniformly (this the undivided

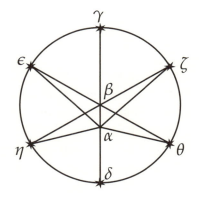

and unimpeded forces of the moving soul provide), and to exhibit all the distances $\alpha\gamma$, $\alpha\epsilon$, $\alpha\eta$, $\alpha\delta$, in such order that they follow by a geometrical law from the eccentric $\beta\gamma$. To this end, the mover should also know how much longer $\alpha\gamma$ is than $\alpha\delta$; that is, by how much the path which it will traverse is eccentric from the body at α around which it is to go. The planet's mover will thus be occupied with many things at once. To escape this conclusion, one must assert that the planet pays attention to the point β, entirely empty of any body or real quality, and maintains equal distances from that point.

10

---

[7]Kepler uses a plural demonstrative pronoun here, although there does not appear to be any possible plural antecedent.

The prior model is explained physically thus. Let a motive power be conceived which, seated on the concentric *B* and itself without body, moves around the body at *A* with a uniform exertion of forces, maintaining equal distances from that point. Let there be another power in the body of the planet *C*, capable of holding its attention on the incorporeal power at *B*, estimating and maintaining its distance from that power, and moving uniformly around it. Thus, as before, this power again will have numerous tasks. But it is also incredible in itself that an immaterial power reside in a non-body, be moved in space and time, but not have a substrate to move itself (as I said) from place to place. And I am making these absurd assumptions in order to establish in the end the impossibility that every cause of a planet's motions inhere in its body or somewhere else in its orb, and to build a path for other less difficult forms of motions that are more readily persuasive.

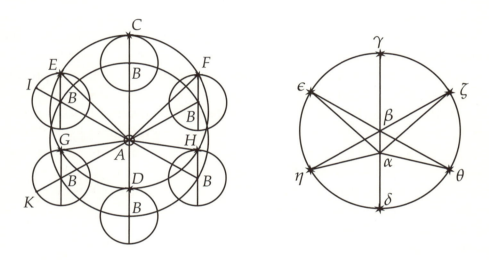

I have presented these models hypothetically, the hypothesis being astronomy's testimony that the planet's path is a perfect eccentric circle such as was described. If astronomy will discover something different, the physical theories will also change.

In this equivalence of hypotheses, not only the equality of the apparent angles at *A* and *α*, but also the actual paths themselves of the planets through the surrounding aether, each remain the same. For in both shape and size, the planet traverses an arc from *C* to *E* through angle *CAE* that is also the same as it traverses from *γ* to *ε* through the equal angle *γαε*.

CHAPTER 3

On the equivalence and unanimity of different points of observation and of hypotheses, for laying out one and the same planetary path.

Next, I must show how one and the same planetary motion, while in itself remaining the same, can display one or another appearance, and how the pair of forms are equivalent here.

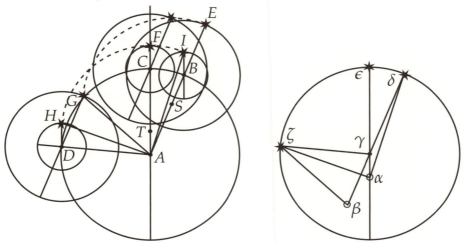

About centers A and γ, with equal radii AC, γε, let the circles CD, εζ be described with CA, εγ drawn through the centers parallel to one another, and other lines AB, γδ, and AD, γζ, through the centers inclined to the former, both pairs likewise parallel. Also, about B let an epicycle be described, with radius BE, and likewise 11 *about D with radius DG equal to BE. Let the planet be placed at E and G, so that DG and AB are parallel. On the line δγ let a segment equal to BE be set out on the side opposite δ, and let it be γβ. Let G be joined with A and ζ with β. The hypotheses will therefore be equivalent, by the preceding chapter, and to an observer placed at A and β, EAG and δβζ will be equal. EA and δβ will also be equal, as well as GA and ζβ. And finally, the arcs EG and δζ will be equal.*

Now let a smaller epicycle be described upon BCD with radii BI, CF, DH, and let AC be extended to F, and BI and DH be parallel to CF. And let the planet be on IFH. Again, by ch. 2, the circle IFH will be equal to the circle δζ.

Next, extend the arc IF from the point δ, so as to end at ε, and from ε through γ draw εγ, such that εγ is parallel to CA. And let a magnitude equal to CF be set out on the line εγ, and let it be γα, on the side opposite ε. Let I and H be joined with A, and also δ and ζ with α. Again, therefore, the hypotheses will be equivalent by the preceding chapter, and to an observer placed at A and α, FAH and εαζ will be equal, as well as FAI and εαδ. FA and εα will also be equal, as well as HA and ζα, and IA and δα. And finally, the arcs FH and εζ will be equal and similar, as are also FI and εδ, by construction.

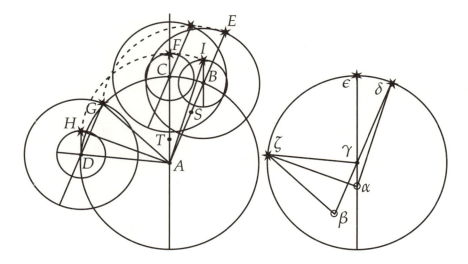

Therefore, if the path of the planet remain the same while the observer is moved from β to α, different appearances will be produced at the same moments of time. For the same places, δ and ζ, are viewed in different ways from β and α. On the other hand, if the observer remain at A, and the quantity of the planet's path EG, IH remain the same while changing its place, the planet will again appear in different places, even when at the same place on its path, because the entire path has been shifted. Accordingly, since the planet, whether viewed from α or β, is at δ, or at ζ, at the same moment in each observation, and the hypotheses are entirely equivalent, it must also be said that I and E, which are positions of different epicycles, are occupied by the planet at the same moment. The same is true of G and H. The only difference is that in the first diagram the planet's path is shifted by altering the epicycle in its position, the observer remaining in the same place, while in the second diagram, for the planet's path the position also stays the same while the observer's position is changed by the same amount in the opposite direction. If required, however, it is possible to keep the path fixed in the former and the observer fixed in the latter fixed by shifting what is now fixed, in accordance with the demonstrations of the preceding chapter.

12   This demonstration will be put to use below. For surely, if the first inequality of the superior planets could be accounted for by the simple hypothesis of the second chapter, no difficulty would arise as to whether one should examine the first inequality at mean opposition to the sun, or at apparent opposition. For the actual path would in fact remain the same, and in both models the planet would be at the same points of the path at any given moment. Only the position of this path, in the first model, would be altered through the space of the sun's eccentricity, while in the second, although the path would stay fixed, the point whence the eccentricity is reckoned would also be shifted by the same amount.

In the physical account, the above characteristics remain unchanged. Only their quantities change as the motive powers are intensified.

CHAPTER 4

On the imperfect equivalence between a double epicycle on a concentric, or eccentric-epicycle, and equant on an eccentric.

That is how it would be if there were scope for the simple hypothesis of ch. 3 to account for the first inequality of the superior planets. However, for demonstrating the first and simple inequality of the planets, Ptolemy makes use of a more elaborate hypothesis.

About center B let an eccentric DE be described, with eccentricity BA, so that A is the place of the observer. The line drawn through AB will indicate the apogee at D and the perigee at F. Upon this line, above B, let another segment BC be extended, equal to BA. C will be the point of the equant, that is, the point about which the planet completes equal angles in equal times, even though it lays out the circle around B rather than around C.

In Book V ch. 4, as well as Book IV ch. 7[1], Copernicus marks this hypothesis among other things in this respect, that it offends against physical principles by instituting irregular celestial motions. *For let a point E be chosen on the circle which the planet is bodily traversing, and let it be connected with C, B, and A. Now let DCE be a right angle, as well as ECF. Now since these angles are equal (for they are set up in equal times), and the exterior angle DCE is equal to the interior angles CBE and CEB, therefore, when the part CEB is subtracted, the remainder CBE or DBE will be less than DCE. Con-*

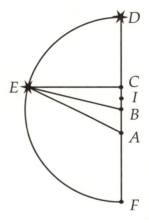

sequently, FBE will be greater than DCE or FCE. But the arc DE measures the angle DBE, and the arc EF measures the angle EBF. Therefore, DE is smaller than EF, and the planet passes over them in equal times. Therefore, the same solid orb (Copernicus believed in them)[2] in which the planet inheres is slow when the planet borne by the orb proceeds from D to E, and fast when the planet goes from E to F. Therefore, the 13 entire solid orb is now fast, now slow. This Copernicus rejects as absurd.

Now I, too, for good reasons, would reject as absurd the notion that the moving power should preside over a solid orb, everywhere uniform, rather

---

[1] The passages in which Copernicus presents the argument to which Kepler refers are in Book IV ch. 2 and Book V ch. 2.

[2] Again, it should be remarked that it is probably anachronistic of Kepler to ascribe this opinion to Copernicus. Although it is true that Copernicus believed in the physical truth of his hypothesis, and consequently in the reality of the orbs, it does not follow that the orbs had to have the terrestrial properties of hardness and impenetrability. On the contrary, natural philosophers contemporary with Copernicus generally held (with Aristotle) that the heavens are devoid of such qualities. For a thorough discussion of Copernicus's orbs, see N. Jardine, "The Significance of the Copernican Orbs", *Journal for the History of Astronomy* 13 (1982), especially pp. 174–180.

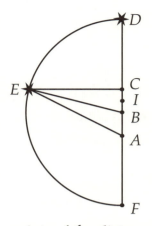

than over the bare planet. But because there are no solid orbs, consider now the physical coherence of this hypothesis when very slight changes are made, as described below. This hypothesis, it should be added, posits two motive powers to move the planet (Ptolemy was unaware of this). It places one of these in the body *A* (which, in the reformed astronomy will be the very sun itself), and says that this power endeavors to drive the planet around itself, but possesses an infinite number of degrees corresponding to the infinite number of points of the distance from *A*. Thus, just as *AD* is the longest, and *AF* the shortest, so the planet is slowest at *D* and fastest at *F*, and in general, as *AD* is to *AE*, so is the slowness at *D* to the slowness at E^3, as will be demonstrated at great length in Part III below. The hypothesis attributes another motive power to the planet itself, which has the capacity to adjust its approach to and recession from the sun, either by the strength of the angles or by inspection of the increase or decrease of the solar diameter, and to make the difference between the mean distance and the longest and shortest equal to *AB*. Therefore, the point of the equant *C* is nothing but a geometrical short cut for computing the equations from a hypothesis that is clearly physical. But if the planet's path is a perfect circle, as Ptolemy certainly thought, the planet also has to have some additional perception of the swiftness and slowness by which it is carried along by the other external power, in order to adjust its own approach and recession in such accord with the power's precepts that the path *DE* itself is made to be a circle. It therefore needs both a comprehension of the circle and a desire to effect it. Also, the ratios of its own slowness and swiftness must differ from the gradations of the external power. However, if the demonstrations of astronomy, founded upon observations, should testify that the path of the planet is not quite circular, contrary to what this hypothesis asserts, then this physical account too will be constructed differently and the planet's power will be freed from these rather troublesome requirements.

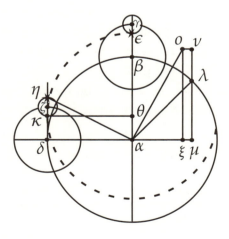

But let me return to Copernicus. In avoiding the absurdity explained just above from his own opinions, he substituted another epicycle for the equant, in the following way. *About center α with radius αβ equal to BD let the concentric βδ be described, so that the observer is at α; let αβ, parallel to BD, be extended in both directions; and let the angle βαδ be set up equal to DCE. Now let BC*

---

[3]Kepler measures the degree of slowness by finding the increments of time over a certain arc or path element. See "mora" in the Glossary, and the beginning of ch. 40.

be bisected at I, and about the centers β and δ with radii βγ and δζ equal to AI let the first or greater epicycle be described, and let δζ be parallel to αβ. Next, about centers γ and ζ, but with radii γε, ζη equal to IC, let the second epicycle be described, and let its motion be eastward, with twice the speed of the motion of the first. And let the westward motion of the first epicycle be equal to the motion of the eccentric. And since γ is on αβ, let the planet be at ε, the point nearest β. And since βαδ is right, let the planet be at η, the point farthest from the center of the greater epicycle δ. This particular hypothesis of Copernicus is also followed by Tycho Brahe religiously, in all particulars.

14

Physically considered, this hypothesis is in any event valid if you grant solid orbs. However, if you remove these, as Brahe does with good reason, it says something practically impossible. For it attaches three movement-producing minds to a single planet; and besides, the other two [minds] will be thrown into confusion by the motion and impulse of one of them towards the body at α. For that any of them should pay heed to its own center, which is not distinguished by any body and is mobile besides, cannot be represented even in thought. Further, while Copernicus strives to outdo Ptolemy in the uniformity of motions, he is in turn outdone by him in the perfection of the planetary path. For, in Ptolemy, the planet bodily traces out a perfect circle in the aethereal air. Copernicus, on the other hand, says in Book V ch. 4 that for him the path of a planet is not circular, but goes outside the circular path at the sides. This is easily demonstrated in the present diagram.

If from ε, the planet's position at apogee, you extend the distance αβ, the semidiameter of the orb, to θ, and from θ draw θκ parallel to αδ, the circle εκ described about θ will indeed go through ε and the perigeal point opposite, but since it touches the straight line δη only at κ, while the planet goes through η, it does not stay on the circle εκ, but strays outside this track. To this excursion of the planetary path from the perfection of the circle Ptolemy might well have objected against Copernicus, but I do not. For below, in Part IV, it will be demonstrated that by the agency of two physical powers, simple in capability, acting in concert to move the planet, it necessarily happens that the planet turns aside from the circle for a short time, though not by running outside of it, as in this Copernican hypothesis, but in the opposite direction, towards the center; that is, by making an incursion.

Besides, should Copernicus retain that liberty he had of setting up the ratios of the epicycles, it can happen that the planet's path would come out twisted, higher before and after apogee than at apogee itself, and lower before and after perigee than at perigee itself. This happened to Tycho in his lunar theory, inasmuch as he followed Copernicus.

That these two forms of hypothesis are not simply equivalent, I shall demonstrate numerically.

In the Ptolemaic form it can be computed more simply than Ptolemy did in the following manner. *First, in triangle CBE, given the mean anomaly ECB or DCE, the side CB — the eccentricity of the equant — is also given, as well as the radius of the orb, BE. Therefore, as the radius of the orb is to the sine of ECB, so is CB to the sine of CEB. And since ECD is equal to the two opposite interior angles CEB and CBE taken together, therefore CEB subtracted from DCE leaves CBE. Therefore, in triangle EBA, the angle at B is given, together with the sides about it. For BA is the*

Terms:
The mean anomaly is the time elapsed since the planet was at apogee, expressed according to an arbitrary rule: the whole time in which the planet makes one return from apogee to apogee is divided into 360 degrees, like a circle.

The true anomaly is the arc of the zodiac between the position of the apogee and the apparent position of the star, viewed from the center of the zodiac.

The equation is the difference between the two anomalies.

15

eccentricity of the eccentric, while EB is the radius of the orb. Therefore, following the rule for this form of triangle, the angle BEA is given. But CEB was given before. Therefore, the whole equation CEA will be given.

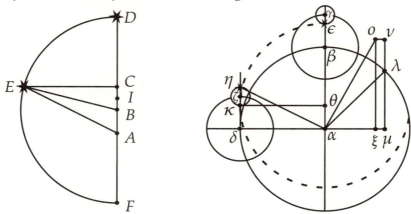

We shall now make use of numbers belonging to the motion of Mars. Although Ptolemy made *CB* and *BA* equal, Copernicus, freed from this rule, nonetheless also adopted other ratios, which Tycho Brahe undertook to imitate. *Let CB be 7560, BA 12,600, where BE is 100,000; and, first, let DCE be 45°, whose sine is 70,711. Therefore, as 100,000 is to 70,711, so is 7560 to 5346, the sine of the arc of 3° 4' 52", which is CEB. Subtracting this from 45° leaves CBE, 41° 55' 8", whose half is 20° 57' 34", the tangent to which arc is 38,304. And since EB is 100,000, while BA is 12,600, the difference, 87,400, multiplied by the radius and divided by the sum, 112,600, gives 77,620. Multiply this by the tangent found above (38,304). The product, that is, 29,732, is the tangent of the arc 16° 33' 30", which, subtracted from the half of CBE, found above, leaves 4° 24' 4", which is the angle BEA. Therefore, the whole, CEA, is 7° 28' 56", in the Ptolemaic form.* In the Copernican form, although the ordinary means of finding the equation is clearly presented in Tycho's lunar tables in Vol. I of the *Progymnasmata*, and in Copernicus himself, let me nonetheless follow a different, less usual procedure, which is adapted to an anomaly of 45°. *Let βαλ be 45°, and λν or βγ be 16,380, γε or νο be 3780, and ονλ be right, that is, twice βαλ. Now let νλ be parallel to βα, and let νλ and δα be extended, so as to meet at μ. From o let οξ be dropped parallel to νμ. Therefore, λαμ is 45°, and consequently αμ, and also μλ, are 70,711. Add λν, 16,380, and μν or οξ will be 87,091. And because γε, νο, and ξμ are equal, subtract ξμ from αμ. The remainder, αξ, is 66,931. Therefore, as οξ is to ξα, so is the whole sine[4] to the tangent of αοξ or οαβ, 76,852, giving an angle of 37° 32' 37", which differs from the arc 45° by 7° 27' 23". Therefore, the difference between the Copernican and the Ptolemaic equations at this position is 1' 33", a very small difference indeed.*

Again in the Ptolemaic form, let DCE be 90°. Therefore, since ECB is right, and EB is 100,000, BC will be the sine of the angle CEB, or 4° 20' 8". Therefore, EBC is 85° 39' 52", and EC is 99,713. Now, as EC is to CA, so is the radius to the tangent of CEA, 20,218. Hence, the equation, CEA, is 11° 25' 48". But in the Copernican form, the whole magnitude ηδ, equal to CA, becomes the tangent, because ηδα is right and δα is the radius. Therefore, ηαδ is 11° 23' 53". The difference is 1' 55".

Terms:
The equation of the eccentric is in the first inequality.
The equation of the orb is in the second inequality.
Likewise the annual equation of the center.

---

[4]That is, the radius.

Thus you see that, as far as the eccentric equation is concerned, there is something very slight lacking that prevents the two forms of hypothesis from being equivalent.

They are different, however, in the distances of the planet from the observer at α, and as a consequence, in the annual equations of the center as well. For, in the Ptolemaic form, as the sine of the angle AEC is to AC, the whole sine is to AE, which becomes 101,766 when DCE is 90°. But in the Copernican, $\eta\alpha$ is the secant of angle $\eta\alpha\delta$, that is, 102,012. The difference is 246 parts, and this can have a somewhat greater effect upon the equation of the center for the annual orb, as will be clear below in Part IV. We can eliminate even this extremely slight difference in the equations by positing 20,103 as Mars's eccentricity in the Ptolemaic form where Brahe, in the Copernican form, found it to be 20,160. However, the distances in the Copernican form cannot be made equal to those in the Ptolemaic unless the equation be altered by 43′. In a certain equivalence I tried out in Tycho's hypothesis of the lunar tables, I transposed those two Copernican epicycles into such a Ptolemaic eccentric with an equant point. Nevertheless, I added yet another epicycle on account of another inequality, peculiar to the moon.

Finally, in accord with ch. 2, the greater epicycle with its concentric in the Copernican form can, by virtue of its complete equivalence, be transformed into an eccentric whose eccentricity is equal to the semidiameter of the greater epicycle. Therefore, when a smaller epicycle is added to this Copernican eccentric, an eccentric with an epicycle will be created which matches the double epicycle on a concentric to a hair, and which differs from the Ptolemaic eccentric with an equant by no more than does this double epicycle.

16

CHAPTER 5

The extent to which this arrangement of orbs, using either an equant or a second epicycle, while remaining entirely one and the same (or very nearly one and the same), can present different phenomena at one and the same instant, according to whether the planets are observed at mean or at apparent opposition to the sun.

17   This is done in two ways: one, in which the Ptolemaic and Copernican forms are equivalent, and another which is peculiar to the Copernican form. This latter, as it is further from our enterprise, we shall explain first, for it remains more allied to itself than the other.

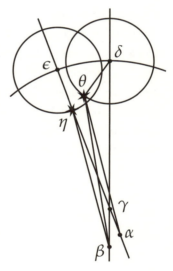

About center γ, with radius γδ, let an eccentric be described, upon which, in the first instance, let αγ be the line of apsides, with α the observer, and let this line be extended to ε. And let γα be the magnitude of the eccentricity, or the radius of the greater Copernican epicycle (the equivalence of the two was discussed at the end of Ch. 4, preceding). Next, about center ε, with radius εη, let the smaller epicycle be described, and since its center is at ε, let the planet be at η, falling on the line εγ, so that it is not the star but the center of the epicycle bearing the star that runs along the eccentric εδ. Thus, by Ch. 4, it is the Copernican form that is expressed here. By Ch. 3, we shall set up another that is equivalent to it in reality or in the indication of the exact planetary path, but is different in appearance. This we shall do by moving the observer from α. By what was said at the end of Ch. 3, we could do the same thing even if the observer were to remain at α, by moving the eccentric while keeping lines parallel, in such a way that the size of the eccentric remains constant and only its position changes. But we shall carry it to completion as we have begun it. Adopting a position for the observer not on the prior line of apsides (let it be β), such that βγ has a magnitude different from that of αγ (that is, a new eccentricity or new radius of the greater epicycle), let us draw a new line of apsides βδ through βγ, and at δ let us describe an epicycle equal to the former. Here, although the center of the epicycle is at the apsis δ, nonetheless we shall not place the planet at the point nearest γ, as before, but, taking the measure of the angle εγδ, we shall set out the angle θδγ twice that size, in the direction of ε, and shall place the planet at θ, when the epicycle is at the apsis δ. For this is where the planet would be, were the observer at α and the epicycle at δ. In this way, the true compound planetary path remains the same to a hair's breadth, while the appearance is altered. For when the lines of sight are inclined to one another, as βθ and αθ here, or βη and αη, they fall upon different positions beneath the fixed stars.

Term:
The word "eccentric" has a peculiar denotation here.

You may object that even when the lines of sight are parallel, they fall upon different positions beneath the fixed stars, and it is therefore not essential that they be inclined to one another. I answer, this is indeed true, but in that case the space of the sphere of the stars intercepted between the two lines is not perceptible to the power of sight unless the distance between the parallels is perceptible in relation to the radius of the sphere of the stars.

In the physical account, this must be posited in addition to what was said in Ch. 3, in order to establish the identity of the path while the appearances are altered: the mind to which the smaller epicycle is committed pays attention to a point on the circuit different from the one regarded by the mind of the greater epicycle. For in the second position, the greater epicycle or the eccentricity returns to its starting point on the line $\beta\delta$, while the smaller epicycle does so on the line $\alpha\epsilon$ which does not pass through the place of the observer. This is because, in the second position, the observer is located at β, while in the first position, where the observer is located at α, both epicycles return to their starting points on the same line $\epsilon\alpha$. The form of the hypothesis thus does not remain simply the same physically in such a way that that the planet keeps the same path. But suppose you were going to simulate the same path in the second position by having both epicycles return to their starting point at the same line of apsides $\beta\delta$. If so, while the eccentric as well as the epicycle stay the same in both instances, the position of the planet on the epicycle will be different at every single moment. Thus, although in the second instance the form of Ptolemaic hypothesis presented is the same to a hair's breadth, the actual path of the planet will be altered. Hence, it will be inferred below, even though the first inequality of the planets may be entirely accounted for by the compound hypothesis of Ch. 4, it cannot happen that the first inequality have the same measure at the planet's mean opposition to the sun as at apparent opposition, unless the planet's orbit be moved from its own location at the same time (unlike the circles in the theory of the sun), or the Ptolemaic form of Ch. 4 be changed.

Maestlin made use of this form of transposition in constructing the table in Ch. 15 of my *Mysterium cosmographicum*. For when Copernicus transformed the Ptolemaic [models] into his own general form of hypotheses, he supposed the observer to be stationed at some nearly motionless point near the sun, distant from the sun's own body by the entire eccentricity of the solar orb. I, however, in adapting Copernicus to the subject matter of that book, needed a different fiction. The observer was to be imagined as transported from that point to the very center of the solar body, and from there (that is, from the body of the sun) the departures of the planetary bodies were to be computed, moving on the same path which the suppositions of Copernicus fashioned. But, as has just now been shown, by reason of the particular times my translation of the line of apsides did not effect exactly the same path. The difference, however, was very slight, and was clearly of no importance in that little book. For there the question concerned only the location of the path, which this procedure did not affect.

As for the rest, in order to avoid confusion in what follows, I shall no longer make use of this Copernican eccentric, described by the center of the epicycle rather than the star. It differs from the planet's true path, which is higher

Term:
What the word
"eccentric" will signify
henceforth.

at perigee and lower at apogee. The term "eccentric" from now on we will use only to designate the actual path of the planet, or of the point to whose motion the first inequality belongs. In proceeding in this way, it is appropriate that we imagine only the Ptolemaic eccentric, or something like it. For it was shown in the fourth chapter that our computation of the equation, based upon the Ptolemaic form, will differ from the Copernican by only two minutes at most. Then, too, the procedure for computing the first inequality is easier in the Ptolemaic form than in the Copernican. Finally, as was said, this Ptolemaic form of the first inequality is better accommodated to nature herself and to our speculations that follow in the third and fourth parts. However, because of the equivalence, anyone who so chooses will always be able to supply in thought

19   the Copernican eccentric-cum-epicycle considered so far in this fifth chapter.

I now proceed to the prior procedure for setting up the two sorts of equivalence that I have proposed—the one common to the particular hypotheses of the authorities. I shall demonstrate this first in the Ptolemaic form.

About center β let the Ptolemaic eccentric ιζη be described, with ιβ the line of apsides, α the observer, and γ the equalizing point.[1]

Now when I say that the observer is at α, I mean it either as a fiction or as truth. Physically speaking, it is not so much the observer which is to be placed at α as the power which renders the planetary circuit around itself slow or swift according to the ratios of its proximity to α, as was said above. *Let some point on the circumference not on the line of apsides (say, η) be connected with γ, β, and α. Let it be that about as many angles ιαη may be computed by this hypothesis throughout the entire circuit as are observed from α, and after certain periods of time, which the angle ηγι measures uniformly. Later, in the second part, it will be shown how one can find, through astronomical observations, how great the angle ηαι should be for any given ηγι. Again, let the observer or moving power be at some point not on the line ια, and let this be δ. Also, let it be granted us that the apparent angles about δ be apprehended by astronomical observations at certain times; that is, how much the planet appears to move forward in sidereal position in a given time when seen from δ. Let this be granted as well: that these appearances at δ square with a hypothesis in conformity with the previous one, with only the magnitude of the eccentricity altered. But since it is certain that at one and the same time the planet traverses one and the same path in the heavens, not one seen to an observer at δ and another to an observer at α, it is also certain, as a consequence, that the planet cannot appear to both observers (both the one at α and the one at δ) to be equally moved in the same time. For let ιη be a portion of the planet's true path, and let the planet traverse this in a given time, say twenty days. Now since α is nearer ιη than δ is, ιη will appear greater at α than at δ, by what is demonstrated in optics. Therefore, during the same twenty days the planet will appear to make greater progress to one who is at α than to one who is at δ. And since any planet maintains a fixed and constant number of days in which it is returned to the same sidereal position, the slowness has to be compensated by a contrary speed. Therefore, since in the portion ιη the planet appears slower to one at δ, it will in some other portion appear swifter to the one at δ than to the one at α. Hence it happens that it appears slowest to the one at δ in one place, and to one at α in another. Nevertheless,*

20   *the planet itself can be truly slowest in but one place on its orbit.*

---

[1]*Punctum aequatorium.*

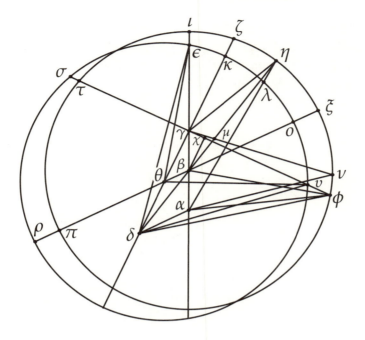

With these as preliminaries, the question is raised whether one and the same true path of the planet in the heavens (this is presupposed) can present two sets of appearances each proper to an observer, one at δ and the other at α, and each such as comply with and admit the Ptolemaic form of computation.

If the planet were of equal speed at all parts of its orbit, the answer, according to what was said in Ch. 3, is yes. But since in terms of real and true increment of time[2] the planet is slowest at one point on the eccentric and fastest at the opposite point, the answer must therefore be, clearly not.

The reason for this is that the two slowings are intermingled, one real and physical, occurring at one place on the eccentric, and the other optical and apparent and not occurring at a single place, but in the place most distant from whatever position is chosen for the observer. *Now, when the observer α lies upon a line drawn through the center of the eccentric β and the center of the equant γ on the side of β opposite the center of the equant γ, then both slowings verge toward the same sidereal position ι. But when the observer departs from this line, as at δ, then a straight line drawn from δ through the center of the circle β marks the place of the optical slowing, η, while the true and physical slowing is at ι. Furthermore, each of these inequalities or slowings diminishes the other, and they are accumulated at an intermediate point between ι and η, as would be if a line were to be drawn from δ through γ to the point ζ. Consequently, were one to adopt a form of computation in which δβ were the line of apsides of the eccentric and βγ the line of the equant's eccentricity, then even though the planet's true path ιη remained the same, it would be represented differently at δ than at α. For to the observer at δ the planet would be slowest at ζ, and to the one at α it would be slowest at ι. But no such thing would be represented at δ that ought to have been represented by a hypothesis of the same form as the previous one, according to our presuppositions above. For the forms of the hypotheses differ in that in the former, β is the midpoint between α and γ (as physical considerations require, if the moving power is in α), while in the latter the center of the eccentric β would not be the midpoint between δ and γ, nor would the line of the*

---

[2]*Mora.* See the Glossary.

equant's eccentricity pass through the observer δ, as before. Even if it did pass through δ, as δγ, it would not cut the eccentric into two equal parts, because it would not cut it at the center β, and it would not allow the planet to appear on one side fastest and on the other slowest, at opposite places.

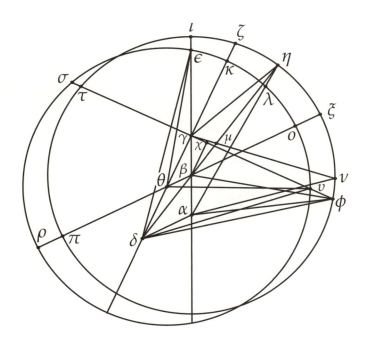

It is thus established that when a planet's path in the heavens remains in all respects unchanged, the form of hypothesis cannot persist entirely unchanged. This raises the further question, how much the path of a planet is changed from its prior position if the same form of hypothesis is set up about δ, and how much this newly established hypothesis at δ will be at variance with the appearances as seen at α. *First, if the center of the equant be transferred from γ to the line δβ, and βμ be made equal to βγ, the position of the planetary path is quite unchanged, but the planet is slowest, in physical terms, at η rather than ι. This*

21 *changes what cannot be changed in the planet's path, because, unlike the optical slowing, the physical slowing is independent of the observer's viewpoint. Even though the planet would traverse the same path ιη in twenty days (which path appears greater at α and smaller at δ), nevertheless, if you consider the parts of this time, their ratio in comparison with the parts of this path will be violently perturbed, and much more so at other places, not between ι and η. In particular, for the observer at α the quantity of his equations will be changed noticeably if, for the observer at δ, you will prevent the planet's being slowest at ι; that is, if you transfer the equant from γ to μ. For if you draw a straight line through γμ to the point ν on the circumference and connect αν, this equation ανμ alone will be equal to the prior, ανγ. Above ν the equations about μ will be smaller, and below ν, greater. For example, at η the angle μνα is much less than γηα.* But then what we proposed to do has not been done, for the prior form of hypothesis has not yet been quite established. *For αβ is not to βγ as δβ is to βμ, since βμ is equal to βγ, while δβ is greater than αβ. But if, on the other hand, you make δβ be to βμ as αβ is to βγ, βμ will become greater than βγ.* Whence it follows that for the observer at α, his equation will be more in error, even when it is at a maximum, on account of the increased eccentricity. Not

only will the planet be slowest in a different place than before, but in also by a different and in fact greater measure of its true slowness. It is clear, therefore, that the equivalence we have been seeking cannot be established by drawing the line of apsides from δ through the center of the eccentric β. And since it has at the same time become clear how important it is to keep the same equant point γ, a breakthrough must in all events be made here, or nowhere.

But what will happen if a new line of apsides be drawn from δ through the old equant point γ, and a new hypothesis of the same form as the old be set up? That is, if the center of the eccentric be transferred from β to the line $\delta\gamma$, and $\delta\theta$ be made to $\theta\gamma$ as $\alpha\beta$ is to $\beta\gamma$, θ thus being the center of the eccentric? Obviously, the result will be this: the path of the planet in the heavens does not remain entirely the same. *About θ let the eccentric $\epsilon\kappa\lambda$ be described, equal to the previous one, and through $\theta\beta$ let a straight line be extended to the circumference, on one side to ζo and on the other to $\rho\pi$. Therefore, ζo and $\rho\pi$ are both of the same magnitude as $\theta\beta$, and the planet is this much closer to β at o, and more remote at ρ, than it would have been had it traversed the previous eccentric. However, the planet is slowest in a different region, for previously the apsis was at ι, and now it is at κ.* Through this mutual tempering it is brought about that the observer previously stationed at α has his observations pretty much unchanged, which is the only thing sought for here. But now we shall prove it with numbers belonging to Mars's motion, although Brahe recorded somewhat different ones. This should prove no impediment to us, since we are only performing a preliminary exercise[3] here.

Let the magnitudes on the line $\delta\gamma\alpha$ be taken as follows: let $\delta\alpha$ be the quantity of the sun's eccentricity, 3584; $\delta\gamma$ the eccentricity of Mars, $30,138$ of the same parts, and the angle $\alpha\delta\gamma$ $47°\ 59\frac{1}{4}'$, which is the angular difference between the sun's and Mars's apogees. Now, from these three given quantities, $\gamma\alpha$, Mars's new eccentricity, will also be given, and will be $27,971$, while the angle $\delta\gamma\alpha$ is $5°\ 27'\ 47''$. On the supposition that the old apogee of Mars $\delta\gamma$ is positioned at $23°\ 32'\ 16''$ Leo, Mars's new apogee $\alpha\gamma$ will fall at $29°\ 0'\ 16''$ Leo.

22

Now let $\beta\xi$ be $100,000$ and $\alpha\gamma$ be $18,034$ of the same parts. Before, it was $27,971$, in units of which $\delta\gamma$ was $30,138$. Therefore, in these units, $\delta\gamma$ will be $19,763$. Next, let both be divided by the letters θ and β in such a ratio that $\delta\theta$ is to $\theta\gamma$, and $\alpha\beta$ to $\beta\gamma$, as 1260 is to 756: $\delta\theta$ will be $12,352$, $\theta\gamma$ 7411; and $\alpha\beta$ $11,271$, $\beta\gamma$ 6763. In this way, a Ptolemaic hypothesis for the first inequality may be set up about both δ and α. Then, in the former units, of which $\delta\alpha$ is 3584, $\theta\beta$ or $o\xi$ will be 1344, but in units of which $\beta\xi$ is $100,000$, $\theta\beta$ or $o\xi$ will be 880. These should be kept in mind.

To find the basis of a computation whereby we may investigate, for the observer at δ, how much his appearances are changed by transposing the eccentric from $\rho\theta o$ to $\pi\beta\xi$, we must proceed as follows. *Since γ is the common center upon whose circle the times are indicated, let the line $\gamma\epsilon\iota$ indicate the same moment in both hypotheses. Therefore, if the planet is traversing the eccentric ϵo, it will at that moment be at ϵ with equation $\delta\epsilon\gamma$, but if it is traversing $\iota\xi$, it will be at ι with no equation, since the line of apparent motion $\alpha\iota$ coincides with the line of mean motion $\gamma\iota$. Again, after a certain amount of time, whose measure shall be $\iota\gamma\zeta$ or $\epsilon\gamma\kappa$,*

---

[3]Kepler used the Greek word προγυμναζόμεθα here

whose vertical angle is δγα, just found to be 5° 27' 47", let a common moment be taken, represented by γκζ. At that time, the planet traversing the eccentric εο will be at κ with no equation, while the planet traversing ιξ will be at ζ with equation γζα. Thus, in both instances, the planet is always on a line drawn from γ, at the point where that line cuts one or another eccentric. If the observer were at γ, there would be no difference in the appearances, whether the planet were at κ or at ζ. But since in the present model the observer is placed at δ by the practitioners[4] and at α by myself, the question arises at what point on the circumference the distance between the eccentrics is perceptibly a maximum for the observer at δ. *In order that this difference become perceptible, three factors must concur. First, the distance itself must be large (just as it is a maximum around οξ and ρπ). Second, as nearly as possible it should be presented directly to an observer at δ (just as it vanishes at ζκ and the point opposite) according to optical principles. Thus, it appears greatest at the intermediate regions, below ξ and above ρ. Third, it must be close to δ (just as it is closer above ρ than below ξ, because the center of the other eccentric β lies off to the right of δ). If we construct a right angle to point γ of the line γδ, this perpendicular from γ to the circumference will bring us as close as possible to the place where the apparent magnitude is greatest. Let σφ be drawn through γ perpendicular to δγ, intersecting the eccentric about θ at σ and υ, and the other at τ and φ, and to this let the perpendicular βχ be dropped. Therefore, at the moment γσ the planet will be at σ and τ, and at the moment γφ, at υ and φ.* First of all, the quantity υφ must be found. *Let θυ and βφ be joined. In θυγ, θυ is given as 100,000, because θ is the center of the eccentric υ. The magnitude θγ is 7411, and θγυ is right. Therefore, γυ is 99,725. The same is to be done in βγφ. But first, βχ[5] has to be found. This will be revealed by the triangle βγχ, in which βχ is parallel to θγ, the angle at χ is right, and γβχ is equal to θγβ (5° 27' 47") and βγ is 6763. Hence, the sides are found to be: γχ 644, βχ 6732. Therefore, in the right triangle βχφ, since βφ is 100,000 (β being the center of the eccentric φ) and χβ is 6732, χφ will be 99,773. Add χγ, 644, and γφ is given, 100,417. But γυ was 99,725. Therefore, the quantity sought, υφ, is 692.*

With υ and φ joined to the position of the observer δ, the magnitude of angle υφδ is found as follows. *Above, δγ was 19,763 in the new units, and the angle at γ right. Therefore, as δγ is to γφ and γυ, so is the whole sine to the tangents of the angles γδφ and γδυ. These come out to be 78° 51' 54" and 78° 47' 30". The difference of these angles, the angle υδφ, is 4' 24". The angle σδτ will be much less, because στ, as it is closer to the intersection of the eccentrics, is smaller than υφ.*

You see, then, how nearly the appearances remain unchanged for the observer at δ, despite the substitution of a new planetary path in the heavens by transposition of the observer and change of hypothesis. Moreover, it is still within the practitioner's power to vary somewhat the mean motion and the ratio of the eccentricities both among themselves and in relation to the radius of the orb, should this become useful, for the purpose of obliterating this discrepancy of some five minutes.

---

[4]*Artifices* (singular, *artifex*). For a discussion of the meaning of this term, see the Glossary.

[5]Incorrectly denoted as βγ in all editions.

This equivalence pertains chiefly to the first inequality—that is, to the appearances at δ, close to the center of the eccentric. However, in the second inequality (the equation for the annual orb), it makes a big difference whether the planet goes around on ξπ or oρ, as was also noted above in the other equivalence. And there we could not ignore the 246 units (the difference between the Ptolemaic and the Copernican hypotheses). Much less can we overlook these 880 units, which are 1344 of the old units. In the next chapter we shall see how much of a difference in Mars's apparent position this would occasion.

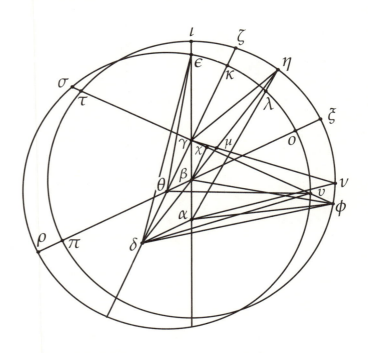

Hitherto, we have transposed the observer from δ to α. Let it now be demonstrated that very nearly the same things happen if the observer remains fixed while the point of the equant is transposed. This we do to make it clear that the same thing can be done in this chapter with an eccentric that has an equant as could be done on a simple eccentric (at the end of Chapter 3 above). In the earlier case, the result was the same whether the observer or the center of the eccentric were transposed, while here, likewise, the result is almost the same whether the observer or the center of the equant be transposed. However, it necessary to adapt this demonstration to this divergence, on account of the great dissimilarity of the opinions followed by the practitioners in demonstrating the second inequality of the planets, which opinions will be keeping us in court in the next chapter.

24

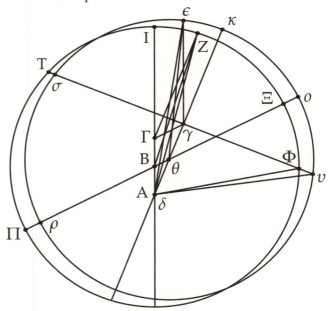

Let the points α and δ merge into one, so that the observer remain in the same place. Let δ, θ, and γ remain the same, but let the line γβα of the previous diagram be deleted, and replaced by ΑΒΓ parallel to it and passing through the point δ or Α. Let the segments ΑΒ, ΑΓ be equal to the previous segments αβ, αγ. Therefore, Γγ will be the transposition of the equalizing point γ, equal to the previous transposition of the observer αδ. Once more, two eccentrics or planetary paths through the aethereal air will be described, about Β and θ. All the letters on each circle will be carried over along with them, and the magnitudes of the lines will remain precisely the same. The only difference is that the two points on the two eccentrics at which the planet is to be placed at a given moment are no longer determined by a single line, but by parallel lines drawn from the two equant points Γ, γ, each extended to its own eccentric. *For example, when the eccentric θκ has its planet at κ, the eccentric BI will have it at Z, where γκ and ΓZ are parallel. And when the former has the planet at ε, the latter will have it at I, where γε and ΓI are again parallel. The rest is clear from the diagram without demonstration.*

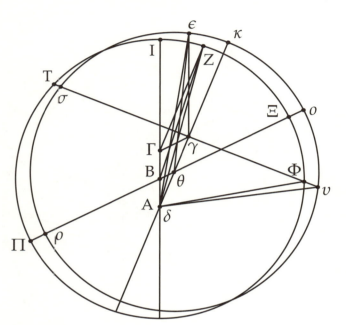

Now, suppose it is not permissible to shift the observer (and it is not permitted by those who make the earth the center of the world, as will be remarked in the next chapter), and that the planet has been observed in several positions on the zodiac, always at opposition to the sun's mean position, and that the practitioner will use the positions and the intervals of time between them to construct this sort of hypothesis, with the observer at δ, δθ the eccentricity of the eccentric θκ, θγ the eccentricity of the equant, and κ the apogee. Comes now Kepler, who would change the observed positions and times; that is, he would observe the moments and points at which the planet is at opposition to the apparent position of the sun rather than its mean position. From these positions and times he will have come up with another hypothesis, in which the observer would be left unmoved at δ or A, but where the eccentricity would come out to be AB in a new eccentric BI, and the eccentricity AΓ of a new equant Γ, and there would be a new apogee I. The question now is whether, if the prior practitioner combines the new eccentric BI with his original equalizing point γ, the computed equation and sidereal position of the planet will turn out much different from what he had formerly found using his eccentric γκ. (It is the first inequality that is in question; this discussion is not concerned with the second inequality, and the nature and magnitude of the changes which this procedure would effect therein.) The answer, arising from the equivalence of transpositions, is that the discrepancy will be extremely small. Its maximum, reached in the neighborhood of the points υ and Φ, will not exceed five minutes, exactly as before when the observer was transposed, except that now the line υΦ is closer to the observer δ than is its endpoint υ. Consequently, the angle υδΦ, which previously was 4′ 24″, is now 4′ 43″. The opposite happens at σT.

It has thus been demonstrated in a Ptolemaic eccentric what sort of disturbances would arise if one should transpose either the observer or the orb and construct a new eccentric, making use of the planet's oppositions to the sun's apparent position.

I do not think there is any need to repeat the arguments and demonstrate the same equivalence in the Copernican or Tychonic form, which makes use of two epicycles. I shall only show, by what was established at the end of Ch. 3, how to delineate both the eccentric-cum-equant that suits the planet, and its transformation into different magnitudes and different positions for the observer, in terms of the Copernican double epicycle. This is done in such a manner that, while the observer is transposed, the path of the planet through the aethereal air is invariant, as nearly as possible, in accord with what has been said in this fifth chapter. (This is exactly the possibility adumbrated in Ch. 3).

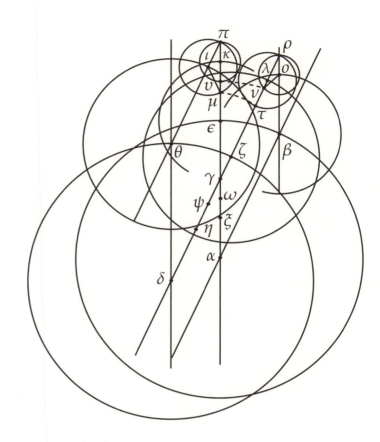

Let the triangle δγα be constructed equal to the previous one, with corresponding lines parallel, and through α let αβ be drawn parallel to δγ, and through δ, δθ parallel to αγ. And about centers α, δ let two concentrics be described, equal to the previous eccentrics δθ, αβ. Let δγ be extended to ζ and λ, and αγ to ε and κ, and let δζ and αε be semidiameters, as before, and lines of apsides (since both go through the same point γ). Now let δγ and αγ be cut at η and ξ in the same ratio as before, and let ηγ and ξγ be bisected at ψ and ω. Then, with radius δψ and upon centers θ and ζ let the epicycles ι and λ be described, and let ζλ be parallel to θι. Then, about centers ι and λ, with radius ψγ, let epicyclets be described through πμ and ρτ.

Again, let the epicycles κ and ο be described about centers ε and β with radius αω, and with βο parallel to εκ. Next, about centers κ and ο with radius ωγ, let epicyclets be described through πυ and ρυ, and let θιμ and βου be double δγα. Let 26 the planet on epicyclet κπ be at υ nearest ε, and the planet on epicyclet λρ be at ν nearest ζ. Therefore, according to the hypothesis deriving from δ the planet falls upon τμ, while according to the hypothesis deriving from α the planet falls upon νυ. Here you see that the points μ and υ, as well at τ and ν, hardly differ when seen from δ and α, respectively, when the planet is near the apsides. But around the middle elongations these points will be separated from each other only by the amount that separated υ and Φ in the previous diagram. The magnitudes will all be nearly equal and the demonstrations entirely the same. For if θι and εκ be extended to meet at π, and ζλ, βο to meet at ρ, the triangles θπε and ζρβ will be congruent to triangle δγα, and corresponding sides will be parallel.

These demonstrations will be perplexing enough in themselves, so it is not really advisable to make them more involved by a heaping up of Copernican or Brahean epicycles. Therefore, in what follows, we shall decree that the

Copernican or Brahean form count as belonging to the first inequality. For the procedure for treating the hypotheses of the second inequality, since it is always going to concern each of the three, will furnish us with a great abundance of things to do.

I now immediately state it as a postulate that whatever we shall demonstrate using the Ptolemaic equant-cum-eccentric also be taken as demonstrated in the Copernican or Brahean concentric-cum-double-epicycle, or eccentrepicycle. For in Ch. 4 above, the difference was found to be very small.

CHAPTER 6

On the equivalence of the hypotheses of Ptolemy, Copernicus, and Brahe, by which they demonstrated the second inequality of the planets, and how each one varies when accommodated to the sun's apparent position and when accommodated to its mean position.

The discussion so far has concerned the hypotheses of the planets' first in- equality, which completes its cycle each time the planet returns to the same sign of the zodiac. Now we pass on to the other inequality, which completes its cycle not at a single constant sign of the zodiac but with the sun's opposition to, or conjunction with, the planet. People have wondered exceedingly at this, different ones proposing different reasons why a planet in conjunction with the sun is made swift, direct, high, and small; and opposite the sun, retrograde, low, and large; while in between it becomes stationary and of a medium size.

The Latin authors considered that in the sun's aspects and rays there is a force by which the other planets are in fact attracted. Their opinion cannot be shown numerically, because it is not astronomical. But it is also improbable, now that the true causes have been found, and manifestly false, since Saturn begins to retrogress at quadrature with the sun, or beyond; Jupiter, at trine; Mars, at biquintile or before sequiquadrate, and all at variable distances.[1]

Ptolemy said that at a determined point on the planetary circle that serves for the first inequality, there is fixed, not the planet itself, but the center of an epicycle bearing the planet fixed upon its circumference, which is in turn borne by the planet's chief circle. The motion has the following form: if the center of the epicycle be in conjunction with the sun, the planet is also at the highest point of the epicycle and is moved along with the sun in the same direction, [and] when the sun, which is faster, departs from the center of this epicycle, the planet simultaneously descends on the epicycle. But since the epicycle's motion about its center is faster than the motion of its center about the earth, it hence happens that when the planet traverses the lower parts of the epicycle while the epicycle's center is at opposition to the sun, the compounding of motions makes it actually retrograde. Thus Ptolemy made his opinions correspond to the data and to geometry, and has failed to sustain our admiration. For we still seek the cause that connects all the epicycles of the planets to the sun, so that they always complete their periods when their centers are in conjunction with the sun.

Copernicus, with the most ancient Pythagoreans and Aristarchus, and I

---

[1]Quadrature is 90° of angular difference; trine, 120°; biquintile, 144°; and sequiquadrate, 135°.

along with them, say that this second inequality does not belong to the planet's own motion, but only appears to do so, and is really a by-product of the earth's annual wheeling around the motionless sun. In this way, just as in Ch. 1 the diurnal motion was separated from the motion proper to the planets, the second inequality of the planets is likewise now separated from the first by Copernicus, and in quite the same way. For some practitioners admit that the first motion is extrinsic to the planets, but still think it is in fact in the planets, inserted into them, so that the planets, too, are moved with the same motion. Copernicus holds that it is neither intrinsic to the planets nor inserted, but only attached to them through an optical illusion. For while the earth rotates upon its axis from west to east, it appears to our eyes that the rest of the world rotates from east to west. It is, I claim, in just the same way that Copernicus asserts that the planets do not really become stationary and retrograde, but only appear so. For he says that since the earth is in addition carried along by another motion, which is annual, in a very large circle (which he calls the *orbis magnus*), those who believe that the earth is at rest think that the planets and the sun are carried in the opposite direction; and he says that when the sun is between the planet and the earth, in the appearance the motions of the earth and the planet in are added, whence the planet appears to be swift; and when, on the other hand, the earth is between the sun and the planet, the planet is apparently left behind and thus retrogresses, owing to the earth's being swifter than the planet.

Tycho Brahe holds something in common with the Latins: although the sun does indeed not attract the planets through the aspect, the planets do fawn upon the sun. For he says that they strive to keep the sun (although it is moving) nearly in the middle of their circuits, and indeed, that they arrange their real paths around the sun as if it were motionless. Thus any given planet, besides its own path, traverses the sun's path in the aethereal air, and out of these motions compounded with one another there is produced exactly what Ptolemy had (that is, a spiral), as described in Ch. 1. In astronomical terms, Ptolemy put epicycles on eccentrics, while Brahe put eccentrics on a single epicycle, which is the sun's orb itself.

I, in the demonstrations that follow, shall link together all three authors' forms. For Tycho, too, whenever I suggested this, answered that he was about to do this on his own initiative even if I had kept silent (and he would have done it had he survived), and on his death bed asked me, whom he knew to be of the Copernican persuasion, that I demonstrate everything in his hypothesis.

Furthermore, we shall demonstrate, both right here and through the entire book (though while doing other things), that these three forms are absolutely, perfectly, geometrically equivalent. For the present we must carry out what we set out for ourselves and what is to be demonstrated, namely, that there is a very great error indeed in the second inequality if the apparent position of the sun is replaced by its mean position, to which the planet is at opposition at the beginning of this second inequality.

I shall begin with the Copernican opinion.[2] *About center β let the earth's*

---

[2]The diagram here differs somewhat from the one in the 1609 edition. It has been redrawn to

eccentric γυ be described, such as Copernicus, putting his trust in Ptolemy, imagined. Let γβ be its line of apsides, κ the position of the motionless sun, and β the point about which the earth's motion is uniform.

Through β and perpendicular to βγ let υβσ be drawn, intersecting the circumference at points υ and σ, and let υ and σ be joined to κ.

Copernicus, intending to transfer the Ptolemaic numbers into his own form of hypothesis, reckoned the eccentricities of the planets from the conjectural center β of the earth's uniform motion rather than from the sun κ. *For if lines be drawn from β (as βγ, βυ, βσ), whenever a planet and the earth lie upon one of these, the planet was supposed to shed the second inequality, to which it was subject on account of the earth's motion, as, for example, if the earth were at υ and the planet were found on βυ extended.*

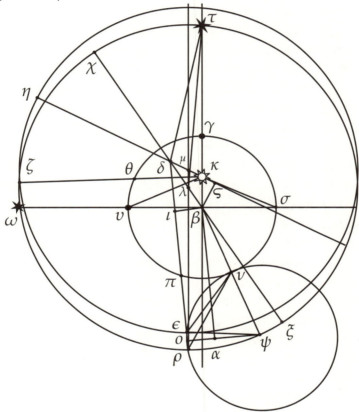

In effect, by adopting this procedure, Copernicus established a fictitious   29 observer at the point β. *For provided that the planet is on the line βυ, it makes no difference for the purpose of designating its sidereal position whether it be viewed from σ or from β. The same may be truly said of the lines βγ, βσ, and all the other infinite lines intersecting at β.* Therefore, β is the point of intersection of all the lines of vision, and is thus the fictitious common point for all observers. In fact, however, the point of vision is the earth, our home, which is found at one or another point on the circle σγυ at various times.

So Copernicus believed the planet was freed from the second inequality

---

fit the demonstrations as clearly as possible. The most conspicuous change is the omission of a circle that appears in Kepler's original diagram but is never referenced. It appears to be the same size as the earth's orbit (γθυ) but passes through κ, φ, and σ. Its center is marked on a line through ν that appears to be parallel to κβ extended.

whenever the planet and the earth were found on any one line passing through
β. Consequently, he endeavored to find the apparent sidereal position of a
planet at those moments of opposition to the mean position of the sun through
mathematical instruments. For if the planet's position was found on one of the
nights near the planet's opposition to the sun, and if at that time the mean position of the sun was found by calculation to be at the point precisely opposite,
then that was the moment of time desired. However, if there was still a little
distance between them on that night, then he tracked down this moment of
time, and the point or position held by the planet at this moment, by a comparison of two or more nights and of the diurnal motions of Mars and the earth
over the interval. When he had done this as many times and in as many places
30   on the zodiac as he thought necessary (suppose, for example, $\beta\gamma$, $\beta\upsilon$, $\beta\sigma$), the
practitioner now began to use these known sidereal or zodiacal positions of
the planet, $\beta\gamma$, $\beta\upsilon$, $\beta\sigma$, to investigate the hypothesis of the first inequality. This
involved finding the magnitude of the planetary circle's eccentricity from the
selected point β, and the parts of the zodiac into which the apogee points,
by comparing the angles which the observed positions set up about the point
of observation β, with the time intervals between them. The method of this
undertaking, however, I shall present clearly in its proper place, below.

Suppose these things already done, giving a line of apsides $\beta\delta$, eccentricity
of the equant point $\beta\delta$, and the center of the eccentric on this line at the point
λ, and let this hypothesis correspond to all positions observed at the moment
of opposition to the sun's mean position.

Now, Kepler, what more could you ask of Copernicus? Are you denying
that this hypothesis corresponds entirely to observations or to astronomers'
experience?[3] That is indeed not in question at present. Nor was I, when I first
entered upon this undertaking, tempted by the observations to take up a different opinion. But it is this that I have been wanting: *Let $\beta\delta$ be extended so as
to intersect the eccentric at χ and ξ, and near χ let some point τ on the eccentric be
chosen, and lines be drawn from τ to δ and λ. Now $\chi\tau$ is the measure of the angle
$\chi\lambda\tau$, while the angle $\chi\delta\tau$ is greater than the angle $\chi\lambda\tau$ by the amount $\delta\tau\lambda$, and δ is
the point of temporal uniformity.[4] Therefore, the time designated by $\chi\delta\tau$ is greater,
with respect to the whole periodic time designated by four right angles, than is the arc*
31   *$\chi\tau$ with respect to the whole circumference. The planet, then, is actually slower over
the arc $\tau\chi$ (this is not just an optical illusion), and fast over the opposite arc; and at χ
it is slowest, and at ξ it is fastest. Nevertheless, it is not farthest from the sun κ when
it is at χ, nor is it nearest κ at ξ. But by all the arguments, even the testimony of*

---

[3]*Experimentis.* The distinction between *experimentum* and *experientia* is not the same as our distinction between "experiment" and "experience". The Latin words mean very nearly the same
thing, rather like our "observation", although *experimentum* is more nearly related to proving
something, while *experientia* may denote the skill gained through practice or experience. *experimentum* can also mean something close to the English "example," as in Kepler's letter to
Maestlin of 3 October 1595 (letter no. 23 in *KGW* 13, p. 38 line 184), where he writes, "*Capiamus
a luce experimentum,*" that is, "Let us take an example from light."

[4]Kepler's Latin contains a malapropism: *temporaria* ("temporary") instead of *temporalis* ("temporal").

the very hypothesis set up upon β which I am refuting, it is fitting that this real slowing down of the planet arises from its moving away from the body of the sun, and the speeding up from its approach to the sun itself, seated at κ. On the other hand, it is impossible even to conceive of how a force could inhere in point β, which has no body, rather than in κ, quite nearby, in which there is the sun, the heart of the world, which force would drive the planet around more swiftly or slowly according to its approach and recess. Furthermore, even if one who is not prepared to admit that the slowings and quickenings arise physically from the close interconnection of the eccentrics, should consequently assert that these affects of motion are naturally under the control of the motive faculties residing in the body of the planet, we will again maintain the same probable conclusion. For what would be the reason why those

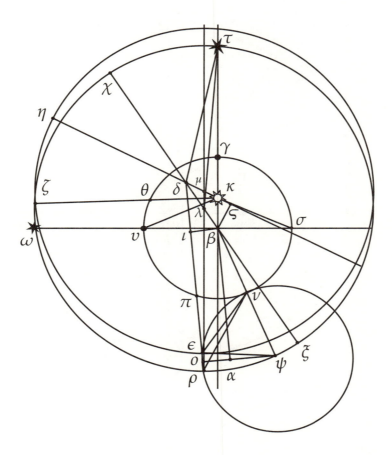

minds would bypass the point κ (which has a geometrical affinity for motion, being invested with a body of no small magnitude) and pay attention to the point β, only four semidiameters (diameters, according to the authorities)[5] of the solar body distant from the sun, empty, and propped up by nothing but imagination alone? Even Copernicus himself admits, in Book 5 Ch. 16, that the sun is really fixed at κ, wherefore the eccentricity $\delta\kappa$ is constant, while he holds

---

[5]Kepler is referring to his reduction of the solar eccentricity, introduced in Part III of the present work. See especially Ch. 27–31.

that the point β, which he takes as the center of the annual orb, is displaced over the ages, thus making βδ shorter. Thus either β is no longer in the center of the world today, or it was previously not there. But it is probable either that the motion originates from the center of the world, or that the moving minds pay attention to the center of the world, and thus not to β but to κ, which Copernicus says is fixed, as the center of the world should be.

Led by these probable arguments, I concluded that the line of apsides taken in order to produce the planet's first inequality, ought to go right through κ itself rather than β. But we shall obtain this result when we make use of the sidereal positions which the planet has at the moment when the sun's apparent position and its own are opposite.

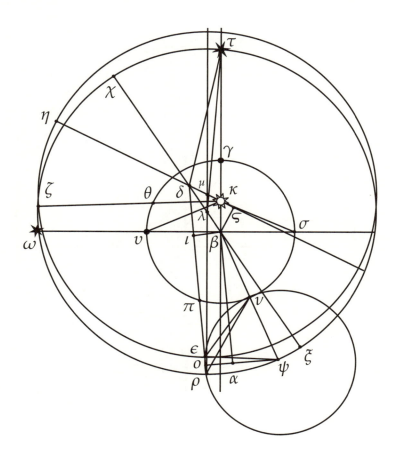

And when the points κ and β are collinear with the earth γ, and when the planet itself falls on the same line at the same time, as at τ, it is at the same moment at opposition to both the sun's mean position and its apparent position. Its position is the same whether it be designated by βτ or by κτ extended out to the fixed stars, and the planet is truly stripped of its second inequality, whether it depends upon the earth's apparent position or its mean position. But when the earth comes to the side of its eccentric, or rather the middle elongations,[6] a fairly large difference arises between them. For let the earth have traveled from γ to υ (that is, let the sun, directly opposite, have moved from perigee in Capricorn to Aries), and let the line of the sun's mean

---

[6]For this odd term, see the Glossary.

vβ be found in Aries, while the line upon which the planet is observed in Libra be
exactly opposite, namely, vω. Now since vκ is farther eastward beyond vβ, the sun's
apparent position is beyond opposition to the planet. And since v is the observer's
home, the earth, and ω is the planet, and both are going down towards ζ, the earth v
more swiftly, the line vω will at a later time be still more inclined to the line of the   32
sun's visible position vκ. Therefore the apparent opposition precedes the mean. So, at
a time preceding the moment designated by βv (call it βθ), the planet will fall upon a
line drawn from κ through θ, that is, at ζ. And then the planet's line of vision θζ
(as the less experienced should diligently note) lies farther eastward beneath
the fixed stars than the line *vω* of the later time. This is because, although
θζ precedes the line *vω* in being farther west, it is nonetheless exactly as if *θ*,
v, and absolutely all points on the earth's circle were a single point and were
the center of the sphere of the fixed stars. Therefore, it is not the distance of
the endpoints *θ* and *v* but the inclination of the lines θζ and *vω* that causes the
lines to strike upon different zodiacal positions, since they would be perceived
as coinciding if they were parallel. *But that ζ is inclined towards ω is clear from*
the supposition that the planet is moved from ζ to ω in the same time that the earth
is moved from θ to v. For the earth is swifter than the planet. Therefore, the earth
traverses a greater space θv than does the planet along ζω.

But it can be shown even more easily that at an earlier time the planet is
farther east, since at opposition it is retrograde, as is clear to everyone. It is
therefore clear what is altered in the positions stripped of the second inequality, in this transition from the mean to the apparent positions of the sun.

For, at τ and the point opposite, the original positions remain; at ζ or ω an addition
is made to the observed position, since θζ (as was said) is farther eastward than vω.
A subtraction is made in the intervening time, because θζ is the line of vision at an
earlier time than is vω. At the opposite position the outcome is the contrary, that
is, an addition is made to the time and a subtraction from the position. Accordingly,
these positions of the planet differ considerably from the original ones. And therefore,
the operation set up in this new way produces quite different results. That is, since
we have transferred the fictitious point of observation to the sun *κ (by virtue*
of our having viewed the planet when it was at τ and ζ and the earth was on the lines
κτ and κζ, namely, at the points γ and θ), the eccentricity will now originate at
the point *κ*. But in Ch. 5 above, *it was shown that when the observer is shifted from*
β to κ and a line is drawn from κ through the original point of uniformity δ, although
this new hypothesis does result in a new eccentric, it is one that, for the observer at
β, leaves nearly all his appearances undisturbed. So, by joining δκ, dividing it at μ
so that δλ is to δμ as δβ is to δκ, tracing out a new eccentric ηε about μ equal to the
previous one ξχ, and drawing a new line of apsides through κδ, a new hypothesis will
be formed, whose apsis is at η. Previously, however, we had improperly called χ
the apogee, because the Copernican center β on the line χβ was the successor
to the Ptolemaic position of the earth. Now, following my own notions, we
shall call η the aphelion (since we are in the Copernican hypothesis), and the   Terms:
point opposite it perihelion, because the sun's distance from η is a maximum.[7]   What are aphelion and
                                                                                    perihelion?

---

[7]The terms "aphelion" and "perihelion" were coined by Kepler following the analogy with

It has been said how this pair of opinions, mine and those of the authorities, differ as regards physics. It has also been shown how each is to be constructed geometrically in the Copernican form. Third, it has been driven home that in astronomical terms those opinions[8] do not differ in any important way at the moments of conjunction and opposition[9]. The next thing for me to do is to demonstrate what remained unexplained in Ch. 5 above, that there is a considerable difference between the two hypotheses if you are required to use
33   them to compute the planet's position outside the acronychal location.[10]

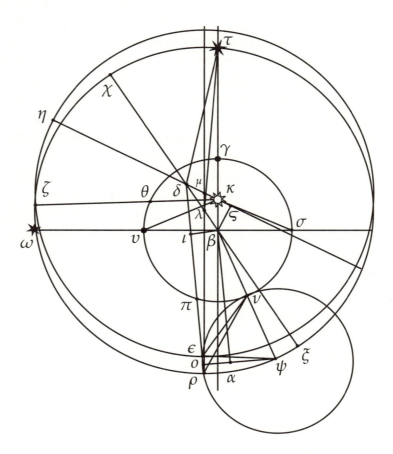

If a line be drawn through the centers of the eccentrics λ, μ, parallel to βκ, and extended to intersect each eccentric in two points, above and below, there will be set up, below, the maximum space ερ between the two, equal to λμ. But because it is lines from δ, not those from λ, that designate certain unchanged moments of time, which is what we need here, let δρ be drawn intersecting the eccentrics at ε and ρ, so that at one and the same moment the planet on one might be at ε and on the other at ρ. When

---

Ptolemy's "apogee" and "perigee."

[8]Reading *illas* for *illos*.

[9]Kepler clearly shows here for the first time the three distinct areas in which he requires a comprehensive theory to operate: the physical, the geometrical, and the observational. The entire work can fruitfully be viewed as a thoroughgoing effort to bring these three into harmony with one another. For more on this "Keplerian synthesis", see the translator's introduction.

[10]For an explanation of this term, see the glossary.

the earth is on the line δρ, at π, the planet, whether it is at ε or ρ, will in either case be seen at the same place on the zodiac. For, optically considered, the line ερ appears as a point. But when the earth departs towards either side of this line, the magnitude of the line ερ appears greater and greater, since it is viewed obliquely.

It is required to find the point on the earth's orb from which the lines of vision passing through ε and through ρ are at their greatest distance from one another and form the greatest angle of vision, and at which the error would be greatest if the planet were placed at ρ when it should have been placed at ε.

First, the angle will be greater down at ε than up near τ because the earth's orbit, described about β, moves the observer nearer to ερ than to τ. Next, since δρ is beyond τβ, ερ is seen more obliquely from the left side than from the right. It will consequently appear less at the former place than at the latter even when the distances of the earth from the line δρ are equal. Therefore, the point we are to find is on the right side. I say   34
that ερ subtends a maximum angle of vision when the observer is stationed at the point where a circle drawn through ε and ρ is tangent to the earth's circle. *For let such a circle be described through ερ tangent to the circle vσ on the side towards σ, and let the point of tangency be v. From ε and ρ let lines extend both to the tangent point v and to several other points of circle vσ before and after the point of contact. Now since circles touch one another in one and only one point, the sides of all angles extending from ε and ρ and meeting with points on circle vσ will therefore be cut by the circle through ερ, with the exception of those that terminate at the tangent point of the circles, v. But the sides from ε and ρ that are cut by the circle ερ before they intersect would form a greater angle had they intersected at either of the points of section,[11] by Euclid's* Elements *I. 21. And by Euclid III. 21, all angles at the circumference set up on segment ερ are equal. Therefore, the one at v (the point of contact) is greater than all the others.* q.e.d.

Next, to investigate its magnitude in appropriate numbers, we need to find ερ and also the perpendicular from β to δρ.

We shall find both by solution of the triangles δλρ and δμε. Now we said above[12] that in δλρ, δλ is 7411[13] where λρ is 100,000, and ρλβ is 47° 59′ 16″. This gives ρδλ a value of 44° 59′ 10″, and δρ 105,123. Therefore, in εδμ, since εδλ is 44° 59′ 10″, and λδμ earlier came out to be 5° 27′ 47″, the whole εδμ is 50° 26′ 57″, and δμ above was 6763 where με is 100,000. Therefore, in εδμ, with three magnitudes 35
given the rest are given: εμκ is 53° 26′ 17″, and through this, δε is 104,170. But earlier, δρ was 105,123. Therefore, the remainder, ερ, is 953. Above, λμ was 880, to which ερ would be equal were ε and ρ on the line μρ. But because here ε is

---

[11]That is, points of intersection with the circle ερ

[12]In Ch. 5, p. 99

[13]As a result of a slight error in the passage cited in the previous footnote, this magnitude is too large. In particular, the angle δγα (in the diagram in Ch. 5) should have been 5° 29′, not 5° 27′ 47″, and so the magnitude of θγ (or δλ in the present diagram) should be 7314, not 7411. However, as the intention of the present calculation is to show only that the two hypotheses give significantly different results, this error is of no importance. Kepler says that ρυε is 1° 3′ 32″, although it really is 0° 55′ 12″: in any event, it is about one degree, which is a difference that cannot be overlooked.

on the line $\delta\rho$ which is inclined to $\mu\rho$, you should not be surprised that $\epsilon\rho$ is longer than $\mu\lambda$[14]. *Now let $\beta\iota$ be drawn from β perpendicular to $\delta\rho$. In triangle $\delta\beta\iota$, the angle at ι is right, and $\beta\delta\iota$ is 44° 59′ 10″, and $\beta\delta$ was found to be 19,763, above. Therefore, the required perpendicular, $\beta\iota$, is 13,971, and $\delta\iota$ is 13,978. Consequently, $\iota\rho$ is 91,145. It is also necessary to figure out the quantity of the radius $\beta\upsilon$ in the same units. Above, when the line corresponding to the present $\beta\kappa$ was taken to be 3584 parts, $\beta\upsilon$ was assumed as 100,000. Now, however, $\lambda\rho$ is assumed as 100,000, and $\lambda\rho$ is to $\beta\upsilon$ (taken above) as approximately 61 to 40, whence the other ratios are extrapolated. Thus 60 is to 41 as 100,000 is to 65,656 $\frac{1}{2}$, the appropriate magnitude for $\beta\upsilon$.*

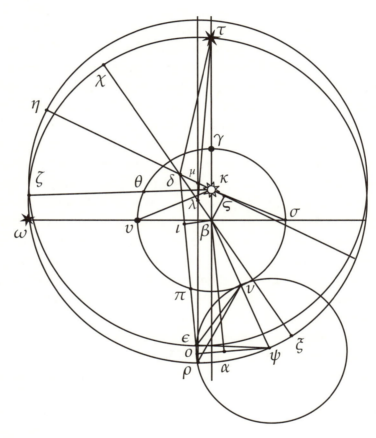

Next, let a circle passing through ϵ and ρ touch the circle $\beta\upsilon$ at point υ, and, $\epsilon\rho$ being bisected at o, let ψo be set up perpendicular to $\iota\rho$, and $\beta\upsilon$ be extended so as to intersect $o\psi$ at ψ. The center of the circle will be ψ. For the center of the circle is on the line passing through the center of one of the tangent circles and the point of tangency, by Euclid III. 11: hence, it is on the line $\beta\psi$. Again, by Euclid III. 3, the center of the circle is on the perpendicular bisector of the chord $\epsilon\rho$, which connects the points of intersection ϵ and ρ. Therefore, the center is on the line $o\psi$, and hence is at the point ψ common to the two lines. Let $\epsilon\psi$ be joined, and from β parallel to $\iota\rho$ let $\beta\alpha$ be drawn intersecting $o\psi$ at α. Therefore, $\beta\alpha$ is equal to ιo, and αo is equal to $\beta\iota$.

---

[14]Since the angle at ρ is only 3°, the inclination of the lines is not enough to account for $\epsilon\rho$'s being longer than $\mu\lambda$. However, if Kepler's errors are corrected, the computed magnitudes of both are the same, 870.

But βι was just found to be 13,971, while ιο is known through ιρ and ερ. And ιρ, above, was 91,145, and ερ was 953. But ορ is half of ερ, and therefore, ορ is 476$\frac{1}{2}$. So, when ορ is subtracted from ιρ, the remainder ιο or βα is 90,668. Now since α is a right angle, βψ is the power[15] of the two, βα, αψ. However, βψ is composed of βν, which is known (65,656), and νψ. But because ο is a right angle, νψ, that is, εψ, is the power of the known εο (476$\frac{1}{2}$) and οψ which is composed of the known οα and αψ, which is unknown but was also noted previously. Therefore, οψ must be made long enough that when you add the powers of ψο and οε, the side εψ or ψν will be just so long that when the power of the sum of βν and νψ is diminished by the power of βα, it leaves the power of ψα of such a magnitude that when it is compounded with αο the sum is equal to οψ that was taken at the beginning.[16]

I take ψο as a figured unit [x]. Its square will also be a figured unit [x^2]. Add the square on εο, 227,052, and the sum of the two will be the square of ψε or ψν. But the square of βν is 4,310,747,475. If you add this to the square of ψν and complete the rectangle, the result will be the square of the whole ψβ. Then each rectangle is the root of

$$4,310,747,475x^2 + 978,763,835,536,363.$$

And thus the square on βψ is obtained for the first time.[17] [18]

Now since αο is 13,971, ψα will be represented by the figured unit diminished by 13,971. Its square will be

$$x^2 - 27,942x + 195,188,841.$$

Add to this the square of βα, 8,220,686,224, so that the square on βψ may be

---

[15]*potentia.* For an explanation of this term, see the glossary.

[16]In algebraic terms,
$$(\beta\nu + \sqrt{(\epsilon o^2 + o\psi^2)})^2 - \beta\alpha^2 = \alpha\psi^2,$$
where $\alpha\psi = o\psi - o\alpha$.

[17]Here and in the following mathematics, the text and equations are set in continuous paragraph style in the Latin editions. For the sake of clarity, the translator has reformatted the equations or expressions where appropriate. However, where Kepler uses words instead of symbols, the translation retains them; for example, "*a* augmented by twice *b*," rather than "$a + 2b$."

[18]The argument makes use of the geometrical "algebra" of Euclid II. 4.

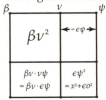

The square on βψ is made up of the squares on its parts plus twice the rectangle formed by its parts. Each rectangle is:
$$\beta\nu \cdot \epsilon\psi = \beta\nu\sqrt{(x^2 + \epsilon o^2)} = \sqrt{(\beta\nu^2 x^2 + \beta\nu^2\epsilon o^2)},$$
which is what Kepler has here. The whole square on βψ is:
$$\beta\nu^2 + x^2 + \epsilon o^2 + 2\sqrt{(\beta\nu^2 x^2 + \beta\nu^2\epsilon o^2)},$$
or: $4,310,974,527 + x^2 + 2(4,310,747,475x^2 + 978,763,835,536,363)$.

established for a second time:

$$[\beta\psi^2 =]x^2 - 27,942x + 8,415,875,065.$$

Previously it was

$$x^2 + 4,310,974,527$$

augmented by double the root of

$$4,310,747,475x^2 + 978,763,835,536,363.$$

Subtract x^2 from both, and also $4,310,974,529$. In the former, the remainder will be

$$-27,942x + 4,104,900,538,$$

36　　*and in the latter,*

twice the root of $4,310,747,475x^2 + 978,763,835,536,363,$

and these are equal. Therefore, the simple root in the former is equal to

$$-13,971x + 2,052,450,269.$$

And since this is equal to the latter's root, its square will be equal to the quantity itself. This square is

$$+195,188,841x^2 - 57,349,565,416,398x + 4,212,552,106,718,172,361.$$

Subtract from each

$$195,188,841x^2 \text{ and } 978,763,835,536,363,$$

and add to each

$$57,349,565,416,398x.$$

The two will remain equal, the former being

$$4,115,558,634x^2 + 57,349,565,416,398x,$$

while the latter,

$$4,211,573,342,882,635,998.$$

In least terms,

$$x^2 + 13,934x \text{ is equal to } 1,023,329,690.$$

Solving the equation gives the figured unit oψ the value of $25,772$.

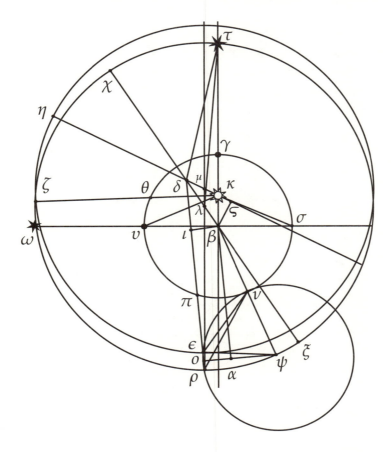

Now that the semidiameter of the circle is known, the angles are easily obtained. *From ψο subtract οα,* 13,971. *The remainder, ψα, will be* 11,801. *And βα is* 90,668$\frac{1}{2}$, *and βαψ is right. Therefore, αβψ is* 7° 30′ 10″. *But above, αβ or ρδ were inclined to ρλ or βκ by* 3° 0′ 6″, *which latter lies at* 5$\frac{1}{2}$° *Cancer. Therefore, ρι or αβ will be at* 8$\frac{1}{2}$° *Cancer, and ψβ is consequently at* 16° *Cancer.* Therefore (assuming these numbers), when the sun is passing through 16° Cancer while the planet is at 8$\frac{1}{2}$° Capricorn in its mean and uniform motion, and in the neighborhood of 27° Scorpio in its apparent motion, ερ appears maximum. If the planet is beyond 8$\frac{1}{2}$° Capricorn, that is, beyond ρε, even though ρε is then diminished, its apparent size can increase when viewed from a point beyond ν owing to the greater nearness of the orbs. This apparent size is now immediately obtained. *For since οψ was found to be* 25,772, *and ορ* 476$\frac{1}{2}$, *οψε will be* 1° 3′ 32″ *But ρνε* 37 *(which is what we have been seeking) is equal to οψε, by Euclid III. 20. This is because the whole angle at the center ρψε is twice the angle at the circumference ρνε, and at the same time οψε is half of ρψε. But if βδ and κδ be bisected [at λ and μ], and λμ were to be taken as half βκ (on which more will be said below), then ρε, and consequently its angle at ν as well, could possibly become greater by one fourth.* So, at last, you see how much my transposition of the hypothesis from the mean to the apparent position of the sun creates a disturbance in the annual parallax.

Therefore, by the observations as well, a door has opened for us to determine what I had deduced *a priori* by consideration of moving causes: that the planet's line of apsides, which is the only line bisecting the planet's path into two semicircles equal both in size and in vigor—this line, I say, does not (as the practitioners have it) pass through some point other than the sun, but right through the center of the solar body. In the conclusion of the work I shall demonstrate this by means of observations, in Parts IV and V.

Now, to the extent that this is possible, I shall deduce the same things in the Ptolemaic hypothesis.

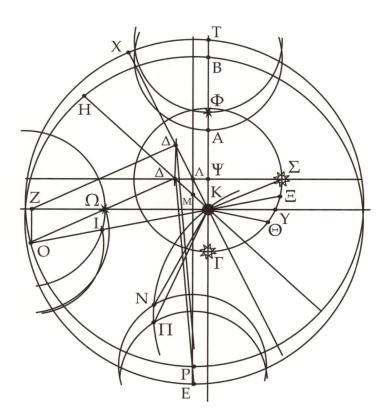

About center Ψ *let the eccentric of the sun* Γ *be described, upon which let* ΨΓ *be the line of apsides, with the motionless earth at point* K *of the line* ΨΓ *on the side nearest* Γ, *and* Ψ *be the point about which the sun's motion is reckoned as being uniform. From* Ψ *and* K *let the perpendiculars* ΨΣ *and* KY *be set up, and let* ΣK *be joined. Let* KΣ *be the line of the sun's apparent motion, and* KY *the line of the sun's uniform* 38 *motion.*

Now Ptolemy measured out the courses of the planets, not on the lines KΣ, but on the lines KY drawn from K parallel to the lines ΨΣ passing through the sun's body. For whenever the planet fell upon these lines KY on the side opposite the sun, it was supposed to have shed its second inequality, which fell to it (in Ptolemy's opinion) because of the epicycle. Then the planet's sidereal position was sought out by means of instruments, and the center of the epicycle was supposed to be found at that time on the same line. This was done a number of times and at various places on the zodiac: say, KΓ, KY, and the points opposite. From three such positions of the planet (or of the center of the epicycle, which, according to Ptolemy, accounts for the second inequality), the practitioner sets out to investigate the hypothesis for the first inequality by comparing these angles which the observed positions form about K, the center of the earth and of observation, with the intervening times. The method for this enterprise is found in Ptolemy Book IX.[19]

Now suppose the treatment done, and let the eccentric's line of apsides come out to be KΛΔX, with Δ the equalizing point, the center of the eccentric upon this line and at point Λ, and the eccentric XZ. And let this hypothesis correspond to all positions observed at the times of the planet's opposition to the sun's mean position.

Here, the objections I raised against Copernicus concerning the ordering of the physical motion are also not clearly in agreement with Ptolemy. For the center of the epicycle which accounts for the second inequality, here just as the planet itself previously, is borne slowly or swiftly according to its approach   39 to or recession from the earth K, performed on the circle XZ. Furthermore, to stipulate that there is in the earth K (just as there was before in Copernicus's sun, the heart of the world) a motive force that drives the centers of this kind of epicycle around, is absurd and monstrous. The hypothesis can also be discredited by another physical argument. For intrinsic to this form of hypothesis, in one way or another, is solidity of the orbs; and since this has been destroyed (by Tycho Brahe's observations of comets), the hypothesis appears, so to speak, to collapse under its own weight. For a moving force would be declared to reside in the center of the epicycle (not in a body, but in a mathematical point), and to rouse itself to move from place to place, by an unequal amount in equal times; but at the same time would draw the planet along with itself at the distance of the epicycle's diameter; all the while causing the planet to gyrate around itself, by equal amounts in equal times. That this enormous variety cannot fall to a single moving mind (unless it be God), Aristotle gives supporting arguments in the *Metaphysics* Book XII Ch. 8: he holds that the individual minds preside over individual, perfectly uniform and simple circular motions. Besides, how will some power sit in a non-body, and flow out of the non-body into the planet? Even if you divide up the tasks, and locate one motive intelligence at the center of the epicycle and another in the body of the planet, the one at the center will use the earth (a body, of course) as its reference, and will move around the earth nonuniformly in a circle, while the one at a point on the circumference (that is, in the planet's body) will move around an incorporeal center and do so uniformly. The question, then, is (as before), by what aids would it move around the incorporeal point? Not by geometrical imagination, for it does not admit of geometrical imagination of itself. Nor can a mobile point subsist in a non-body even in imagination. We humans, in imagining points of this sort, have recourse to the aids of tablets or paper, which we draw upon with our hands or at least remember ourselves to have drawn once. But it likewise cannot happen through a physical flowing of power (which is in the center of the epicycle) out to the circumference and the planet's body. For we have submitted to this outflow of power on the supposition that the tasks of the compound motions are divided between two

---

[19]The general considerations are presented in *Almagest* IX.5–6, while the use of three positions appears in the treatments of the individual planets, in Books IX–XI.

minds. Why not also question whether in the first, eccentric motion, some natural force constituted to produce motion might subsist in some point entirely devoid of any body? And even more, whether this kind of incorporeal power can move itself around the earth, and go from place to place? And most of all, whether it can communicate or transfer another's motion through an efflux from itself, although it is not supported by any body that could serve as its nest? Those sublime considerations of the essence, motion, place, and operations of the blessed angels and of separate minds which some people will want to raise in opposition to me, are irrelevant. For we are arguing about natural things which are far inferior in dignity, about powers that exercise no choice of how to vary their action, and about minds which are not in the least separate, since they are yoked and bound to the celestial bodies which they are to bear. These, then, are the general objections that can be raised against
40   Ptolemy.

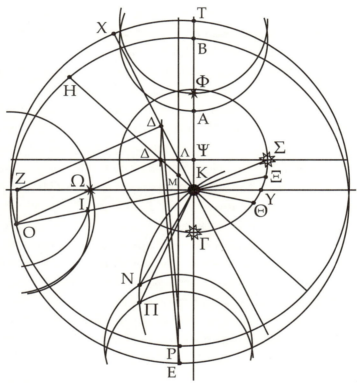

But something further may be said to Ptolemy, on account of which in particular he would wish to abandon his mean position of the sun and embrace the apparent with us. For if the power moving the planet (whether single or double) acts with respect to the sun, so that it places the planet at the lowest point of the epicycle whenever the center of the epicycle is opposite the sun, I ask (as above) why it looks to the imaginary point Y (which sometimes precedes and sometimes follows the sun, here marked Σ, and is sometimes above and sometimes below it), rather than the sun's body itself? Or how it is that that power can have any perception whatever of the motion of Y around the earth K, since there is no body at Ψ? Is it not more probable that the epicycle is restored to the lines KΣ of the sun's apparent position, when these pass through the center of the epicycle?

Let us then see what is changed on the eccentric by substitution of the sun's apparent position. *Again (as before) when the sun Γ, the center of the sun's eccentric*

The epicycle referred to here is the Ptolemaic epicycle that accounts for the second inequality.

Ψ, *and the earth* K *are on the same line, so that the line of the sun's apparent position* ΨΓ *and the line of its mean position* KΓ *coincide, then for the center of the epicycle* T *this place remains the same, whether its sidereal position is designated by* KT *or by* ΨT, *and the planet is really on the line* KT *or* ΨT *and at the lowest point on its epicycle* Φ, *since here it is nearest to both* Ψ *and* K. *The planet is thereby truly divested of its second inequality. But when the sun comes to the side of its eccentric (that is, to the middle elongations), a great enough difference arises. For let the sun have gone from* Γ *to* Σ, *and the line of the sun's mean position* KY *be found in Aries, and the planet's line of vision* KΩ *precisely opposite in Libra, so that* YKΩ *is a single line. Now since* Ptolemy declared that the planet Ω *on this line of vision* KΩ *is divested of its second inequality, he places the center of the epicycle* Z *on the line* KΩ. *But since* KΣ *has gone beyond* KY, *the sun's apparent position is beyond opposition to the planet. And in the last part of the time,* KΩ *is not descending, so as to be opposite* KΣ, *but ascends towards* KΦ, *because the lowest parts of the epicycle* Ω *are retrograde and swifter than the center* Z, *and it is there, of course, that the planet is at opposition to the sun. Therefore, this apparent opposition precedes the mean one.*

Therefore, at some time preceding the moment designated by KY *(let it be* KΘ*) when the sun is seen on the line* KΞ, *the planet will be seen opposite it (suppose it to be at* I*) on* KI *which forms a straight line with* KΞ. *Also, because it is now laid down that it sheds the second inequality at this true opposition, the center of the epicycle will also be seen on this line* ΞK, *say at* O. *And because the planet is retrograde, at the time* KΘ *prior to* KY *the planet is on the line* KI *later than at* KΩ. *But* KI *and* KΩ *are parts of the lines* KO *and* KZ. *Therefore,* KO *too is more to the east than* KZ.

Thus it is clear what would be changed on the line of the epicycle's center by this restoration from the sun's mean position to its apparent position. *For at* T *and the opposite point on the original line the motions of the epicycle's center remain unchanged. At* Z *this line, and the center of the epicycle upon it, is moved forward and a subtraction is made from the intervening time. At the opposite place the contrary happens: an addition is made to the time, and the line of motion of the center of the epicycle is moved back westward. And thus these lines of the center of the epicycle differ much from the original lines.* For this reason also, when we investigate the causes and measure of the first inequality by a new and repeated operation using these several observed positions of the center of the epicycle (that is, using the observed positions of the planet, after which we suppose the center of the epicycle to lie upon the same line), the outcome of this operation differs much from that of the preceding one. *That is, since the time in the semicircle containing the apogee will have been diminished, so that the planet is correspondingly faster, the eccentricity of the equant will come out smaller. And since in the greater quadrant* BZ *of the semicircle containing the apogee the time will have been diminished by an amount equal to the diminution in the other, smaller part, the planet has been made proportionally much faster in the remaining part of the semicircle. Therefore, the perigee has come closer to the latter, and the apogee has descended from* X *towards* Z.

The quantification of the new hypothesis will be made clear thus. *It is presupposed that the planet* Ω *falls on the line drawn from the center of the epicycle* Z *through the earth* K, *only when* KZ *is collinear with* KΣ, *the sun's apparent position. Therefore,* KΣ *and the line drawn from* Z *through the body of the planet always proceed in parallel. Moreover, we have just taken it from Ptolemy that at the time when the line of the sun's mean position was KY, drawn through Ω, the planet was observed on the line KΩ, and nonetheless we do not grant him that the center of the epicycle Z is on KΩ at the same time. Therefore, in accord with our position, let a line be drawn from the planet's position Ω parallel to KΣ, and let this be ΩO. We are supposing that the center of the epicycle is at this moment on the line ΩO, or some other line nearby and parallel, according as Ω (representing the planet) is closer to or farther from K on the line KZ. Let ΩO be drawn from whatever point on the line KZ (now, it is Ω), equal to ΩZ. From O let a line be drawn to ZK parallel to KΨ, and let this be OZ. Now since ZΩO is equal to KΣΨ, and KΣ is imperceptibly greater than ΨΣ*

42   *or ΩO (because KΨΣ is right, and the angle at Σ is not greater than 2° 3′, so that if ΨΣ is 100,000, KΣ is 100,064), OZ is also imperceptibly less than KΨ. Let ZΔ be joined, and let a line be drawn at O parallel to ZΔ. Now the moment of time at which the center of the epicycle is placed at Z by Ptolemy and at O by me, is the same, and is designated by both of us by the line KY in the theory of the sun. In the theory of Mars, that moment should be designated in the former theory by ZΔ, because Δ is the point of equality, while in the new theory it will be designated by a line parallel to ZΔ. Therefore, there will be a new point of equality, about which the time is counted, on this parallel through O.*

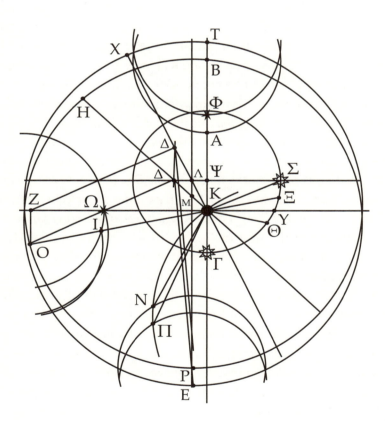

Further, the same things occur when the center of the epicycle (in Ptolemy's account) is at the other end of the line of the sun's mean position in the neighborhood of KY (which, for the sake of brevity, I shall not prove). Therefore, if some line is again *drawn parallel to the Ptolemaic line of the mean position of the center of the epicycle, a line ΔΔ drawn from Δ to where the two new parallels intersect will be parallel to ZO or ΨK, equal to ZO, and very nearly equal to ΨK, and the new Δ will be the common point of equality in the new hypothesis.*

But at the end of Ch. 5 above, it was shown that if through Δ a line ΔΔ be drawn parallel to KΨ, KΨ being equal to ΔΔ, and if the new Δ be joined to K, and the new line KΔ be cut at M in the same ratio in which the previous KΔ was cut at Δ, then by this new hypothesis a new eccentric is constructed, that is, one having a different position from the former, which even if applied in the prior hypothesis would still leave all the appearances to an observer at K very nearly undisturbed. *Let such a new eccentric be described about M, equal to the former, and let KM be extended in both directions. The new apogee will be H, and the center of the epicycle will be on the points B, O, of the new eccentric, with the planet nearer at A and farther at I than before.* However, in positions in which the second inequality is involved, the former observations are thoroughly and vehemently perturbed by this new eccentric introduced into their hypothesis. (This is clearly so if an epicycle equal to the sun's eccentric is attributed to the planet, as it must be if we wish to carry over the force of Copernicus's and Tycho's discoveries into the Ptolemaic form.) *The reason is not that the point of equality Δ does not remain the same, but that near the positions of the sun's apsides the centers of the two eccentrics, the Ptolemaic and ours, are distant by an interval ΛM. Also, from this distance of the centers there necessarily follows an equal distance of the positions of the planetary body.* Furthermore, this discrepancy is not greatest when the center of the epicycle is about the sun's middle elongations. *For it has been said that at those places the position of the center of the epicycle on either eccentric is nearly the same, even though they stand apart by parallels from ΔΔ.* It is therefore greatest near the sun's apsides, *and greater near perigee on MΛ extended so as to intersect the eccentrics at P and E. For PE is of the same magnitude as MΛ. But this one line MΛ does not designate the same moment, since it is not M or Λ that is the point of equality, but Δ. Therefore, let parallels be drawn from ΔΔ towards P and E, which shall designate the same moment, and let them be ΔP and ΔE. Also, let the epicycles N and Π be described about P and E.*

The question now is, where would this discrepancy appear greatest, in terms of the circumference of the epicycle? It is certain that this does not occur at the parts of the epicycle nearest K, the earth, because these parts would be in the same direction as K, nor at the highest parts of the epicycle, because they would be too far away. Therefore, this will occur at the parts of the epicycle near perigee, therefore when the sun, and the planet with it, are not exactly at perigee but nearby: namely (to put it briefly) at the points N and Π corresponding to the same moment of time, such that the small circle through them and K is a minimum. The center of this small circle is, moreover, on the line through K which, extended upward and meeting PΔ likewise extended, makes an angle of $7\frac{1}{2}$ degrees.

Let anyone who disagrees with this adapt the previous demonstration to the present conditions. The numbers are just the same, except for Ptolemy MΛ is greater than μλ, in the numbers taken above. For this reason, the difference in apparent

45   *position,* NKΠ, *is also greater.*

For previously, βκ was to λμ as δβ was to δλ, which is less than half δβ. To Ptolemy, ΔΔ (equal to βκ) would be to MΛ as the whole KΔ is to the half, KΛ.

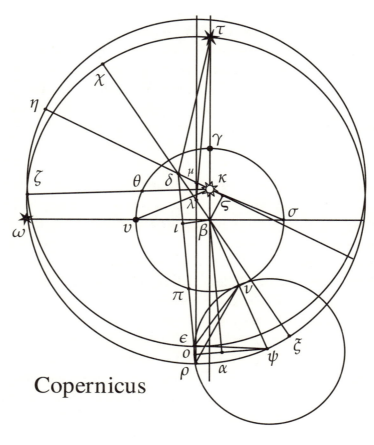

Copernicus

Ptolemy

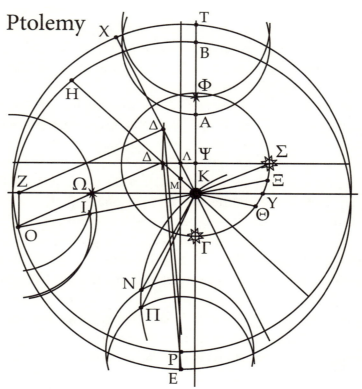

Finally, I shall deduce the same things in the Tychonic hypothesis.

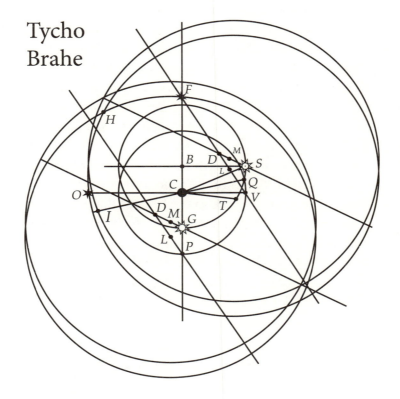

Tycho
Brahe

About center B let the solar eccentric GS be described, with line of apsides BG, C the position of the immobile earth, and B the point of equality, following the opinion of the authorities. For it will be shown in due course that the point of equality and the center of the eccentric are not the same in the theory of the sun. *Upon BC let the perpendiculars BS and CV be erected, and let SC be joined, so that CV may be the line of the sun's mean position, and CS, of its apparent position.*

Now, although Tycho Brahe had never finally decided whether to refer the planets to the lines *CV*, or to *CS* instead, in his initial conception he used the lines *CV*, according to the explanation he left in the *Progymnasmata*, vol. I p. 477 and vol. II p. 188. This is the way shown him by the footprints of Ptolemy and Copernicus. Of this path trodden by Tycho (if we proceed on it with Ptolemy's views), it should be said that whenever a planet is on one of the lines of the sun's mean position *CV*, on the side opposite the sun, the planet is divested of its second inequality, which, in Brahe's opinion, is applied to it on account of the motion of the center of the eccentric around the earth in the same time as the sun does so.

This common point, with respect to which all the planets are said to perform their eccentric motion, and at which the whole planetary system is conceived as being attached to the sun's orbit—this point, I say, is always on the line of the sun's mean position, distant from the earth C by a constant distance *BS*, and describing a concentric *V* equal to the eccentric *GS*. This was Tycho Brahe's opinion—except that he denied solid orbs. So what we said about the attachment of the whole system to the sun's orbit, we said to win over those who believe in solid orbs. *Let VC be extended, and let the planet be on that line*

Although for Brahe the orb of Mars intersects the sun's orb, I have nonetheless found it preferable to exclude this intersection, because I am presenting general considerations in this first part, which belong to all planets. For this would have produced much obscurity in the diagram.

46

beyond C. In this configuration, Brahe will place the point at which the planetary system is attached at V. The planet will be observed on the line VC. So, even though the observer is on the earth, C, it is exactly as if he were at V, the point upon which the first inequality depends. Next, let the planet's sidereal position be taken with instruments whenever it shall be seen at some point on the line CV opposite V beyond C (let this be on the lines CV, CG, and points opposite). Thus the center of the planetary system will be on the circle VP, the sun at S and G, and the body of the planet opposite it at O and F. (Nevertheless, in the theory of Mars, the planet's eccentric is so small in proportion to the sun's eccentric that Mars's eccentric and the points O and F are nearer to the earth C than is the sun S, which was one of the reasons why Brahe denied the solidity of the orbs.) From many positions of this sort, as many, in fact, as he could obtain, Tycho Brahe used to investigate the hypothesis of the first inequality, by removing the magnitude of the orb VP and treating it as a single point, as though in the meanwhile the center VP of the planetary

47   *system, the point at which the system is attached, had remained at rest. So he set up a comparison of the elapsed times with the angles which VO and PF, drawn from a single point (V and P coinciding), would form. These are in fact identical to the angles OCF and VCP.*

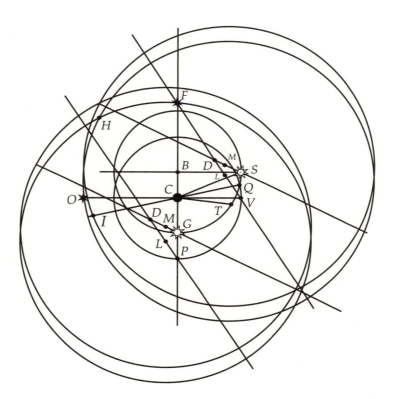

Suppose this treatment done, and let the line of apsides of the eccentric come out to be *VLD* or *PLD*, *D* the equalizing point, *L* the center of the eccentric on this line, and the eccentric *HO* and *FH*. Let this hypothesis correspond to all positions of the planet observed at the moment of the planet's opposition to the sun's mean position.

For the present, I am postponing a more careful examination of whether this sort of hypothesis is in accord with physical principles—a hypothesis, that is, in which the sun, through its motive mind paying attention to the earth, moves around the earth, and of itself (since it lacks an orb) drives itself forward nonuniformly according to its approach to or recession from the earth (unless you would make the earth more important than the sun and ascribe to the earth the sun's motive force); while this same sun (as in Copernicus) sends out a motive force to all the planets, sweeping them around itself with a degree of speed corresponding to their degree of nearness to the sun; the planets meanwhile striving to accomplish their approach to and recession from the sun on a small epicycle, and at the same time to follow the sun, wherever it is supposed to be, in those same tracks not proper to them; and thus any planet, and most of all the sun, attends to many things at once, and the actual trajectories of the planets through the aethereal air (as with Ptolemy) make spirals, such as were depicted in Ch. 1. Whether, I say, these are fitting, we shall consider elsewhere when the occasion arises. Here, let this form of hypothesis be supposed true in its general features. The question is, whether it is also fitting, in particular, that the planets follow the sun's body S, G, or whether they instead follow the point V, P, void of any body, four solar semidiameters (no more) from the center of the sun, which is now above the sun and now below, now before and now behind. And further, whether it is more fitting that the force that drives the planets in an orb around the sun make its nest in the body of the sun S, G, or in some other point such as VP void of body. In brief: if the axle [*axis*] (to derive the sense of the word roughly from a wagon) of the planetary system, to which the orbs of the planets are fastened as with a nail—if this, I say, is near the sun, why not in the sun itself? If this axis or point of attachment travels around the earth both close to the sun and in the very same period, why does it describe its own peculiar path? Why does it not hold to the very same path as the sun?

Term:
Axis or center of the planetary system. Elsewhere, the point or center of attachment.

I therefore wholeheartedly conclude that if, perchance, Tycho Brahe's opinion on the world system be universally true, it must be accepted in such form that the center of the planetary system lies upon SG, exactly on the sun's path, not upon VP; and finally, that it is in the sun itself, and that for liberating the first or eccentric inequality from the second, one should use the planet's oppositions to the apparent position of the sun, not the mean. Brahe himself in his final days embraced this procedure unreservedly. Let us, then, see what is changed on the eccentric. *Once again, as before, when the sun is on the line BC, as at G, and the planet is at F opposite point P, the planet F will be opposite the sun itself, G. Consequently, the sidereal position of the planet will appear to be the same,* 48 *along the line GF, whether the continuation of the line be CP or CG, because both have been made into one line. Therefore, according to either procedure, the planet is divested of the second inequality. But when the sun comes to the side of its eccentric (that is, to the middle elongations), an appreciable difference arises. For let the sun have gone from G to S, let the line of the sun's mean position CV be in Aries, and the planet's line of vision CO be exactly opposite in Libra, so that VCO is one line. Since CS is beyond CV, the sun's apparent position is beyond opposition to the planet. But, because of my alteration, the center of the planetary system is not at V but at S when the planet is seen along CO. Therefore, SO being connected, the earth C will lie outside the line SO, and hence, the planet's apparent position on the line CO still is intermingled with the second inequality. Nor will CO be farther east at a later time, so as to be at opposition to CS. Instead, it will ascend toward CF, because the sun's motion, and with it the motion of the center of the planetary system and of all its parts (and thus those of the planet O and of the center of the eccentric L), is from the line CO upward towards F, and is much faster than the motion of the eccentric or planet at O about L from the point H towards the bottom. Consequently, O is appreciably drawn back westward by a motion extraneous, not proper, to the eccentric; and indeed, it is well known that the planets are retrograde at opposition to the sun. Therefore, at a time preceding the moment designated by CV (let it be CT), when the sun appears on line CQ, the planet will be seen at the point opposite its apparent position, at I. And now, since in the present case it is supposed to have shed the second inequality, QCI will be collinear; that is, the point whence the eccentricity arises will be on the line CQ. But CI, the retrograde planet's line of vision, earlier in time, is beyond the later (and hence more westerly) line of vision CO. Therefore, CQ will also be beyond CV, and Q will be the system's new center, beyond the old, V. Moreover, the line IQ was made to be more distant from the line OV to the east by the angle OCI, while the line of apsides VD or PD (from which the motion begins) remains parallel to itself in the entire circuit. As a result, it appears to be established that the planet goes farther in a shorter period of time around the center Q of the system than it previously did in a greater period of time about the center V of the system.*

49

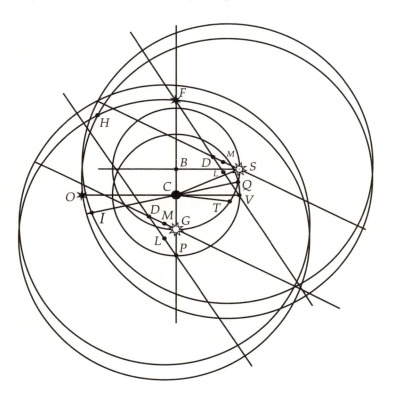

It thereby becomes apparent what is changed in the apparent eccentric motion by this restoration from the sun's mean position to its apparent position. For when the center of the system is at *G* and the point opposite, the line of the apparent eccentric position remains unchanged, at *Q* it is moved forward, and opposite *Q* it is moved back, for at *Q* the time is diminished, and opposite *Q* it is increased. Also, these lines differ greatly from the original ones. Accordingly, when we repeatedly use a new operation, based upon these several apparent positions of the planet (opposite which we are supposing the center of the planetary system to be found; that is, in the sun itself), to investigate the causes and measure of the first inequality, the result of the operation differs greatly from the previous result.

For we have just transposed the point of attachment from the circle VP, upon which Brahe had it move around, to the circle GS, that is, to the body of the sun. This new center always stands on a line parallel to CB, at a distance CB from the original Brahean point; for example, above V and P at S and G. Therefore with the point of equality D fixed (so that CV represents the same moment), in order for it to be possible both for the planet to be at O and for the point of attachment to be at S, a new line of apsides needs to be drawn through D and S or G. Consequently, according to the demonstrations of Ch. 5 (which we have brought forward above, in explaining the Copernican form), with DS or DG drawn and divided at M in the ratio in which DP or DV are divided at L, let there be described about center M, with the same radius as before, a new eccentric. This will not only account for the later observations, upon which it was constructed, but, introduced into the prior hypothesis, it will also save the observations previously cited, with a precision of within five minutes.

However, computations which are carried out upon the previous eccentric and this new one at positions other than night rising will in some places (particularly near the sun's perigee) be able to differ by more than one degree, if we use numbers fitting and proper for Mars, which have been furnished by Brahe.

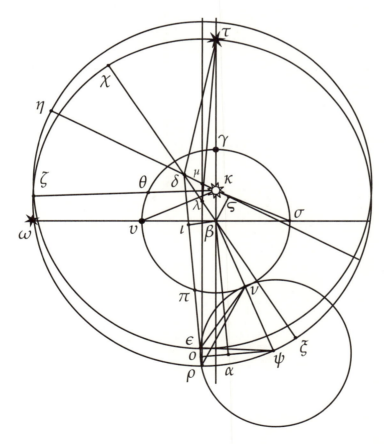

50

There is no need to repeat the demonstration. The drawing is easiest in the Copernican diagram, if from the earth ν you set up a line parallel to βκ, measure off upon it, at an interval βκ above ν, the center of the sun's eccentric, and upon this center set up the Brahean eccentric of the sun through κ, and delete the Copernican center of the earth's eccentric.

The differences of the hypotheses, and their equivalence in the first inequalities but discrepancy in the second, have now been expounded. Let us then conclude the first part, which, as I see it, is the most difficult of the entire work, because of the almost inescapable labyrinths of opinion and the perpetual ambiguities of words and extremely tiresome circumlocutions. What necessity it was, however, that forced me to prefix this body of instruction, will now directly appear in Ch. 7. Anyone who is less energetic can defer the whole part until he understands the easier parts.

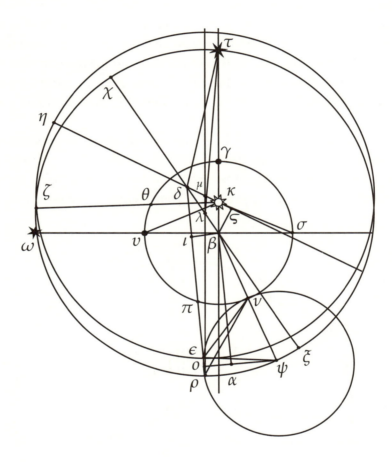

Part Two

on the
First Inequality of the Star Mars
In Imitation of the Ancients

CHAPTER 7

The circumstances under which I happened upon the theory of Mars.

It is true that a divine voice, which enjoins humans to study astronomy, is expressed in the world itself, not in words or syllables, but in things themselves and in the conformity of the human intellect and senses with the sequence of celestial bodies and of their dispositions. Nevertheless, a kind of fate also invisibly drives different individuals to take up different arts, and makes them certain that, just as they are a part of the work of creation, they likewise also partake to some extent in divine providence.

When, in my early years, I was able to taste the sweetness of philosophy, I embraced the whole of it with an overwhelming desire, and with practically no special concern about astronomy. I certainly had enough intelligence, nor did I have any difficulty understanding the geometrical and astronomical topics included in the normal curriculum, aided as I was by figures, numbers, and proportions. These were, however, required courses, nothing that would bespeak an exceptional inclination towards astronomy. And since I was supported at the expense of the Duke of Württemberg, and saw my comrades, whom the Prince, upon request, kept trying to send to foreign countries, stalling in various ways out of love for their country, I, being hardier, quite maturely agreed with myself that whithersoever I was destined I would promptly go.

The first to offer itself was an astronomical position; however, to tell the truth, I was driven forth to take it by the authority of my teachers. I was not frightened by the distance of the place, for (as I have just said) I had condemned this fear in others, but by the low opinion and contempt in which this kind of function is held, and the sparsity of erudition in this part of philosophy. I therefore entered upon this better furnished with wits than with knowledge, protesting loudly that I by no means gave over my right to follow another kind of life which seemed more splendid.[1] What came of those first two years of study may be seen in my *Mysterium cosmographicum*. The additional goads which my teacher Maestlin gave me towards embracing the rest of astronomy, you will read of in the same book, and in that man's prefatory letter to Rheticus's *Narratio*.[2] I had the very highest opinion of what I discovered there, and all the more so when I saw that Maestlin, too, held it in similar esteem. It was not so much his untimely promise to the readers of what he called my "universal uranic opus" that spurred me on, as it was my

---

[1]Kepler had hoped eventually to obtain an ecclesiastical post.

[2]Maestlin, who saw to the printing of the *Mysterium cosmographicum*, took the liberty of appending Georg Joachim Rheticus's *Narratio prima* (1540) to aid the reader in understanding the Copernican hypothesis.

own ardor to seek, through a reworking of astronomy, whether my discovery would support observations of complete accuracy. For it had then been demonstrated in that book that the discovery was consistent with the observations within the limits of accuracy of ordinary astronomy.

53   So from that time I began to think seriously of comparing observations. In 1597, I wrote Tycho Brahe asking his opinion of my little book,[3] and when he, in reply, mentioned (among other things) his own observations, he ignited in me an overwhelming desire to see them. Moreover, Tycho, who was indeed himself a large part of my destiny, did not cease from then on to invite me to come to him. And since I was frightened off by the distance of the place, I again ascribe it to divine arrangement that he came to Bohemia. I thereupon came to visit him at the beginning of 1600 in hopes of learning the correct eccentricities of the planets. But when I found out during the first week that, along with Ptolemy and Copernicus, he made use of the sun's mean motion, while the apparent motion would be more in accord with my little book (as is clear from the book itself), I begged the master to allow me to make use of the observations in my own manner. At that time, the work which his aide Christian Severinus[4] had in hand was the theory of Mars. The occasion had placed this in his hands, in that they were busy with the observation of the acronychal position or opposition of Mars to the sun in 9° Leo. Had Christian been treating a different planet, I would have entered upon it as well.

I therefore once again think it to have happened by divine arrangement, that I arrived at the same time in which he was intent upon Mars, from whose motions alone we will have to come to a knowledge of the hidden secrets of astronomy, or remain forever ignorant of them.[5]

A table of mean oppositions was worked out, starting with the year 1580. A hypothesis was invented which was proclaimed to represent all these oppositions within a distance of two minutes in longitude. (The numbers I used in Ch. 5 were taken from among these, or differed from them only slightly.) The apogee at the beginning of the year 1585 was placed at 23° 45' Leo. The maximum eccentricity, made up of the semidiameters of the two small circles, was 20, 160, in units of which the semidiameter of the greater epicycle was 16, 380. Therefore, in the Ptolemaic form of the first inequality, the eccentricity of the equalizing point was 20, 160 or a little less.

From this hypothesis, a table of eccentric equations for individual degrees was constructed, as well as a table of the corrected mean motion, made by adding $1\frac{3}{4}'$ to the mean motion of the Prutenic Tables.[6] These mean motions, apogees, and likewise the nodes, were extended over a period of forty years, exactly as was done for the solar and lunar motions in Book I of the *Progymnasmata*. It was only in the latitude at acronychal positions and also the parallax

---

[3]Letter to Tycho Brahe, 13 December 1597, letter number 82 in *KGW* 13 p. 154.

[4]Better known under the name "Longomontanus", the Latinization of his birthplace, Longberg.

[5]Mars and Mercury are the only planets whose orbits differ enough from circles for that difference to have an effect observable by Tycho's instruments. Mercury is too near the sun to afford reliable observations of its entire orbit. Therefore, only the observations of Mars could have led Kepler to his "new astronomy."

[6]Erasmus Reinhold, *Prutenicae Tabulae Coelestium Motuum* (Tübingen, 1551).

of the annual orb that Christian got stuck. There was, actually, a hypothesis and table for the latitudes, but they failed to elicit the observed latitude. This result was a problem for him, as he was about to brood over the lunar motions.

Now since I suspected what proved to be true—that the hypothesis was inadequate—I girded myself for the work following opinions that had been preconceived and had been expressed in my *Mysterium cosmographicum*. At the beginning there was great controversy between us as to whether it were possible to set up another sort of hypothesis which would express to a hair's breadth so many positions of the planet, and whether it were possible for the 54 former hypothesis to be false despite its having accomplished this so far over the entire circuit of the zodiac.

I consequently showed, using the arguments presented already in Part I, that an eccentric can be false, yet answer for the appearances within five minutes or better, provided that the equant point be true. As for the parallax of the annual orb, and the latitudes, that prize was still unclaimed, and had not yet been won by the hypothesis of those others. What remains, then, is to find 55 out whether they, with their computation, might not somewhere differ from the observations by five minutes.

I therefore began to investigate the certitude of their operation. What success came of that labor, it would be boring and pointless to recount. I shall describe only so much of that labor of four years as will pertain to our methodical enquiry.

CHAPTER

Tycho Brahe's table of observed and computed
of the sun's mean motion,

Now the table, mentioned above,
An accurate rendering of the motion of the planet Mars on
of positions, carried out sedulously over a period

Uniform Time of Mars					Obs. Long. with respect to Mars's Circle				True obs. Latitude				Long. with respect to the ecliptic		
Year	Mo.	D	H	M	Deg.	m.	s.		Deg.	m.	s.		Deg.	m.	s.
1580	Nov.	17	9	40	6	50	10	Gemini	1	40	0	N	6	46	10
1582	Dec.	28	12	16	16	51	30	Cancer	4	6	0	N	16	46	10
1585	Jan.	31	19	35	21	9	50	Leo	4	32	10	N	21	10	26
1587	Mar.	7	17	22	25	5	10	Virgo	3	38	12	N	25	10	20
1589	Apr.	15	13	34	3	54	35	Scorpio	1	6	45	N	3	58	10
1591	Jun.	8	16	25	26	40 30 P / 42 0 N		Sagit.	3	59	0	S	26	32	0
1593	Aug.	24	2	13	12	35	0	Pisces	6	3	0	S	12	43	45
1595	Oct.	29	21	22	17	56	5	Taurus	0	5	15	N	17	56	15
1597	Dec.	13	13	35	2	34	0	Cancer	3	33	0	N	2	28	0
1600	Jan.	19	9	40	8	18	45	Leo	4	30	50	N	8	18	0

P denotes the Paduan observation carried out by Magini with Brahe's student Gellius Sascerides. N denotes our observation (that is, Brahe's), carried out at Uraniborg.

The amended mean longitudinal motion of Mars at the beginning of 1585 was found to exceed the numbers provided by the Prutenic calculation[1] by at least a minute and a half, or at most $1\frac{3}{4}'$, which by all indications appears more nearly correct.[2] However, the position of its apogee then fell short of the Prutenic calculation at the same time, by 5° 2', both being compared with the first star of Aries in the Copernican manner. Hence, owing to our removal of the vernal equinox westward from that star, which was then 28° $2\frac{1}{2}'$,[3] it is

---

[1] Erasmus Reinhold, *Prutenicae Tabulae Coelestium Motuum* (Tübingen, 1551).

[2] In Kepler's edition, this paragraph, and the following one, were set in lines that ran continuously from the left page to the right, as the chapter title and Brahe's table had done. It is remarkable that Kepler was able to arrange with the printer to have the beginning of this chapter fall on the middle two pages of a signature, so that the facing pages were printed on a single sheet. Although the translation retains Kepler's layout for chapter title and table, it was thought better to have subsequent text flow conventionally down each page, to avoid confusion.

[3] In Tycho Brahe's *Stellarum inerrantium restitutio* (*TBOO* III p. 344), the first star in Aries (Gamma Arietis) had a longitude of 27° 37' Aries at the end of 1600 (that is, at noon on 1601 January 1). Tycho's correction for the interval of 15 years 11 months would be 13' 32", putting the star at 27° $23\frac{1}{2}'$ Aries at the end of January 1585. Hence, Kepler's longitude for the star appears incorrect. Nor is there any other star in Tycho's catalog which would have the stated longitude at that time. This error appears to affect the figures in the column headed "Our precess. of the equin.", as well as the positions of the apogee given in this paragraph. Note that the third entry in the "precession" column is the stated longitude in question, and the other figures in this column reflect only the change required by the Tychonic rate of precession.

8

observations of Mars's oppositions to the line and an examination thereof.

was the following:
its eccentric from trustworthy acronychal observations in a variety
of twenty years (1580 to 1600) using our instruments.

Difference		Simple Long. of Mars[4]				Apogee of Mars[4]				Our precess. of the equin.			Computed		
m.	s.	S.	Deg.	m.	s.	S.	Deg.	m.	s.	Deg.	m.	s.	Deg.	m.	s.
4	10 +	0	27	29	46	3	25	21	40	27	58	50	6	50	40
5	20 +	2	11	34	56	3	25	22	17	28	0	38	16	51	26
0	36 −	3	22	37	46	3	25	22	55	28	2	25	21	9	41
5	10 −	5	3	27	46	3	25	23	32	28	4	10	25	4	50
3	35 −	6	16	53	7	3	25	24	10	28	5	55	3	54	33
10	20 +	8	7	47	30	3	25	24	48	28	7	47	26	40	23
8	45 −	10	10	53	50	3	25	25	26	28	9	40	12	34	36
0	12 +	0	8	26	47	3	25	27	35	28	11	27	17	57	14
6	0 +	1	24	55	47	3	25	29	5	28	13	20	2	32	20
0	45 −	3	6	46	16	3	25	30	6	28	15	5	8	19	57

concluded that Mars's apogee was at 23° 25′ Leo. For the first date it was set at 23° 20′ Leo, and for the last, at 23° 45′ Leo.

Also, the maximum eccentricity, composed of the semidiameters of the two small circles, was found to be 20, 160 parts, of which the semidiameter of the greater epicycle, or distance between centers as given by Copernicus, is 16, 380. In both of these, however, he differs both from himself and from Ptolemy. Care was taken, where appropriate, concerning refraction, using the solar parallax.

THIS, THEN, IS BRAHE'S TABLE

56

We shall begin the examination of the sun's mean motion with the listed instants of equal time, as many as the table presents. For indeed, it is the sun's mean position in opposition to which the table says the star Mars was found, referred to the ecliptic.

---

[4]These are apparently measured from "the first star in Aries, in the Copernican manner", as Kepler puts it below.

Year	Day	Month	H	M	Sun's mean position				Star's obs. pos. on the ecliptic		Difference		
					S	°	′	″	′	″	′	″	
1580	17	November	9	40	8	6	48	32	46	10	2	22	−
1582	28	December	12	16	9	16	50	58	46	10	4	48	−
1585	31	January	19	35	10	21	10	13	10	26	0	13	+
1587	7	March	17	22	11	25	5	57	10	20	4	23	+
1589	15	April	13	34	1	3	53	32	58	10	4	38	+
1591	8	June	16	25	2	26	45	24	32	0	13	24	−
1593	24	August	2	13	5	12	34	36	43	45	9	9	+
1595	29	October	21	22	7	17	56	17	56	15	0	2	−
1597	13	December	13	35	9	2	28	51	28	0	0	51	−
1600	19	January	9	40	10	8	18	43	18	0	0	43	−

You see here that the sun's mean position differs from opposition to Mars's apparent position on the ecliptic by $13\frac{1}{2}'$ in some cases, nearly thrice the error which could arise through a change of hypothesis. Therefore, the exactness of their hypothesis did not prevent my seeking another.

But they permitted this discrepancy advisedly. This is clear from the following: since the nodes are about 17° Taurus and Scorpio, and the limits about 17° Leo and Aquarius, as will be said below, the additions and subtractions are made chiefly at 17° Cancer, 25° Virgo, 4° Scorpio, 27° Sagittarius, and 13° Pisces, intermediate points, and none at 21° Leo and 18° Scorpio, the nodes and the limit. They therefore had reason to believe that a planet is not divested of its second inequality unless the sun's departure from the node is as great as the planet's on its own orbit. Their intentions, moreover, were not consistent. For, in their way of thinking, the difference ought to be greatest at 3° Cancer, because Cancer is closest to the 45th degree,[5] where this difference is generally greatest. But at 17° Cancer they subtracted 5 minutes, while at 3° Cancer, one minute only. Because of this, the following table, comparing the positions (referred to the orbit of Mars) with the mean positions of the sun at these moments, is presented.

Minutiae of the sun's mean position		Minutiae of Mars's apparent position on the orbit		Difference		
′	″	′	″	′	″	
48	32	50	10	1	38	+
50	58	51	30	0	32	+
10	13	9	50	0	23	−
5	57	5	10	0	47	−
53	32	54	35	1	3	+
45	24	42	0	3	24	−
34	36	35	0	0	24	+
56	17	56	5	0	12	−
28	51	34	0	5	9	+
18	43	18	45	0	2	+

---

[5]From aphelion.

Clearly, they did not compensate the whole difference in this way.

We shall discuss this plan of theirs once more a little later.

Now, we shall also examine Mars's mean motion, for the sake of which, see the following table.

57

Minutes and seconds of mean motion.

By my compu-tation from Brahe's tables		Value pre-sented		Difference		
'	"	'	"	'	"	
29	9	29	46	0	37	+
35	26[6]	34	56	0	30	−[7]
37	4	37	46	0	42	+
27	16	27	46	0	30	+
52	33	53	7	0	34	+
46	45	47	30	0	45	+
53	18	53	50	0	32	+
26	5	26	47	0	42	+
54	48	55	47	0	59	+
45	39	46	16	0	37	+

I am therefore missing something small in the mean longitude. For, that nearly everywhere there is half a minute too much, may be so because of my having computed the mean motions from the most recent table, in which something might possibly have been deliberately altered.

There follows a table of Mars's eccentric positions.

By my compu-tation from the Brahean tables		Value pre-sented		Difference		
'	"	'	"	'	"	
49	37	50	40	1	3	+
52	59	51	26	1	33	−
9	47	9	41	0	6	−
4	49	4	50	0	1	+
54	46	54	33	0	13	−
34	45	40	23	5	38	+
33	59	34	36	0	37	+
57	37	57	14	0	23	−
31	48	32	20	0	32	+
45	39	46	16	0	37	+

All positions are tolerably accurate except 27° Sagittarius. Here, for various reasons, a small but appreciable quantity has been accumulated. First, the

---

[6]This should be 34′ 26″, according to the translator's recomputation.

[7]Should be positive.

sun's position is 26° 45′ 24″ Gemini. Now the computed position on Mars's orbit is 26° 24′ 43″ Sagittarius. In the opinion of the table, 10′ 20″ are to be subtracted from the latter to refer it to the ecliptic. Therefore, the computed position on the ecliptic would be 26° 24′ 13″ Sagittarius, a difference from opposition to the sun of 21′ 11″.

CHAPTER 9

On referring the ecliptic position to the circle of Mars.

But it is now time for us to discuss in detail this adjusting to the ecliptic or to the planet's orbit, which serves as a foundation.

First, the table provides us the following information from the observations: the northern latitude takes its rise from 18° Taurus, at which it was five minutes, reaches its observed maximum at 21° Leo, decreased thereafter and became only $1\frac{2}{3}°$ at 3° Scorpio, but right away at 27° Sagittarius it was south and a rather large 4°, and still greater at 13° Pisces. From this one concludes roughly that the ascending node is a little before 18° Taurus, and the descending node far beyond 3° Scorpio. The nodes will therefore be around 17° Taurus and 17° Scorpio, and the limits around 17° Leo and Aquarius. Since the plane of Mars's eccentric is inclined to the plane of the ecliptic, nearly the same thing that happens with the right ascensions of parts of the ecliptic will happen here: the observed arcs of one circle do not correspond to the same observed arcs on the other, except the ones beginning at a node and ending at a limit. I use the term, "observed arcs", because here one must mentally separate out the planet's eccentricity, and proceed as though Mars's path were in the orb of the fixed stars, exactly as is the ecliptic, and as though it really intersected the latter. And indeed, when asked what is the ecliptic position of a planet, astronomers define it thus: it is that point on the ecliptic at which the circle of latitude (at right angles to the ecliptic) passing through the sidereal position of the planet's body intersects the ecliptic.

58

Term:
What is the ecliptic position of a planet? It is contrasted with the position on the orbit, or the position considered with respect to the orbit.

It is therefore clear from the demonstrations in Theodosius's *On the Sphere* that unless this circle passes through the poles of both circles (the ecliptic and the planet's path), its points of intersection will always cut off unequal arcs as measured from the point at which the two circles intersect. And since the circle of latitude is at right angles to the ecliptic, it will always be oblique to the planetary orbit if it does not pass through the poles of the orbit. Consequently, the arc between the planet's position on its orbit and the nearest node is always greater than that between its ecliptic position and the same node.

Now when we observe the planets, we do not feel convinced that we have defined their exact positions until we have referred them to the ecliptic. We do this by indicating the point on the ecliptic at which the circle of latitude passing through the planet is found. The ecliptic position is used, therefore, to aid our memory and comprehension. But when, on the other hand, we compute the planet in its own hypothesis,[1] we are not concerned with the ecliptic to which it is inclined, but with the exact path of the planet. Therefore, to be able to

What does it mean to refer a planet to the ecliptic?

---

[1] In Chapters 13 and 14 below, Kepler will demonstrate that the orbit of Mars is in a single plane that does not librate. Since this has not yet been established, he cannot say "in its own plane," especially since all previous planetary models (which is what he means by "hypotheses," involve motions in a variety of planes.

compare the observed position with the computed position, we must either extend the arc between the ecliptic position and the nearer node, or abridge the arc between the body of the planet and the same node, so that from the former operation the position on the orbit might be given, and from the latter, the ecliptic position. This is actually accomplished by adding or subtracting, according as the node precedes or follows the planet's position.

Such care concerning the planets Ptolemy considered unnecessary. Copernicus did not forego it in treating the moon, and Tycho Brahe diligently embraced it for the sake of precision.

To continue: in the referring process which we have been considering, there are two things I would like to know, both of which I can seek using the same procedure and diagram.

59

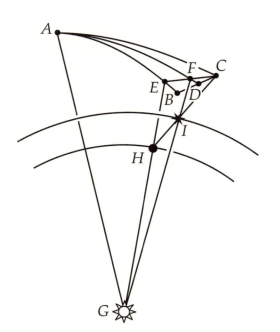

Let A be the sidereal position of the node, AB the arc of the ecliptic, and upon it let AC be set equal to AB, and let the planet be observed beneath C. Further, from C let an arc be drawn perpendicular to the ecliptic, and let it be CE.

Now in the first place the ancients thought that since E is the position on the ecliptic and C the position of the orbit of the planet I, the planet is at the point opposite the sun when the sun is at E, the planet being observed at C. However, as was said above, those who constructed the tables thought that the planet is not exactly at opposition to the sun unless arc AB,

the elongation of the place opposite the sun from the same node, is made equal to AC (the observed distance of the planet from the node).

Now the truth of the matter is quite different. The planet is, indeed, seen in opposition to the sun at that time, but it really is not, and the advantage we seek from the planet's opposition to the sun is more harmed by making *AC* and *AB* equal, than they themselves were hoping it would be corrected. For why are the planets observed at opposition to the sun? In order, of course, that they then lack the second inequality of longitude. And when the point opposite the sun is at *B* and the planet is at *C*, both being between the nodes and the limits, the planet is more wrapped up in the second inequality of longitude than if the point opposite the sun were at *E*, the planet remaining at *C*. *For let G be the sun, the center of the planetary system, at which all the orbs intersect the ecliptic, whether in the Copernican or the Brahean form. Let G be joined to A and E, points on the ecliptic, and let the earth be on the line EG, at H. Let HC be joined, and from H let the sun G be observed opposite E, while from the same point H let the planet be seen at C, its sidereal position along the line HC. Therefore, in this sighting, the planet is really on the line HC. It is, however, far below the fixed stars. Let it be at the point I on the line HC, and let a straight line be drawn from G through I, which will intersect the arc CE. For the whole plane CEHG is beneath the arc EC. Let its point of intersection be F, and let a third arc AF be drawn from A through F to BC, cutting BC at D. It is obvious that the plane of the planet's eccentric viewed from H towards C is set beneath AF, not AC, and that when the point opposite the sun is at E, the planet will really be beneath F, while when the point opposite the sun is located at B the planet is really going to be beneath D, although both do appear beneath C. But AD is shorter than the legs of the isosceles triangle BAC. Therefore, the point B opposite the sun is farther from A than is D, the position beneath which the planet is at the moment they have chosen.* Therefore, the sun really stands beyond the point opposite the planet's true position. This is contrary to what they proposed to do.

But it is likewise false that if the orbit of the planet were beneath AC, AB should on that account be taken equal to AC. *For the orbit's really being beneath AD is likewise not sufficient reason for taking AD equal to AB. For since the planet is observed at the point opposite the sun in order that it might shed the second inequality of longitude, while the longitude is to be reckoned on the planet's true orbit, or on AD which stands above it, surely, unless the point opposite the sun falls upon the arc drawn through the planet at right angles to the orbit (that is, unless ADB is a right angle), the point B opposite the sun will not coincide in longitude with D. But if ADB is a right angle, AB is longer than AD. They are therefore not equal.* Clearly, therefore, the equality between arcs AC and AB that the table aspires to is undermined.

However, for practical purposes, these differences are smaller than can be perceived. I therefore do not hesitate to allow the place opposite the sun to be at E, with AEF right and AFE consequently acute, even though it has just been demonstrated that AFE should be right instead. But it was also necessary to proceed against a new claim of accuracy, by means of accurate arguments. What follows here is the harm arising from this accuracy.

In the second place, then, I wish to establish this: that in the table of adjustment they followed a procedure that is unsound. *For, given Mars's ecliptic position E and apparent latitude EC, they computed the length of AC, and stated that the planet on its orbit was then distant from the node by their quantity AC. Now the orbit of the planet (whose first inequality we are investigating) is not beneath AC, but beneath AD, as was just shown. Therefore, the arc AC has nothing to do with the first inequality, but adulterates the planet's true elongations from A. And furthermore, the apparent latitude is EHC, while the true latitude of the point F, the inclination of the line GF to the ecliptic, is EGF. Thus, although the second inequality of longitude is swallowed up at opposition to the sun, the second inequality of latitude is nonetheless near its maximum there, and its measure is the angle HIG. Therefore, just as the whole latitude EC causes AC to be longer than AE by the arc EB, similarly, the part FC or HIG of this apparent latitude, which is a result of the second inequality, makes AC longer than AF.* So it is longer than it should be. And this error cannot be ignored, as it can be as much as 9 minutes.

This error could also have been perceived in the inconstancy of the angle BAC, which they attributed to the inclination of the planes of the ecliptic and of Mars's orbit. This is clear from the result obtained if you suppose the arc AC to be increased by the amount of the addition expressed in the table, and use this and AC to compute the angle EAC. For the angles come out as in

60

°	′
4	58
4	58
4	0
4	33
5	29
6	20
6	26
4	30
4	22
3	10

the accompanying table, from which it is clear that in the northern semicircle they suppose an angle of maximum northern latitude of 4° 33′, and in the southern, of 6° 26′ south. According to this, at the subtending line connecting the nodes, which passes through the sun or earth, the plane of the eccentric would somehow be bent, since the upper part is less inclined than the lower. Or rather, the whole path or plane of the planet's eccentric would be full of twists and turns, just as is the path described beneath the fixed stars by the observed latitudes of Mars, which is no circle.

However, all this is in conflict with the simplicity of the celestial motions, as many examples from experience will attest.

Therefore, the true procedure for referring the inclinations to the orbit is this: from the planet's position E on the ecliptic, known from the observations, to find the angle of inclination EGF for that position, using the method that will follow below. Then, since the angle E is right, from AE and EF (the measure of the angle EGF) AF is found by trigonometry, or instead of EF the constant angle EAF may be used. And since, from arguments which I shall present below, it is clear that for the star Mars the angle EAF is not greater than about 1° 50′, the adjustment about 45° from the node

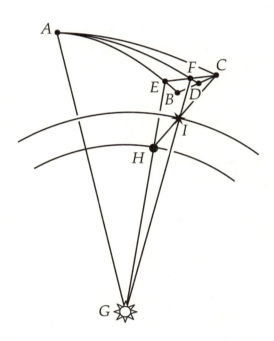

61

(where it is maximum) accordingly does not exceed one minute, for which the table nonetheless has one add 8 or 10 minutes in certain places. So for this reason, too, the hypothesis can be in error by as much as 7 and 9 minutes, since the observations upon which it was founded suffered no little loss through this adjustment. I was consequently subject to much less restraint than before in seeking out a new hypothesis.

CHAPTER 10

Consideration of the observations themselves, with which Tycho Brahe hunted for the moments of opposition to the mean sun.

In an enquiry of such precision, I could not have foregone a deeper inspection of the foundations themselves. And Brahe had given me the opportunity to make use of his observations. This is what I found.

I. On 1580 November 12 at $10^h 50^m$,[1] they set Mars down at 8° 36′ 50″ Gemini,[2] without mentioning the horizontal variations, by which term I wish the diurnal parallaxes and the refractions to be understood in what follows. Now this observation is distant and isolated. It was reduced to the moment of opposition using the diurnal motion from the *Prutenic Tables*.[3]

Term:
What are horizontal variations?

For in Maestlin,[4] on the twelfth at noon, Mars is put at 8° 20′ Gemini, and on the seventeenth, again at noon, it is at 6° 25′ Gemini. Therefore, the motion over five whole days would be 1° 55′. In Stadius,[5] it is 1° 52′. Therefore, on the seventeenth at the same hour of $10^h 50^m$, Mars ought to have been seen at either 6° 41′ 50″ Gemini, or 6° 44′ 50″. At $9^h 40^m$ (which Tycho gives as the moment of opposition), it is 1′ 4″ farther forward, at either 6° 42′ 54″ or 6° 45′ 54″. They put it at 6° 46′ 10″ Gemini.

You see that this opposition (as regards exactness) is a little more uncertain because it makes use of a diurnal motion which is not observed but imported from elsewhere, and about which the different authors differ from one another by three minutes over these five days.

II. On 1582 December 28 at $11^h 30^m$, they set Mars down at 16° 47′ Cancer by observation.[6] The moment of opposition assigned by Tycho comes 46 minutes later, during which the planet retrogressed less than one minute. Tycho therefore puts it at 16° 46′ 16″ Cancer.[7] On an inserted sheet here, an attempt

---

[1]Dates are given "old style," that is, in the Julian calendar. Dates in the Gregorian calendar (which is the one we presently use) were ten days later than the Julian dates. Times are reckoned from noon on the stated day.

[2]This observation is recorded in *TBOO* 10 p. 84, where the position is given as: "longitude 8.36$\frac{1}{4}$ Gemini, latitude 1 16$\frac{1}{6}$ N." Kepler's stated value is more than half a degree too far east.

[3]Erasmus Reinhold, *Prutenicae Tabulae Coelestium Motuum* (Tübingen, 1551). The Canons for Mars begin on fol. 66; diurnal motions would have to be computed from these.

[4]Michael Maestlin, *Ephemerides ab anno 1577 ad annum 1590*, Tubingen, 1580.

[5]Joachim Stadius, *Ephemerides ab anno 1554 ad annum 1606*, Cologne, 1556-1581. Both of these references are provided by Caspar in *KGW* 3.

[6]This position is found in *TBOO* 10 p. 198; however, it was obtained by interpolation between two observations, one on December 26 (recorded on p. 176) and the other on December 30 (p. 178), in a first attempt to find the moment of opposition. This moment was then further adjusted: Brahe's final data for the opposition (p. 199) are: 11 pm, longitude 16° 46′ Cancer.

[7]The time and position differ from those given by Brahe, who stated (p. 199) that the corrected time was 11 pm, longitude 16° 48′ Cancer. Nevertheless, the "inserted sheet" that Kepler men-

was made to correct for a refraction of two minutes. This was, I think, first trial of the theory of refraction then being developed. Nevertheless, he followed the observed value unchanged, thus declining to consider the planet as something which could alter its position. Nor was there any need for correction, since it was in Cancer, beyond the reach of refraction, and was in mid-sky where, in Cancer, there is no longitudinal parallax.

III. On 1585 January 31 at $12^h \, 0^m$, Mars was placed at 21° 18′ 11″ Leo.[8] The diurnal motion, by comparison of observations, was 24′ 15″. The moment of opposition followed at $19^h \, 35^m$, 7 hours and 35 minutes later. To this period belongs 7′ 41″ of diurnal motion westward. Therefore, at the designated moment, it would have been at 21° 10′ 30″ Leo, which is what was accepted. There is no mention of parallax. Nothing had to be done about refraction, because Mars was high and at mid-sky. I therefore find the bit of advice in the table about refraction (properly) ignored.

IV. On 1587 March 7 at $19^h \, 10^m$ they deduced the position of Mars from the observations, which was 25° 10′ 20″ Virgo. This they kept in the table, but changed the time to $17^h \, 22^m$. The difference of $1^h \, 48^m$ multiplied by a diurnal motion of 24′ gives the same number of minutes and seconds (that is, 1′ 48″), no more. It therefore should have been 25° 8′ 32″ Virgo, which also approaches nearer the point opposite the sun. The difference is of practically no importance.[9]

V. On 1589 April 15 at $12^h \, 5^m$ they established the position of Mars very carefully at 3° 58′ 21″ Scorpio, and corrected for longitudinal parallax so as to make it 3° 57′ 11″.[10] There remain $1^h \, 30^m$ until the designated moment of opposition, which, for a diurnal motion of 22′, bring the planet back 1′ 22″, so as to be at 3° 55′ 49″. They took the value of 3° 58′ 10″. The former is closer to the sun's mean position.

VI. On 1591 June 6 at $12^h \, 20^m$, Mars is placed at 27° 15′ Sagittarius.[11] There remained 2 days 4 hours and 5 minutes until the designated moment. In four days it was found to be moved forward 1° 12′ 47″. Therefore, to $2^d \, 4^h \, 5^m$ correspond 39′ 29″. Consequently, at that moment Mars was at 26° 35′ 31″ Sagittarius. There is no need to consider any horizontal variations in longitude, since Mars is at mid-sky and at the beginning of Capricorn. The table has 26° 32′ Sagittarius.

VII. On 1593 24 August at $10^h \, 30^m$ they report Mars as being at 12° 38′ Pisces[12] with an observed diurnal motion of 16′ 45″, and this near the nona-

---

tions in the next sentence is transcribed on this page of *TBOO*. The source of Kepler's time and position is unknown.

[8] *TBOO* 10 p. 388.

[9] The observations for this date are listed in *TBOO* 11 pp. 184–5. The moment of opposition was said to be 5:20 am. However, no longitude is given for any of the entries on this day.

[10] See *TBOO* 11 p. 333. However, no correction for parallax is entered there, nor does it appear in the table of observations on the following page.

[11] *TBOO* 12 p. 142.

[12] The observations of Mars for that day are listed in *TBOO* 12 pp. 288–289; however, these are raw observations and no longitude is given. Brahe collected all the positions of Mars observed in July and August in a table on p. 291, with the time and position given by Kepler. The diurnal

gesimal[13] where there is no longitudinal parallax. The moment designated for the opposition preceded this by 8^h 17^m (for it was at 2^h 13^m),[14] to which corresponds a motion of 5′ 48″ eastward. Therefore, the planet falls at 12° 43′ 48″ Pisces. And the table has 12° 43′ 45″.

VIII. On 1595 October 30 at 8^h 20^m, they found Mars at 17° 48′ Taurus, with a diurnal motion of 22′ 54″.[15] The designated moment preceded by 11^h 48^m, for which is required a motion of Mars of 11′ 7″ eastward, so that it would be at 17° 59′ 7″ Taurus. But it was projected eastward on account of parallax. Therefore, possibly using another observation on the meridian, they put down 17° 56′ 15″ Taurus in the table.

IX. On 1597 December 10 at 8^h 30^m, they first placed Mars at 3° 30′ Cancer, and again at 4° 1′ Cancer, the mean being 3° $45\frac{1}{2}$′ Cancer.[16] The moment of opposition came 3 days 5^h 5^m later, to which, from Magini, corresponds 1° 15′ westward. Therefore, Mars would have been at 2° $30\frac{1}{2}$′ Cancer. In the table, it was put at 2° 28′. The reason for the rough measurement, carried out with a measuring staff, is clear from the date. Tycho had left the island, leaving all instruments but the staff behind. Nevertheless, he did not wish to ignore this opposition completely. But I wish he had still remained on the island, for this opposition was a marvelous opportunity, not often recurring within a man's lifetime, for examining Mars's parallax.   63

X. On 1600 January 13/23[17] at 11^h 50^m, the right ascension of Mars was:

	°	′	″
using the bright foot of Gemini	134	23	39
using Cor Leonis	134	27	37
using Pollux	134	23	18
at 12^h 17^m, using the third in the wing of Virgo	134	29	48[18]
The mean, treating the observations impartially:	134	24	33[19]

Hence, Mars is at 10° 38′ 46″ Leo, at an adjusted time of 11^h 40^m reduced to

---

motion is listed just above this table.

[13]The 90th degree from the horizon on the ecliptic.

[14]Immediately above the table on *TBOO* 12 p. 291, Brahe gives the moment of opposition as 5:04 pm, Mars's longitude being 12° 41′ 40″.

[15]*TBOO* 12 pp. 453–454. Brahe initally gives a longitude of 17° 48′ 30″ Taurus, then adjusts for refraction, bringing it back to 17° 47′ 15″ Taurus. On the following page, he gives a longitude of 17° 48′ 40″ Taurus, at "8-ish."

[16]*TBOO* 13 pp. 112–113 has observations at 7:38, 8:12, 8:30, 8:47, 8:51, 9:11, and 9:26. The longitude of 3° 30′ Cancer was recorded for the last of these times. The other two longitudes do not appear among these observations. In a concluding note, Brahe writes, "The above observations, made with a measuring staff here at Wandesburg, are not of great weight, and do not agree with each other."

[17]The first date is in the Julian calendar, used by Lutherans, and the second is in the Gregorian calendar, used by Catholics.

[18]Gamma Geminorum, Regulus, Pollux, and Beta Virginis, respectively.

[19]This should be 134° 26′ 6″. Kepler took the figures, including the average, straight from Tycho: they appear in *TBOO* 13 p. 221. Apparently, Tycho averaged in pairs, and then instead of averaging the averages, he averaged the first pair's average with the third figure alone (surely unintentionally).

the meridian of Uraniborg. But on January 24/February 3 at the same time it was at 6° 18′ Leo. This gives a diurnal motion of 23′ 44″, and a position on January 19/29 at $9^h 40^m$ of 8° 18′ 45″ Leo, just as they put it.

I have presented these discrepant values for the right ascension in order to show that even in the observations themselves there is an uncertainty of several minutes unless extreme care be exercised and all possible aids used. At that time the instruments (except the largest) had arrived in Bohemia, but they were still not well enough positioned, and were affected by the journey besides. However, even in observations at the island it too often happened that right ascensions measured from two different stars differed by three minutes. When I asked Christian [Longomontanus], on this subject, whether I should consider this an effect of the limitations of observation or vision, he replied, "This is not unusual."

Finally, the reader should be advised that Tycho, in his table, claimed to have made use of the solar parallax in correcting the positions of Mars. But it will now shortly be made clear that the parallax of Mars is a slippery and imperceptible business. However, this does not much affect the certainty of the positions in this table, as Mars can almost always be observed in mid-sky where it has no longitudinal parallax.

CHAPTER 11

On the diurnal parallaxes of the star Mars.

The starting point of my new elaboration and renewal of the motions was where I have just stopped. For it is clear from Part I that the positions of Mars should be taken at the moments of true opposition to the sun. However, it is also clear that not every trace of the second inequality is thus removed, it being also necessary to refer the arc measured on the ecliptic to the planet's orbit. But the planet's orbit must first be investigated by the inclination of the planes and by finding out the nodes. Again, the inclination and the nodes 64 cannot be found without knowing the diurnal parallax, at least if this should turn out to be relatively large. One must therefore start with the parallax. I shall present two ways to find it.

The first way (and the one more familiar to others) will be examined using the Brahean observations.

In 1582, when Mars was opposite the sun in Cancer, I found the observation done with incredible care, with Tycho's manuscript title, "For investigating the parallaxes of Mars", from which you will, however, deduce either no parallax at all, or one exceedingly small. I pass over without saying, that (as is customary) they compared the star Mars with nearby stars on the ecliptic, and frequently with ones at a great distance. Now it is usual to find the parallax of a mobile star (for Mars moves, with a retrograde motion when opposite the sun) by comparing morning and evening observations. As a result, it has happened that almost all the stars by which Mars was observed in the morning are different from those by which it was observed in the evening. For a fixed star which is at hand in the morning (and higher than Mars), if it be near the ecliptic, has either set by evening (when Mars is in the west) or is rendered useless by refraction for this delicate procedure. Another star therefore had to be substituted. But if the fixed stars are substituted for one another, there is always less trust in the procedure than if the same star should be retained.

Brahe, however, announced to the learned world in many places that from the observations of that year, the parallax of Mars was found to be considerably greater than that of the sun. I therefore very carefully scrutinized the whole book in order to examine his operation or computation more deeply. I did indeed find a chapter which professed to offer a procedure for investigating the parallax of Mars from the observations of that year. But here was something really surprising: they fitted the position of Mars found by observation into a Copernican diagram drawn very laboriously and carefully. In this diagram, they took up the immense labor of solving all triangles created by the double epicycle on the concentric, in numbers replete with a great many digits. Finally, this was the goal of the calculation: to issue a pronouncement that the parallax of Mars is indeed greater than the solar parallax. Brahe had thus asked one thing, but his assistants in calculation carried out something else. He wanted them to find out the parallax of Mars by comparing morning and evening observations with one another, but they had in fact found out how much parallax the Copernican diagram would bring about. Whether Brahe's pronouncement on parallax was founded solely upon his trust in his assistants, is unknown to me.[1]

As for us, let us consult the actual observations, insofar as they pertain to our undertaking.

In 1582 on the night between 23 and 24 November, the distances from the fixed stars remained the same at different times. This, then, was a station point.[2]

On the following two days, the motion was 11′ and 15′.

On the night of 26 December,[3] it passed between the second and seventh stars of Gemini,[4] its distance (measured with the staff) from the head of the lower of the Twins (the second star) being 2° 25′ or 2° 26′, but from the seventh, 1° 6′ or 1° 7′, making the latitude about 4° 9′. Then, at $8^h 28^m$, it was 44° 41′ from the eye of Taurus,[5] whose latitude is 5° 31′ south, longitude 4° $12\frac{1}{2}′$ Gemini, in 1600. Hence, Mars's longitude as if the year were 1600 is 17° $53\frac{1}{3}′$ Cancer,[6] or, at the end of 1582, 17° 38′ Cancer, at an altitude of 40° 50′. It is

65

---

[1]Kepler's interpretation cannot be correct, as the computation in question exists in Tycho's own hand as well as in the copy made by assistants. Tycho apparently made no direct attempt to measure Mars's parallax, but concluded that Mars must be nearer than the sun at opposition by measuring its diurnal *motion* at opposition. See Owen Gingerich, "Dreyer and Tycho's World System", *Sky and Telescope*, August 1982, pp. 138–140, and Owen Gingerich and Robert S. Westman, "The Wittich Connection: Conflict and Priority in Late Sixteenth-Century Cosmology." *Transactions of the American Philosophical Society* 18 (1988), Part 7, p. 71.

[2]*TBOO* 10 pp. 174–175.

[3]This set of observations is in *TBOO* 10 pp. 176–177.

[4]Pollux and Kappa Geminorum, respectively. The arrangement of the stars is as in this diagram:

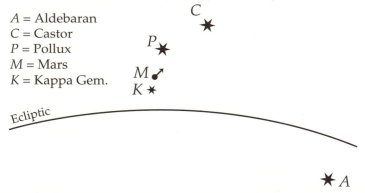

A = Aldebaran
C = Castor
P = Pollux
M = Mars
K = Kappa Gem.

Since κ Geminorum is nearly at the same longitude as Pollux, the distance between them is almost entirely latitudinal, and Mars's latitude can be obtained simply by adding Mars's distance from κ (*KM*) to κ's latitude.

[5]Aldebaran.

[6]In the diagram in the footnote above, arc *MA* is 44° 41′. The difference in longitudes is found by drawing great circle arcs from *M* and *A* through the pole of the ecliptic (call it *N*), and finding

thus beyond the effects of refraction.

Again, at $7^h 15^m$ on the morning of December 27, it was 36° 43′ from Cor Leonis,[7] whose latitude is 0° $26\frac{1}{2}'$; hence, its longitude at the end of 1582 is 17° $28\frac{1}{3}'$ Cancer, altitude 14° 4′, and thus affected by refraction. Therefore, from $8^h 28\frac{1}{2}^m$ in the evening to $19^h 15^m$, an interval of $10^h 46\frac{1}{2}^m$, it was observed to retrogress $9\frac{2}{3}'$.

For the diurnal motion, on the 29th at $7^h 47^m$, the distance of Mars from the southern foot of Erichthonius[8] was 29° $38\frac{1}{2}'$. But on the 30th at $8^h 8^m$ the distance from the same star was 29° $13\frac{1}{2}'$.[9] Therefore, over $24^h 21^m$ it moved 25′. And this diurnal motion remained the same on the 27th. Therefore, to $10^h 46\frac{1}{2}^m$ there should have corresponded $11\frac{1}{2}'$ of arc, but we found only $9\frac{2}{3}'$. Let us consider this.

On the previous evening, when Mars was rising farther to the east (because it was retrograde), parallax moved it eastward, and in the morning, when it was setting and was farther to the west, parallax moved it westward. So, just as the moon's diurnal parallax apparently slows its motion, the same parallax in turn quickens Mars's retrograde motion. Therefore, if parallax is perceived, it is perceived through an excessively enlarged diurnal motion. But this motion is diminished. Therefore, there is no parallax. Again, refraction is perceived as contrary to parallax. Now the refraction at altitude 13° is 4′ from the table of fixed stars, and 8′ from the table of the sun,[10] and only a very small part of this affects the longitude, as Cancer descends quite obliquely. So the refraction in longitude comes to three minutes at most, which, added to $9\frac{2}{3}'$, makes the refraction-free motion $12\frac{2}{3}'$ over $10\frac{3}{4}$ hours, which motion, if it also were free from parallax, ought to have been $11\frac{1}{2}'$. Therefore, the excess of $1\frac{1}{3}'$ is the longitudinal parallax for the two observations—clearly minimal, untrustworthy, and entirely contemptible.

On 1583 January 16 at $7^h 30^m$ in the evening,[11] Mars was 23° 29′ from the bright star in the foot of Erichthonius at an altitude of 51°. The next morning at $5^h 0^m$, it was 43° 58′ from Cor Leonis at an altitude of 15°. And, measured by the straight edge, Mars was perfectly collinear with the two stars. And so, since Mars's motion is carried out along this line, Brahe made a note that the longitudinal parallax is given from the diurnal motion of Mars. This is

---

the angle *MNA*. This procedure is similar to the one detailed in the Translator's Introduction, beginning on p. xxviii.

[7] Regulus. The observation is in *TBOO* 10 p. 177. Since both Mars and Regulus were very near the ecliptic, the longitude can be obtained simply by addition, with a correction for precession.

[8] Beta Tauri, which also serves as the foot of Erichthonius (Auriga, also called Heniochus).

[9] These two observations are in *TBOO* 10 pp. 177–178.

[10] Tycho believed the sun, moon and fixed stars each to have a different angle of refraction (see *TBOO* 2 pp. 64, 136, and 287). Caspar (in *KGW* 3 p. 462) states plausibly that the large solar refraction is a result of Tycho's acceptance of a large solar parallax, since their effects are in opposite directions.

[11] *TBOO* 10 p. 242–243.

obtained here as follows. On January 16 at $10\frac{1}{2}^h$ it was 23° 27′ from the bright star in the foot of Erichthonius. On January 17 at $10\frac{3}{5}^{h\,12}$ it was 23° $12\frac{1}{2}'$ from the same star. Therefore, the diurnal motion would be $14\frac{1}{2}'$. Now in order to comply with Brahe's advice, we must set out the distance between the foot of Erichthonius and Cor Leonis, which is found to be 67° 21′. Subtracting Mars's distance from the bright star in the foot of Erichthonius, 23° 29′, leaves Mars's distance from Cor Leonis, 43° 52′, at $7\frac{1}{2}^h$ in the evening, which at 5^h in the morning was 43° 58′, 6 minutes greater. The time interval is $9\frac{1}{2}$ hours, to which ought to correspond $5\frac{5}{8}'$ of the diurnal motion. Here, therefore, the sum of the two parallaxes is no more than $\frac{3}{8}'$, except that the amount of Mars's longitudinal refraction at 15° is added to it. But this is quite small. For Cancer and Leo are setting extremely obliquely, and Mars's large northern latitude put it at nearly the same altitude as Cor Leonis.

66

On January 17 at $5^h\ 20^m$ in the evening,[13] Mars was 23° 16′ from the foot of Erichthonius. On the following day, the 18th, at $3^h\ 0^m$ in the morning, this distance was 23° 9′, and at $5^h\ 5^m$ in the evening it was 23° $1\frac{1}{2}'$. So the motion over $23^h\ 45^m$ is $14\frac{1}{2}'$, but over $9^h\ 40^m$ it is 7′. This should have been 6′. We are left with a longitudinal parallax of no more than 1′. Refraction does not affect anything, since in both instances the altitude of Mars was about 30°.

Likewise, at $7^h\ 34^m$ its distance from the seventh star of Gemini[14] was 7° 51′. At $4^h\ 52^m$ in the morning, it was 7° 59′ from the same star. Therefore, over $9^h\ 18^m$ it moved 8′. We thus have one minute more than before. Of this star (at the shoulder of Gemini), Brahe wrote, "Note that I am taking Mars's distance from this star because its course passes through it, as it were, so that the morning and evening distances compared might show the parallax." I have reproduced this here so that the reader may rest assured that Brahe did not proceed without a purpose.

On January 18 at $8^h\ 52^m$ in the evening[15] there was 44° 22′ between Mars and Cor Leonis. At $4\frac{3}{4}^h$ in the morning the same distance was 44° $27\frac{1}{3}'$. Therefore, the motion over $7^h\ 53^m$ was $5\frac{1}{3}'$. On January 19 following, at $7^h\ 3^m$, this distance was 44° $32\frac{1}{2}'$. Therefore, for $22^h\ 11^m$ the motion is $10\frac{1}{2}'$. And for 8 hours there would be less than 4′ of motion. Our profit is about $1\frac{1}{2}'$ of parallax.

But now let us calculate, for January 17, how much of an increase in the hourly motion should result from a parallax greater than the solar parallax as usually accepted. Since we consider the sun's parallax to be three minutes, let Mars have four.

---

[12]*TBOO* 10 p. 243–244.

[13]*TBOO* 10 p. 243–244.

[14]*κ* Geminorum.

[15]*TBOO* 10 p. 244–245.

1583 January 17	5h 20m °	'		15h 0m °	'	
Position of the sun	7	22	Aquarius	7	31	Aquarius
Its right ascension	309	47		309	56	
Add the hour angle[16]	79	0		225	0	
Right ascension of meridian	28	47		174	56	
Degree of the meridian	0	56	Taurus	24	29	Virgo
Declination	11	50		2	12	
Oblique ascension risen	118	47		264	56	
Degree rising	19	41	Leo	26	0	Scorpio
Nonagesimal degree from rising....	19	41	Taurus	26	0	Leo
Between deg. of mer. and 90th	18	45		28	29	
Between deg. of mer. and zenith....	44	5		53	43;	and hence,
Between zenith and nonagesimal ...	40	40		47	41;	that is,
Altitude of the nonagesimal	49	20		42	19	
Corresponding long. horiz. parall. .		2	36 sec.		2	58 sec.
And because Mars is about	10	0	Cancer	10	0	Cancer, therefore,
Between Mars and the nonagesimal	50	19		46	0	
Corresponding parallax in longitude		2	0 sec. east		2	8″ west

It follows that the motion of Mars over those hours should have appeared 4′ greater than what follows proportionally from the diurnal motion. Since this is repudiated by the observations, Mars's parallax is not so large.

There exist similar observations from 1585, 1595, and others, from which an exceedingly small parallax is found—often none at all. There was even an occasional note in Brahe's hand saying, "It strayed over to the wrong side." So this is the first way of investigating Mars's parallax.

I shall now add the other way, because of its beauty; I cannot use the Brahean observations in it. Therefore, in using my own observations, I am going to give you a clown show, and will show by example why Brahe needed such diligence, precision of instruments, assistants, and other equipment.

I have two instruments, which I use through the generosity of Baron Friedrich Hoffmann, L.B.:[17] an iron sextant and a brass azimuthal quadrant. The latter is two and a half feet in diameter and the former three and a half, and both are calibrated in one-minute divisions.

Now at this very time, 1604, at which I am considering parallax (whether that of the sun more than Mars's is hard to say, for my *Hipparchus*[18] requires Mars's aid even in the lunar eclipses), a very suitable occasion for observation arose, if only the climate zone had been different and Mars had moved a little higher. For about 19/29 February[19] of this year 1604, Mars was stationary both in longitude and in latitude at the same time. This occurred in Libra, and therefore from Mars's rising to the sun's rising the angle of the horizon with the ecliptic continually decreased. Consequently, according to Ch. 9 of

---

[16]This is the angular diurnal motion of the sun from noon to the stated time. For the first observation, this should have been 80°, not 79°.

[17]Johann Friedrich Hofmann von Grünpichl und Strechau (1540–1607) was one of Kepler's closest friends and patrons in Prague. See R. J. W. Evans, *Rudolf II and his world* (Thames and Hudson, 1973), p. 153.

[18]A projected work on the sizes and distances of the heavenly bodies, which Kepler never published although his work on its composition extended over many years. The surviving manuscripts relating to the *Hipparchus* have been published in *KGW* 20.1 pp. 181–268.

[19]Dates presented in this format show the "old style" or Julian calendar date first, and then the "new style" or Gregorian calendar date. For more on dates, see "Calendars," in the Glossary.

the *Astronomiae Optica*[20] the latitudinal parallax, if any, continually increases. But from the increment (found through the columns of the parallactic table[21] opposite the initial and final angles of the ecliptic with the horizon), the whole horizontal parallax (at the front of the column) is known.

THERE FOLLOWS THE SERIES OF MY OBSERVATIONS.[22]

On the night between Thursday and Friday, which was February 17/27, while Corvus was on the meridian, there was 9° 44′ between Mars and Spica, and between it and the Northern Pan,[23] 17° 41′; and between Mars and Arcturus, 29° 13′. Also, to test the sextant, we measured the interval between Arcturus and Spica as 32° 57′, which should have been 33° 1′ 45″, as is clear if you calculate it using the right ascensions and declinations, or latitudes and longitudes, which Tycho assigned to these stars in Book I of the *Progymnasmata*.[24] Therefore, my distances are smaller than the true distances by $4\frac{3}{4}$ minutes. I applied this correction to the distances of Mars from the fixed stars, so as to make it 9° 48′ 45″ from Spica, 17° 45′ 45″ from the Pan, and 29° 17′ 45″ from Arcturus.[25]

I then used the quadrant to obtain the meridian altitude of Mars, 32° 4′, and of Spica, 30° 50′. Since the latter has a declination of 9° 2′, Mars is left with a declination of 7° 48′. However, the altitude of Spica showed that my perpendicular was not well enough set up. For the altitude of the equator, at my location, is 39° 54′. Accordingly, the meridian altitude of Spica is 30° 52′, and of Mars, 32° 6′. Now, from the declination of Mars and its distance from the fixed star, its right ascension comes out:

	°	′	″
From Spica	205	57	36
From the Pan	206	3	17
Difference	0	5	41
Resultant mean	206	0	26[26]

---

[20]Kepler considers diurnal parallax in *Optics*, Ch. 9 Sect. 4, English trans. pp. 326–329. There is no passage describing the present configuration; however, Kepler's statement here may be deduced from the principles set forth in the *Optics*.

[21]*KGW* 2, following p. 240; also contained in the pocket inside the back cover of the English translation of the *Optics*.

[22]The following series of observations and computations is presented in detail, with commentary, in the Translator's Introduction, beginning on p. xxviii. Kepler's central aim in the initial observations is to show that Mars's latitude is constant over the two-week period: hence, there are many observations of the distance between Mars and Arcturus. Since they are very approximately at the same longitude, any significant change in the distance between them would indicate a change in Mars's latitude.

[23]Beta Librae.

[24]Brahe's star catalog appears in *TBOO* 2 pp. 258–280. Brahe also published separately a larger catalog, which is in *TBOO* 3 pp. 344–377.

[25]This last distance is corrected from 29° 17′ 43″, obviously a typographical error.

[26]The right ascensions in the Latin text are all 100° too large, evidently a typographical error that has been corrected here.

I am not certain whether, when (as happened a few times) the clamps holding the arm loosened and it (being a heavy piece of iron) fell precipitously and
hit hard, it might not have changed the position of the sights, since they are re-
movable and subject to dislocation. But from this right ascension, I first select,
from Tycho's table of right ascensions, the degree of the right sphere that was
rising at the same time, 28° 1′ 0″ Libra,[27] whose declination, from another of
that author's tables,[28] is 10° 48′ 30″ [South]. But that of Mars is 7° 48′ [South].
Therefore, Mars is distant from the ecliptic, the oblique path, by 3° 0′ 30″ on
the circle of declination. But the angle which the circle of declination makes
with the ecliptic is 68° 59′ from the appropriate table. Its complement is 21°
1′. And in my table of parallax,[29] under the heading of 60′, I find, opposite 68°
59′, the entry 56′ 1″. Under 30″, however, I find 28″. But since I have thrice
60′ in this distance of Mars from the ecliptic (which I call the base of the latitude), I multiply what I extracted under 60′ by 3.[30] This gives me a latitude of
2° 48′ 31″. The same operation opposite 21° 1′ shows me what has to be subtracted from the place rising at the same time, namely, 1° 5′ 4″. Accordingly,
Mars's position will be 26° 56′ Libra. I come within a minute of this using a
computation whose fundamentals I am going to be presenting in this work.

From the right ascension and declination of a star, to find its longitude and latitude without calculation, with the help of tables.

Term: What is the base of the latitude?

To test the latitude of Mars, I also consulted the distance from Arcturus,
through the star's latitude and longitude provided from Tycho, and Mars's
longitudinal position just found, and it replied to me that Mars was at a latitude of 2° 47′ 48″. Before, it was at 2° 48′ 31″.

On February 19/29 we had moved the sight, and began to observe Mars
rising. Its distances from Arcturus were noted, and were these:

I think we are ten minutes too high. For the wind was
blowing so hard that it was only by a glowing coal that we
could cast light upon the scale so as to read it. And the altitude
of Mars was then 11°. Later, the back of Leo[31] culminated at
altitude 62° 37′, according to a corrected plumb line. Thus
the altitude of the equator was shown to be 39° 55′: almost correct. At that
moment, Mars's altitude was 23°. We next reinvestigated the former distance
[to Arcturus], which came out to be:

°	′
29	$22\frac{1}{2}$
	24
	20
	22

°	′			
29	14	Therefore without	$12\frac{1}{2}$
	19	doubt the previous	14
	13	distance was	10
	18		12

For first, when Mars was near the horizon, refraction moved it toward Arcturus, later letting it drop down when Mars had acquired some altitude. It
was, however, the cold and the extremely biting winds that occasioned so
much variety in observations made at the same time. For it was impossible to

69

---

[27] *TBOO* 3 p. 74.

[28] *TBOO* 3 p. 71.

[29] For the use of this table, see the Translator's Introduction, pp. xxvi–xxviii, and p. xxxii.

[30] Relying on the approximation $\sin 3x = 3 \sin x$, for small angles.

[31] Delta Leonis.

handle the iron and close the clamps with bare hands, and with gloves the arm was not securely enough clamped to be read to the minute. Vindemiatrix[32] showed a meridian altitude of 53° 5′, a little greater than it should have had. But Spica's 30° 54′ was correct within one minute. The altitude of Mars at culmination was 32° 6′, the same as it had been two days before, and of Arcturus, a correct 61° 13′. Hence, by computation, it is concluded that the distance of Mars and Arcturus was 29° $18\frac{1}{3}'$. Now since, according to the *Prutenic Tables* and to my computations, Mars was stationary in longitude at this time, there could be no change in meridian altitude resulting from its wandering about on the ecliptic. For this reason, since the meridian altitude remained quite the same (for my instrument allows an uncertainty of one minute), neither did any change in the latitude occur during this time.

On February 22 or March 3 we tested the sextant, just as we had done above, and found 26° 2′ between Canis Minor and Orion's higher shoulder,[33] which the calculation shows to be 26° 2′ 15″. So also, between Canis Minor and Palilicium[34] was found 46° $22\frac{1}{2}'$, which Tycho, in his *Epistolae*, indicates is 46° 22′.[35] Therefore, when the fifth star of Leo[36] culminated, with the arm of the instrument fixed above 29° 17′, the distance of Mars and Arcturus was less, but with it fixed above 29° $13\frac{1}{2}'$ the distance was now more, while no error could be found at 29° 15′. Then the whole sky unexpectedly clouded over. It became clear again on the morning of March 4, and now, when Antares culminated, with the arm fixed at 29° 19′, both stars were seen evenly [with the sights]. However, it seemed that something needed to be added, but at 29° 20′ too much had been added. When the observation was finished, Saturn preceded the meridian by less than Jupiter preceded Saturn.

On the night following February 29 or March 10, the instrument having meanwhile been displaced, this distance was, first, between 29° 9′ and 29° 10′, half an hour before the culmination of the Heart of Hydra.[37] Later, it appeared to the investigators to be between 29° 12′ and 29° 13′, as it was now higher and free from refraction. For at the end of this observation it had an altitude of $19\frac{1}{6}'$. But a little afterward—I do not know whether the sights were disturbed—it would not allow that much, for it appeared to be 29° $9\frac{1}{2}'$. The Tail of Leo[38] was about half a degree from the meridian. The altitude of Mars was then $24\frac{3}{4}°$. The Tail of Leo, when culminating, had the correct altitude of 56° 44′, within one minute. When one third of the distance between Mars and Spica had crossed the meridian, it [i.e., the distance to Arcturus] first appeared to us to be 29° $9\frac{1}{2}'$, but the cylinder was not well enough applied, as it was too long. Thus, a little later, this could not be accepted, but 29° $10\frac{1}{4}'$ or a little less

---

[32] Epsilon Virginis.

[33] Procyon and Betelgeuse, respectively.

[34] Aldebaran.

[35] *Epistolarum Astronomicarum Libri* (Uraniburg, 1596), p. 51, in *TBOO* 6 p. 79.

[36] Zeta Leonis.

[37] Alpha Hydrae.

[38] Beta Leonis.

seemed to be required. And Mars was seen on both sides of the cylinder.

Then, between Mars and Spica, there was 9° 26', and less than 9° 27'.

Mars culminated at an altitude of 32° $19\frac{1}{2}'$.[39]

There was then 18° 25' between Mars and the Northern Pan [of Libra].   70

For the investigation of the sextant, the distance between Spica and the Pan was taken as 27° 39', although it should have been 27° 34'.[40] Also, the distance between Spica and the northern star in the head of Scorpius[41] was 39° $32\frac{1}{2}'$, which should have been 39° $26\frac{1}{2}'$. So the sextant read five minutes high. Furthermore, calculation of Mars's position also provides evidence of this. For unless you diminish Mars's distance from the fixed stars by 5 minutes, the right ascension measured through Spica and through the Pan will be discrepant by 10 minutes, but if (as required by the examination of the sextant) the five minutes be subtracted, they will coincide exactly, and will be 205° 27' 10'', with declination 7° $35\frac{1}{2}'$. Therefore, the position is 26° 18' 48'' Libra, with latitude 2° 47' 20''. You see that the latitude is manifestly the same, though meanwhile the planet had retrogressed 38' in longitude. If, from the position of Mars found thus, you find its distance from Arcturus, it will come out to be 29° $9\frac{1}{6}'$, while using the faulty instrument it was 29° 14'.

Now, with the Heart of Scorpius[42] culminating, our distance [to Arcturus] was 29° $13\frac{1}{2}'$, the instrument in the meantime having been displaced and later repositioned. We next tested the sextant again, which showed 44° 45' between Polaris and the Tail of Cygnus.[43] This should have been 44° $39\frac{1}{2}'$. Therefore, the instrument was in its pristine state. But when Saturn had passed the meridian by one degree, the distance could not admit 29° $13\frac{1}{2}'$; nevertheless, it was greater than 29° $12\frac{1}{2}'$, about 29° 13'.

So this is the series of observations. I would be crazy if I tried to use them to build something of great precision. Therefore, I am presenting an example to another more diligent and successful observer, rather than an argument. I also hope that the nausea evoked by these uncertain observations will lead readers to desire all the more fervently the extremely certain Tychonic ones. Now, on with the example.

The first and second days, agree only in showing the station in latitudinal motion. In both, Mars was 29° 18' from Arcturus, and in both, its meridian altitude was 32° 7' or 6'. Those days were busying me in preparation for meeting the following days properly, should the instruments be needed.

But on March 3, with the Mouth of Leo[44] culminating, the distance was 29° 15', and with the Heart of Scorpius[45] culminating, 29° 19' plus. Therefore the distance

---

[39] Corrected from 30°, which is consistent with the given altitude of the equator and the declination.

[40] Brahe printed a list of distances between various prominent stars in *Epistolae Astronomicae*, from which this number and the next are taken. See *TBOO* 6 pp. 78–79.

[41] Beta Scorpii.

[42] Antares.

[43] Deneb. The correct distance is again from *Epistolae*, in *TBOO* 6 p. 79.

[44] Lambda Leonis.

[45] Antares.

changed about $4\frac{1}{4}'$ over the interval. And since Arcturus and Mars have very nearly the same longitude,[46] this change of distance bespeaks a variation of latitudinal parallax. I am not unaware that $29°\ 19'$ is hardly different from $29°\ 18'$, and that by analogy with the previous day the latter ought to be the distance at about the same time if Mars is standing still. I also know that when the Mouth of Leo is on the meridian Mars is $12\frac{1}{2}°$ high and is somewhat affected by refraction. But of these we shall speak afterwards. For now, let us entirely ignore them, so as not to complicate our example. Now the altitude of the nonagesimal was $57\frac{1}{3}°$ (about) when the Mouth of Leo was culminating, but finally, after the Heart of Scorpius had culminated, it was $20\frac{1}{3}°$. Let me therefore look through the parallactic table to find the column in which, from a distance from the zenith of $32\frac{2}{3}°$ to a distance of $69\frac{2}{3}°$, the entry in the table would change $4\frac{1}{4}'$. I find that this happens in the column whose heading is $9'$.[47] Therefore, the maximum parallax of Mars would be $9'$. And since, on that day, the distance of Mars and earth was to the distance of Mars and the sun as 28 is to 60 (given approximately by an anticipatory acquaintance with the hypotheses of Tycho and Copernicus), the ratio of the parallaxes will be the inverse of this, and the

71   *maximum solar parallax will be about $4'\ 24''$. It is set at $3'\ 0''$.*

But now let us consider that at an altitude of $12\frac{1}{2}°$ Mars would be subject to refraction, if the table of fixed star refractions constructed at Hven is valid for Prague. At this altitude, it was $4'\ 20''$, $2'\ 18''$ of which is owed to the latitude, by which Mars is caused to be closer to Arcturus. If, however, we apply the sun's refraction to Mars (as it often appears we should), it is $8'\ 45''$ at this altitude, twice as great. Therefore, the latitudinal parallax would also be twice as great, $4'\ 36''$. In this way, the whole difference which observation imposes upon itself, at these two different moments, would be due entirely to refraction. Reckoning it in the former way would leave a latitudinal parallax of $2'$—the difference in parallax in the column whose heading is $5'$. This would give the sun only $2'\ 25''$ of maximum parallax. So refraction makes our third day, too, suspect and doubtful, and ultimately quite worthless. I know that since Arcturus and Mars are 9 degrees apart,[48] which is one third of the amount by which Arcturus's latitude exceeds Mars's, it comes about that not all the latitudinal refraction is subtracted from the distances from Mars, and that the parallax changes Mars's latitude more than it changes this distance from Arcturus. But since this is very small, I have, with greater fear, considered it as something to be left buried. Let him observe it who is versed in more subtle matters.

Now on the fourth day, what seems to have been accomplished is nothing other than the total destruction of Mars's parallax. The meridian distance ought to have been $29°\ 9\frac{1}{2}'$ with an accurate instrument, and consequently $29°\ 14'$ with the faulty one. But at the end it was found to be $29°\ 13\frac{1}{2}'$, when the latitudinal parallax (had there been any) ought to have been greater, and hence the distance from Arcturus greater. Hence, from that time when Mars attained an altitude of $19°$, the distance was found to be $29°\ 12\frac{1}{2}'$, and one minute greater at the end. This would be a very

---

[46]They are actually about $9°$ apart, but this is close enough to ensure that any latitudinal parallax will be evident in an apparent change in the distance.

[47]An error. This difference in this column of the table is $3'\ 33''$. The difference under $11'$ is $4'\ 20''$, and under $12'$ is $4'\ 45''$. Hence, the correct value would be about $11\frac{1}{2}'$.

[48]In longitude.

small parallax. And what is this ratio? When its altitude was 9 degrees (when Hydra was culminating), the distance was 29° 9′ with a faulty instrument, and still subject to refraction. Later, at an altitude of 25° and near the meridian, it was again 29° 9′, measured twice at different moments. Could the refraction have been zero initially, that the arc thus remain constant? Or should it rather be said that I (though to myself I might have seemed most diligent) erred in observing? Especially on account of the length of the cylinder.

Still, it is at least established by these observations, whatever their quality, that Mars's parallaxes of latitude are not greater than 4′, which is the amount of uncertainty in the instrument. It is more credible that the parallax is very small. In Ch. 64 below, you will find further discussion of this point.

That the parallaxes of Mars are greater than the sun's, on the other hand, is argued by the proportion in the Tychonic and Copernican hypotheses; and from this proportion Mars's parallax could easily be computed if we were certain of the sun's parallax. Is, then, the procedure for finding the sun's altitude and parallax from eclipses uncertain? It is indeed relatively uncertain quantitatively, but as concerns the thing itself it is perfectly certain. The sun is not nearer than 230 semidiameters of the earth, but it is not an infinite number of semidiameters away. But between 700 and 2000 semidiameters (of which the first quantity is from my *Mysterium cosmographicum*, while the other two are given as the upper and lower limits in observations of eclipses) it does not appear that any indisputable number has yet been demonstrated, as I shall prove in my *Hipparchus*.

CHAPTER 12

Investigation of Mars's Nodes.

Means are not wanting of investigating the planets' first inequality through observations, even when these are entangled in the second inequality. Nevertheless, in this second part, I prefer to follow the footsteps of the authorities and make use of acronychal observations, in order to establish my credibility. For I want to be sure that later, when I bring forth something contrary to accepted opinion, no one may complain that I am hiding behind the briar-path of my own method.

Furthermore, it is now clear that nothing of any importance is wanting in Mars's diurnal parallax as taken by Tycho. I shall therefore move on, little by little, towards the reduction of the observed positions of Mars to the point opposite the sun's apparent position.

As a first step, we must find the nodes. Tycho Brahe used to investigate them thus:

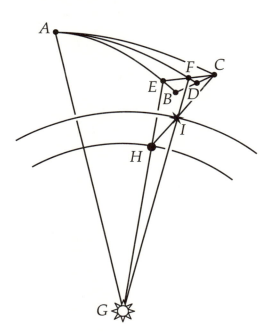

In the diagram of Ch. 9, let A be the position of the node, E the planet's position on the ecliptic in 1595, C the observed sidereal position of the planet, 17° 56′ 5″ Taurus, EC the observed latitude, 0° 5′ 15″ north. Now it is presupposed that the angle EAC is very nearly 4° 34½′, the same as the maximum northern latitude observed in the same way in 1585. In the right triangle CEA (or the isosceles triangle CBA; the difference is unimportant in this procedure), from side CE and angle EAC he sought the length EA, the distance of the ecliptic position from the node. There is nothing wrong with this operation, since EC is small and near the node. However, the need for accuracy[1] in this demonstration commends another. *For it was said in Ch. 9 that the angle EAC is not constant; hence, through different latitudes of different oppositions they will also show different positions for the node. Nor is EAC as great as the maximum apparent latitude, because AC is a curved arc; and neither is it AC, but some path inside (AF, say), that is the planet's path as it would be seen from the center of the sun. Therefore, A will not necessarily be the node, at least as found through this operation.*

---

[1] Kepler uses the Greek ἀκρίβεια here.

I, therefore, investigated the nodes differently, using observations on the day on which they were at the node. Even though this method depends upon some preconceptions, and the subject will be treated more accurately below in Part 5, we ought to have a first taste of it here, if only to give confirmation.

My presupposition was that when the planet, in its eccentric position, is truly at the node, it can by no disposition of the earth or sun be made to appear elsewhere than at the node. For in the Copernican hypothesis, this in itself is in agreement with the nature of things: that the moving faculty of any star is not bound to observe a star foreign to it (including the earth), but has its own laws governing its circuit. In the Ptolemaic hypothesis, this would be exactly as if one were to say that the epicycle pays heed, not to a line from the sun through its own center, but to certain positions beneath the fixed stars, beneath which 73 it places the planet in the plane of the ecliptic. In the Tychonic system, the same will be said of the eccentric.

Further, that which I presupposed, I have found to be true using the following observations:

I. On 1590 March 4 at $7^h 10^m$ in the evening, Mars's declination was 9° 26′ N, and the right ascension was 22° 35′ 10″.[2] Hence, its position comes out as 24° 22′ 56″ Aries, and its latitude as 3′ 12″ south. Parallax and refraction were opposite and approximately equal, and are therefore ignored.

II. On 1592 January 23 at $10^h 15^m$ in the evening, Mars was at 11° 34′ 30″ Aries with latitude 0° 2′ south.[3] The altitude of Mars was 25°, and therefore (from the table for the fixed stars),[4] there was no refraction. The parallax was about as much as the sun's, because Mars and the sun were sextile, and were therefore about equally distant from earth.[5] Nearly all of this was latitudinal. Therefore, about two minutes of latitude must be added northward to Mars in order to free it from parallax, and it thus will fall upon the ecliptic. For on February 6 it was already at about 7′ northern latitude.[6]

III. On the evening of 1593 December 10, Mars was observed at the ascending node. For after correction of the horizontal variations it retained no more than 0° 0′ 45″ north latitude.[7]

IV. On 1595 October 27 at $12^h 20^m$, Mars's true latitude after the removal of parallax was 0° 2′ 20″ south.[8] On the 28th, when the parallax had similarly

---

[2]The raw observations are in *TBOO* 12 p. 44. Kepler uses them to calculate the right ascension and the ecliptic position. For right ascension and declination, and conversion to latitude and longitude, see the Glossary and the Translator's Introduction.

[3]The position was again computed by Kepler from Tycho's raw observations (*TBOO* 12 p. 219).

[4]Tycho's table appears in *Stellarum inerrantium restitutio*, folium 28 v, in *TBOO* III p. 377. Brahe believed that the light from the fixed stars is refracted differently from sunlight; hence, he had two tables of refraction.

[5]That is, about sixty degrees apart (see the Glossary for the various planetary aspects. When Mars is sextile to the sun, it is over 1.7 times as far from earth as is the sun; therefore, it is only very roughly the same distance.

[6]Cf. *TBOO* 12 p. 220; however, Tycho gives a latitude of 10′ 30″ N., which is consistent with his right ascension and declination.

[7]*TBOO* 12 pp. 296–297. The latitude is given by Tycho.

[8]*TBOO* 12 p. 450. Kepler derived the latitude by computation from Tycho's transit altitude.

* This approximate
argument is sufficient for
the present enterprise. In
Ch. 61 and 67 below, with
everything considered
more carefully, it is found
to have been at the node
at 15^h on the 29th.

been removed, the latitude was 0° 0′ 25″ north.[9] Therefore,* in the meanwhile, it was at the ascending node.

Counting backwards 687 days, the number of days in Mars's revolution on its eccentric, starting from noon on 28 October, the end will fall on 1593 December 10, while on the preceding night Mars was observed near the node. Count back another 687. This will end up at 1592 January 23, when the planet was observed right at the node. If you do the same a third time, you will come out at 1590 March 7, while on the fourth day preceding, Mars had some southern latitude, which it would make up over the four days, so as to fall at the node about the seventh.

From this it is known that it makes no difference where the earth is, either in sidereal position or with respect to Mars. In the Ptolemaic system, it makes no difference where the sun is with respect to the center of Mars's epicycle and where Mars is on the epicycle, and in the Tychonic system, it makes no difference where the center of the eccentric or the sun is located with respect to the line from Mars through the earth, so drawn that Mars lies in the plane of the ecliptic. For the diameter of the nodes is always the same in Copernicus and Ptolemy, and in Tycho it is always parallel to itself, except that over the ages the nodes move slightly. This motion was not perceived over these six years.

Now let us find the opposite node.

I. On the morning of 1595 January 4, when Mars was observed at $7^h 10^m$ at altitude 8°, with reference to Spica Virginis and Cor Scorpii,[10] it was observed at latitude 0° 3′ 46″ N, and it was itself at 13° 36′ 40″ Sagittarius.[11] The paral-
74   lax is small, because Mars was more than twice as far from the earth as is the sun. Refraction, on the other hand, is large: 6′ 45″ from the table for the fixed stars, or 11 minutes from the table for the sun.[12] This is nearly all latitudinal, owing to the low altitude of the nonagesimal.[13] Therefore, Mars (reckoning by the refraction displayed by the sun) was really at a few minutes south latitude—about 2 or 3, or even more.

II. On the night of 1589 April 15 Mars's observed latitude was 1° 7′ north.[14] The parallax of the annual orb was drastically increased, owing to the approach of Mars and earth. After 21 days, the latitude decreased to a paltry $6\frac{2}{3}′$ north.[15] And thus, although on May 6 it decreased somewhat more slowly,

---

[9]*TBOO* 12 p. 452; Kepler again used the transit altitude.

[10]Antares.

[11]*TBOO* 12 p. 441.

[12]See footnote 1 in this chapter.

[13]The nonagesimal is the point on the ecliptic 90° westward from the point of intersection of the ecliptic and the horizon (the ascendant). See the Glossary for this and related terms, and see also the translator's introduction for an explanation of the role of the nonagesimal in the computation of refraction and parallax.

[14]*TBOO* 11 pp. 332–3. No latitude is given for April 15th; however, on p. 334 Tycho gives the latitude on April 14th as 1° 7′ 26″, which appears to be computed from planetary models.

[15]May 6 is 21 days after April 15, and the observations for that day appear in *TBOO* 11 p. 334.

since the star was moving away from the earth, we shall still not be far wrong if we extrapolate proportionally: we make 21 days be to the number of days after which Mars falls upon the ecliptic, as 60 minutes of diminution is to the $6\frac{2}{3}$ minutes remaining. This rule shows it to be two and one third days, so that Mars was at the node on May 9. Once again, counting thrice 687 days thence, we will come out on 1594 December 30. On this day, Mars ought to have been at the node, and for the next 5 days—up to the morning of January 4—it should have been moving off southward. And indeed, from observations of it on January 4, we have given it a few minutes of south latitude. Mars was not observed more frequently at this eccentric position. It is enough that we have this 1595 observation, provided that it does not disagree with us. In 1589 there is nothing that we can bring into question. Nor should it disturb you that in 1589 we assigned a latitudinal motion of $6\frac{2}{3}'$ to $2\frac{1}{3}$ days, while we do not allow so much over a 5-day period around 1595 January 4. For, as will appear in the course of this work, the latitude is most attenuated at conjunction with the sun (as in 1595), owing to the parallaxes of the annual orb, and at opposition (as in 1589) it is augmented. It is therefore fitting that the diurnal motion in latitude appear less in 1595, and greater in 1589.

Now, how are the sidereal positions of the two nodes found? Thus: one finds an approximate value for the mean position of Mars at each place, using tables for Mars (which, accordingly, we presuppose for this purpose). Whether you do this with the help of the Prutenic or Tychonic tables, taking into account the true precession of the equinox, you will find that on the morning of 1594 December 30, the mean position of Mars is $27° \ 14\frac{1}{2}'$ Scorpio, and on the morning of 1595 October 28 it was at $5° \ 31'$ Taurus. It therefore appears that the diameter of the nodes does not pass through the center of equable motion,[16] but far beneath it. For from $5° \ 31'$ Taurus to $27° \ 14\frac{1}{2}'$ Scorpio is more than from the latter to the former.

If, on the other hand, you make use of the Tychonic equations, $11° \ 17'$ must be added to the latter figure and $11° \ 30'$ subtracted from the former. Accordingly, the one comes out to be $15° \ 44\frac{1}{2}'$ Scorpio, and the other, $16° \ 48'$ Taurus, which are Mars's equated eccentric positions.[17] As you see, the nodes are nearly opposite one another at about $16\frac{1}{3}°$ Taurus and Scorpio, when viewed from the center of the planetary system, which Ptolemy described as a point very near the earth, and Copernicus and Tycho as a point very near the sun.

Further, it will be seen in Part 5 below how much we are going to change these positions of the nodes when we change the equations by transposing the theory of the sun from the sun's mean position to its apparent position.

---

[16]The equant. See the Glossary.

[17]That is, Mars's positions as if observed from the sun's mean position (Ptolemy and Tycho), or the center of the earth's orbit (Copernicus).

CHAPTER 13

Investigation of the inclination of the planes of the ecliptic and of the orbit of Mars.

Terms:
Inclination and *Latitude* are understood differently. "Inclination" concerns the angle at the sun or center of the planetary system, which is (for Copernicus) formed by lines extended to the body of Mars and to its position on the ecliptic. "Latitude" is the angle under which any inclination appears when viewed from the earth. For Ptolemy, the inclination is the angle between the straight lines drawn from the earth through the center of the epicycle and through the epicycle's position on the ecliptic. The latitude is the angle made by straight lines through the earth, one through the body of the planet, the other through the place on the ecliptic which corresponds to it.

In the previous chapter, the nodes and limits have been found quite accurately according to the opinions of Tycho Brahe and myself. Now it is to be inquired what exactly the inclination of the plane of Mars's orbit is to the plane of the ecliptic.

It is not so evident how to deduce this from the observations themselves. For the angle of this inclination is set up around the center of the planetary system, which for Copernicus and Tycho is the sun.

But the eye cannot be placed at the sun so that this inclination might thence appear beneath the fixed stars and be measured, and the maximum distance of the limit from the ecliptic, seen from another place, will also be seen under another angle. In the Ptolemaic form there might appear to be a more direct procedure, but this is not so. For it will be demonstrated that the plane of the epicycle always remains parallel to the plane of the ecliptic. Therefore, place the center of the plane of the epicycle at either limit, and let the planet lie on the same line of longitude passing from the center of vision through the center of the epicycle. The planet will then either be more distant than the center of the epicycle from the observer, and thus its distance from the ecliptic will appear less than the distance of the center of the epicycle from the same ecliptic, or it will be nearer to the observer, and will thus appear greater than what we seek.

In this difficulty we may take consolation from this one circumstance: that the purpose for which we seek to know the inclination as one of our principles is not such as to require the highest accuracy. It will consequently permit us to use those means which furnish indirect evidence of the quantity of the inclination: we shall offer three of these.

Now it is apparent from what has just been said that it will be most directly helpful to us if we find an observation of the star Mars at that moment at which Mars is reported to be equidistant from the sun and the earth and on the line drawn from the sun to 16° or 17° Leo or Aquarius (the positions of the limits). In the Ptolemaic form, this is where the center of the epicycle is at 16° or 17° Leo or Aquarius and Mars is as far from the earth as is the center of the epicycle. In Mercury alone this problem does not occur.

Let B be the sun, A the earth, and upon AB let the isosceles triangle ABC be set up, with C the planet's position in the plane of the ecliptic; and, CE being drawn perpendicular to the orbit of Mars, let the body of Mars be at E. EC will therefore appear the same whether seen from B the sun or A the earth—this is immediately evident.

But in order that it be known at what position Mars is equidistant from the sun and the earth, observe that when the lines from Mars at C and the earth at A falling upon the sun at B make the angle CBA right, then CB is shorter than CA. Consequently,

BA, the position opposite the sun's, should make with BC, the eccentric position of Mars, an angle less than 90°, in order that CAB and CBA be equal. Therefore, if BC is directed toward 17 Leo, the sun ought to be beyond 17 Taurus and before 17 Scorpio. Or if on the contrary BC is directed toward 17 Aquarius, the sun should be beyond 17 Scorpio and before 17 Taurus. We use these circumlocutions to denote morning   76 risings or evening settings, with Mars and the sun sextile or quintile.[1]

In the Ptolemaic form, if C be the earth, A the center of the epicycle, and B Mars, CAB will not be able to be right, as CA and CB are to be made equal. So the anomaly of commutation[2] ought to be more than 90° or less than 270°.

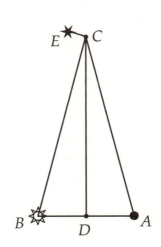

If you wish to work a little more precisely, take from Copernicus or from an anticipation of Tycho's reformation the approximate ratio 1525 : 1000 as (in Copernicus) the ratio of the orbits of Mars and earth, (in Tycho) of Mars and the sun, or (in Ptolemy) of the eccentric and the epicycle. At 16° or 17° Leo, this is about 5 : 3, and at 16° or 17° Aquarius, about 11 : 8.

So the triangle ACB is isosceles with sides AC, CB equal, and (with AB = 1000), BC is $1666\frac{2}{3}$ when directed toward 17° Leo. Therefore (CD being dropped perpendicular to AB), where AD, which is half AB, is 1000, AC will be $3333\frac{1}{3}$. Looking this up in a table of secants, we find the angles CAD and CBD to be 72° 33′. So also, at 16° or 17° Aquarius, with AB = 1000, AC is 1375, so if AD = 1000, AC is 2750, showing 68° 40′ in the table of secants.

Therefore, with BC directed towards 16° or 17° Leo or thereabouts, the apparent position of Mars, AC, ought to be $72\frac{1}{2}°$ from the apparent position of the sun AB. And with BC at 16° or 17° Aquarius, these ought to be $68\frac{2}{3}°$ from one another. And since the sum of the two (CAB, CBA) at 17° Leo is 145°, ACB will be 35° at 17° Leo. Wherefore, Mars, which lies on the line AC, ought to be seen at 22° Virgo (the sun being on AB at 5° Sagittarius), or at 12° Cancer (the sun being at 30° Aries).

Similarly, at 17° Aquarius, since the sum (CAB, CBA) is $137\frac{1}{3}°$, ACB will be $42\frac{2}{3}°$, wherefore Mars, lying on the line AC, ought to be seen at $24\frac{1}{3}°$ Sagittarius[3] (the sun being on AB at 16° Libra) or at 0° Aries (the sun being at 9° Gemini).

Something approximating this could have happened, first, in November of 1586 or 1588; again, in April of 1581, 1583, 1596, and 1598; third, in September or October of 1587 and 1589; and fourth, in May or June of 1580, 1582, 1595, and 1597. In the last instance, suitable observations are lacking, since Mars in Aries, on account of its small ascension (where the sun in Gemini makes the

---

[1] That is, the apparent angle between Mars and the sun would be in the region of 60° to 72°.

[2] *Anomalia commutationis.* This denotes the motion of the planet upon the epicycle, as measured from the epicycle's apogee; hence, in the heliocentric system, it is the relative motion of the planet and the earth.

[3] As Caspar points out (*KGW* 3 p. 463), this is clearly wrong. Instead of "at $24\frac{1}{3}°$ Sagittarius (the sun being on *AB* at 16° Libra)", the text should read "at $4\frac{1}{3}°$ Capricorn (the sun being on *AB* at 26° Libra)."

nights bright) can hardly be observed, or even be seen at all.

Accordingly, on 1588 November 10 at 6^h 30^m in the morning the planet Mars was seen at 25° 31′ Virgo, with a latitude of 1° 36′ 45″ north, the sun being at 21 Scorpio.[4] The sun is thus only $62\frac{1}{2}°$ from Mars, although it should have been 72° from Mars in order to make the triangle isosceles, as the problem requires. Therefore, Mars is then farther from the earth than from the sun. Consequently, its latitude at that place appeared less than was the true inclination.

On December 5 following, at 6^h in the morning, Mars was seen at 9° $19\frac{2}{5}′$ Libra, with a latitude of 1° $53\frac{1}{2}′$ N., the sun being at 23° Sagittarius.[5] Therefore, since the sun was $73\frac{1}{2}°$ from Mars, the digression of the point which Mars then occupied on its orbit was a little less than 1° $53\frac{1}{2}′$ (since there should have been 77 72° between Mars and the sun). Since the present angle is greater, the distance of Mars from the earth turns out to be less than the distance of Mars from the sun. The apparent magnitude of the inclination—at least, of this point from the plane of the ecliptic—is consequently greater. But since on December 5 the planet in its eccentric motion was already several degrees beyond the limit, again diminishing its true digression from the ecliptic, this was consequently greater right at the limit. And since these two effects cancel one another, the maximum inclination of the planes will be about 1° 50′.

Similarly, on 1586 October 22 at 6^h in the morning, about dawn, there was about 6° 9′ eastward between Mars and Cor Leonis.[6] The declination of Mars from the equator was 13° 0′ 40″ north. Hence, its apparent longitude is found to be 0° 7′ Virgo,[7] latitude 1° 36′ 6″ N.[8] The sun stood at 8° Scorpio, 68° from Mars. It should have been farther. Consequently, the line between Mars and the earth was longer than that between Mars and the sun. And so the apparent latitude was less than the true digression of the planet from the ecliptic, and this was, in fact, long before it reached the limit.

But on November 2 at $4\frac{2}{3}^h$ in the morning, with the sun at $19\frac{2}{5}°$ Scorpio, Mars was seen at 5° 52′ Virgo, with latitude 1° 47′ N.[9] The sun was $73\frac{1}{2}°$ from Mars, by a nearly exact measure. But Mars was a few degrees before its northern limit which is at about 16° 17′. Therefore the latitude at this position appeared about right, although exactly at the limit it is reckoned to be greater than 1° 47′, namely, about 1° 50′.

On December 1 following, at $7\frac{1}{2}^h$ in the morning, the equatorial distance[10] between Cor Leonis and Mars was 25° $12\frac{1}{4}′$, and Mars's declination was

---

[4]Should be 28 Scorpio. The raw observation is in *TBOO* 11 p. 275; presumably Kepler computed the longitude and latitude.

[5]*TBOO* 11 p. 276. However, Tycho gives the position as 9° 40′ Libra, latitude 1° 57′ N. Perhaps Kepler recalculated the position from Tycho's raw observations.

[6]Regulus. The observation is in *TBOO* 11 p. 64.

[7]0° $2\frac{1}{2}′$ Virgo, by the translator's computation.

[8]1° 37′, by the translator's computation.

[9]*TBOO* 11 p. 64 (raw observation only).

[10]The "distantia aequatoria" is the difference in right ascensions; that is, the distance measured parallel to the equator.

$6° 2\frac{1}{4}'$.[11] Hence is found its longitude, 20° 4' 30" Virgo, and latitude, 2° 16' 30",
with the sun at 18° Sagittarius, which is 88° from Mars. It should have been
only $72\frac{1}{2}°$. Therefore the line between Mars and the earth is made less than that
between Mars and the sun, and because of the lesser distance, the digression
appeared greater than it really was. At this point, therefore, the digression
from the ecliptic was less than 2° $16\frac{1}{2}'$. Indeed, it was much less, but not
thereby much greater than 1° 47'. So here the magnitude of the maximum
inclination is indirectly confirmed to be 1° 50'.

On the other hand, on 1583 April 22, at $9\frac{3}{4}^h$ in the night, an interval of
20° 58' was observed between Mars and the Dog[12] and 22° $47\frac{1}{2}'$ between Mars
and Cor Leonis.[13] Hence, the position of Mars is found to be 1° 17' Leo, with
latitude 1° $50\frac{2}{3}'$ north. The sun was at 11° Taurus, 80° distant from Mars. This
should have been $72\frac{1}{2}°$. Accordingly, Mars is closer than it should be. Therefore, its observed latitude is greater than its true digression from the ecliptic.
But Mars is more than twenty-one degrees beyond the northern limit. So at
the limit, is digression will again be greater. Therefore, the opposite causes
again cancel one another, and the maximum inclination is 1° 50'.

Likewise, at 8^h in the evening on 1596 March 9, it was observed at 15° 49'
Gemini, with latitude 1° $49\frac{2}{3}'$ north.[14] The sun was at 30° Pisces, 76 degrees
from Mars. It should have been a little closer. Therefore, Mars's true digression from the ecliptic was a little less than the observed latitude. However, this
digression was not at its maximum, since Mars had not yet approached within
about 25 degrees of the limit. So once again, indirect support is provided for
a maximum digression of about 1° 50' at the limit.

Now, at the other limit, at 17° Aquarius, although observations are rarer,   78
there is one available.

On 1589 September 15 at $7\frac{1}{4}^h$ in the evening, Mars was observed at 16° $47\frac{1}{3}'$
Sagittarius with 1° $41\frac{2}{3}'$ southern latitude.[15] But when the correction for refraction of light which it underwent at this low altitude is applied, its position
was 16° $45\frac{2}{3}'$, with latitude 1° $52\frac{1}{3}'$ south. The sun was at 2° Libra, $74\frac{1}{3}°$ distant
from Mars. It ought to have been only $68\frac{2}{3}°$. Therefore, the observed latitude
is a little greater than the digression of its position from the ecliptic. However, that is not the most distant point, as it is several degrees before the limit.
Therefore here, too, the effects cancel.

On November 1 following, at $6\frac{1}{6}^h$, it was seen at 20° $59\frac{1}{4}$ Capricorn, with
latitude 1° 36' south, the sun being at 19° Scorpio.[16] While it was then no
more than 62° from Mars, it should have been $68\frac{2}{3}$. Therefore, the apparent
latitude is less than the true digression from the ecliptic. But at the same time,
the digression at this point is less than the digression at the limit, because this

---

[11]*TBOO* 11 p. 66. However, Tycho gives the distance as 25° $12\frac{1}{2}'$.

[12]Procyon.

[13]*TBOO* 10 p. 250.

[14]*TBOO* 10 p. 47. However, the latitude is given as 1° 47' 40" N; possibly the 0 was misread as 9.

[15]*TBOO* 11 p. 335 (raw observations).

[16]*TBOO* 11 p. 335. The longitude is not given.

point is beyond the limit. Therefore, the maximum inclination is much greater than 1° 36′, and by all indications is about as great as the apparent latitude on September 15, namely, 1° 50′, approximately.

I have carried through one method, in which a knowledge of the approximate proportion of the orbs is presupposed. The observations followed this method within the limits of calculation, indicating readily enough the maximum inclination of the planes.

I shall now present another method, for which rarer, more select observations are required. If these are to be had, what we are seeking is found without prior knowledge of the ratio of the orbs and this without the encumbrance of a laborious computation.

When two planes cut one another, any two lines drawn in the respective planes to the same point on the line of intersection and at right angles to that line always include one and the same angle.

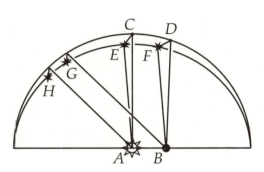

Let the plane of the ecliptic be ACDB, the plane of Mars's orbit AEFB, and let them intersect one another in AB. Let the sun be at A, the earth at B, and from A and B, perpendicular to AB, let AC and BD be set up in the plane of the ecliptic, and AE and BF in the orbit of Mars. Let the planet be at F. The inclination EAC of the limit E will be equal to the apparent latitude of the planet at F, namely, FBD. You will therefore note that if ever there is a perfect quadrature of the sun and Mars, with the line BA, that is, the sun, at 16° or 17° Taurus or 16° or 17° Scorpio—where between the line BA from earth to sun (which in this instance is also the line of intersection of the planes), and the line BF from earth to Mars, there is 90° or one quadrant intervening—then, whatever the apparent latitude of Mars FBD will be there, that will also be the maximum inclination of the planes EAC, although at that place F, Mars is not as far from the ecliptic as at E.

The first such day fell on 1583 April 22, which only just now I had under consideration. The sun was at 11 Taurus, five or six degrees below the node. The earth, therefore, was above the line of intersection towards Mars. On this account, the apparent latitude will be greater than the truth, since it is seen from nearer. On the other hand, since there are not 90° between the sun and Mars, the apparent latitude will on this account be less than the truth. On the supposition that these opposite deviations cancel one another, the inclination of the planes will approximately equal the observed latitude. The observed latitude was 1° 50⅔′. Therefore, the inclination of the planes is approximately that much.

On 1584 October 30, there was a select occasion, but no observation is available. However, on November 12 following, at $1\frac{1}{2}^h$ in the night, when the sun had already fallen about 14° or 15° below the diameter of intersection, the earth having risen that much (for Copernicus), or the diameter of intersection having fallen that much toward the earth (for Tycho), Mars was seen at 23° 14′

Leo with latitude 2° 12$\frac{2}{5}'$ north, while the sun was at 1° Sagittarius.[17] This angle is somewhat diminished owing to the inclination of Mars's line of vision to the line of intersection. But it is greatly augmented by its approach toward the earth. Therefore, the inclination is much less than 2° 12′, namely, 1° 50′.

On 1585 April 26 at 9^h 42^m, Mars was seen at 21° 26′ Leo with latitude 1° 49$\frac{3}{4}'$ north.[18] The sun was at 16° Taurus, right near the node. Mars's line of vision was a little inclined, since Mars was beyond 16° Leo. Therefore, the angle of maximum inclination of the planes was only a little greater than 1° 49$\frac{3}{4}'$; that is, 1° 50′ or a little greater.

Similarly, near the other limit, on 1591 October 16 at 6^h 30^m in the evening, Mars was seen at 1° 27$\frac{1}{3}'$ Aquarius with latitude 2° 10$\frac{5}{6}'$ south, decreasing.[19] (For on October 10 preceding, the latitude was 2° 18$\frac{2}{3}'$,[20] and on October 2 it was 2° 38$\frac{1}{2}'$.) The sun was at 2$\frac{1}{2}$° Scorpio, above the node. The earth was therefore below the node towards Mars. So because of this proximity, the observed latitude was greater than the inclination of the plane of the ecliptic. Fourteen days later, when the sun was at the node, if it were again to have decreased 28 minutes (the amount of decrease in the previous 14 days), there would remain 1° 45′. But the ratio of decrease does not remain the same when the earth departs from a star, or vice versa. For at a greater distance the decrease is always less. Therefore nothing can be adduced here against a maximum inclination of 1° 50′. On the contrary: it is indirectly confirmed.

The demonstration can be extended farther. *Let B A be a line drawn from the earth through the body of the sun at the place of the node, 17° Scorpio or Taurus, and let the planet be observed at any point whatever on the zodiac. Now the latitude which it appears to have measures the inclination of a point on the plane truly removed from the limit by an amount equal to Mars's apparent removal from the limit. Let Mars be observed on BG. Draw AH parallel to it. The apparent latitude of point G seen from B will be the same as the inclination of point H. And BG and AH are directed towards the same sidereal degree, because they are parallel.* For example, in the observation of 1585 April 26, the sun was at 16° Taurus, and Mars was observed at 21° 26′ Leo, with latitude 1° 49$\frac{3}{4}'$. Therefore, the inclination at the eccentric motion of 21° 26′ Leo is 1° 49$\frac{3}{4}'$. And since 21° 26′ Leo is 5° from the limit, and the sine of 85° is $\frac{1}{250}$ less than the whole sine,[21] the maximum inclination here will be greater by $\frac{1}{250}$ of it, that is, about 1° 50$\frac{1}{2}'$.

In the Ptolemaic hypothesis, the demonstration of this theorem proceeds   80 on the following basis.

---

[17]*TBOO* 10 p. 321.

[18]*TBOO* 10 p. 399 (raw observation).

[19]*TBOO* 12 p. 146. The position given is slightly different: 1° 27′ 48″ Aquarius with latitude 2° 10′ 52″ south.

[20]Brahe gives 2° 24′. The October 2 position, however, is stated as Brahe gave it.

[21]That is, the sine of 90°.

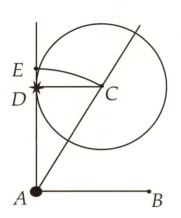

Let A be the earth, AB the line through the sun and the point opposite, at 17° Taurus or Scorpio, AD Mars's line of vision, D Mars, and BAD a right angle. AD will accordingly be at 17° Leo or Aquarius. And because D is Mars, a line drawn from D parallel to BA will pass through the center of the epicycle C (since the motion of Mars on its epicycle follows the motion of the sun in its orb). Take a point E on AD such that AE equals AC. Therefore, since AC will not be at 17° Leo or Aquarius, it will not be so far from the ecliptic as the northern limit E. D will likewise not stand so far from the ecliptic as E, because CD and all the points of the epicycle are equally distant from the ecliptic, since the plane of the epicycle, in order to make the hypothesis equivalent, is supposed always to remain parallel to the plane of the ecliptic. But proportionally as D or C is less distant from the ecliptic than E, D is closer to A than E, with the result that the distance D may be seen from A as proportionally greater, and both may be seen under the same angle from A. *Now, according to spherical trigonometry, as the distance of C from the ecliptic is to the distance of E from the ecliptic, so is the sine of arc CB (that is, AD) to the whole sine AE, because ECB is a circle inclined above AB. But C and D are equally distant from the ecliptic, as was just said. Therefore, AD is to AE as the distance of D (or the perpendicular drawn from D to the ecliptic) is to the perpendicular from E. Therefore, the triangles ADD and AEE will be similar, since they have right angles at points D and E on the ecliptic, and the sides are proportional. Also, they will be concurrent, since the sides (AD, AE) are drawn in the plane of the ecliptic through the same point A, and are directed toward the same point of longitude, 17° Leo or Aquarius. Therefore, the lines AD, AE in the orbit are also concurrent; that is, the line extended from the earth A through Mars D at this position will hit upon E, the center of the epicycle, when it is at the limit. Thus the angles of maximum inclination and of observed latitude of Mars will be the same at this place.*

A third way depends upon computation and upon prior knowledge of the ratio of the orbs. We shall have a taste of this one only for the sake of confirmation. A true and accurate treatment is reserved for Part V and ch. 63, and is not required here.

In Tycho's table of oppositions, the apparent latitude at 21° 16′ Leo was 4° 32$\frac{1}{6}$′.[22]

Let A be the sun, B the earth, C Mars on the eccentric. Thus the line AE running out through the earth B among the fixed stars will intersect the ecliptic, and AC will intersect the orbit of Mars. And since Mars is at 21° Leo, near the limit, the angle EAC is near its maximum. I track this down as follows. Let BA be 1000, AC 1664,

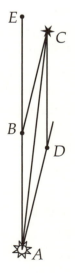

---

[22]This is the observation for January 1585, given in the table at the beginning of Ch. 8, above. However, Kepler has changed the longitude from Brahe's value of 21° 9′ 50″ Leo.

and *EBC* 4° 32$\frac{1}{6}$$'$. *Therefore, as AC is to EBC, so is BA to BCA,* 2° 43$'$ 27$''$.[23] *This,* subtracted from *EBC,* leaves the required angle *BAC,* 1° 48$'$ 43$''$; hence, right at the limit, it would be about 1° 49$'$ and is somewhat altered if the ratio *BA* to *AC* is altered (more on this below). In this manner, from any given acronychal observation with a comparatively great latitude, the inclination is found, first, of that point on the orbit, and then at its maximum, by considering the distance from the node or limit. For example, on 1593 August 24, the apparent latitude at opposition to the sun comes out to be 6° 3$'$ south.[24] Mars was at 12$\frac{1}{2}$° Pisces.   81 *Now let BA be* 1000 *and AC* 1389, *from our prior knowledge. As CA is to the sine of CBE, BA is to the sine of BCA,* 4° 21$'$ 10$''$. *Subtracting from CBE, this leaves the required angle BAC,* 1° 41$'$ 50$''$. *However, this position is about* 26° *from the limit,* 64° *from the node. So, as the sine of* 64° *is to this digression from the ecliptic of* 1° 42$'$, *so is the whole sine to the maximum inclination of the planes,* which comes out to be 1° 53$'$. We need not be concerned about the three minutes' excess, for they arise from the assumed ratio, for which see Part IV below.

In the Ptolemaic form, *A will be the earth, C the center of Mars's epicycle, and D the lowest point on the epicycle, since Mars is situated at opposition to the sun. And because the sun's line EA is on the ecliptic, while the plane of the epicycle is set up parallel to the plane of the ecliptic, CD will be parallel to EA. Therefore, BAC and ACD, the inclinations of the eccentric and the epicycle, are equal. But, owing to the full equivalence of the hypotheses, CD is equal to BA; that is, as AB is to AC in Copernicus, so is the semidiameter of the Ptolemaic epicycle DC to CA, the line from the earth to the center of the epicycle. Therefore, CDA and CBA are equal, and EBC and BAD are equal: the apparent latitude.*

---

[23]This is correct if one replaces the angles with their sines.

[24]See the table of observations in Ch. 8 above.

CHAPTER 14

The planes of the eccentrics are non-librating.[1]

The convolutions of Ptolemy's hypothesis forced him to accumulate many monstrosities in the theory of latitudes. For when he decided to make the plane of the epicycle tip every which way (it not being immediately clear, through the mists of his hypothesis, that the plane of the epicycle is parallel to the plane of the ecliptic), he contrived the latitude from three components, and in order that contraries counterbalance one another,* utterly wrenched his epicycle from its parallel position. He did not choose to find average values, whether on the grounds that his observations were not closely spaced, or that he distrusted those that were so, and hence accepted extreme values which were in error.

* See Maestlin's *Epitome astronomica*, on the final page of his account of the outer planets.

As a consequence, you may see that in the usual computation (e.g., in Magini's *Ephemerides*) there is no conjunction whatever of Mars and the sun that is not, as they say, "through the body". If this were true, it would been in vain that nature devised latitudinal temperings, which prevent the excessive arousal of the sublunar powers that often repeated physical conjunctions would cause.

Efficient cause of the latitudes.

Copernicus, ignorant of his own riches, ever took it upon himself to express Ptolemy, not the nature of things, to which, nonetheless, he of all men came closest. (In this regard, see Rheticus's *Narratio prima*.) For although he rejoiced to find that when the earth approaches a celestial body, the apparent latitudes are increased, he still did not dare to reject the remaining Ptolemaic increases in the latitudes (which this approach of the earth would not bring about). Instead, in order to reproduce these as well, he fabricated librations of the planes of the eccentrics, in which the angle of inclination (which for Ptolemy was fixed) would be varied. Moreover, in a manner close to being monstrous, this happens not according to the laws of motion[2] of its own eccentric, but to those of the earth's orb, clearly foreign. See Copernicus Book VI Ch. 1.

82

Armed by my own skepticism, I always opposed this gratuitous connection of diverse orbs as a cause of motion, even before seeing Tycho's observations. In this I more greatly congratulate myself that the observations were found to stand in agreement with me, as has happened with many other preconceived opinions.

But lest anyone deem me untrustworthy on this very account, claiming that I would treat the observations with prejudice, let him now witness that I

---

[1]Kepler used the Greek word ἀτάλαντα, derived from the verb ταλαντεύω, "sway to and fro."

[2]*Leges motuum.* This term was frequently used by astronomers of the sixteenth century to denote the principles governing planetary motion. For more on this topic, see the entry *"lex"* in the Glossary.

have most solidly demonstrated that there are no librations in the inclination of the eccentric. For three ways of investigating the maximum inclination were proposed. In the first, the sun was near Mars's sextiles and quintiles,[3] that is, about as close to Mars's conjunction as it could be and still allow Mars to be conveniently seen and observed. In the second, it was near quadrature with Mars, and in the third it was right opposite Mars. But when the sun was at all three places, Mars exhibited one and the same maximum inclination (1° 50′, about), northwards, at the same place on its eccentric, and the same amount southwards at the opposite position. Likewise, in Ch. 12, it appeared that when Mars was near the nodes of its motion on the eccentric, no matter what position the sun occupied on its orb (whether near Mars or far from it), Mars was never seen to have any latitude. And in the fifth part below it will be proved in many ways that the declination of Mars's orbit from the ecliptic is constant at any particular position on its orbit.

And so let us conclude with great certainty that the inclination of the planes of the eccentrics to the ecliptic does not vary at all. (For why should I not form a general conclusion, seeing that there is no reason why it should exist in only a single planet? Even so, I have demonstrated the same for both Venus and Mercury from the observations.) And a follower of Ptolemy may learn from this that the plane of the epicycle is always parallel to the plane of the ecliptic. For this is already demonstrated where the center is near the limits, and it was proved above in Ch. 12 that when the center is near the nodes, the epicycle lies entirely on the ecliptic.

Now who will set me up a fountain of tears, by which, for his deserts, I might bewail the pathetic industry of Apianus, who in his *Opus Caesarium*,[4] following his trust in Ptolemy, spent so many hours and wasted so many ingenious meditations trying to express, by means of spirals and corollae and helices and volutes and a vast labyrinth of the most intricate curves, a human figment which the nature of things clearly disowns? But that man shows us that he was easily capable of equaling nature by the divine talents of his most perspicacious wits. Apart from this, he entertained his mind with these tricks (in which he rivaled nature herself), which were thoroughly mastered and assembled in his models, and he has consequently won the prize of undying fame,[5] whatever diminishment fortune herself might have in store for the works themselves. But what are we to say of the empty artistry[6] of those who made the devices?[7] For they make six hundred, nay rather twelve hundred little wheels, so they can triumph in the presentation of the latitudes (that is, human figments thereof) in their works, and claim the consequent reward.

Peter Apianus' Opus Caesarium

---

[3]That is, in the region of 60°–72°.

[4]Ingolstadt, 1540. A lavishly produced book consisting mainly of "volvelles" (movable graduated paper wheels, like circular slide rules), by which one could predict planetary positions.

[5]*Famae perennis.* This may be an allusion to Ovid, *Amores* I.15 lines 7–8: "I seek undying fame, that I may ever be sung in the whole orb," though the phrase is also used by Virgil, *Aeneid* IX.79.

[6]Kepler coined a Greek word, κενοτεχνία, from κενός (empty) and τέχνη (art, skill).

[7]*Automatopoeorum.* This word, taken from the Greek, is a modification of the title of a work by Hero, on the making of marionettes.

CHAPTER 15

Reduction of observed positions at either end of the night to the line of the sun's apparent motion.[1]

Now, with that investigation carried through to its conclusion, and the positions of the nodes, the inclination of the planes, and its constancy all demonstrated, all of which were necessary for the coming reduction, we shall now define the positions which a planet may occupy on its orbit when the sun itself is diametrically opposite it. The years 1580 and 1597 may be omitted from the argument, as they present no suitable evidence owing to uncertainty of the observations.[2]

I. Suppose, however, that on 1580 November 12 at $10^h 50^m$ Mars was observed at 8° 37′ Gemini,[3] and the motion over five days was 1° 55′. Since at the given time the sun stood at 0° 45′ 36″ Sagittarius, and its motion over five days is 5° 5′, the sum of the two motions will come to 7° 0′. But the sun is 7° 51′ 24″ removed from opposition to Mars. Of this, seven degrees exactly are traversed in 5 days, or 120 hours. So, according to the same ratio, the remaining 51′ 24″ will be traversed in 14 hours 41 minutes. Therefore, the moment of opposition was November 18 at $1^h 31^m$. Its position was 6° 28′ Gemini on the ecliptic. Now this is 20° away from $16\frac{1}{2}°$ Taurus. I want to know how much longer this makes the arc on the orbit extended from the node to the arc of latitude through 6° 28′ Gemini. So I turn to Philip Lansberg's trigonometry.[4] I mention him out of honor and gratitude, for he has supplied me in abundance with the finest axes, best adapted for building astronomical foundations, from nearby, and at little expense of time. Without him, these would have had to be sought from afar with a great deal of trouble and toil, and the handles would

Lansberg's trigonometry.

---

[1] The positions in this chapter are, with three exceptions, modifications of those obtained from Brahe's observations, presented in Chapter 10, with which they should be compared (the exceptions are Obs. IV, taken from a different day, and Obss. XI and XII, which were made by Kepler). It is remarkable how much the raw observations had to be modified and extrapolated before Kepler could even begin constructing a theory; and even more remarkable that the resulting theory, which gives longitudes with unprecedented accuracy, is to be proved false immediately thereafter!

Since these acronychal observations are the foundation of Kepler's Mars theory, it was thought worthwhile to provide modern counterparts, as computed by the best available algorithms. They were calculated by Yaakov Zik with Guide 9 using the JPL DE 430 ephemeris, rounded to the nearest arc second. Time used is local apparent time for the location of the observations.

[2] Kepler believed the observations for 1580 to be uncertain because the diurnal motion was obtained from tables rather than from observations. But in fact, the observation itself was off by 20′. And ironically, although the observations for 1597 were made with a crude instrument, they were very nearly correct. For more of Kepler's comments, see Ch. 10, observations I and IX.

[3] Modern value: 8° 37′ 37″. Observation from *TBOO* 10 84.

[4] Philip Lansberg (or Lansbergius, or van Lansbergen) *Triangulorum geometriae libri* IV, Leyden 1591.

not have been so well fitted. From Lansberge, then, the tangent of the 20° side multiplied by the secant of the angle of inclination, 1° 50′, the last five digits being dropped,[5] gives an increase of only $18\frac{1}{2}$ of the smallest units, to which corresponds about 35 seconds. Mars therefore, standing opposite 6° 28′ Gemini, is 35″ further along on its orbit. It should therefore be placed at 6° 28′ 35″ Gemini, a tiny and quite unnecessary correction. The latitude is 1° 40′ north.

II. On the night following 1582 December 28, at 11^h 30^m, Mars was observed at 16° 47′ Cancer,[6] while the true position of the sun was 17° 13′ 45″ Capricorn. The moment of opposition had therefore passed. Now the sun's diurnal motion was 61′ 18″, that of Mars 24′, and their sum, 85′ 18″. At this moment, the distance between the stars was 26′ 45″. Therefore, as 1° 25′ 18″ is to 24 hours, so is 26′ 45″ to 7 hours 32 minutes. Subtracting this from 11 hours 30 minutes gives December 28 at 3^h 58^m after noon as the moment of true opposition. Its position on the ecliptic was 16° 54′ 32″ Cancer, and by reduction to the orbit (a 50″ correction), 16° $55\frac{1}{2}′$ Cancer. The latitude was 4° 6′ north, as given by Brahe's table of oppositions. For among the observations I find various latitudes: on the night following December 26, 4° 6′ or 4° 2′, while on the night following December 29, 4° 8′ or 4° $6\frac{1}{2}′$.

III. On 1585 January 31 at 12^h 0^m, Mars was observed at 21° 18′ 11″ Leo.[7]   84
The sun was at 22° 21′ 31″ Aquarius. The true opposition had therefore passed. The distance was 1° 3′ 20″. The sun's diurnal motion was 61′ 16″, that of Mars 24′ 15″, and their sum 85′ 31″. Now as 1° 25′ 31″ is to 24 hours, so is 1° 3′ 20″ to 17 hours 46 minutes, to which correspond about 18′ of Mars's motion. Therefore the time was January 30 at 19^h 14^m, and Mars's ecliptic position was 21° 36′ 10″ Leo. For reduction, some very small quantity is subtracted, because Mars is then beyond the limit. Therefore, the extension of the arc on the orbit from the following node is directed westward. But because Mars was only 4 or 5 degrees from the node, the subtraction is rendered quite imperceptible. The latitude, on the authority of the Tychonic table, was 4° 32′ 10″ north. For the observation on January 31 at 12^h gave 4° 31′. They added the remainder to the Tychonic figure, on account of diurnal parallax.

IV. On the night following 1587 March 4, at 1^h 16^m past midnight, the position of Mars was found to be 26° 26′ 17″ Virgo,[8] from Cor Leonis[9] and Spica Virginis, with an observed latitude of 3° 38′ 16″ north. But because Mars was elevated $37\frac{1}{2}°$ above the horizon, diurnal parallax comes into the reckoning, and subtracts some small quantity from the longitude, thus making it 26° 26′ Virgo, with a slightly greater latitude. For since the sun is nearly twice as far away from earth as is Mars, Mars's parallax will consequently be nearly twice

---

[5]Since decimal notation had not yet come into use, multiplication of trigonometric functions yielded numbers with many extra digits, which had to be removed. This was done, in effect, by dividing by the number of units in the radius. See the translator's introduction for a more complete explanation.

[6]Modern value: 16° 48′ 18″. Observation from *TBOO* 10 176–7.

[7]Modern value: 21° 18′ 47″. Observation from *TBOO* 10 388.

[8]Modern value: 26° 27′ 7″. Observation from *TBOO* 11 184 (under 5 March). Note that this observation is different from the one used in Ch. 10.

[9]Regulus.

the sun's. On the supposition that the sun's is 3′, Mars's will be about 5′. Now when 9° Sagittarius is rising, the nonagesimal[10] is 55° from the zenith. Opposite this number in our parallactic table,[11] under the column headed 5′, the latitudinal parallax 4′ is shown. Therefore, the latitude observed from the center of the earth would be 3° 42′ 22″ north. In Part V below, this will be useful to us in a more accurate examination of Mars's parallaxes, where a determination will be made of the precise inclination and of the absolutely certain distance of Mars for this position.[12] The sun's true position was 23° 59′ 11″ Pisces. The true opposition was therefore still to come. The stars were 2° 26′ 49″ apart. The sun's diurnal motion was 59′ 35″, that of Mars 24′, and their sum 1° 23′ 35″. As this is to 24 hours, so is 2° 26′ 49″ to 1 day 18 hours 7 minutes, to which corresponds 42′ 7″ of Mars's motion. Therefore, the time of true opposition was March 6 at 7^h 23^m. Mars's position on the ecliptic was 25° 43′ 53″ Virgo. For the reduction to the orbit, 55″ must be subtracted. Therefore, on the orbit it was at 25° 43′ Virgo. The latitude was decreasing. It was therefore somewhat less than 3° 38′ N., or 3° 42′ corrected for parallax.

V. On the night following 1589 April 15 at 12^h 5^m, the planet was found at 3° 58′ 20″ Scorpio,[13] with latitude 1° 4′ 20″ north, decreasing. Mars's altitude was $22\frac{1}{5}°$, where refraction from the table for the fixed stars was zero, and from the table for the sun, $3\frac{1}{2}′$. But the parallax was about twice as great as the sun's, that is, 6 minutes at the horizon. The degree rising was 24° Sagittarius. Therefore, the nonagesimal was 64° from the zenith, giving a diurnal latitudinal parallax of 5′ 24″. Whether it really was that much will become apparent below, through a careful consideration of latitudes.[14] For there, the northern latitude, free from diurnal parallax (and if there is no refraction), would come out to be 1° 9′ 45″ north. And because the altitude of the nonagesimal is 26°, the longitudinal parallax[15] at the horizon is 2′ 38″. But Mars is 40° from the

---

[10]The nonagesimal is the point on the ecliptic 90° westward from the rising point (the intersection of the ecliptic with the eastern horizon). The great circle drawn through the zenith and the nonagesimal is perpendicular to the ecliptic; hence, it represents the shortest distance from the zenith to the ecliptic, and is of use in determining parallax. In particular, at the nonagesimal the parallax is entirely latitudinal, longitudinal parallax being reckoned by finding the distance from the nonagesimal. Since at this time (unlike the next observation) Mars is close to the nonagesimal, no longitudinal correction is necessary. For an account of Kepler's treatment of parallax, see the translator's introduction.

[11]The parallactic table, which is essentially a multiplication table, is contained in Kepler, *Optics* (1604), W. H. Donahue, trans. (Green Lion Press, 2000), in the back pocket. For its use, see the translator's introduction.

[12]See Ch. 64, p. 465

[13]Modern value: 3° 59′ 49″. Observation from *TBOO* 11 333.

[14]See Ch. 64, p. 466.

[15]The longitudinal parallax (P_{long}) is expressed by the following equation:

$$P_{\text{long}} = P \sin A \sin d,$$

where P is the total parallax at the horizon;
   A is the altitude of the nonagesimal;
   d is the distance of the planet from the nonagesimal.
   The column mentioned in the following sentence is found in the parallax table in Kepler's *Optics*.

nonagesimal, counting from 4° Scorpio to 24° Virgo, which, under the column   85
headed 2′ 38″ shows a true longitudinal parallax of 1′ 42″. That hastens Mars
further forward than when viewed from the center of the earth, and this is so
on the assumption that it underwent no refraction. But to me it is more probable that it undergoes the same refraction as the sun, greater, that is, than that
of the fixed stars, because the opposition of the sun and Mars stirs up the air,
while the fixed stars are observed when the air is as calm as possible. Still, let
there be no refraction at all, and let Mars be placed at 3° 57′ Scorpio. At that
moment the sun was at 5° 36′ 20″ Taurus. At this time, therefore, Mars was
1° 39′ 20″ past opposition to the sun. Mars's diurnal motion, as is clear from
comparison with April 13, is 22′ 8″; the sun's, 58′ 10″; the sum, 1° 20′ 18″. As
this is to 24 hours, so is 1° 39′ 20″ to 1 day 5 hours 42 minutes. Therefore, the
moment of opposition was April 14 at $6^h 23^m$ PM. Its position was 4° 24′ 30″
Scorpio, or a little past, if refraction were applied or if the previously assumed
value for the diurnal parallax was too great. For reduction to the orbit, some
imperceptible quantity must be subtracted, since it is barely 12 degrees from
the node. This would be about 24 seconds, which are of no importance, and
Mars would be at 4° 24′ Scorpio, with latitude three minutes greater than before. For from that position, the latitude was decreasing from the eighth of
March; nor was it a maximum at opposition.

    VI. On the night following 1591 June 6 at $12^h 20^m$, Mars was found at 27°
14′ 42″ Sagittarius,[16] with latitude 3° 55$\frac{1}{2}$′ south. Refraction was of course provided for (from the table for the fixed stars), since it was large, in that Mars
had no more than 6° altitude on the meridian. There was, however, no mention of parallax. But at that time Mars was distant from the earth by half the
solar distance. Therefore, the horizontal parallax is greater than 6 minutes,
on the supposition that the sun's parallax is 3′. I omit it nonetheless, partly
because the refraction is supplied from the table for the sun (which, as I said,
is the more probable), exceeding that which Brahe took here by 4$\frac{1}{2}$′, which
almost completely cancels out the parallax; and partly because Mars was on
the meridian and near the winter solstice point, and thus had no longitudinal
parallax. Of the latitude, on the other hand, it will have to be seen below in
Part IV whether it might not be a few minutes less, since the parallax projects
the planet too far south.

    The sun was at 24° 58′ 10″ Gemini. The difference in position between
the stars was 2° 16′ 10″. The sun's diurnal motion was 57′ 8″, and Mars's
(for four days) was 1° 12′ 24″, since on June 10 at $11^h 50^m$ it was at 26° 2′ 18″
Sagittarius. For one day, therefore, 18′ 12″. The sum of the diurnal motions is
1° 15′ 20″. This corresponds to 1 day 19 hours 24 minutes, which, added to the
6th at $12^h 20^m$ (because opposition was yet to come), shows [the moment of
opposition to be] the 8th at $7^h 43^m$. Mars's position was 26° 41′ 48″ Sagittarius,
to which are added 52″ for reduction to the orbit, so as to make it about 26° 43′
Sagittarius. The latitude was six minutes greater than on 6 June, because, by
the observations, the latitude here was increasing until the fortieth day after
opposition, and increased nearly thirteen minutes between the 6th and the

---

[16]Modern value: 27° 13′ 41″. Observation from *TBOO* 12 142.

10th of June. Therefore, ignoring parallax and keeping the same refraction, it would be $4° \, 1\frac{1}{2}'$.[17]

86   VII. On 1593 August 24 at $10^h \, 30^m$, the ecliptic position of Mars was found to be 12° 38′ Pisces,[18] with latitude 6° 5′ 30″ south. The altitude was great enough that horizontal variations canceled one another. On the August 29th following, at $10^h \, 20^m$, Mars was observed at 11° 15′ 24″ Pisces, with latitude 5° 52′ 15″ south. It was decreasing precipitously. For before August 10 it was maximum, fourteen days before opposition. The motion for the five days was 1° 22′ 36″, and for one day, 16′ 31″. The sun's position on August 24 at $10\frac{1}{2}^h$ was 11° 2′ 31″ Virgo. The stars were 1° 35′ 30″ apart. The sun's diurnal motion was 58′ 20″. The sum of the diurnal motions was 1° 14′ 51″. This requires 1 day 6 hours 57 min. until opposition, so that this will be August 26 at $5^h \, 27^m$ in the morning. Mars's position was 12° 16′ Pisces. Its latitude was 6° 2′ south, approximately, if the horizontal variations do indeed cancel.

VIII. On 1595 October 30 at $8^h \, 20^m$, the planet was found at 17° 47′ 15″ Taurus,[19] not far from the nonagesimal. We may thus be sure of the parallax, although we must take it into account. The latitude was 0° 5′ 10″ north. The sun's position was 16° 50′ 30″ Scorpio. The distance between the stars was 56′ 45″. The sun's diurnal motion was 1° 0′ 35″; that of Mars, 22′ 54″, as appears by comparing the nearby observations. The sum of the diurnal motions was 1° 23′ 29″. If the distance between the stars be divided by this, it comes out to 40′ 47″ of a day, or 16 hours 19 min. Therefore, the true opposition was $0^h \, 39^m$ PM on October 31. Mars's position was 17° 31′ 40″ Taurus. This needs no reduction to the orbit, as it is nearly at the node. The latitude was about 0° 8′ north. But comparison with the preceding and following days shows a latitude of about 5′ north.

IX. Let, at any rate, Mars's position on 1597 December 10 at $8^h \, 30^m$ be 3° $45\frac{1}{2}'$ Cancer (as above).[20] The sun's position was 29° 4′ 53″ Sagittarius. The distance between the stars was 4° 40′ 37″. The sun's diurnal motion was 61′ 20″, that of Mars 23′ 40″ (for in 1580 the diurnal motion in Gemini was 23′, and in 1582, at 17° Cancer, it was 24′). Therefore, the sum of the diurnal motions was 1° 25′ 0″. These data show that the time of true opposition followed 3 days 7 hours 14 min. later, on December 14 at $3^h \, 44^m$ in the morning. Mars's position was 2° $27\frac{1}{3}'$ Cancer. The reduction to the orbit (quite ridiculous here, since the observation itself has an uncertainty of several minutes) requires the

---

[17]The computations for this reduction are among the Kepler manuscripts in the archives of the Russian Academy of Sciences in St. Petersburg (vol. XIV 213v–214v, transcribed in *KGW* 20.2, pp. 263–265). They contrast dramatically with what is presented here: he actually used a dozen observations between May 13 and July 16 in a much more laborious procedure than the present one. Here, as often elsewhere, Kepler has spared the reader much of the agony of his search for the true orbit. (On this volume of the St. Petersburg Kepler manuscripts, see Gingerich, "Kepler's Treatment of Redundant Observations", *Internationales Kepler-Symposium Weil der Stadt 1971* (Hildesheim, 1973), pp. 307-314.), reprinted as "The Computer versus Kepler Revisited," in Gingerich, *The Eye of Heaven* (American Institute of Physics, 1993), pp. 367–378.

[18]Modern value: 12° 37′ 48″. Observation from *TBOO* 12 288.

[19]Modern value: 17° 48′ 32″. Observation from *TBOO* 12 453.

[20]Modern value: 3° 44′ 20″. This was the "doubtful" observation made at Wandesburg with a backstaff: see Observation IX in Ch. 10, and *TBOO* 13 112.

addition of about 52″. Therefore, the corrected position was 2° 28′ Cancer. The latitude, from the table, was 3° 33′ north.

On the same night, the one following December 10, at $12\frac{1}{6}^{h}$, Fabricius in East Frisia found Mars's position to be 3° $40\frac{1}{4}'$ Cancer with latitude 3° 23′ N. In this observation the longitude comes out nearly the same. For the motion over 3^{h} 40^{m} is $3\frac{1}{2}'$, so that in the Brahean observation too, at $12\frac{1}{6}^{h}$ Mars could have been at 3° 42′ Cancer—two minutes beyond the Fabrician position.

X. On 1600 January 13/23 at 11^{h} 40^{m}, the time being adjusted to Uraniborg time, the planet was observed at 10° 38′ 46″ Leo.[21] The sun's position was 3° 26′ 30″ Aquarius. The stars were 7° 12′ 16″ apart. The sun's diurnal motion for the next few days was 1° 1′ 3″; that of Mars, 23′ 44″. The sum: 1° 24′ 47″. The opposition therefore followed 5 days 2^{h} 22^{m} after; that is, on January 19/29 at 2^{h} 2^{m} in the morning, before dawn. Mars was at 8° 38′ Leo. There is no need for reduction, since it is near the limit. The latitude, from the table, was 4° 30′ 50″ N.

XI. On the evening of 1602 Febr. 18/28 at 10^{h} 30^{m}, using the Tychonic in-   87
struments (with the help of the learned Matthias Seiffard,[22] bequeathed us by Tycho), I took the distance of Mars from the middle star of the tail of Ursa Ma-jor[23] to be 52° 22′. And since the distance between Cor Leonis and Procyon was 37° 22′ 20″, which should have been 37° 19′ 50″,[24] we know that the sex-tant reads $2\frac{1}{2}$ minutes high. Therefore, the corrected distance of Mars from the tail of Ursa was 52° $19\frac{1}{2}'$. And since the latitude of the fixed star is 56° 22′, the remainder, by subtraction, is 4° $2\frac{1}{2}'$—supposing that Mars was at precisely the same longitude as the fixed star. But because there was a difference of $3\frac{3}{4}$ degrees between them (as is clear from the following observations), a slight correction is required.

For let *AB* be 3° 43′ 30″ on a parallel close to the ecliptic, *B* Mars, *C* the fixed star, and *BC* 52° 19′ 30″. Dividing the secant of *BC* by the secant of *AB* gives the secant of *CA*, 52° 14′, which, subtracted from 56° 22′ (the latitude of the fixed star) leaves 4° 8′ north as Mars's observed latitude. At the same time, we found 19° 23′ between Mars and Cor Leonis (19° $20\frac{1}{2}$ corrected), and 21° 20′ between Mars and the bright star in the wing of Virgo[25] (21° $17\frac{1}{2}'$ corrected). From these two distances (using the latitudes of the stars and Mars), Mars's lon-gitude is found to be 13° 19′ 6″ Virgo, by consensus of all measurements.

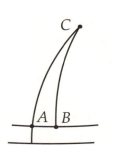

The reasoning behind this approximation [Ἀτεχνία] is given in the book on the star in Serpentarius.

---

[21] Modern value: 10° 40′ 16″. Observation from *TBOO* 13 221

[22] For Seiffard, see John Robert Christianson, *On Tycho's Island* (Cambridge University Press, 2000) pp. 356–357. He served Kepler for a number of years as assistant and secretary.

[23] Mizar.

[24] This distance is obtained from Brahe's *Epistolarum Astronomicarum Liber* (Uraniburg 1596) p. 50, in *TBOO* 6 p. 78.

[25] Gamma Virginis. There is no star in the standard catalogs with the name that Kepler gives. Other possible candidates are β Virginis and ε Virginis (Vindemiatrix). However, the distance given is not consistent with these stars, while computations from γ Virginis yield the longitude given by Kepler.

Alternatively, the meridian altitude of Mars was found at $12^h\ 40^m$, using two quadrants, to be 50° 19', while the tail of Leo[26] was 56° 45'. Therefore, from the declinations and right ascensions of the fixed stars and our distances, Mars's position is determined as 13° 19' 30" Virgo, latitude 4° 7' 55".[27] This is the Tychonic procedure. I have included the other for the sake of showing a consensus, and also that it might be evident that despite the lack of absolute perfection in the demonstration, short cuts either in computation or in our understanding can under certain circumstances be applied. For in that previous procedure there is less in the actual work than in the reporting of it. At Prague, 5° Scorpio was rising.[28] Therefore, the nonagesimal was about $32\frac{1}{2}°$ from the zenith. And since Mars's distance from the earth was rather more than half the sun's distance, the resulting parallax of about 5' opposite $32\frac{1}{2}°$ in our parallactic table, shows a latitudinal parallax of 2' 41". Thus the latitude as seen from the middle of the earth would be $4°\ 10\frac{2}{3}'$ north. And because the altitude of the nonagesimal was $57\frac{1}{2}°$, the longitudinal parallax at the horizon was 4' 13". But since Mars was 38° from the nonagesimal, the longitudinal parallax corresponding to this position is 2' 36", and if this be eliminated, Mars would be placed at about 13° 18' Virgo.[29] At that moment, the sun's position was 10° 16' 42" Pisces. The distance between the bodies was 3° 1' 18". The sun's diurnal motion was 1° 0' 4", and that of Mars, 24' 5" (for it was 24' 18" at 21° Leo in 1585, and 24' at 26 Virgo in 1587). The sum of the diurnal motions was 1° 24' 9". The true opposition therefore followed 2 days 3 hours 43 minutes later, on 21 February / 3 March before dawn at $2^h\ 13^m$, Mars being at 12° 27' 35" Virgo. Forty seconds must be subtracted to reduce the position to the orbit, putting Mars at 12° 27' Virgo, with slightly less latitude than before, since the latitude was then decreasing. It was therefore about 4° 10', or $4°\ 7\frac{1}{3}'$ if the parallax be neglected.

88   But because, since Tycho's death, we have not made frequent observations nor continued them over several days, it would be best, for certainty's sake, also to make use of those observations which David Fabricius of East Frisia, a sedulous practitioner of astronomy, has communicated to me.

On February 16, old style, at 5^h in the morning, he took the planet's distances from the tail of Leo (for the latitude), from the neck of Leo,[30] and, on the other hand, from the bright star in the southern wing of Virgo, so as to check its longitude by working it out twice.

I could make use of Tycho's line of reasoning, which he routinely adopted

---

[26]Beta Leonis, or Denebola.

[27]These are the longitude and latitude resulting from the distance from γ Virginis; the position resulting from the distance from Regulus is somewhat different.

[28]This was at the time of the previous observation, namely, $10^h\ 30^m$.

[29]Modern value: 13° 19' 41". The procedure is vitiated by Kepler's excessively small estimate of the sun's distance, resulting in too large a value for the total horizontal parallax. However, Kepler's practice—exemplified at the end of this paragraph—of giving alternative data with the parallax correction omitted, shows that he suspected the correction to be too large (see also Ch. 64). Oddly enough, when he applied the correction to the longitude, he appears to have subtracted only one minute rather than $2\frac{1}{2}'$, thus improving the accuracy of this figure.

[30]Gamma Leonis.

in volume I of the *Progymnasmata* when (as here) the declination of the planet was not known. But because that method extends to ten operations, I prefer for the sake of brevity to proceed as I did before with my own observations. Here there are no hidden dangers.

First, the star in the wing of Virgo, adjusted to our time, is at 4° 36′ 30″ Libra with latitude 2° 50′ north. Fabricius found that Mars is 20° 18′ westward from it on the zodiac, which puts Mars at about 14° 18′ 30″ Virgo. This is a preliminary, approximate figure: the longitude will shortly be corrected. Now the tail of Leo[31] is at 16° 4′ Virgo, with latitude 12° 18′ North, and Mars was found to be 8° 17′ distant from the Tail. What is sought is the distance of its parallel from the Tail, since the difference in longitude is 1° 45′. Dividing the secant of 8° 17′ by the secant of 1° 45′ gives the secant of 8° 6′, the arc sought. This subtracted from the fixed star's northern latitude of 12° 18′ leaves Mars's latitude, 4° 12′ north. This I now take as determined, and compare it with the latitudes of the fixed stars according to the laws of trigonometry. From the Wing of Virgo, I find Mars's longitude to be 14° 19′ Virgo; from the Neck of Leo, 14° 23′ 36″. The mean of these two is 14° 21′ 18″ Virgo. And since the sextant gave distances larger than the truth, the latitude would come out at 4° 14′ north.

On the night following February 23, at 12^h, he observed Mars in relation to five fixed stars: the Tail of Leo and Arcturus for latitude, and for the longitude, in the first instance, Spica (which followed it), and, in the second, the Neck of Leo and Cor Leonis (which preceded it).

By a rough estimate, I foresee that Mars is going to fall at $11\frac{1}{4}°$ Virgo. It was found to be 9° 24′ from the Tail of Leo, and hence, its latitude comes out at 4° 6′.[32] And now, through this and the latitudes of the fixed stars, together with their distances (17° 26′ from Regulus, 17° 51′ from the Neck of Leo, 37° 28′ from Spica, 44° 15′ from Arcturus), Mars's position comes out at 11° 21′ 23″ Virgo (from Regulus), 11° 20′ 52″ Virgo (from the Neck of Leo), and 11° 17′ 40″ (from Spica). Again (as you see) the distances err in being too great. For from the Heart and the Neck [of Leo], Mars is pushed a little eastward, and from Spica and Arcturus a little westward, and more from Arcturus, owing to its greater northern latitude.[33] The mean (ignoring Arcturus), 11° 19′ 20″ Virgo, is very nearly true. Also, the latitude is greater, namely, 4° 7′ 40″ north. Now from February 15 at 17^h to 23 February at 12^h, a period of 7 days 19 hours, Mars moved 3° 0′: 180 minutes in 187 hours. That is, about one minute per

---

[31] Denebola.

[32] According to the translator's computations, the latitude consistent with the given longitude and distance, together with Brahe's position for the Tail of Leo, would be 4° 13′. This is the first of several odd discrepancies in Kepler's treatment of Fabricius's observations; others will be noted as they arise. It is interesting to note that Kepler changes the latitude below.

[33] Recomputation (using Kepler's questionable latitude) shows agreement with Kepler's result for Regulus; however, the longitudes computed for the others come out about 7′ lower—that is, more to the west. Kepler does not give a position computed from Arcturus (which would be less trustworthy owing to that star's high latitude), but the translator finds that the given numbers result in a position that is nearly a degree *east* of the other computed longitudes, contrary to what Kepler says about this position. Altogether, the treatment of these observations is puzzling.

hour. And if you wish to take it into account, the parallax (if any) is subtracted from the longitude on February 16 and a little is added on February 23.

89   Because the last observation follows the time of observation found by me by 2 days 21 hours 47 min., add the motion corresponding to this time, 1° 7′: the position will come out 12° 26′ Virgo. The agreement is thus very good—it could not be better—given that both of us [that is, Kepler and Fabricius] work independently and do not rely on the facilities used by Tycho Brahe.

As for the latitude, on the 16th it was 4° 12′, and on the 23rd it was $4° 7\frac{2}{3}'$. It is therefore fitting to set it at 4° 9′ on the 21st, which comes between the other two dates. Correction for parallax makes it somewhat greater. And I, too, was putting it at a little less than $4° 10\frac{2}{3}'$, or 4° 10′.

XII. Finally, in 1604, when I had published my previously written Ephemerides, in which on the night between March 29 and 30 / April 8 and 9 the planet was placed on a line from Arcturus to Spica, that very thing appeared. For on the evening of April 8 it was slightly east of that line, but already on April 9 it was to the west. At that time, with the help of Johann Schuler and using Hofmann's sextant, I found 33° 4′ between Arcturus and Spica, which ought to have been $33° 1\frac{1}{2}'$. Therefore, the reading was $2\frac{1}{2}'$ too great. Immediately after, between Arcturus and Mars there was $29° 43\frac{1}{2}'$; therefore, correctly, 29° 41′. And since the latitude of Arcturus is $31° 2\frac{1}{2}'$ north, this left $2° 21\frac{1}{2}'$ as the latitude of Mars.[34]   There was then $54° 8\frac{1}{2}'$ between Cor Leonis and Mars, and at the same time, the same amount between Cor Leonis and Spica. This should, however, have been 54° 2′. There were therefore $6\frac{1}{2}$ minutes too much, while before the excess was only $2\frac{1}{2}'$. The origin of this uncertainty of four minutes could not be ascribed to the intervention of obstacles, as we were unable to eliminate it while observing. But let us suppose that, as before, the excess was $2\frac{1}{2}'$, making the distance between Mars and Cor Leonis 54° 6′. The error could then be in Spica's position, possibly because Mars was mistaken for Spica, since they were close to one another. Hence, Mars's latitude comes out to $2° 21\frac{1}{2}'$, longitude 18° 25′ Libra. The hour is known since, at the time of observation, the Back of Leo was culminating, whose right ascension is 163° 13′. Now the sun's position at noon is 18° 56′ 24″ Aries, whose right ascension is 17° 27′ 55″. Hence the difference in ascensions is 145° 45′, which resolves into 9 hours 43 min. The rising point was $22\frac{1}{2}°$ Scorpio. Therefore, the nonagesimal was 39° from the zenith, and the distance of Mars from the earth was a little greater than half that of the sun from the earth. So the parallax was about $5\frac{1}{2}'$, and its latitudinal part, 3′ 28″. Therefore the latitude without parallax was 2° 25′ (whether this correction was rightly made, we shall consider below). And because the altitude of the nonagesimal was 51°, and Mars's distance from the nonagesimal was 56°, the longitudinal parallax was 3′ 32″.

---

[34]Since Arcturus and Spica have nearly the same longitude, while Mars is collinear with them, the latitude of Mars can be found simply by subtracting its distance from Arcturus from the latitude of Arcturus. However, the numbers that Kepler gives result in a latitude of $1° 21\frac{1}{2}'$, 1° less than the stated latitude. Since Mars is at opposition, it is easy to find what Mars's observed latitude should be, given the Mars-sun and earth-sun distances and the inclination of the planes and distance from the node. This computation confirms the latitude given by Kepler, as does Kepler's table in Ch. 62. It is therefore evident that the correct distance between Arcturus and Mars was 28° 41′, not 29° 41′: the latter number may have been the result of a printer's error.

Therefore, Mars would be at 18° 21½′ Libra.[35]   At our chosen moment, the sun's position is 19° 20′ 8″ Aries. The two celestial bodies are 58½′ apart. The sun's diurnal motion is 58′ 38″, and that of Mars, 22′ 36″. (For in 1587 in Virgo it is 24′, and in 1589 at 4° Scorpio it is 22′ 8″). The sum of the diurnal motions is 1° 21′ 14″. From all these beginnings, it follows that the true opposition preceded the observation by 17 hours 20 min., namely, on 29 March/8 April at 4^h 23^m in the morning. Mars's position was 18° 37′ 50″ Libra. For reduction to the orbit, subtract about 39 sec., making Mars's position 18° 37′ 10″ Libra. The latitude was slightly greater than 2° 25′, but, when parallax is ignored, it is 2° 22′ north.

Now these twelve eccentric positions of Mars (so called because the longi-   90
tudes are freed from the effects of the second inequality) have been established with all possible care. If in this prickly business something has escaped me somewhere (and it did escape sometimes for a period of as much as eighteen months: I relied upon a false foundation—false, that is, for the applied observation—and all that work was in vain), I am entirely at a loss to imagine what it could be.

I shall therefore set out all the positions in the following table, with the addition of the mean longitudes from Tycho. I could have gotten these from the Prutenic tables, or from the computation upon which Ptolemy based his demonstrations and which he designed for that purpose, but this would be unnecessary. For if the mean motion needed correction, it will be corrected later. For the present, it will serve to measure the time intervals without any appreciable error.

	Date, Old Style					Longitude				Latitude			Mean Long.			
	Year	D.	Month	H^{36}	M	D	M	S	Sign	D	M		S	D	M	S
I	1580	18	Nov.	1	31	6	28	35	Gemini	1	40	N.	1	25	49	31
II	1582	28	Dec.	3	58	16	55	30	Cancer	4	6	N.	3	9	24	55
III	1585	30	Jan.	19	14	21	36	10	Leo	4	32⅙	N.	4	20	8	19
IV	1587	6	Mar.	7	23	25	43	0	Virgo	3	41	N.	6	0	47	40
V	1589	14	Apr.	6	23	4	23	0	Scorpio	1	12¾	N.	7	14	18	26
VI	1591	8	Jun.	7	43	26	43	0	Sagitt.	4	0	S.	9	5	43	55
VII	1593	25	Aug.	17	27	12	16	0	Pisces	6	2	S.	11	9	55	4
IIX	1595	31	Oct.	0	39	17	31	40	Taurus	0	8	N.	1	7	14	9
IX	1597	13	Dec.	15	44	2	28	0	Cancer	3	33	N.	2	23	11	56
X	1600	18	Jan.	14	2	8	38	0	Leo	4	30⅚	N.	4	4	35	50
XI	1602	20	Feb.	14	13	12	27	0	Virgo	4	10	N.	5	14	59	37
XII	1604	28	Mar.	16	23	18	37	10	Libra	2	26	N.	6	27	0	12

---

[35]Modern value: 18° 22′ 34″.

[36]Hours are reckoned from noon; hence, 19^h 14^m on January 30 is the same as 7^h 14^m on the morning of January 31.

CHAPTER 16

A method of finding a hypothesis to account for the first inequality.

Ptolemy, in Book 9 chapter 4[1] of the Great Work, where he is about to take up the first inequality, made by way of preface a somewhat cursory declaration of the suppositions of which he wished to make use. It is, in summary, as follows: We see that a planet spends unequal times on opposite semicircles. As, although from $2\frac{2}{3}^\circ$ Cancer through Leo to $26\frac{3}{4}^\circ$ Sagittarius is less than a semicircle, and from 26° Sagittarius through Aquarius to Cancer is more than a semicircle, nonetheless the planet is found to spend longer on the former than on the latter, although a law of uniformity[2] would require the contrary. For from a mean longitude of $2^S\ 23^\circ\ 18'$ to $9^S\ 5^\circ\ 44'$ is $6^S\ 12^\circ\ 26'$, more than a semicircle, that is, more than half of the planet's periodic time.[3] So from $12^\circ\ 16'$ Pisces through Leo to $12^\circ\ 27'$ Virgo is about a semicircle and 11 minutes. But if the mean longitude of the former position ($11^S\ 9^\circ\ 55'$) be subtracted from the longitude of the latter ($5^S\ 14^\circ\ 59'$), the difference is seen to be $6^S\ 5^\circ\ 5'$, which is $5^\circ\ 5'$ more than half. The planet consequently takes a proportionally shorter time from Virgo through Aquarius to Pisces. Now if you examine adjacent positions one at a time and compare the intervening arcs with the times or with the arcs of mean longitude, you will see that the planet is slowest at one fixed point on the zodiac, and swiftest at the opposite point, and that at the intermediate points its motion gradually increases or decreases, according to its proximity to one or the other.

These things reveal first of all that the motion of a planet (however irregular it may appear) is governed according to cycles, and that the present cycle is the successive modification of motion and a return to its same state. For if the planet moved in straight lines joined by angles (such as if it should move around a pentangle—I was once engaged in such ideas), its motion would sometimes suddenly change from swifter to slower in an evident manner, according to the relationship of the lines, and this would happen not in one but in many places on the zodiac, according to the number of lines. However, since so great an inequality still remains in the planet's motion, after the removal of

[1] Book IX Ch. 5, in modern editions.

[2] *Lex aequalitatis.* The phrase is odd enough to suggest the use of an awkward translation. It would be clearer to say "pattern of uniformity", but the Latin equivalent for that would be *ratio aequalitatis* or perhaps *species aequalitatis*. Kepler's use of the word *lex*, "law", should be noted, for while it is commonplace to speak of "laws" in a scientific context today, the term was then only beginning to acquire its modern signification.

[3] For an explanation of mean longitude as a measure of time, see the Glossary.

the inequality that depends upon the sun, it therefore will be incapable of being either governed or demonstrated by the supposition of a simple circle (one set up at the center of observation). This can, however, be done by composition of several circles, or the equivalent (as Ptolemy said in his preliminaries to Book 3). The simplest ways of doing this are two: by using either an eccentric circle or a concentric with an epicycle.

Thus Ptolemy chooses an eccentric for the first inequality, for the sake of distinguishing between the two and providing an aid to comprehension, since an epicycle would be required for the second inequality. Then, thinking over this general description, he denies that a mere eccentric suffices the planets. For he first considered closely what would duly follow from the simultaneous revolution of an epicycle (to account for the second inequality) and an eccentric (for the first inequality), and it was then evident, by comparing observations, that the center of the epicycle approaches much nearer to the earth at apogee, and flees farther from it at perigee, than the simple eccentric that accounts for the first inequality allows. From this discovery, by a continuous train of thought, he alights on the measure of this approach, and relates that he discovered that the center of the eccentric that carries the center of the epicycle is exactly at the midpoint between the center of observation, the earth, and the center of uniformity or of the eccentric accounting for the first inequality. And, without a single demonstration, he nevertheless relies upon this principle for the three superior planets.

Copernicus, as he frequently did on other occasions, here too followed his master religiously, his form of hypothesis being accommodated to this measure.

On this point, see the marginal note to Ch. 19.

Not without reason, astronomers have wondered about this, and I among them (using Maestlin's voice), as you see in the *Mysterium cosmographicum* Ch. 22 p. 79. Despite my having opined, in that passage of the said book, that Ptolemy used blind guesswork to establish this, the truth is the opposite. For he was able to prove it with a perfectly good demonstration given a suitable observation, as I shall demonstrate below. One finds fault with the theorist only in that he did not transmit to posterity those observations along with the demonstration.

92

And so, since I thought then that this was altogether too much to assume,[4] and also saw it pointedly called into question by Copernicus when he argued for a change in Mars's eccentricity on the basis of his figures which were not in accord with this bisection of the eccentricity, I envisaged a method which would lead me to a knowledge of the ratio of the two eccentricities (because, as I said, it is not indubitably 2 : 1). And since Ptolemy used three acronychal[5] observations and this preconceived opinion of the ratio of the eccentricities to find the position of the apogee, the correction of the mean longitude, and finally, the magnitude of the eccentricity, I saw that in order to weaken the

---

[4]μέγα λίαν αἴτημα.

[5]ἀκρονυχίοις.

sinews of the problem (once the axiom of the ratio of the eccentricity is taken away), it will be indeterminate and not having a single case, and thus I would need in addition the support of a fourth acronychal observation. And so, in the year 1600, having acquired knowledge of this art, I came to Tycho, and was happy to learn that he too did not assume this ratio, but made an investigation, as his figures indicate. For he makes the center of the (Copernican*) eccentric distant from the center of vision by 13, 680 units, while the point of equality in turn is another 3, 780 of these units beyond that. In the Ptolemaic form, this would be as if he were to make the distance of the centers of vision and of the eccentric 9, 900, and the remaining distance between the center of the eccentric and the point of equality 7, 560.

*Whose definition is at the beginning of Ch. 5 of this book.

Now I myself could also have taken the bisection of the eccentricity as certainly established, and with better reason than Ptolemy, because in Ch. 22 of my *Mysterium* I had brought forward a physical cause for the bisection. Indeed, it was for this very reason that I had come to Tycho, that I might use his observations to inquire further into my opinions expressed in that book. I of course did this without prejudice, and continue to do so. And if I survive to see astronomy achieve its purity and perfection, so that a verdict can be given in the case which I have brought before her tribunal in that book, I promise the reader that I shall retract that book and, upon confirming what is seen as true, will faithfully reveal the remainder that has turned out not to be so.[6]

On the *Prodromus*, or *Mysterium Cosmographicum*.

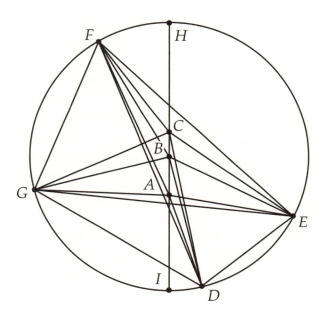

But back to the argument. *About center B let the eccentric FG be described, and on it, through B, the diameter of the apsides HI, taken as if immutable over any number of years. If there be danger of error in this assumption, we are not lacking ways of taking it into account. On this line below B let A be the observer, and above B let C be that center about which angles are made proportional to amounts of time, since (as was just said above) these are not proportional about A. Now let F, G, D, and E be four observations distributed about the circumference of the circle, so situated that the planet, stripped of the second inequality, would appear there as if* viewed from point A. For indeed, according to Ptolemy, A is the true center of vision

---

[6]Kepler kept this promise by publishing a second edition of the *Mysterium cosmographicum* in 1621, with extensive notes in which he frequently repudiated or modified his earlier opinions.

or the center of the earth, while according to Tycho and Copernicus vision takes place along the lines FA, GA, DA, and EA, and A is the sun.[7] *But it was said above that in either way the planet is likewise shorn of its second inequality. Now let each point be connected with each of the others, and let AF be at 25° 43′ Virgo, AG at 26° 43′ Sagittarius, AD at 12° 16′ Pisces, and AE at 17° 31$\frac{2}{3}$′ Taurus. Hence, the four angles about A are given: FAG is 91° 0′, GAD is 75° 33′, DAE is 65° 15$\frac{2}{3}$′, and EAF is 128° 11$\frac{1}{3}$′. These must be corrected somewhat on account of the precession of the equinoxes. For in relation to the fixed stars the planet is not so far forward at E (the last observation) as is indicated by these numbers. Wherefore FAE is a little greater, and the others smaller by the same amount. In the same way, by subtraction of the [mean] longitudes, the angles about C are also obtained.*

Proposition. It is now required to select values for angles *FAH* and *FCH* such that, once these are supposed, the points *F*, *G*, *D*, *E* stand on one circle, and that center *B* of that circle lie between the points *C* and *A* on the line *CA*.

The solution is not geometrical, at least if algebra is not geometrical, but proceeds by a double iteration.[8] For algebra, too, forsakes us here, because terms communicated by strictly straight lines through straight line do not extend to angles,[9] unless perchance one would wish to cram the entire theory of sines into this one operation.

But behold what we are required to do. If we were to assume a value for the angle *FAH*, then since the line AF has a certain sidereal position, the other leg *AH* would also be assumed to have a certain sidereal position. But let *AH* be the line of the apogee, or the line of the aphelion in the Copernican and Tychonic notion. We are thus required to assume and posit that which is sought. For it was in order to learn the position of this aphelion that we embarked upon this path. In the same way, since the sidereal position of *AH* (that is, *CH*) was arrived at through this assumption of ours, and it passes through *C* the center of our equant circle (and therefore also through the starting point from which the numbering of its parts has its beginning, namely, the apsis which is conceived as being above *H*); and since we are also required to assume the angle *FCH*, the line *CF* therefore also will acquire its position on the circumference of the equant. And indeed, this is the mean longitude corresponding to the observed position of the planet at *F*. And we were seeking to

---

[7] In the first edition, the diagram was printed with the letters *E* and *G* exchanged, and all subsequent editions and translations (except the present one) have repeated the error. (This error was discovered independently by D. T. Whiteside and the translator.)

[8] *Falsam positionem.* This is not, as one might guess, a *reductio ad absurdum,* but rather a procedure of supposing a trial value, knowing that it will turn out to be false, which is in turn displaced by a new trial value, and so on until the stated conditions are met. As Kepler says, there are actually two such iterations. In the "inner" iteration, a direction is found for the line *ACH*, and in the "outer" iteration, an adjustment is introduced that makes the center of the eccentric *B* fall on that line. A fine investigation of this iteration is Owen Gingerich's pair of articles, "The Computer Versus Kepler," and "The Computer Versus Kepler Revisited," in *The Eye of Heaven: Ptolemy, Copernicus, Kepler* (American Institute of Physics, 1993), pp. 357–378.

[9] What Kepler may be trying to say here is that there is no algebraic way, using arithmetical operations on variables, to express trigonometric functions. He seems to be hampered by a lack of established terminology for this problem.

know what this mean longitude is. Therefore, in addition to the apogee, we are assuming yet another thing among those which we were seeking.

At the same time, however, it is not unusual, whether in geometry, or arithmetic, or dialectic, to use a form of argument which leads to an impossibility, so that if something absurd is seen to follow from the assumptions, they are rejected as false; and this is carried out until the consequent removal of excesses and defects unveils the exact truth (which in the mathematical disciplines lies hidden in the middle between the two). In the present case this comes about in the following manner.

Let the line CA be taken as the nominal standard (and thus, let it be given). Since the angles FCH and FAH are assumed, and hence also the inclinations of the remaining lines to HCA, and AC is the common side of the four triangles CFA, CGA, CDA, CEA, whose angle are given, therefore the four lines AF, AG, AD, and AE will be given in relation to the length AC. And since in the four new triangles FAG, GAD, DAE, and EAF, the sides are already given with the angles at A between two sides,
94   *the individual angles at the bases of the several triangles (that is, AFG, ADG, ADE, and AFE), will not be a matter of ignorance. But AFG and AFE are parts of the angle GFE. And in the quadrilateral DEFG (if, indeed, it is inscribed in a circle, which is one of the hypotheses here), it is a consequence that two opposite angles (as GFE, GDE) taken together are equal to the sum of two right angles. Therefore, when the four angles which we have just found are combined, if their sum differs from the measure of two right angles, we shall pronounce the assumptions false, whether the falsehood be in one or the other of the assumed values, or both.*

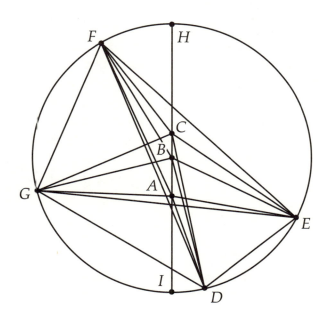

Then, one angle, FCH, being kept the same, and the other, FAH, being changed,[10] *a return will be made to the beginning, and the sum of the four angles will once more sought. If this sum differs more than the previous sum from two right angles, it suggests that FAH was altered in the wrong direction. Therefore, the opposite must be done: if you had added something to it, you now would diminish it, or vice versa. But if, on the other hand, you have come closer to the correct measure, then you will be sure that you are on the way. And then by comparing the original discrepancy with that which still remains, you will carry on in that same*

----

[10]This is not what Kepler does in the example that follows. Instead, he changes only the position of the aphelion, letting the mean longitudes remain fixed. That is, he changes *FAH* and *FCH* by the same amount. Then later, when the center of the circle does not fall upon the line of apsides, he changes the mean longitudes and then repeats the process of adjusting the aphelion until the four points lie upon one circle. Again, the center of the circle is found, and its new position will show how to adjust the mean longitudes so as to bring the center back to the line of apsides.

proportion by increasing or decreasing the angle FAH.

But it is still not certain that this second correction will directly reconcile your four angles with the exact measure. For the rate of increase of circular variables is not the same as that of straight ones. Your labor will have to be repeated again and again, until your sum for the angles in question is 180° or very nearly as much (you may safely ignore very small discrepancies).

When you have carried this out until the angles F and D (and therefore the remaining angles G and E) truly stand upon the same circumference, now, in turn, an enquiry must be made into the other matter that it is fitting to pursue, and that is, whether the center of that circle B lies between C and A on the same line. For on this point it was said above that Ptolemy assumed it outright, and physical considerations demand that the slowest motion occur where the star is at its greatest distance from A, the sun, as at H. This can happen in no other way than if A, B, and C are on the same line. To find this out, let the known angles GAD, DAE, be taken as one, so that the angle GAE may be known, and in GAE from this angle and sides GA, AE, let the side GE be sought. Now in the triangle GFE the angle GFE stands upon the circumference. Therefore, GBE, the angle at the center, is its double. But the angle GFE was previously found through its parts GFA, AFE. So, again, in the isosceles triangle GBE the angle GBE and the side GE are given. Consequently, the angles at the base will not be unknown, as well as GB the radius of the circle, in proportion to AC, the eccentricity taken at the beginning. And because BG and BGE are now had, and AG and AGE were had before, therefore, by subtracting AGE from BGE (or vice versa, as the case requires), the remainder will be AGB. Next, in triangle AGB, AG, BG, and the included angle AGB are given. [Hence, the other sides and angles will not be unknown, and thus the angle BAG will be given].[11] *If this differs from CAG, which was taken at the beginning, it shows that B itself, contrary to what should happen, does falls outside the line CA. So again we shall pronounce the assumed magnitudes of angles FCH and FAH to be false. But if we keep FCH fixed and change FAH we fall into yet another absurdity, namely, that the positions D, E, F, G, do not fall upon* 95 *a circle (just as they did not above, before we had finally established the magnitude of FAH). Therefore, it is obvious that FCH, too, has to be changed. Let it be changed, then; that is, let another quantity be taken at will for the angle FCH, and keeping that constant, let the angle FAH be adjusted four, five, or six times, until once again the four angles at F and D add up to two right angles. Then let an attempt be made at a second enquiry, using the triangles GAE, GFE, GBE, and BGA, to find BAG, in comparison with CAG as it has now most recently been established. Here you will again see whether you have departed farther from the truth, or have in truth come closer, and according to the qualities of excess or defect and ratios of the additions you will thence return to the beginning until you find BAG equal to the value you had assumed during that trial for CAG or HAG. When you have arrived at this point, then finally, in triangle BGA, you will assign a round number (100,000) to BG as a standard, and, in the same ratio (through the mediation of the angles), you will seek out BA the eccentricity of the eccentric and CA the eccentricity of the equant. Whence, by subtracting BA, CB remains. Then you will issue the pronouncement, concerning both the position of the apogee and the correction of the mean motion (which you had*

---

[11] This sentence has been supplied by the translator to fill an obvious lacuna in the text.

assumed in the final operation), that they are well established, at least as far as pertains to this form of hypothesis.

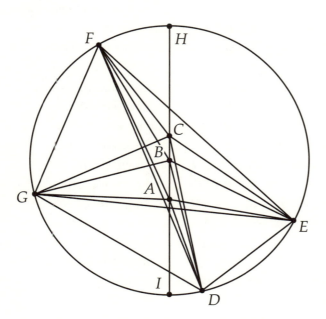

If this wearisome method has filled you with loathing, it should more properly fill you with compassion for me, as I have gone through it at least seventy times at the expense of a great deal of time, and you will cease to wonder that the fifth year has now gone by since I took up Mars,[12] although the year 1603 was nearly all given over to optical investigations.

There will arise subtle geometers such as Vieta who will think it something great to show up the contrived nature[13] of this method. For in this matter, Vieta did object to Ptolemy, Copernicus, and Regiomontanus.[14]   Let them therefore go forth themselves and solve the figure geometrically, and they will be to me great Apolloes.[15] For me it is enough to draw four or five conclusions from a single argument (which includes four observations and two hypotheses); that is, in getting from the labyrinth back to the highway, to show, instead of a geometrical light, a contrived thread,[16] which nonetheless will lead you to the exit. If this method is difficult to grasp, the subject is much more difficult to investigate with no method at all.

There now follows an example of this instruction based upon the four proposed observations.

Because of precession, all positions are reduced to [the time of] the first observation. Here, the apparent longitude was 25° 43′ Virgo, the mean longitude 6^S 0° 47′ 40″, and the annual motion of the fixed stars is 51 seconds, as Brahe has demonstrated in the *Progymnasmata*. Therefore, from 1587 March 6 to 1591 June 8 is 4 years 3 months, to which corresponds a motion of precession of 3′ 37″. Therefore, we must set the apparent position in 1591 at 26° 39′ 23″ Sagittarius, and the mean longitude at 9^S 5° 40′ 18″. Similarly, from 1587 March 6 to 1593 August 25 is 6 years $5\frac{1}{2}$ months, to which corresponds a motion of precession of 5′ 30″. And so Mars is to be placed at 12° 10′ 30″ Pisces, with mean longitude 11^S 9° 49′ 34″. Finally, from 1587 March 6 to 1595 October 31

96

---

[12]Kepler bet Longomontanus at the time of taking up the problem that he would have the orbit within a week!

[13]ἀτεχνίαν.

[14]For the relevant passage, Caspar refers the reader to *Francisci Vietae Apollonius Gallus seu exsuscitata Apollonii Pergaei περι επαφων geometria*, Paris 1600, p. 11.

[15]Virgil, *Bucolica*, Ecl. 3, 104: "Dic quibus in terris, et eris mihi magnus Apollo,…" (Tell in what lands, and you will be the great Apollo to me).

[16]ἄτεχνον.

is nearly 8 years 7 months, to which corresponds a motion of 7' 18''. And so Mars is to be placed at 17° 24' 22'' Taurus, with mean longitude 1^S 7° 6' 51''.

Now first, we shall assume that the apogee or aphelion in 1587 is at 28° 44' 0'' *Leo, and second, we shall assume that the mean longitudes should be increased by 3'* *16'',[17] so that the mean longitudes are 6^S 0° 50' 56'', 9^S 5° 43' 34'', 11^S 9° 52' 50'',* *and 1^S 7° 10' 7''.*

			°	'	''	
And because	CH	is	28	44	0	Leo
and	CF		0	50	56	Libra
	FCH	will be	32	6	56.	
Also, because	CH	is	28	44	0	Leo
and	CD		9	49	34	Pisces[18]
	HCD	will be	168	54	26;	
	its supplement,		11	5	34.	
Also, because	CH	is	28	44	0	Leo
and	CG		5	40	18	Capricorn
	HCG	will be	126	56	18;	
	its supplement,		53	3	42.	
Also, because	CH	is	28	44	0	Leo
and	CE		7	6	51	Taurus
	HCE	will be	111	37	9;	
	its supplement,		68	22	51.	

For the angles of the equations.

CF	0°	50'	56''	Libra		CG	5°	43'	34''	Capricorn
AF	25	43	0	Virgo		AG	26	39	23	Sagittarius
CFA	5	7	56			CGA	9	4	11	
CD	9°	52'	50''	Pisces		CE	7°	10'	7''	Taurus
AD	12	10	30	Pisces		AE	17	24	22	Taurus
CDA	2	17	40			CEA	10	14	15	

---

[17]Kepler makes this adjustment with the benefit of hindsight. He was aware that a small adjustment in the mean longitudes requires a large change in the position of the aphelion. Thus, although his iterative process could begin with nearly any arbitrary aphelion and mean longitudes, a large error, especially in the mean longitudes, could result in more iterations.

One might wonder why the mean longitudes need to be changed, if they are merely arbitrary and relative indications of the time between observations, expressed in terms of the planet's periodic time. The answer is that the mean longitudes also represent actual angles about the equant point *C*, measured from the vernal equinox. This is convenient in that the mean longitude and the observed longitude are the same when the planet is at either apsis.

[18]It is striking that, immediately after giving the corrected mean longitudes, Kepler puts *CD*, *CG*, and *CE* at the previous, uncorrected mean longitudes, and then, in finding the angles of the equations, returns to the corrected forms. Such errors make it seem hardy remarkable that it took him many trials to establish a usable line of apsides and eccentricity.

It should also be remarked that there are many errors in the following computations. These have not been noted in the translation because the chief value of Kepler's example lies not so much in the numbers he uses as in the procedures he demonstrates. Repetition of the procedure will, of course, yield different values (the translator, for example, would put the aphelion at 28° 50' 37'' Leo, having increased the mean longitudes 3' 26''). Those who wish to find errors will find many of them noted by Caspar in *KGW* 3 pp. 157–167.

For the lines from A.

Let AC be taken as the standard, with magnitude 10,000. Now as the [sines of the] angles of the equations are to AC, so are the [sines of the] angles at C to the lines from A. Therefore, the sines of the angles at C, multiplied by 10,000, are to be divided by the sines of the angles of the equations.

Sin. *FCH*	53163	**AF**
Sin. *CFA*	8945	
	44725	5
	84380	
	80505	9
	3875	
	3578	4
	297	
	268	3
	29	3

Sin. *GCH*	79928	**AG**
Sin. *CGA*	15764	
	78820	50
	11080	
	11035	70
	45	3

Sin. *DCH*	19240	**AD**
Sin. *CDA*	4004	
	16016	4
	3224	
	3203	80
	208	
	200	5
	8	2

Sin. *ECH*	92966	**AE**
Sin CEA	17773	
	88875	5
	40910	
	35546	2
	5364	
	5333	30
	31	2

97

For the angles at *A*.

AF	25°	43′	0″	Virgo	*AG*	26°	39′	23″	Sagittarius
AG	26	39	23	Sagittarius	*AD*	12	10	30	Pisces
FAG	90	56	23		*GAD*	75	31	7	
Supplement:	89	3	37			104	28	53	

AD	12°	10′	30″	Pisces	*AE*	17°	24′	22″	Taurus
AE	17	24	22	Taurus	*AF*	25	43	0	Virgo
DAE	65	13	52		*EAF*	128	18	38	
Supplement:	114	46	8			51	41	22*	

For the angles at *F* and *D*

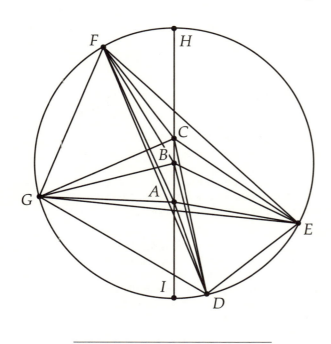

The angles AFG, AFE, ADG, ADE, are approximately the halves of the supplements of the angles at A.[19] *Those at F, however, are smaller, because the lines AG (50,703) and AE (52,302) have been found to be smaller than AF (59,433); and those at D are greater, because those lines AG and AE are longer than AD (48,052). And since those four angles about A are equal to four right angles, the sum of their supplements consequently will also equal four right angles, because four semicircles are equal to eight right angles. Therefore, half of the sum of the supplements is*

---

[19]Because the triangles are nearly isosceles.

equal to two right angles, which is what we wish the sum of GFE and GDE to be. Consequently, the angles at D ought to exceed [the halves of] their supplements by the same amount that those at F fall short of theirs. But the tangents of the differences of the angles at the bases in this kind of triangle are found if you divide the differences of the sides by the sums of the sides, and multiply the quotient by the tangents of the halves of the supplements.[20] *Therefore, if the differences of the two angles at F together equal the sum [of the two differences] at D, angle at F plus the angle at D will equal two right angles.*[21]

		FAG			GAD			DAE			EAF	
Halves		44° 31′ 48″			52° 14′ 27″			57° 23′ 4″			25° 50′ 41″*	
Tangents		98373			129093			156271			48438	
	AF	59433		AG	50703		AD	48052		AE	52302	
	AG	50703		AD	48052		AE	52302		AF	59433	
Differences		8730			2651			4250			7131	
Sums		110136			98755			100354			111735	
		770952	7		197510	2		401416	4		670410	6
		102048			67590			23584			42690	
		99123	9		59253	6		20071	2		33520	3
		2925			8337			3513			9170	
		2203	2		7900	8		3016	3		8938	8
		722	6		437	4		497	5		232	2
Quotients		7926			2684			4235			6382	
Tangents		98373			129093			156271			48438	
		6886	11		2581	86		6250	84		2906	86
		885	33		774	54		312	54		145	34
		19	66		103	20		46	86		38	72
		5	88		5	16		7	81			96
Tangents		7797			3465			6618			3142	
Differences	F	4° 27′ 30″		D	1° 59′ 4″		D	3° 47′ 10″		F	1° 47′ 59″	
					3 47 10						4 27 30	
		Sum of the two at	D	5 46 14			Sum of the two at	F	6 15 29			

From this it is clear, therefore, that the sum of F and D is less than two right   98 *angles, because the difference to be subtracted exceeds that to be added.*

The amount wanting is 24′ 15″. Now I know from many repetitions of this task that by adding 3′ 20″ to the aphelion, the sums come together. This I shall prove.

The angles of the equations and their sines will stay the same, as well as the tangents of the halves of the supplements of the angles at A.

---

[20]This is, of course, the "law of tangents", which, stated algebraically, is:

$$\tan \frac{1}{2}(180 - C)\frac{a-b}{a+b} = \tan \frac{1}{2}(A - B).$$

Kepler neglected to state that it is the tangent of *half* the difference that is found in this way.

[21]That is,

$$(GDA - DGA) + (EDA - DEA) \text{ should equal } (FGA - GFA) + (FEA - EFA)$$

But HCF 32° 3′ 36″			GCI 53° 7′ 2″			DCI 11° 2′ 14″			ECI 68° 19′ 31″		
Sines	53081	AF		79986	AG		19145	AD		92929	AE
Sin. CFA	8945		Si. CGA	15764		Si. CDA	4004		Si. CEA	17773	
	44725	5		78820	5		16016	4		88875	5
	83560			11660	0		3129			40540	
	80505	9		11035	7		28028	7		35546	2
	3055			625			3262			4994	
	2683	3		630	4		2803	8		3555	2
	372			5	0		459	1		1439	
	358	4					10	5		1244	8
	14	½								195	1
										8	

AF	59341		AG	50740		AD	47815		AE	52281	
AG	50740		AD	47815		AE	52281		AF	59341	
	8601			2925			4466			7060	
	110081			98555			100096	4		111622	
	770567	7		197110	2			4		669732	6
	89533			95390				6		36268	
	88065	8		88700	9			2		33486	3
	1468			6690						2782	
	1101	1		5913	6					2232	2
	367			777	8					550	5
	330	3									
	37	3									

Tangents	98373			129093			156271			48438	
	7813			2968			4462			6325	
	6886	11		2581	86		6250	84		2906	28
	786	96		1161	81		625	08		145	29
	9	83		77	40		93	72		9	68
	2	94		10	32		3	12		2	40
	7686			3831			6973			3064	
	F 4° 23′ 41″			D 2° 11′ 37″			D 3° 59′ 10″			F 1° 45′ 18″	
							2 11 37			4 23 41	
			Sum at	D 6° 10′ 47″		Sum at	F 6° 8′ 59″				

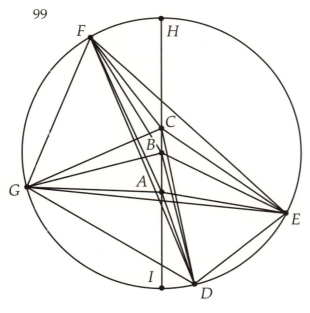

99

Here the sums differ by no more than 1′ 48″. So we have now moved the apogee too far forward, and must move it back by another 12″. But not much care is needed with such a small difference. We shall make it up "from the equal and the good" [that is, by interpolation], in order to be able to carry on with our method. Before, when we had erred in defect by 29′ 15″, the sum of the differences at F and D was 12° 1′ 44″. Now, when we have erred in excess by 1′ 48″, this sum comes out to be 12° 19′ 46″. And so, since 31′ will produce 18′ in the sum of the differences, therefore, $1\frac{4}{5}′$ produces about 1′, so that the exact sum is 12° 18′ 44″, whose half, 6° 9′ 22″, is the sum at either F or D.

For the Triangles *GFE, GBE*.

In FAG half the supplement was		44°	31′	48″.
In FAE		25	50	41.
	The sum	70	22	29.

From this subtract the sum of the differences	6	9	22		*And because GAE is*	75°	31′	7″
The remainder is GFE	64	13	7		*and DAE*	65	13	52.
Therefore, its double, at GBE, will be	128	26	14		*Therefore, GAE is*	140	44	59
The supplement of which is	51	33	46		*its supplement,*	39	15	1 .
Its half,	25	46	53					
Also, in the first trial, GA was			50703		*and AE,*			52302
In the second,			50740					52281
Difference			37					21
Therefore it is now			50739					52282

Next, GE is sought, from the sides GA, AE, and the angle GAE.

GA	50739		*Half the supplement of GAE*	19° 37′ 30″
AE	52282		*Tangent*	35658
Difference	154300			1497
Sum	103021	1		356|58
	51279			142|63
	41208	4		32|08
	10071			2|49
	9272	9		534| 0° 18′ 21″
	799	7		*Half the supplement* 19 37 30
			AGE	19 55 51

As the sine of AGE is to AE, so is the sine of GAE to GE. [22]

Sine GAE	63271			*3307935	*GE*
AE	52282		*Sine AGE*	34088	
	3163550			306792	9
	126542			240015	
	12654			238616	70
	5062			1399	
	127			1363	4
	3307935*			36	1

Therefore, in GBE, as [sine] GBE is to GE, let [sine] BGE be to BE.   100

43494	*Sine BGE* [†]	4218701		
97041	*GE*	78327		*Sine GBE*
3912460		391635	5	
304458		302351		
1740		234981	3	
43		67370		
4218701		62662	8	
		4708		
† *Incorrectly labelled GBE.*		4699	6	
		9	0	

[22]When Kepler marks a number with an asterisk, as in the computation below, he is alerting the reader that the number is carried over to or from somewhere else nearby.

And because AGE was	19° 55′ 51″	*BG*	53860
But now BGE is	25 46 53	*AG*	50739
BGA will be	5 51 2	*Difference*	312100
Supplement	174 8 58	*Sum*	104599
Half	87 4 29		209198 |2
Tangent 1957200			102902
2984 *			94140 |9
39144			8762
17615			8368 |8
1564			394 |4 *
78			

58401 | 30° 17′ 8″
      87 4 29
117 21 37   BAG

In the last trial we had moved the aphelion forward a total of 3′ 8″.

Therefore, because AH was 28° 47′ 8″ Leo,
And AG was   26 39 23 *Sagit.,*
HAG or CAG was   117 52 15.

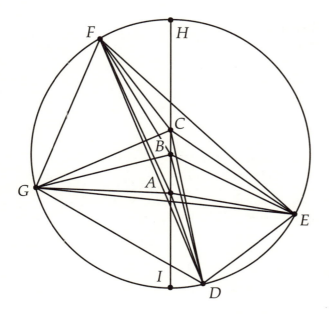

Therefore B is a little off the line CA on the side of G, since CAG is greater than BAG by 30′ 38″. But I know from many trials that by the addition of one half minute to the mean longitude, B is made to lie upon the line CA. But at the same time, to keep the quadrangle on the circle, the aphelion has to be moved forward 2′. It is worthwhile seeing this through, at the same time demonstrating the eccentricity. So since 30″ are added to CF and its counterparts, and 2′ to CH, HCF will be diminished by 1′ 30″. Therefore,

HCF is 32° 2′ 6″; GCI 53° 8′ 32″; DCI 11° 0′ 44″; and ECI 68° 18′ 1″.

But the angles of the equations are increased and decreased by 30″.

So CFA 5° 8′ 26″		*CGA* 9° 4′ 32″		*CDA* 2° 17′ 10″		*CEA* 10° 13′ 46″	
Sine HCF 53044		80012		19102		92913	
Sine CFA 8960	*AF*	15758	*AG*	3989	*AD*	17758	*AE*
44800 |5		78790 |50		15956 |4		88790 |5	
8244		12220		3146		4123	
8064 |9		11030 |7		27923 |7		35516 |2	
180		1190		3537		5714	
179 |2		1103 |7		3191 |8		5327 |3	
01 |0		87 |5		346		387	
|1				319 |8		355 |2	
				27 |7		32 |2	

AF 59201		AG 50775		AD 47887		AE 52322	
AG 50775		AD 47887		AE 52322		AF 59201	
8426		2888		4435		6879	
109976		98662		100209		111523	
769832	7	197324	2	4 836	4	669138	6
72768		91476		42664		18762	
65986	6	88796	9	40084	4	11152	1
6782		2680		2580	2	7610	
6599	6	1973	2		6	6691	6
183		707				919	
110	1	690	7			892	8

The tangents remain:

98373		129093		156271		48438	
7661		2927		4426		6168	
6886	11	2581	86	6250	84	2906	28
590	22	1161	81	625	08	48	44
59	02	25	82	31	25	29	06
	98	9	03	9	36	3	87
7536		3779		6917		2988	
4° 18′ 36″		2° 9′ 52″		3° 57′ 24″		1° 42′ 41″	
		3 57 24				4 18 36	
The one sum		6 7 16		*The other sum*		6 1 17	

We have six minutes left over, which are removed by moving the aphelion back 38″. So, since it was at 28° 49′ 8″ Leo, it will now be at 28° 48′ 30″ Leo.

102

Test

HCF 32° 2′ 44″		GCI 53° 7′ 54″		DCI 11° 1′ 22″		ECI 68° 17′ 23″	
53060		80001		19120		92905	
8960		15758		3989		17758	
4480	5	78790	50	15956	4	88790	5
8260		12110		3164		4115	
8064	9	11031	7	27923	7	35516	2
196		1080		3717		5634	
179	2	945	6	3591	9	5327	3
170	1	135	9	126		307	
79	9			120	3	178	1
				6	1	129	7

The numbers have the same derivations as just above.

59219		50769		47931		52317	
50769		47931		52317		59219	
8450		2838		4386		6902	
109988		98700		100248		111536	
769916	7 Prev. 7	1974	2 Prev. 2	400992	4 Prev. 4	669216	6 Prev. 6
75084		864		37608		20984	
65993	6 6	7896	8 9	30074	3 4	11154	1 1
9091		744		7534		9830	
8799	8 6	691	7 2	7017	7 2	8922	8 6
292	3 2	53	5 7	517	5 6	908	8 8
	Diff. 21		*Diff.* 52		*Diff.* 51		*Diff.* 20

	98373	129093	156271	
	21	52	51	48438
	98373	2 | 58168	1 | 56271	| 20
Increase of	19 | 6746	64 | 5465	78 | 1355	9 | 68760
tangent	21	67	80	10
Increase of arc	41″	2′ 14″	2′ 39″	19″
		2 39		41
	Prev. 6° 7′ 16		*Prev.* 6° 1′ 17″	
	Now 6 2 23	*Note the Equality*	*Now* 6 2 17	

103   *And so, with the quadrangle contained in the circle, let it again be enquired whether B lies upon the line C A. And from the sum of 70° 22′ 29″ established above, subtract the difference of 6° 2′ 20″ just found. The remainder is*

GFE	64° 20′ 19″	*The tangent of half the supplement of GAE remains*	35658
Its double, GBE	128 40 18		1502**
Supplement	51 19 42		356 | 58
BGE	25 39 51		178 | 29
In the last trial, GA was	50769		| 71
AE was	52317	535½ 18′ 24″	
	154800	19° 37 30	
	103086 | 1	*AGE* 19 55 54	
	51714		
	51543 | 50		
	171 | 2**		

Sin GAE	63271	3310148*	
AE	52317	34089	*Sin AGE*
	3163550	306801 | 9	
	126542	242138	
	18981	238623 | 7	
	633	3515	
	442	3409 | 10	
	3310148*	106 | 3 *GE*	

BG	53866	
GA	50769	
	309700	
	104635	2
	209270	
	100430	9
	94172	6
	6258	0

AGE	19° 55′ 54″	
BGE	25 39 51	
BGA	5 43 57	
Suppl.	174 16 3	
Half	87 8 1½	
Tangent	1997100	
	2960	
	1198	26000
	17973	9
	39942	
	59114	

Aphelion	28° 48′ 30″	*Leo*	30° 35′ 22″		
AG	26° 39′ 23″	*Sagittarius*	87 8 1		
	117° 50′ 43″	*CAG*	117 43 23	*BAG*	
			62 16 37		

B still lies 7′ 20″ from the line CA towards G.

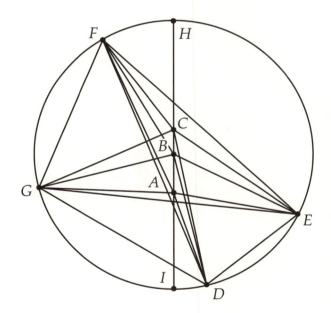

   Whence we see that because before, by adding 30″ to the mean motion and 82″
to the aphelion, we moved forward 23′ 18″, we will take up the remaining 7′ 20″ by
adding 9″ to the mean motion and 25″ to the aphelion. Therefore, the total addition
to Tycho's longitude is 3′ 55″,[23] and the aphelion is placed at 28° 48′ 55″ Leo.

   And with such a small error, one incurs no disadvantage by finding BA from the
given angles and sides of triangle CAG as if B lay exactly on the line CA.

---

[23]That is, he began by adding 3′ 16″ to the mean longitudes, then added another 30″, and finally,
9″, the total being 3′ 55″.

Sine BGA	998800000	
Sine BAG	8852	1
	11360	
	8852	1
	2508	2
	17704	
	7376	8
	7082	
	294	3

Therefore BA is 11283
where BG is   100000.

But as BG, 53,866, is to 100,000, so is 100,000 to AC.

BG	53866	1
	46134	
	430928	8
	30412	5
	26933	
	3479	6
	3232	4

Therefore AC is   18564
And BC is   7281
where BG is 100000

But to exclude all error, let us interpolate.

At first, *BG* was	53860	*AG* 50739	*BGA* 5° 51' 2"	*BAG* 62° 38' 23"
Now it is	53866	50769	5 43 57	62 16 37
Difference	6	30	7 5	21 46
We must proceed with				*BAG* 62 8 37
one third more:	2	11	2 25	5 41 32
BG corrected	53868	*AG* 50780	*BGA* 5° 41' 32"	67 50 9

100000	
BG 53868	1
46132	
430428	8 [17]
30392	
26933	5
3459	
3232	6
227	4

Sine BGA	99190	
Sine BAG	88414	1
	11776	[18]
	8841	1
	2935	
	2652	3
	283	
	265	3
	18	2

So the whole eccentricity remains   18,564
but that of the eccentric   11,332
and of the equant   7,232

---

[24] This number should have been 430944; however, unlike the error cited in the following footnote, this one has no effect on the quotient.

[25] Because of a subtraction error, this number should have been 10776. The resulting eccentricity for the eccentric is 11,219, slightly smaller than Kepler's stated eccentricity. This makes the eccentricity of the equant 7,345, somewhat greater than Kepler's. This error changes only the position of point *B* relative to *A* and *C*, which has a negligible effect on the equations: for example, at a mean anomaly of 90°, the total equation is increased by three arc seconds.

In the Copernican and Tychonic form, the diameter of the small epicycle would be 3,616, and of the greater, 14,948. Or, following what was said at the end of the fourth chapter, the tangent may be used instead of the sine, in this manner.

Let the maximum equation be investigated at the ninetieth degree. Let HCG be 90.   105
BC will be the sine of the angle BGC, 4° 8′ 51″. And GBC will be 85° 51′ 9″. And GC 99,738. But in the Copernican form, with C at the center of the concentric, GC will be 100,000. Therefore, in order that CGA, the angle of the equation, remain unchanged, the same 18,564 is to be increased in the same ratio for Tycho and Copernicus:

1856400000	
99738	1
85902	
79790	8
6112	
5984	6
128	1
99	3

The composite Copernican-Tychonic eccentricity. And in the tangents, this shows 10° 32′ 38″ as the common angle of the equation at 90° of anomaly.

Therefore the corrected diameter of the smaller epicycle is   3,628.
of the greater,   14,988.

Compare all this with Ch. 5, where I transposed the Tychonic rendition from the mean to the apparent motion of the sun, and see how slight the difference is.

So this is the method by which the hypothesis of the first inequality was investigated using four acronychal[26] positions of Mars. In this, with Ptolemy, I have supposed that all positions of the planet throughout the heavens are arranged on the circumference of one circle; also that the planet moves most slowly where it is at its greatest distance from the center of the earth (according to Ptolemy) or of the sun (according to Tycho and Copernicus); and that the point about which this slowing is measured is fixed. Everything else I have demonstrated, although the demonstration is a "reduction to the impossible." But whether the things I assumed in the demonstration are in fact so, or the opposite, will become clear in what follows.

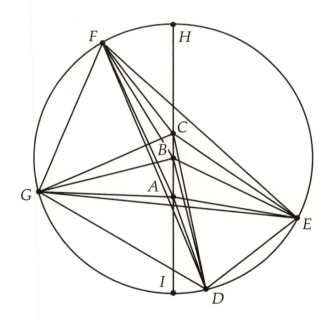

I shall now test the remaining eight positions against this hypothesis, for   106 the sake of consensus. But in order that the test be universal and legitimate, I shall also throw in the motion of the apogee. So I shall take this up first.

---

[26] ἀκρονυχίοις.

CHAPTER 17

A superficial[1] investigation of the motion of the apogee and nodes.

This investigation will be of the same degree of certainty as the observations (or rather, those things handed down by Ptolemy). If that practitioner had not existed, we would know even less today about those very slow motions. For besides him there is no one at all to be found, from the earliest records of civilization to the present, who could help us here.

We lay down here those suppositions found in Ptolemy, which are not in all respects perfectly certain. First, that the fixed stars have remained exactly in the zodiacal positions in which Ptolemy placed them (Ptolemy book 7). Second, that Ptolemy's figure for the sun's eccentricity was correct: 4153, where the semidiameter of the orbit is 100,000 (Ptolemy book 3 Ch. 4). Third, that the sun's apogee was fixed at $5\frac{1}{2}°$ Gemini (in the same chapter). Fourth, that Mars's apogee (when its motion is adjusted to the sun's mean motion) was found to be at $25\frac{1}{2}°$ Cancer (Ptol. book 10 Ch. 7). Fifth, that the eccentricity of Mars was 20,000 where its semidiameter is 100,000 (in the same chapter). Sixth, that the ratio of the epicycle (in Ptolemy) or of the annual orb (in Tycho and Copernicus) to the orb of Mars was as 100,000 to 151,900. Hence, where the semidiameter of the sun's orb (or the earth's) is 100,000, Mars's eccentricity will be 30,380 (Ptolemy book 10 Ch. 8).

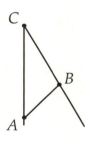

We shall proceed as in Chapter 5. *Let A be the point about which the earth's orb is described, C Mars's equalizing point, B the center of the sun's orb. And because AB is at $5\frac{1}{2}°$ Gemini, while AC is at $25\frac{1}{2}°$ Cancer, CAB is 50°. By supposition, AB is 4153, while AC is 30,380 of the same units. Therefore, since two sides and the included angle are given, the angle CBA is given as 123° 27'. And because BA is directed towards $5\frac{1}{2}°$ Sagittarius, the direction of BC (by subtraction of the angle 123° 27') will be about 2° 3' Leo, for Ptolemy's time. Also at that time, the eccentricity of the equant, after transposition to the sun's true motion, was 18,353. I discovered this above by transposition of the Tychonic value, 18,342, with one change: for the size of Mars's orbit, instead of 151,386 I took 152,500, which is nearer the truth. But this is tangential to the matter at hand,* to which we return.

107

---

[1]*Superficiaria.* In classical Latin, this term means "on ground belonging to someone else." However, by the seventeenth century this word came to be used as synonymous with *superficialis,* "relating to the surface, superficial." In other writings, Kepler used it to denote a property of the surface of something (as opposed to its interior), or a property of a surface in a mathematical sense. It therefore appears that it should be read as "superficial." Nevertheless, Kepler may also have had the classical meaning in mind, since he begins by stating that this investigation is built on "things handed down by Ptolemy."

Table of the motion of the aphelia and nodes

Years	Aphelion		Limits & nodes	
	M	S	M	S
1	1	4	0	40
2	2	8	1	21
3	3	12	2	1
4	4	16	2	42
5	5	20	3	22
6	6	24	4	3
7	7	28	4	43
8	8	32	5	24
9	9	36	6	4
10	10	40	6	45
11	11	44	7	25
12	12	47	8	6
13	13	51	8	46
14	14	55	9	27
15	15	59	10	7
16	17	3	10	48
17	18	7	11	28
18	19	11	12	9
19	20	15	12	49
20	21	19	13	30
21	22	23	14	10
22	23	27	14	50
23	24	31	15	31
24	25	35	16	11
25	26	39	16	52
26	27	43	17	32
27	28	47	18	12
28	29	51	18	53
29	30	55	19	33
30	31	59	20	13

Months	Aphelion	Limits & nodes
	Seconds	Seconds
1	5	3
2	11	7
3	16	10
4	21	13
5	27	17
6	32	20
7	37	23
8	43	27
9	48	30
10	54	33
11	59	37
12	1 4	40

On the motion of the aphelia

Because the precession of the equinoxes was exceedingly high around Ptolemy's time, while before and after there remains not the least suspicion of any such thing, I shall exclude this, and relate the position of the apsis to the fixed stars. At that time, Cor Leonis was at 2° 30′ Leo. The apsis or aphelion of Mars therefore preceded this star by 27′, in about A. D. 140. In our times, in 1587, Tycho Brahe found this star at 24° 5′ Leo. Since the aphelion progressed to 28° 49′ Leo, it was 4° 44′ east of Cor Leonis. If you add the above mentioned 27′ to this, the sum (5° 11′) is the motion over the 1447 years between 140 and 1587. Therefore, the annual motion is very nearly 13″: 6′29″ every 30 years. If, in turn, you add to this the Tychonic value for the motion of the fixed stars or of precession, which is quite uniform and is the same for all times (Ptolemy's alone excepted), namely, 25′30″ in 30 years, you will get the sum of 31′59″. Therefore, the annual motion of Mars's aphelion with respect to the equinox is 1′4″ in our time.

On the motion of the nodes

Although it is not really necessary, we shall consider this subject here because it is related to the motion of the aphelia. And because Ptolemy says in Book 13 Ch. 1 that Mars's northern limit is "near the end of Cancer, and almost at its apogee,"[2] it will therefore be at 29 Cancer, that is, $3\frac{1}{2}°$ before Cor Leonis. Despite this, in book 3 Ch. 6 Ptolemy put the northern limit exactly at the position of the apogee ($25\frac{1}{2}°$ Cancer) because it made calculations easier. And today it is at about 16° 20′ Leo, about 7° 45′ before Cor Leonis. By subtraction of 3° 30′, the northern limit, and conse-

---

[2] περὶ τὰ τελεθταῖα τοῦ Καρκίνου, καὶ σχεδὸν περὶ τὸ ἀπογηότατον· See Ptolemy, *Almagest*, ed. Heiberg II p. 526 lines 10–11.

quently the nodes, are found to have retrogressed 4° 15′ from Cor Leonis. This accords well with the motions of the moon, whose apogee likewise has a progressive motion with respect to the fixed stars, while its nodes retrogress. So the annual westward motion is 10″34‴, or 5′17″ in 30 years. Subtract this from the motion of precession, 25′30″. The remainder is 20′13″. And the nodes of Mars are moved the same number of minutes with respect to the equinoctial point in 30 of our present years, likewise eastward.

Chapter 18

Examination of the twelve acronychal positions using the hypothesis we have found.

I shall use that form[1] of calculation which I explained above in Chapter 4, 108–109
because it is more succinct. Also, it is indubitable that not a minute and a half
(actually somewhat less) would be gained or lost by using the Copernican or
Tychonic form, as I noted in the same place.

You see, then, O studious reader, that the hypothesis found by the method
developed above, is able in its calculations not only to account, in turn, for the
four observations upon which it was founded, but also to comprehend all the
other observations within two minutes—a magnitude which this star, when in
its acronychal position, always fills and even exceeds in the size of its body.[2]
This shows that if anyone were to repeat the above method, taking in turn 110
various sets of four observations, the same eccentricity with the same division,
an identical aphelion, and very nearly the same mean motion, would always
result. I therefore proclaim that the acronychal positions displayed by this
calculation are as certain as the observations made with the Tychonic sextants
can be. As I have said before, these observations are subject to some degree
of uncertainty (at least two minutes), owing to Mars's small but appreciable
diameter, and refraction and parallax, which are not yet known with complete
certainty.[3]

Finally, you see how nothing prevented the transposition of acronychal
observations from the mean to the apparent motion of the sun, so as to keep
me from, not just imitating, but even surpassing, the certitude of the Tychonic
calculation, which was raised as an objection against my abandoning the sun's
mean motion.

---

[1] Reading *forma* instead of *firma*.

[2] Mars actually subtends only about 18″ at opposition.

[3] This hypothesis theoretically gives results accurate within one minute, provided that its constants are correctly established.

	1580	1582	1585	1587	1589	1591
	s ° ′ ″	s ° ′ ″	s ° ′ ″	s ° ′ ″	s ° ′ ″	s ° ′ ″
Aphelion for 1587	[4] 28 48 55 Leo	4 28 48 55	4 28 48 55	4 28 48 55	4 28 48 55	4 28 48 55
Motion over the intervening years	6 42	4 28	2 14	0	2 15	4 32
Aphelion in the year written above	4 28 42 13	4 28 44 27	4 28 46 41	4 28 48 55	4 28 51 10	4 28 53 27
Mean longitude	1 25 49 31	3 9 24 55	4 20 8 19	6 0 47 40	7 14 18 26	9 5 43 55
Add	3 55	3 55	3 55	3 55	3 55	3 55
Corrected mean longitude	1 25 53 26	3 9 28 50	4 20 12 14	6 0 51 35	7 14 22 21	9 5 47 50
Therefore, the angle C is	87 11 13	49 15 37	8 34 27	32 2 40	75 31 11	126 54 23
Its sine	99880	75767	14909	53058	96823	79961
Eccentricity of the equant	7232	7232	7232	7232	7232	7232
	65088 / 6509 / 579 / 58	50624 / 3616 / 506 / 43 / 5	07232 / 2893 / 651 / 6	36160 / 2169 / 36 / 6	65088 / 4339 / 578 / 14 / 2	50624 / 6509 / 651 / 43 / 1
Part of the equation	7223 ; 4° 8′ 33″ ; 91 19 46[1]	5479 ; 3° 8′ 26″	1078 ; 0° 37′ 4″	3837 ; 2° 11′ 57″	7002 ; 4° 0′ 55″	5783 ; 3° 18′ 55″
The angle B	88 40 14	46 7 11	7 57 23	29 50 43	71 30 16	123 35 28
Its half	44 20 7	23 3 36	3 58 42	14 55 21	35 45 8	61 47 44
Tangent	97706	42572	6955	26650	72002	186464
Quotient resulting from the division of the difference of the sides by the sum	79643	79643	79643	79643	79643	79643
	716787 / 55750 / 5575 / 48	318572 / 15929 / 3982 / 557 / 16	47786 / 7168 / 398 / 40	159286 / 47786 / 4779 / 398	557501 / 15929 / 16	796430 / 637144 / 47786 / 3186 / 478 / 32
Tangent	778160 ; 37° 53′ 22″	33906 ; 18° 43′ 47″	5539 ; 3° 10′ 13″	21225 ; 11° 59′ 0″	57344 ; 29° 49′ 54″	148506 ; 56° 2′ 40″
	44 20 7	23 3 36	3 58 42	14 55 21	35 45 8	61 47 44
Angle at A	82 13 29	41 47 23	7 8 55	26 54 21	65 35 2	117 50 24
Aphelion	148 42 13	148 44 27	148 46 41	148 48 55	148 51 10	148 53 27
Mars's position is	6 28 44 Gem.	16 57 4 Can.	21 37 46 Leo	25 43 16 Vir.	4 26 12 Sco.	26 43 51
Should be	6 28 35	16 55 30	21 36 10	25 43 0	4 24 0	26 43 0
Difference	0′ 9″	1′ 34″	1′ 36″	0′ 16″	2′ 12″	0′ 51″

---

[3]This is the sum of angle C and the part of the equation. The angle at B is its supplement.

Values are given in the form sign · degrees minutes seconds (s ° ′ ″).

	1593	1595	1597	1600	1602	1604
Aphelion for 1587	4 28 48 55	28 48 55	4 28 48 55	4 28 48 55	4 28 48 55	4 28 48 55
Motion over the intervening years	6 48	9 14	11 30	13 43	15 56	18 11
Aphelion in the year written above	4 28 55 43	4 28 58 9	4 29 0 25	4 29 2 38	4 29 4 51	4 29 7 6
Mean longitude	11 9 55 4	1 7 14 9	2 23 11 56	4 4 35 50	5 14 59 37	6 27 0 12
Add	3 55	3 55	3 55	3 55	3 55	3 55
Corrected mean longitude	11 9 58 59	1 7 18 4	2 23 15 51	4 4 39 45	5 15 3 32	6 27 4 7
Therefore, the angle C is	11 3 16	111 40 5	65 44 34	24 22 53	15 58 41	57 57 1
Its sine	19174	92934	91171	41280	27528	84759
Eccentricity of the equant	7232	7232	7232	7232	7232	7232
(partial products)	07232 6509 072 51 3	65088 1446 651 22 3	65088 0723 072 51 1	28928 0723 145 58	14464 5062 362 14 6	57856 2893 506 36 6½
Part of the equation	1387; 0° 47′ 42″	6721; 3° 51′ 14″	6593; 3° 46′ 50″	2985; 1° 42′ 40″	1991; 1° 8′ 26″	6130; 3° 30′ 52″
The angle B	11 50 58	107 48 51	61 57 44	22 40 13	14 50 15	54 26 9
Its half	168 9 2[3]	53 54 26	30 58 52	11 20 6	7 25 8	27 13 5
Tangent	963600	137171	60045	20046	13021	51433
Quotient resulting from the division of the difference of the sides by the sum	79643	79643	79643	79643	79643	79643
(partial products)	7167870 477858 23893 4779	796430 238929 55750 0796 557 08	477858 00318 40	159286 319 48	079643 23893 159 8	398215 07964 3186 239 24
Tangent	767440; 82° 34′ 30″ 84 4 31	109247; 47° 31′ 49″ 53 54 26	47822; 25° 33′ 30″ 30 58 52	15965; 9° 4′ 14″ 11 20 6	10370; 5° 55′ 14″ 7 25 8	409628; 22° 16′ 32″ 27 13 5
Angle at A	166 39 1	101 26 15	56 32 22	20 24 20	13 20 22	49 29 37
Aphelion	148 55 43	148 58 9	149 0 25	149 2 38	149 4 51	149 7 6
Mars's position is	12 16 42 Pis.	17 31 54 Tau.	2 28 3 Can.	8 38 18 Leo	12 25 13 Vir.	18 36 43
Should be	12 16 0	17 31 40	2 28 0	8 38 0	12 27 0	18 37 10
Difference	0′ 42″	0′ 14″	0′ 3″	0′ 18″	1′ 47″	0′ 27″

---

[3]This is obviously the supplement, not the half.

Chapter 19

A refutation, using acronychal latitudes, of this hypothesis constructed according to the opinion of the authorities and confirmed by all the acronychal positions.

Who would have thought it possible? This hypothesis, so closely in agreement with the acronychal[1] observations, is nonetheless false, whether the observations be considered in relation to the sun's mean position or to its apparent position. Ptolemy indicated this to us when he teaches that the eccentricity of the equalizing point is to be bisected by the center of the eccentric bearing the planet. For here neither Tycho Brahe nor I have bisected the eccentricity of the equalizing point. Now for Copernicus* it was a matter of religion not to neglect this anywhere. For he made very little use of observations, perhaps thinking that Ptolemy used no more than are referred to in his Great Work. Tycho Brahe balked at this. For in imitating Copernicus, he set up this ratio of the eccentricities, which the acronychal observations required. But when this was gainsaid not only by the acronychal[2] latitudes (for these still underwent some increase arising from the second inequality) but also, and much more forcefully, by observations of other positions with respect to the sun which are affected by the second inequality, he stopped here and turned to the lunar theory, and I meanwhile stepped in.

Now the method by which the whole theory of Mars could easily be acquitted of error, if the premises were correct, and by which it is demonstrated to be incorrect, is this.

First, through the latitudes at acronychal positions. *In the Copernican form, let the line DE be set out in the plane of Mars's eccentric, upon which let A be the sun, D the northern limit, E the southern limit, or the point nearest it. And through A let the straight line HL be drawn, lying in the plane of the earth's eccentric orb. Now let AH and AD be conceived as lying in the one plane of a circle of latitude, and likewise AL and AE. And let the earth in 1585 lie at B on line AH, and in 1593 let it be at the point C on line AL. Now AB and AD are directed toward 21° Leo, so that the sun A seen from B appears at 21° Aquarius. And on the other hand, E and C are at 12° Pisces, so that the sun A seen from the earth C appears at 12° Virgo. But 12° Virgo is nearer to the sun's apogee than is 21° Aquarius. Therefore, BA is shorter than AC.* I shall take these lines from vol. I p. 98 of Tycho Brahe's *Progymnasmata*,[3] and shall suppose them to be correct, although below (by a method developed for the purpose) I am going to be showing them to be slightly different.[4] In Tycho

* In Saturn and Jupiter he bisected it simply; that is, the Copernican form attributes the quadrant to the semidiameter of the epicycle. In Mars, however, since he had attributed to the epicycle the quadrant of the Ptolemaic eccentricity, he argued that in our time the whole Ptolemaic eccentricity must be diminished, but left to the epicycle its original quantity. And so he moved the center of the eccentric (to speak Ptolemaically) 40 units closer to the center of the annual orb than to the center of the equant circle. Book 5 Ch. 16. See also Ch. 16 of this book.

111

---

[1] ἀκρονυχίοις.

[2] ἀκρονύχιοι.

[3] *TBOO* 2 p. 83: *"Tabula distantiarum solis a terra."*

[4] This "method" is the introduction of an equant into the theory of the earth's motion, presented

BA is shown as 97,500, and *AC* as 101,400, while in the correction which is to follow, *BA* turns out a little longer and *AC* a little shorter; they are not equal, however. *Now because in Ch. 13 above, through two procedures independent of the present enquiry, the angle BAD was found to be about 1° 50′ at the limit (about 16° Leo), therefore, at four or five degrees from the limit it will be 1° 49½′. But HBD, the apparent latitude in 1585, was 4° 32′ 10″. Hence, given the angles HBD and BAD, their difference BDA of 2° 42′ 40″ is also given. Now as the sine of BDA is to the known length BA, so is the sine of DBA to DA. So that if BA is taken to be 97,500, DA comes out to be 163,000. But if BA is 100,000, DA will be 167,200.*

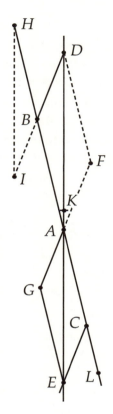

Likewise, since in 1593 C and E are in Pisces, and Mars is 26° from its limit, 64° from the node, consequently, the whole sine is to the sine of the maximum inclination 1° 50′ as the sine of 64° is to the sine of CAE, the inclination at this position. CAE is therefore 1° 39′. But the apparent latitude LCE was 6° 3′. Therefore the angle AEC is 4° 24′. And now, as before, as the sine of AEC is to the known length AC, so is the sine of ACE to AE. So that if AC is taken to be 101,400, AE comes out to be about 139,300. But if AC is 100,000, AE comes out to be about 137,380. Now since 21° Leo is about 8° from the aphelion, the line AD will be about 150 parts longer at aphelion (as will be clear to anyone who computes the distances from the hypothesis we have found and substitutes these numbers); that is, either 163,150 or 167,350. And since 12° Pisces is about 13° from the perihelion, AE when right at the perihelion will be about 300 parts shorter; that is, either 139,000 or 137,080. So we have the lengths of the lines AD and AE when right at the apsides, when they are parts of the same straight line DE. So let them be added:

DA	163,150	*or*	167,350.
AE	139,000	*or*	137,080.
So the whole, DE, is	302,150	*or*	304,430.
the half, DK,	151,075	*or*	152,215.
Resulting eccentricity AK	12,075	*or*	15,135.

Let these numbers be substituted for the original ones, where the radius of the eccentric was 100,000.
Thus, as 151,075 is to 100,000, so is 12,075 to 8000,
and as 152,215 is to 100,000, so is 15,135 to 9943.

112

Therefore, the exact eccentricity of the eccentric, as indicated by the acronychal latitudes, is somewhere between 8000 and 9943, where the radius of the eccentric orb is 100,000. But our hypothesis based upon the acronychal observations of longitude resulted in an eccentricity of the eccentric of 11,332, far different from that which is near the mean between 8000 and 9943. Therefore, something among those things we had assumed must be false.   But

---

in Ch. 29. This change reduced the eccentricity of the eccentric to half the size used by Brahe.

what was assumed was: that the orbit upon which the planet moves is a perfect circle; and that there exists some unique point on the line of apsides at a fixed and constant distance from the center of the eccentric about which point Mars describes equal angles in equal times. Therefore, of these, one or the other or perhaps both are false, for the observations used are not false.

The same demonstration also holds against that hypothesis established by observations adjusted to opposition to the sun's mean position, because the latitudes remain about the same over the time interval between the two moments. Thus they show an eccentricity of the eccentric of 9943, while in Ch. 5 above, 12,600 was taken from the Brahean revision, and 12,352 in the Ptolemaic equant, where the whole eccentricity of the equalizing point was 20,160 or 19,763.

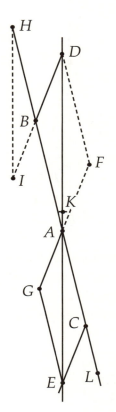

For the transformation of our diagram to the Ptolemaic form, *let DE be the line of apsides, A the earth, D and E the center of the epicycle at the highest and lowest apsis, and from both points D and E let straight lines be drawn toward the earth A parallel to the plane of the ecliptic BC. On these, let DF, EG be taken as radii of the epicycle, equal to BA, AC, and let the planet be at F and G. Therefore, the inclination FDA will be equal to the inclination BAD, and the line of vision AF will be parallel to the original line BD. Therefore, the observed latitude, HAF and HBD, is the same. The same must be said of the congruent triangles ACE and EGA. And thus the demonstration and the magnitudes of the corresponding lines are the same.*

It will occur to the reader to question why I make the semidiameter of Mars's epicycle unequal to itself, namely, *DF* equal to *BA*, which is longer, and *EG* equal to *CA*, which is shorter. I answer from Part I, that this happens because of the transposition of observations from opposition to the sun's mean position to opposition to the sun's apparent position. If we should stay with the sun's mean motion—and the present line of argument prevails even then—*DF* and *EG* will remain equal at least up to this point. But for this see Part I Ch. 6.

For the Brahean form, *leaving one of the triangles (DBA, say) the same, so that B may be the motionless earth and A the sun in 1585, let AB be extended so that BH is equal to AC, and let H be the sun in 1593, at 12° Virgo, and let HI be made equal and parallel to AE in the same direction, so that Mars at perigee is at I, and at apogee at D; HBA the ecliptic, BHI, BAD the inclination; IBA the latitude at perigee, and DBH at apogee. So again the sum of DA and HI will come out the same, whose half is DK, and the eccentricity will be KA.* The only difference is this, that for Ptolemy 113 the plane of the epicycle, and for Tycho the plane of the eccentric, is moved from north to south and back, remaining parallel to itself, while in Copernicus both stay in the same place.

Meanwhile, consider this also. In chapter 16 I had found the combined eccentricity to be 18,564, whose half, 9282, is just about the mean between 8000 and 9943. And Ptolemy too, as was remarked above, had taught us that half of the eccentricity found by acronychal observations is to be assigned to the eccentricity of the eccentric. So it was no lack of reason that so moved him, and we should not rashly reject this bisection, since the observed latitudes support it.

Now if, on the other hand, we bisect the eccentricity of 18,564 that we found, we shall indeed represent the acronychal positions near the middle elongations on the eccentric accurately enough, but not so well the positions around the octants and towards the apsides.

Take, for example, the opposition in 1593. From the preceding chapter, the simple anomaly was 6ˢ 11° 3′ 16″. I multiply the sine of 11° 3′ 16″, namely, 19,174, by 9282 (before, it was to be multiplied by 7232). This gives 1780, the sine of the arc 1° 1′ 12″, which is part of the equation. When this is added to 11° 3′ 16″, the sum is the semi-equated anomaly, 6ˢ 12° 4′ 28″. The supplement of this is 167° 55′ 32″, whose half is 83° 57′ 46″. The tangent of this is about 945,500, which, multiplied by the perihelial distance of 90,718 and in turn divided by the aphelial distance of 109,282, gives a tangent of 784,880, whose arc is 82° 44′ 20″. This, subtracted from the 83° 57′ 46″ found earlier, leaves 1° 13′ 26″, which is the other part of the equation. If this be added to the semi-equated anomaly, and the sum be added to the aphelion, it puts the planet at 12° 13′ 37″ Pisces, which differs from the former hypothesis by three minutes, and is more distant from the observed position. For it should have been 12° 16′ Pisces.

This appears more clearly at 17° Cancer in 1582. For with the eccentricity bisected, Mars falls at 17° 4$\frac{3}{4}$′ Cancer, and this calculated value differs from ours by 7$\frac{2}{3}$′ at about 45° from aphelion, but by about 9′ from the observation.

And from this difference of eight minutes, so small as it is, the reason is clear why Ptolemy, when he made use of bisection, was satisfied with a fixed equalizing point. For if the eccentricity of the equant, whose magnitude the very large equations in the middle elongations fix indubitably, be bisected, you see that the very greatest error from the observations reaches 8′, and this in Mars, which has the greatest eccentricity; it is therefore less for the rest. Now Ptolemy professes not to go below 10′, or the sixth part of a degree, in his observation. The uncertainty or (as they say) the "latitude" of the observations therefore exceeds the error in this Ptolemaic computation.

Since the divine benevolence has vouchsafed us Tycho Brahe, a most diligent observer, from whose observations the 8′ error of this Ptolemaic computation is shown in Mars, it is fitting that we with thankful mind both acknowledge and honor this favor of God. For it is in this that we shall carry on, to find at length the true form of the celestial motions, supported as we are by these proofs showing our suppositions to be fallacious. In what follows, I shall myself, to the best of my ability, lead the way for others on this road. For if I had thought I could ignore eight minutes of longitude, in bisecting the eccentricity I would already have made enough of a correction in the hypothesis found in   114
Ch. 16. Now, because they could not be ignored, these eight minutes alone will have led the way to the reformation of all of astronomy, and have become the material for a great part of the present work.

* Nevertheless, in the equations of the center for the annual orb, these 8 minutes in some places increase to as much as 30 minutes.

CHAPTER 20

Refutation of the same hypothesis through observations in positions other than acronychal.

I shall now proceed to the other argument whereby the eccentricity of the eccentric as found in Chapter 16[1] is proven false, despite its providing true longitudinal motions. This argument is based upon observations of Mars at positions with respect to the sun other than opposition, when the planet was observed in the region of the eccentric's apsides.

On 1600 March 5/15 about midnight, Mars was observed at 29° 12$\frac{1}{2}$′ Cancer with latitude 3° 23′ north. Its mean longitude, corrected by our addition,[2] was 4^S 29° 14′ 58″, while the aphelion was at 4^S 29° 2′ 45″. Therefore, the anomaly was 0^S 0° 12′ 13″, requiring an equation of 2′, to be subtracted, according to the hypothesis of eccentric positions established above. Therefore, Mars's eccentric position was 29° 13′ Leo, and the sun's position, 25° 45′ 51″ Pisces.

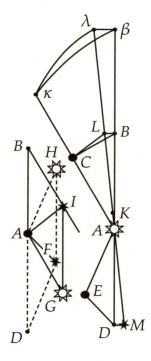

In the diagram, let A be the sun, B Mars, and C the earth. By subtraction of CB (29° 12$\frac{1}{2}$′ Cancer) from AB (29° 13′ Leo) the angle CBA will be 30° 0′ 30″, while by subtraction of CA (25° 45′ 51″ Pisces) from CB (29° 12′ 30″ Cancer), BCA will be 123° 26′ 39″. But as [sin] CBA is to CA, so is [sine] BCA to BA. But CA, the distance of the sun from the earth, is 99,302 from Tycho's table. (Although this is incorrect, the true value is nevertheless between this and 100,000, as we shall learn below in Ch. 30). Therefore, AB is between 165,680 and 166,846.

At perihelion, let the observation be taken that was made on the night following 1593 July 30 at 1^h 45^m.[3] Mars was found to be at 17° 39$\frac{1}{2}$′ Pisces with latitude 6° 6$\frac{1}{4}$′ south. Mars's mean longitude was 10^S 26° 16′ 38″, aphelion 4^S 28° 55′ 43″, so Mars was 2° 39′ 5″ from perihelion, to which corresponds an equation of 32′, to be subtracted, in accordance with the above hypothesis, making Mars's eccentric position 10^S 25° 44′ 30″, and the sun's apparent position 17° 3′ 0″ Leo.

In the diagram let BA be extended to D, and let AD be at 25° 44′ 30″ Aquarius and ED at 17° 39′ 30″ Pisces. Therefore, EDA is 21° 55′ 0″. And because ED is at 17° 39′ 30″ Pisces and EA at 17° 3′ Leo, AED is therefore 149° 23′ 30″. Now

---

[1]Incorrectly given as XVII in all Latin editions.

[2]This is the 3′ 55″ that Kepler said should be added to the mean longitudes (Ch. 16 p. 199)

[3]That is, July 31 at 1:45 am.

as [sin] EDA is to EA, so is [sine] AED to AD. But EA, the sun's distance from the earth, is 102,689, from Tycho's table, an erroneous figure, to be sure, but it is surely greater than 100,000. Therefore, AD is between 140,080 and 136,409. But since the star Mars is $2\frac{2}{3}$ degrees from perihelion, AD will be shorter by about 15 at the perihelion itself, that is, between 140,065 and 136,394. Distances for both apogee and perigee must be increased, because they were computed using observations related to the ecliptic. Thus AD and AB are lines in the plane of the ecliptic. On which point, take this

115

PROTHEOREM[4] TO BE USED FREQUENTLY BELOW

By observations of the star Mars related to the ecliptic, and by lines in the plane of the ecliptic found through those observations, to show the length of lines corresponding to them and next to them in the plane of Mars's own orbit.

Let the line BAD be set out in the plane of the ecliptic, and through A, which denotes the sun or center of the world, let the straight line LAM be so drawn in the plane of the orbit that the star be at L and M. Now let the earth be at C, and the triangle CAB be part of the plane of the ecliptic, to which the plane of the triangle LBA is to be understood to be perpendicular. Let the points C, L, and B be joined, and lines be extended to the surface of the sphere of the fixed stars: AB to β, AL to λ, AC to κ; and let κβ be an arc of the ecliptic, βλ an arc of the circle of latitude, and κλ a transverse arc. Thus the observation of the star's position beneath the fixed stars is referred to the ecliptic, by means of an arc of the circle of latitude drawn at right angles to the ecliptic κβ through the observed position of the star, and the triangle CLB is part of the plane of that circle. But λβ is also by supposition the circle of latitude, perpendicular to the ecliptic κβ. Therefore, the planes CLB and LBA of the two circles perpendicular to the same ecliptic intersect each other at the line LB. Therefore, by Euclid XI. 19, the line of intersection LB will be perpendicular to the plane of the ecliptic CBA and to the line BA contained in it; that is, LBA will be a right angle. Therefore, once the length of BA on the ecliptic is found, and the angle LAB is known, it will be impossible for the length of LA which is sought to fail to be known. QEF

Now in the matter at hand, since the inclination, or angle *LAB*, is 1° 48′ at this position, *LA* is 82 parts longer than *BA* (in the present units), and *AM* 72 parts longer than *AD*.

So the corrected apogees *AL* will be	165,762	or	166,928
Perigees *AM*	140,137	or	136,466
Sums *LM*	305,899	or	303,394
Halves *KL*	152,950	or	151,697
Eccentricity *KA*	12,812	or	15,371[5]

In new units, taking *KL* or *KM* as 100,000, the eccentricity of the eccentric is between 8377 and 10,106. But our hypothesis postulated 11,332, which exceeds both of these. Therefore, it postulated something false.

---

[4]See the note at the beginning of Ch. 59, p. 431.

[5]This should be 15,231, or 10,040 where *KL* is 100,000.

You should not let it disturb you that the second number, 10,106, which was arrived at through the assumption that *AC* and *AE* are equal, comes rather close to 11,332. For since I have related these observations to the sun's apparent positions, constructing the eccentricity from the center of the sun's body, *AC* and *AE* will therefore not be equal. Consequently, this eccentricity will be much less than 10,106, and would in fact be 8377 if the sun's distances were correctly given as 99,302 and 102,689, which the requirements of this demonstration leads us to take as 100,000 and 100,000. But since these Tychonic distances will be corrected below, and will be brought closer to the average radius, the eccentricity being sought here certainly lies between these two limits (8377 and 10,106). In fact, it is approximately half the total eccentricity found previously (18,564); that is, 9282.

116

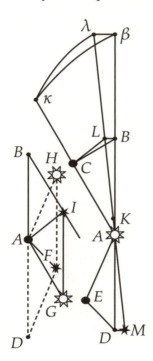

To go through the same demonstration also in the Ptolemaic hypothesis of the second inequality, proceed as in the previous chapter. *Draw AI, BI, AF, DF parallel to CB, CA, ED, EA in the larger diagram, and fix the earth at A, and the center of the epicycle (or, more correctly, the point about which the epicycle is rotated, distant from the center of the epicycle by the whole of the sun's eccentricity) at D and B; the sun at H and G; in such a manner that AH is equal and parallel to EA and AG to CA, so that the angles of equated anomaly of commutation are HAD and GAB; let Mars be at I instead of B or L and at F instead of D or M; and the lines AG, AH parallel to BI and DF (the lines of the planet's position on the epicycle) will be* the sun's position. The rest is obvious.

For the Tychonic form and hypothesis of the second inequality, *let A remain the earth, H and G the sun; and let HF, GI be drawn parallel and equal to AD, AB, so that Mars is again at F and I. The lines of vision, AF and AI, will therefore also be the same as in Ptolemy, and will be parallel to ED, CB, the lines of vision in the larger diagram. Therefore, they point in the same direction from the sun, and the sum of the lines HF, GI will equal the prior BD. Because the lines are parallel, the proof will be the same as the one at the beginning of the chapter.*

Now, as in the previous chapter, I shall accommodate the proof that the eccentric's eccentricity has been falsely determined to the Brahean revision as well, which depends upon the sun's mean motion. This is done so that no one will think the discrepancy is a result of my having wrongheadedly transposed the observations from the sun's mean motion to its apparent motion.

On 1600 March 5, Mars's mean longitude was, in Tycho's reckoning, 4^S 29° 11' 3", apogee 23° 41' Leo. Therefore, the simple anomaly was 5° 30', which, in his reckoning, requires an equation of 1° 7' 11", to be subtracted, so as to give Mars an eccentric position of 4^S 28° 3' 52", with the sun's mean position 23° 44' 31" Pisces. *In the above diagram let A be the point of the sun's mean motion, distant from the center of the sun by the whole of the sun's eccentricity. Therefore, the angle CBA is 28° 51' 22", and BCA is 125° 28' 0". Also, this demonstration requires that AE and AC be assumed equal, namely, 100,000,* retaining those suppositions made by Tycho and the ancients. These will be given an airing in Part III below, where it will be shown that the distance of the earth from the point of the sun's mean position is somewhat less; that is, that the Ptolemaic epicycle or the Copernican-Tychonic annual orb is not placed evenly about that point about which equal angles are traversed in equal times. But for now let us hold to the fundamentals as given: *and let CA be* 100,000; *therefore, AB will be* 168,760.

117

At perigee, on 1593 July 30, since (in Brahe's reckoning) Mars's [mean] longitude was 10^S 26° 12′ 43″ and the apogee was at 23° 34′ Leo, the simple anomaly was 182° 38′ 43″, which requires an equation of 35′ 52″, to be added. Therefore, Mars's eccentric position was 10^S 26° 48′ 35″, and the sun's mean position 18° 24′ 31″ [Leo]. *Therefore, in the diagram, EDA will be* 20° 50′ 55″, *and AED will be* 158° 45′ 0″.[6] *Let EA again be* 100,000, *although below (as has just been remarked) it will turn out to be somewhat greater. AD is therefore* 137,300. *This you shall diminish by* 15 *so as to fit right at perigee: let it be* 137,285. *The other you shall increase by about* 100, *so as to fit exactly at the apogee, so it will be* 168,860. *But we shall increase both (as before) because of the inclination of the planes,* 82 *being added at apogee and* 72 *at perigee. The final values will be:*

AB	168942
AD	137357
BD	306299
BK	153150
KA	15792

,

the eccentricity from the point of the sun's mean motion, or (in the Ptolemaic form) on the line of apsides drawn through the center of the epicycle.

Now where BK is 100,000, *KA is* 10,312. *However, the Tychonic revision based upon acronychal observations and presented in Chapter 8 required BK to be greater, namely,* 12,352.

It has therefore been shown that the Tychonic revision is also subject to the same incongruity, that the eccentric has one eccentricity that results from acronychal observations, and a different one that results from the others.

And meanwhile, the observations in this Tychonic rendition lead the way to bisection. For Tycho's figure for the whole eccentricity of the equalizing point is 20,160, half of which is 10,080, or in the form of the Ptolemaic equant, 9882. And here we have found it to be 10,312, which closely approximates half the Tychonic value. Indeed, it will approach much nearer, decreasing to a value less than the Tychonic (that is, to a very exact 9282), when *AC* in the greater diagram (*BI* in the smaller, on the left) is diminished, and along with it *AB* or *GI* (the distance at apogee); and, in turn, when *AE* of the right diagram (and its equal and equivalent *DF* of the left) is increased, and along with it *AD* or *HF* (the distance at perigee). For when the lesser part is increased, and the greater diminished, the difference between the two is decreased.

The blame for this discrepancy among the different ways of finding the eccentricity (I am repeating this over and over so that it will be remembered) falls entirely upon the faulty assumptions deliberately entertained by me, in

---

[6]This should be 150° 45′.

common with Tycho and all who have ever devised hypotheses. For the necessary consequence of this enquiry is that there is no single fixed point on the planet's eccentric about which the planet always sweeps out equal angles in equal times. We would instead have to make such a point reciprocate up and down along the line of apsides—if, indeed, we could keep the other assumption of a circular orbit. And how such a reciprocation could be reconciled with natural principles, I do not see.

118

But in fact the other assumption will be demolished, in Ch. 44 below; that is, the orbit of the star is not a perfect circle, but an oval, and its greatest diameter is the line of apsides, while its least is that passing through the center at the middle elongations. No wonder, then, that the other observations at points not at opposition to the sun do not accord with the hypothesis constructed in Chapter 16, since we have made two false assumptions in it.

CHAPTER 21

Why, and to what extent, may a false hypothesis yield the truth?

I particularly abhor that axiom of the logicians, that the true follows from the false, because people have used it to go for Copernicus's throat, while I am his disciple in the more general hypotheses concerning the system of the world. I therefore considered it particularly worth while now to show the reader how it does happen here that the true follows from the false.

First, you have already seen that what has followed is not completely true. For the path of the planet through the single plane of the ecliptic was considered in two ways: first, in respect to its longitude beneath certain degrees and minutes of the circle of the zodiac, and second, in respect to its altitude or distance from the center of the world about which it moves, which it shows to be different by means of other zodiacal positions. Therefore, our false supposition, although it does put the planet in the right longitudinal position at the right time, does not give it the right altitude. So what follows from this false hypothesis is not completely true.

Further, even concerning the longitude alone, the fact that the result appears identical to the senses does not prove that the as yet unknown true hypothesis and the false one assumed by us have an identical result. For there can be a very small discrepancy which the senses do not perceive.

There are, however, occasions upon which a false hypothesis can simulate truth, within the limits of observational precision, with respect to the longitude. These I shall now demonstrate.

Through the center of the world A let the straight line MP be set out, falling upon opposite parts of the zodiac (29° Leo and Aquarius, say). And let it so be that according to some true hypothesis the planet spends half its time between lines AM and AP on the left and the other half on the right, so that after successive halves of its periodic time it is always alternately on the lines AM and AP. And let it be assumed that this particular effect of the true hypothesis is expressible by some other hypothesis that has been discovered. And so let a circle of any kind or some other

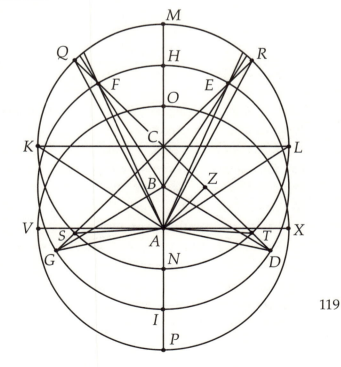

119

curvy line be described about a center taken on the line MP, with the sole provision that it go around the center of the world A and that it be cut into two equal parts by the line MP. What is proposed will happen if the planet traverse the circle with a uniform motion (one which is regular about any one particular point on the line MP, whether fixed or movable); as, for instance, if the circle OP were described about center A and moved uniformly about it. So all these circles and other figures have something in common through which what was proposed occurs, namely, that they move around the center of the world, and move regularly around some point on the line *MP*. Now the figure, whether this or that circle, whether one or another point of uniform motion, out of all those comprehended under the same genus, can be false. But we have brought about what was proposed, not through this false specific model, but through that which, within this false one, was comprehended within a general truth.

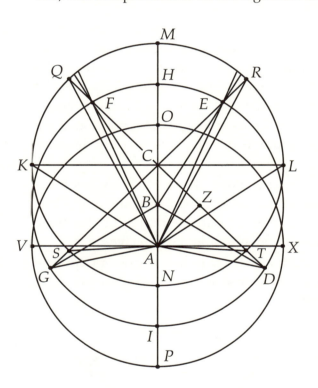

Now let us continue, *and let it happen that after successive quarter periods the planet lies on the lines AM, AK, AP, AL, the angles MAK, MAL, being less than right angles. Here, then, the former circle OP will be in error at the sides. For since the motion was supposed regular about A, a straight line drawn through A perpendicular to MP (namely, VX) will make the angles MAV, MAX measures of quarter periods. And accordingly, this hypothesis would put the planet on the lines AV, AX, when it should have been on AK, AL.*

Now experience testifies that the planets' motion closely emulates circularity (although it may perchance not exactly attain it), and it is the nature of motions of this kind to undergo gradual intensification and remission, admitting nothing sudden. *Therefore, the error of this hypothesis of the circle OP will begin little by little from the line AM, will grow continually greater, becoming a maximum at AK, and will again gradually decrease and vanish at AP. Therefore, the uniform and concentric hypothesis OP will never be more in error than it is at AK, AL, by the angles KAV, LAX, which, for Mars, are $10\frac{1}{2}°$.*

So let there now be another hypothesis which, in addition, also shows us the lines *AK, AL.* But again, there can be several hypotheses that do this. *For we might connect the points where AK and AL intersect the circle OP, and where this straight line intersects the straight line MP we may place the center of uniform motion of the circle OP, so that the motion of the circle OP becomes nonuniform. We would then also have [the planet on] the lines AK, AL [at the appropriate times].* But since we have a certain inclination towards choosing the simplest and most regular, we shall therefore seek out that circle that moves uniformly about its own center while effecting for us what is proposed. *Therefore, beginning from A, mark off equal lengths AK, AL upon the lines AK, AL; let the points K and L be joined by a straight line intersecting MP at C; and about C with radius CK let the eccentric circle MN be described, whose motion shall be regular about its center. This hypothesis will represent the planet in the correct position, on the four lines AM, AN, AK, AL.* But it is not this hypothesis alone, but many others as well, that could have this effect. For they have this feature that is general, and is indeed perfectly true, that the point of uniform motion is on the line that connects the   120 positions of the planet falling upon the lines *AK, AL,* and at that point upon it where the line intersects *MP. Now it follows from the premises that this hypothesis has absorbed the entire maximum error of the former hypothesis OP, namely, KAV, LAX, at about the quarters of the period, nor does it commit a new error (since at AM, AP it is equivalent to the former). Therefore, if this hypothesis is still in error, that error will be much smaller than KAV. And since it has done its job at CM, CN, CK, CL, the error (if any) will retreat to the four regions intermediate to those just mentioned, and will occur at the eighths of the period, since the time is measured about C. Therefore, angles MCK, KCN being bisected, let two new lines be drawn through C intersecting the circumference at Q, T, R, S. The maximum error, if any, will be about these points. But the hypothesis will also place the planet on the lines AQ, AR, AS, AT, at the eighths of the period. Now suppose (as is true for Mars) that after the eighths of the periodic time the planet should not appear on the lines AQ, AR, AS, AT, but should instead be, for the former two, on the lines AF, AE, higher up, and for latter two on the lines AG, AD, lower down. Therefore, if the former error KAV was* $10\frac{1}{2}°$*, the present error QAF will hardly amount to a few minutes.* For Mars, the magnitude of *QAF* or *RAE* is observed to be about 9′, while *SAG* or *TAD* is about 28′.

Now as a third step, let this hypothesis too be corrected. As this can happen in a variety of ways (specifically, by a reciprocation of the point *C* along the line *CA*), we are not prevented by any scruple from *keeping the point of uniform motion C fixed at distance CA, on account of the angle KAV, and also keeping the planet's path circular. These three, taken by choice and not forced by demonstration, will compel us to move the center of the eccentric downwards to B from the point of uniform motion C. Hence, HI is substituted for MN, and the body of the planet departs from the points Q, R, S, T, nevertheless remaining on the lines CQ, CR, CS, CT (because the measure of time stays at C), and arrives at the points marked F, E, G, D. And QF, ER, SG, TD would be such as to make QAF, EAR 9′ and SAG, TAD 28′.* With this done, that error at the eighths of the period will also be absorbed, and the hypothesis will exhibit the longitude perfectly accurately at eight places. Thus if again some error remains, it will be at the sixteenths of the period, the points in between. Also, since this third eccentric *HI* is equivalent to the first at positions *AM, AP,* as well as to the second at the additional positions *AK, AL,* it introduces no new error. And because the error of the second was greatest at the eighths of the period, and this is now absorbed, at the sixteenths there will therefore be a much smaller error remaining from the old error. Let us estimate it proportionally: just as the error of the first eccentric was $10\frac{1}{2}°$ while that of the second was 9′ or 28′, that is, one seventieth or one twenty-fifth of the former, let us now make the errors of the second that many

121   times the errors of the third. Plainly, already at the sixteenths of the period, we will have driven the business down to within the limits of observational accuracy.

It is at least now clear to what extent and in what manner the truth may follow from false principles: whatever is false in these hypotheses is specific to them and can be absent, while whatever endows truth with necessity is in general aspect wholly true and nothing else.[1]

Further, as these false principles are fitted only to fixed positions throughout the whole circle, so the truth does not invariably follow outside those very positions, except to the extent that happens in this procedure, that the difference can no longer be appraised by the acuteness of the senses.

Also, this same dullness of the senses hides the following additional small error which remains at the eighths of the period. That there is such a remainder I prove thus:

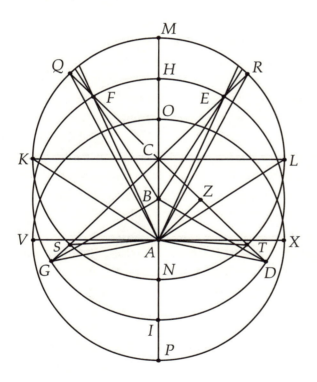

Once again, let a perfect eccentric be described about B so that BD, BE, BF, BG are equal, and let us have made BC such that the angle QAF is of the required magnitude. Now it is not likewise left to our discretion how great we want angle SAG to be, since it will be completely determined. From A draw a perpendicular to QT, and let this be AZ. Now, as above, let AC be 18,564 where CQ is 100,000. And because ACZ is 45°, AZ or ZC will be 13,127 (both in the same units). Therefore, ZQ is 113,127, and AQZ 6° 37′ 5″, and QAZ 83° 22′ 55″, whose tangent is 864092.[2] Now let FAZ be taken as 9′ less: its tangent FZ will be 844900.[3] But where AZ is 13,127, ZF will be 110,910.

Therefore, QF will be 2217. Now QF is larger than TD, which I prove thus: QT is the diameter of the circle, and is therefore equal to the two semidiameters FB, BD

---

[1]Perhaps what Kepler means is this: if a hypothesis by deductive necessity yields a true statement, there is some general feature of the hypothesis that is itself true. For example, both the Ptolemaic and Copernican hypotheses predict sunrise. Although the former is false in placing the diurnal motion in the sun, it is nonetheless true in positing a relative motion of the sun with respect to the surface of the earth. If this relative motion were denied, sunrise would not occur; hence, it is necessarily true. The sun's motion, on the other hand, is specific to the Ptolemaic hypothesis, and could be denied without affecting the phenomena of sunrise.

Throughout the preceding discussion, Kepler makes use of the terms *species* and *genus*. These terms are hinted at, but not fully introduced, in the translation.

[2]This is actually the tangent of 83° 23′ 55″. The correct value is 861,896.

[3]This is the tangent of 83° 15′. The correct value is 842621.

taken together. But BF, BD taken together are greater than FD. Therefore, QT is also greater than FD. Let the common part FT be subtracted. The remainder QF is then greater than TD. And yet, over and above what is required, we shall allow them to be equal. Let CZ, 13,127, be subtracted from CT, so as to leave ZT, 86,873. Now from AZ, AT, ATZ is known, and it is 8° 35′ 33″. So ZAT is 81° 24′ 27″. And because ZT is 86,873, I shall add to it a magnitude equal to QF, as if it were TD, namely, 2217. This will makes ZD 89,090. But where AZ is 100,000, ZD will become the tangent of the angle ZAD, 686,291. Thus this angle is 81° 42′ 35″. But ZAT was 81° 24′ 27″. Therefore TAD, or SAG, is less than the difference, 18′ 8″, since TD is less than 2217.

This then is the required angle TAD, which ought to have been $27\frac{3}{5}'$. And so if you make QAF 12′ instead of 9′, TAD becomes 24′. And in both places the planet is made to be 3 minutes higher than it should be. The equation therefore will be seen to be too large, and thus the eccentricity [of the equant] is too large. It will straightaway be diminished, then, so that the planet is about $1\frac{1}{2}'$ lower at the lines AK, AL, and the same amount (that is, $1\frac{1}{2}'$) higher at D, E, F, and G.[4]

This mutual tempering of various influences causes one error to compensate for another, brings the calculation within the limits of observational precision, and makes it impossible to perceive the falsity of this specific hypothesis. And so this sly courtesan cannot gloat over the dragging of truth (a most chaste maiden) into her bordello. Any honest woman following the lead of this prostitute would stay closely in her tracks owing to the narrowness of the streets and the press of the crowd, and the stupid, bleary-eyed professors of the subtleties of logic, who cannot tell a candid appearance from a shameless one, judge her to be the prostitute's maidservant.

122

This is without doubt the reason for the remaining discrepancies of one or two minutes in Ch. 18, in Cancer, Leo, Scorpio, and several other places. But the error is not easy to see, since the observations used do not fall at the apsides and at the quarters and eighths of the period.

Conclusion of Part II

Up to the present, the hypothesis accounting for the first inequality (in which Brahe and Copernicus are in agreement, both differing somewhat in form from Ptolemy) has been presented using the sun's mean motion, which all three authors had substituted for the sun's apparent motion. Thereafter, it was shown that whether we follow the sun's apparent motion and the hypothesis found in Ch. 16, or the sun's mean motion and the hypothesis proposed in Ch. 8 according to Brahe's revision, in both instances there result false distances of the planet from the center, whether of the sun (for Copernicus and Brahe) or of

---

[4]Reading "*D. E. F. G.*" instead of "*DE, FG*," which makes no sense although it stands thus in all editions.

An elucidation of Kepler's procedure here may be helpful. The angle QAF retains its adjusted magnitude of 12′, which places the planet 3′ too high. However, the eccentricity of the equant (C in the diagram) is decreased so as to diminish this error to $1\frac{1}{2}'$, introducing a $1\frac{1}{2}'$ error in the opposite direction at points K and L. Here, as often elsewhere, when Kepler says "up" and "down" he means "toward aphelion" and "toward perihelion", respectively.

the world (for Ptolemy). Consequently, what we had previously constructed from the Brahean observations we have later in turn destroyed using other observations of his. This was the necessary consequence of our having followed (in imitation of previous theorists) several things that were plausible but really false.

And this much of the work is dedicated to this imitation of previous theorists, with which I am concluding this second part of the *Commentaries*.

Part 3

of the

COMMENTARIES

ON THE MOTIONS OF THE STAR MARS

INVESTIGATION OF THE SECOND INEQUALITY

THAT IS, OF THE MOTIONS OF THE SUN OR EARTH

OR THE KEY TO A DEEPER ASTRONOMY

WHEREIN THERE IS MUCH

ON THE PHYSICAL CAUSES OF THE MOTIONS

CHAPTER 22

124

The epicycle, or annual orb, is not equally situated about the point of equality of motion.

This, then, is the way our predecessors measured the first inequality. With this calculation established, which would represent the planet's eccentric position at any desired moment, they next turned to exploring the second inequality (which depends upon the sun), comparing the observed or apparent position with that which the eccentric and the planet's first inequality alone would assign.

When I was on this same path and was confronted with this equivocal fork in the road (in Chapters 19 and 20 above), and the observations (most faithful guides) were seen to be at war with observations, I had to give thought to altering completely the way the path was set out, using the method which follows.

First, in this third part I shall approach the second inequality. Here I shall use unquestionable observations to demonstrate, with either a confirmation or a refutation, all that I have hitherto supposed as principles but had doubts about. Once this is found it will be like a key: the rest will be opened up. Afterwards, in Part IV, I shall proceed to the first inequality.

In Chapter 22 of the *Mysterium cosmographicum*, when I was giving the physical cause of the Ptolemaic equant or of the Copernican-Tychonic second epicycle, I raised an objection against myself at the end of the chapter: if the cause I proposed were true, it ought to hold universally for all planets.[1] But since the earth, one of the celestial bodies (for Copernicus), or the sun (for the rest) had not hitherto required this equant, I decided to leave that speculation open, until the matter were clearer to astronomers. I nevertheless entertained a suspicion that this theory might perchance also have its equant. After I gained the recognition of Tycho, this suspicion was confirmed in me. For in a letter to me in Styria in 1598 Brahe said the following:

> The annual orb according to Copernicus, or the epicycle according to Ptolemy, does not appear always of the same size, in comparisons made to the eccentric, but introduces a perceptible alteration in all three superior planets, so much so that for Mars the angle of difference reaches one degree 45′.[2]

He also touched upon this point at the same time in the appendix to his

---

[1]*Mysterium Cosmographicum: The Secret of the Universe*, A. M. Duncan, trans. (Abaris Books, 1981), p. 219.

[2]Brahe to Kepler, April 1, 1598 (old style), letter number 92 in *KGW* 13 pp. 197–202.

Mechanica,[3] an account of his studies. Also, his words in volume I p. 209 of his letters[4] are not much different, where he states the opinion that as an effect of the solar eccentricity a certain amount of nonuniformity is also mixed in with the eccentric equations and the acronychal positions. This is, in fact, refuted in Part I: it is not reflected in the acronychal positions, or at least very little. But it appears that this needs to be understood through a certain correction regarding Mars at 90° from the sun.

125   Now when I heard that the annual orb grows and shrinks, an inspiration said to me that this illusion arises thus: Copernicus's annual orb, or Ptolemy's epicycle, is not everywhere distant from that center about which by supposition it is sweeping out equal angles in equal times. For what physical cause could make the circuit of the center of the planetary system (Tychonic) or of the circuit of the earth (for Copernicus) or of the epicycle bearing the star (for Ptolemy) grow and shrink? What, I ask, is this novelty unprecedented in astronomy, this unlikely absurdity? Wouldn't it seem more worthy of belief that the sun (for Copernicus) or the center of the planetary system (Tychonic) or the body of the planet (for Ptolemy) would in certain places be farther from, and in others nearer to, the selected point of uniform motion (at rest for Copernicus and Tycho, and moving around on the circumference of the eccentric for Ptolemy)—and this especially on the line of apsides? And for this, that suspicion of mine arising from my *Mysterium cosmographicum*—that an equant might be introduced into the theory of the sun (or, as I call it, the theory of the Ptolemaic epicycle)—seemed to provide a convenient occasion.

Term:
The center of the planetary system is the common point of intersection of the lines drawn through the apsides of the individual planets. And that point is either near to the sun's body, as Brahe originally thought, or in the sun's very center, as I would say, correcting him.

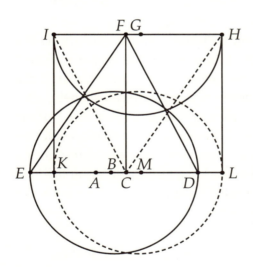

Let us suppose that the second inequality starts from the line of the sun's mean motion, as the practitioners have hitherto been pleased to hold (lest anyone hold suspect here my innovation of using the sun's apparent motion), *and in the present diagram let the planet's eccentricity for Copernicus originate not from the sun's center A, but from the point C about which the earth's motion is supposed regular. But let that point C not be the center of the earth's orb DE but only of uniformity of motion, and let its distance from the sun A be greater than that of B, the center of the earth's orb ED.* I say that, these things being granted, such observations will be produced from which one might suspect the annual orb of growing and shrinking. *Let a line CF be drawn from C perpendicular to DE, and let the star Mars be at F twice:*

---

[3]*TBOO* V p. 115. Brahe had also adumbrated this matter in correspondence with Wilhelm, Landgrave of Hesse, and Magini; cf. *TBOO* VI p. 236 and the endnote to that passage; also Magini's letter quoted in *Mechanica* fol. G4ʳ, *TBOO* V p. 126.

[4]*Epistolae astronomicae* p. 209 (another letter to the Landgrave), in *TBOO* VI p. 239.

once when the earth is at D and again when it is at E; and let F be joined with the points D and E. Now because C is the point of uniformity of the earth's motion on DE, FCD and FCE will be the anomaly of commutation, and (as we suppose) equal on both sides. Now if CD and CE were equal (as has hitherto been thought) then the angles DFC and EFC, parallaxes of the orb, would be equal on both sides, for both anomalies of commutation. But because CE is greater than CD, the angle CFE will also appear greater than the angle CFD. Therefore, anyone not noticing that this growth occurs only at or about E, and that the contrary diminution occurs only at the contrary position D, will think that the entire annual orb sometimes gets larger, with radius CE, and sometimes smaller, with radius CD. This is because such a person presupposes, along with astronomy as it has hitherto been practiced, that the point of equal motion C is at the same time also the center of the circle DE.

In the Ptolemaic form, let the earth be at C, and the lines of the sun's mean motion be CK, CL, in place of DC and EC in the preceding Copernican arrangement. And let the center about which the epicyclic motion is regular be at F, and IH be equal and parallel to ED so that, CI being drawn, it is parallel to DF and CH to 126 *EF. For when the earth (or the observer) is transferred to the center of the world C, as Ptolemy has it, Mars at F is likewise transferred to H. Similarly, owing to the translation of D to C, F is transferred to I. Now Ptolemy, thinking that the point F, about which the motion of the epicycle IH is uniform, is also the center of the epicycle IH, supposed FI and FH to be exactly equal. Consequently, for both of the equated anomalies HFC as well as IFC (that is, at both 90° and 270°, in this diagram) he posited one and the same equation of the epicycle, namely, the equal angles HCF and ICF. So if observation affirms that HCF is greater than ICF, then the center of the epicycle will not be at the point of uniform motion F, but at G, in the direction of H. Further, on the supposition that F be nevertheless considered the center of the epicycle, the epicycle will appear distinctly enlarged at anomaly 90° at H, and diminished at 270° at I, while in both instances Mars, in its eccentric position (that is, on the line CF), is in the same position with respect to the fixed stars.*

In the Tychonic form, let C remain the earth, DE the sun's circle with center B, but let the center of uniform motion be A. And let the lines along which the planet is seen (namely, CI and CH) be the same as in Ptolemy. Accordingly, let HL, IK descend from H and I parallel to FC. In order that K and L be the center of the planetary system, let the center of its circuit be M, which is in the direction of the sun's perigee, so that the point M, the center of the circuit KL (in which is found the point from which Mars's eccentricity originates) descends as far below C as the point B, the true center of the sun's circuit (contrary to common opinion), descends below A, the putative center of the same circuit of the sun. And let AC and BM be equal. The line of equated motion on the eccentric (that is, KI, LH) will be parallel to itself after an integral number of returns of the planet. Therefore Tycho, thinking that the earth C is in the middle of the circuit KL bearing the planets' eccentrics, will make angles CIK, CHL equal when the angles of commutation CLH, CKI are equal. But if these are perceived to be unequal, CHL being the greater, CL will be longer than CK: and the orb KL, the deferent of the center of the system, will appear to grow at L and to shrink at K, because M, the center of the orb which is the planetary system's deferent, is not believed to be elsewhere than at the earth C, about whose center the motion of that orb is uniform.

Now what greatly contributes to obscuring the true cause of this difference, namely, to freeing the sun's eccentricity from suspicion, is the fact that thereby* the distance CK of the center of the system from the earth becomes short just where the distance CE of the sun from the earth becomes long; and conversely, that the former, CL, becomes long where the latter, CD, becomes short.

The reason why the apsides have thus been reversed is this. For Copernicus, the earth traverses the regions opposite the Tychonic sun and the Ptolemaic epicycle, and also DC, CE, the distances of the earth from the sun, of the sun from the earth, and of Mars H or I from the epicycle's center of uniform motion F, subtend angles of the same magnitude in all three forms of hypothesis. Therefore, it also happens that the Copernican distances of the sun and the earth will be transferred to the opposite sides by Brahe and Ptolemy; that is, CE to CL or FH, and CD to CK or FI.

* Here is a notable incongruity [ἀποσδόκητον]. If the general Ptolemaic or Brahean hypothesis concerning the world system is true, and if at the same time we make use of the sun's mean motion, then the epicycle in the former or the deferent circle of the planetary system in the latter becomes an eccentric whose apogee is exactly opposite the sun's apogee, while its eccentricity (as will appear below) is exactly equal to the sun's true eccentricity, that is, half of what it has hitherto been thought to be.

127

Next, in order either to confirm or undermine this speculation by observations, this is the road upon which I set out. Since the sun's apogee is at $5\frac{1}{2}°$ Cancer, I enquired whether there might exist an observation in which Mars, reckoned by the first inequality, would be twice at $5\frac{1}{2}°$ Libra or Aries, while the sun would be at $5\frac{1}{2}°$ Cancer at one time, and then at $5\frac{1}{2}°$ Capricorn. As it turns out, this is not possible within such a short space of time (20 or 30 years). For the periodic motions of Mars and the sun are incommensurable, nor do they ever coincide at 90° from one another, or at opposition, after a certain number of complete periods, or quarter or half periods, of either. I therefore had to choose the next best thing, which was to find many days throughout those 20 years on which the planet was observed, and in which the equated anomaly of commutation was 90° or 270° or nearly that much, with Mars at 6 Aries or Libra or thereabouts. Afterwards, it was necessary to look up all those dates in the catalog of observations of Mars, so I could see whether it had been observed at those moments. Had the indefatigable Tycho Brahe not observed Mars very frequently, the selection would have been so exclusive that I would not have been able to accomplish what I wished. Now since Tycho put the apogee of Mars at $23\frac{1}{2}°$ Leo, while the required position of Mars, corrected by the eccentric equation, was $5\frac{1}{2}°$ Libra, an equated anomaly of 42° was required. And from his table, to an equated anomaly of 42° there corresponds an equation of $8°\ 15\frac{3}{5}'$; therefore, a mean anomaly on the eccentric of 50° 16' was required, and through this I was shown twelve points in time in the twenty years between 1579 and 1600.

Thus, what had to be skillfully tracked down is whether for any of these times there was an equated anomaly of commutation that was at one time

90° and again 270°, or if the former were greater or less, the latter would be correspondingly less or greater.

One revolution of Mars has 687 days, and two of the sun have $730\frac{1}{2}$. The difference is $43\frac{1}{2}$ days, to which corresponds 42° 54′ 23″ of the sun's mean motion. This, therefore, is how much the anomaly of commutation changes at the end of any revolution of Mars. Therefore, when in any two year period one seeks two anomalies of commutation that are equal, with Mars at the same eccentric position, each angle of commutation should be 21° 27′. Over four years 42° 54′ is required; over six years, 64° 22′; over eight years, 85° 49′. And we were supposing 90°, if it were possible. Therefore, we had to look for our two observations eight years apart. However, a team of two such observations is not to be found in the catalog of observations we had.

I next turned to the interval of six years, and found at length that from 1585 May 18 and 1591 January 22 suitable observations exist. For they corresponded to 1585 May 30 at 5^h and 1591 January 20 at 0^h.[5] For both, the mean longitude of Mars was 6^S 22° 43′. The Tychonic equation was 9° 14′ 52″, to be subtracted. Therefore, Mars's eccentric position was 13° 28′ 16″ Libra. The equated commutation for 1585 was 8^S 4° 23′ 30″, by which it was shown, Ptolemaic style, that the planet was 64° 23′ 30″ beyond the perigee of the epicycle. Similarly, the equated commutation for 1591 was 3^S 25° 36′ 30″, by which it was shown that the planet was 64° 23′ 30″ before the perigee of the epicycle. Therefore, both the angles of commutation, *FCD* and *FCE* (or *CFI*, *CFH*) in the diagram, are equal. However, in 1585 the sun was in 18° Gemini, 18° before apogee, and in 1591 in 9° Aquarius, 33° beyond perigee, and this inequality could not be avoided.

Now to the observations: on 1585 May 18 at $10\frac{1}{2}^h$ at night Mars was observed at 0° 50′ 45″ Virgo with latitude 1° 19′ 30″ north.[6] Magini puts it at 1° 5′ Virgo,[7] 14′ or 15′ too much. Therefore, when on the 30th at 5^h in the evening he puts it at 6° 48′ Virgo, we shall again subtract the discrepancy of eleven days previously. So he will be left with 6° 34′ Virgo. Here we will assume that some very few minutes are in error because the deduction over 12 days is too great, and the diurnal motion is not exactly the same as that obtained here from Magini. For consider that on April 18 preceding at 10^h Mars was found at 17° $37\frac{1}{2}′$ Leo, while Magini puts it at 18° 0′ Leo. The difference is $22\frac{1}{2}′$. Over the 33 days elapsed to May 18 this difference was diminished to the measure of $14\frac{1}{4}′$. So if we extrapolate, since over 33 days eight minutes vanished, in the same ratio, over the 12 following days 3 minutes will vanish. Therefore, on May 30 the difference will be $11\frac{1}{4}′$. So, more accurately, Mars is at 6° 37′ Virgo.

128

---

[5]The conditions that must be satisfied are: 1. that Mars have the same eccentric position at the two times, and 2. that the two angles between Mars and the earth about the center of the earth's orbit be equal. These conditions are satisfied on 1585 May 30 and 1591 January 20 (old style). The observations which Kepler had were not made on those dates; however, they were near enough that adjustments could readily be made.

[6]*TBOO* 10 p. 402.

[7]*Ephemerides coelestium motuum Io. Antonii Magini Patauini, ad annos XL. Ab anno Domini 1581. Vsque ad annum 1620.* Venice, 1582.

Similarly, on 1591 January 22 at 7^h in the morning, Mars was 34° 32′ 45″ from Spica with declination 17° 25′ south, at an altitude of 16°.[8] Therefore, after correction for horizontal variations, the declination was 17° 30′. Hence, the right ascension was 230° 23′ 12″, longitude 22° 33′ Scorpio, latitude 1° 0′ 30″ north. Now this time differs from ours by 1 day 19 hours, and the diurnal motion, from Magini, is 33′. Therefore, 59′ are required for the intervening time. Therefore, the remainder is the position of Mars on January 20 at 0^h (which, as we said, corresponds to the other time): 21° 34′ Scorpio.

And since from Tycho's rendition, CF is	13°	28′	Libra (fairly accurately)
while DF or CI in 1585 is	6	37	Virgo
Therefore DFC or FCI will be	36°	51′.	

Likewise, because CF in 1591 is again	13°	28′	Libra*
while EF or CH is	21	34	Scorpio
Therefore EFC or FCH will be	38°	$5\frac{1}{2}′$.	

*The precession over the intervening time does not amount to five minutes; hence, it is neglected.

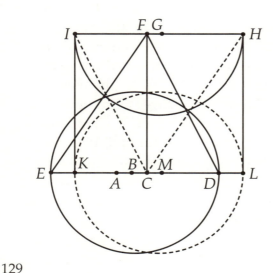

129

Witness the great difference in the equations on the annual orb, despite the promise that the two anomalies of commutation would be equal. The Copernican hypothesis shows us the cause. The earth at *D* and *E* was considered to be equally distant from the point of uniform motion *C*, but it was found to be at unequal distances, in such a way that the center of its circuit is at *B* towards the sun *A*. And by equivalence, in the Ptolemaic form the epicycle *HI* is not equally situated about point *F*, whose eccentric path the acronychal observations have been describing to us, and about which the epicycle's motion is regular. And the center of the epicycle *G* draws towards *E*, on the same side as the solar perigee. Similarly, in the Tychonic form the deferent *KL* of the planetary systems does not go around the earth *C* at a constant distance, although its motion is regular about this point, but the center *M* of its circuit draws towards the region of the sun's perigee.

---

[8]*TBOO* 12 p. 138.

CHAPTER 23

From the knowledge of two distances of the sun from the earth and of the zodiacal positions and the sun's apogee, to find the eccentricity of the sun's path (or the earth's, for Copernicus).

From this it is also not difficult for us to measure the line *BC* tentatively. *Let FC be* 100,000. *And because DFC is* 36° 51′ *and FCD is* 64° 23′ 30″, *therefore the remaining angle FDC is* 78° 45′ 27″. *And as the sine of this angle is to FC* (100,000), *so is the sine of DFC to DC,* 61,148.

In the same way, because EFC is 38° 5½′ *minus a little, and FCE is* 64° 23′ 30″, *FEC will be* 77° 31′ 0″ *plus a little. Therefore, EC is* 63,186 *minus a little.*

Let the earth's orb NED be set out, and on it let CBN be the line of apsides, and N perihelion, R aphelion, B the center, C the point of equality of motion, E and D the positions of the two observations, and let these be joined to C and to B. Now EC and CD are known in the same units, and angle ECD is known, viz. 128° 47′ 19″. *Let EC be extended, and let DO be dropped perpendicular to it from D; and also to DE let two perpendiculars CP, BQ be dropped from C and B. DCO is therefore* 51° 12′ 41″, *and CDO is* 38° 47′ 19″. *Therefore, where DC is* 61,148,

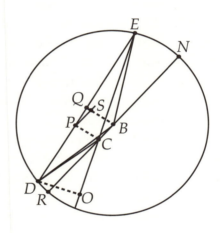

DO will be 47,660 *and CO* 38,305. *And this placed on the end of CE makes EO* 101,491. *But from the two given sides DO, OE about a right angle, the magnitude of DEO is obtained:* 25° 9′ 20″. *Therefore DE is* 112,125. *The half of this,* 56,062½, *is the magnitude of DQ, since DB, BE are equal. And because DEC was* 25° 9′ 20″, *EDC or PDC will be* 26° 3′ 21″. *Therefore, where DC is* 61,148, *CP will come out to be* 26,858 *and PD is* 54,932. *Subtract this from QD, and the remainder, PQ, is* 1130½. *And now from the known inclination of the lines ED and NC the length CB is easily obtained. For since CR is the line of the aphelion, at* 5° 30′ *Capricorn, while CD is at* 17° 52′ *Sagittarius (because the sun is at* 17° 52′ *Gemini), DCR will be* 17° 38′. *But EDC was* 26° 3′ 21″. *Therefore, after subtraction, there remains the inclination of the lines in question:* 8° 25′ 21″. *From P let PS be drawn parallel to CB. This will be equal to CB, and CP will be equal to BS. So in right triangle PQS, as the whole sine* 130 *is to the tangent and secant of the angle QPS,* 8° 25′ 21″, *so is the known magnitude of PQ to QS,* 167, *and SP,* 1143, *which is CB. And because PC and SB are equal, with magnitude* 26,858, *add QS, and QB will come out to be* 27,025. *So in the right triangle DQB, given the sides about the right angle, DB will also be given,* 62,237. *Therefore, the ratio of DB to BC (the radius to the eccentricity, which is being sought) is the same as* 62,237 *to* 1143. *And as* 62,237 *is to* 100,000 *so is* 1143 *to* 1837. *This, at last, is the eccentricity sought. It would have been less if we had accounted for the*

precession of the equinoxes, for then CE would have been less.

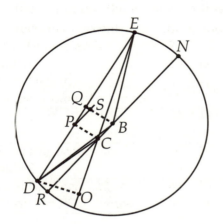

And so from these two observations and the accepted true position of the sun's aphelion [*sic*] there is provided the distance of our equalizing point *C* or *F* (which we were considering as center) from the orbit's true center *B* or *C* or *M*, which is 1837 where the radius of its orbit is 100,000. In contrast, Tycho Brahe found the sun's eccentricity, that is, the distance of the equalizing point *C* from the center *A* of the solar body (in Copernicus) or the distance of the equalizing point *A* of the solar motion from the center of the earth *C* (in the Tychonic-Ptolemaic supposition) to be 3584, whose half, 1792, is only slightly different from 1837. It is therefore fitting that the halving of the eccentricity hold in the theory of the sun, which halving previously held for the eccentric of Mars (in Chapter 19 and 20). For owing to the large corrections and the use of a controverted value for the diurnal motion, the observations which I have presented are not exact enough to allow anything to be concluded from 45 parts in one hundred thousand, not to mention the ignoring of precession in Mars's and the sun's eccentric motions for the time interval involved.

What has been demonstrated here of the circuit of the earth can be demonstrated in exactly the same way concerning the Ptolemaic epicycle and the Tychonic deferent of the system, provided only that in the diagram the apsides be reversed.

I have supposed here that the sun's apogee established by Tycho was in the right place, and that the orbit of the sun (or earth) which it bodily traverses is arranged in a circle. Although in Ch. 44 below the analogy with other planets will declare something different, the small breadth of the deflection nonetheless does not in the least vitiate our demonstration.

CHAPTER 24

A clearer proof that the epicycle or annual orb is eccentric with respect to the point of uniformity.

Such, then, was the beginning of this enquiry, timid and encumbered with so many concerns that the anomaly of commutation be equal on both sides.

Now that we have once made a hazard of this, we are buoyed by audacity 131 and will begin to be more free on this battlefield. For I shall seek out three or more observed positions of Mars with the planet always at the same eccentric position, and from these find by trigonometry the distances of that number of points on the epicycle or annual orb from the point of uniform motion. And since a circle is defined by three points, I shall use triplets of such observations to investigate the position of the circle, its apsides (previously taken as a presupposition), and its eccentricity with respect to the point of uniform motion. If a fourth observation will be at hand, it will serve as a test.

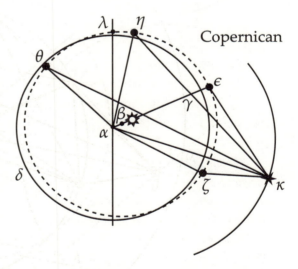

Let the first time be 1590 March 5 at 7^h 10^m in the evening, since then Mars had hardly any latitude, and thus no one looking at the demonstration could be troubled by any irrelevant suspicions about the intermingling of latitude. To this there correspond these moments, in which Mars returns to the same sidereal position: 1592 Jan. 21 at 6^h 41^m; 1593 Dec. 8 at 6^h 12^m; 1595 Oct. 26 at 5^h 44^m.[1] For the first of these times, Mars's [mean] longitude, according to Tycho's revision, is 1^s $4°$ $38'$ $50''$, and for subsequent times $1'$ $36''$ greater for each. For this is the motion of precession corresponding to the periodic time of one return of Mars. And since Tycho places the apogee at $23\frac{1}{2}°$ Leo, its equation will be $11°$ $14'$ $55''$, and consequently the equated anomaly in 1590 will be 1^s $15°$ $53'$ $45''$.

Now at the same time, the commutation, or difference of the mean motions of the sun and Mars, is reckoned to be 10^s $18°$ $19'$ $56''$, so the equated [commutation], or the difference between the sun's mean motion and Mars's equated eccentric motion, is 10^s $7°$ $5'$ $1''$.

We shall present this first in the Copernican form, since it is simpler to perceive.

---

[1]It is remarkable that Kepler does not explicitly state Mars's sidereal period anywhere in *Astronomia Nova*. However, it is clear from the dates and times given here that the period used is 29 minutes less than 687 days.

Let α be the point of uniform motion of the earth's circuit, and let this be considered to be the circle δγ described about α, and let the sun be on the side β, such that the line of the sun's apogee αβ lies in the direction of $5\frac{1}{2}°$ Cancer, even though we are going to investigate this degree freely, as if unknown, in Ch. 25. And let the earth be on αθ in 1590, αη in 1592, αε in 1593, and αζ in 1595. And the angles θαη, ηαε, εαζ are equal, since α is the point of uniform motion and the periodic times of Mars are presupposed equal. And let the planet at these four times be at κ, and its line of apsides be αλ. Thus the angle θακ, according to the measure of the equated anomaly of commutation, is 127° 5′ 1″.[2]

132

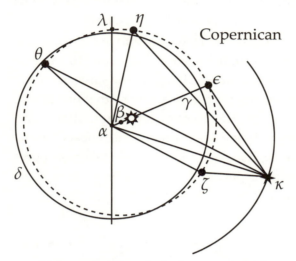

As for the observed position of Mars, at the same time on the preceding day, the fourth,[3] it was at 24° 22′ Aries.[4] Its diurnal motion for the day would be 44′. Therefore, at our time it was seen at 25° 6′ Aries, which is the position of the line θκ. But ακ is directed towards 15° 53′ 45″ Taurus. Therefore, θκα is 20° 47′ 45″. So the remainder αθκ to make up two right angles is 32° 7′ 14″.

Now as the sine of αθκ is to ακ, which we shall say is 100,000 units, so is [the sine of] θκα to θα, which is what is sought. Therefore, θα is 66,774.

Now if the remaining lines ηα, εα, ζα, will turn out to be of the same length, my suspicion will be false, but if they are different, my triumph will be complete.

Second, then, in 1592 at our moment the equated longitude was 1ˢ 15° 55′ 23″; the equated commutation was 8ˢ 24° 10′ 34″ — that is, the angle ηακ is 84° 10′ 34″. It was observed on January 23 at 7ʰ 15ᵐ at 11° $34\frac{1}{2}′$ Aries, with the correction for parallax.[5] And its motion over two days was 1° 25′. Therefore on the 21st at 7ʰ 15ᵐ it was seen at 10° $9\frac{1}{2}′$ Aries. Let the remaining parts of an hour deduct the half minute. Therefore, the angle ηκα is 35° 46′ 23″, and αηκ is 60° 3′ 3″, and αη is 67,467, now longer than αθ. This is doubtless because the sun has descended towards perigee, and the earth has been moved from θ to η; thus, in this region the sun is found beyond β at a nearer point.

133

Third, in 1593 at our moment the equated longitude was 1ˢ 15° 56′ 56″, the equated commutation was 7ˢ 11° 16′ 16″, which makes εακ 41° 16′ 16″.

---

[2]This is simply the equated commutation as given above, except that 180° is subtracted because of the change of viewpoint from the earth to the sun (θ to α, in the diagram).

[3]This was March 4, 1590.

[4]*TBOO* 12 p. 44. This is a raw observation in the usual Brahean manner: a declination and a distance from a fixed star (Aldebaran, in this case). To find the longitude, Kepler had to use the observation to find Mars's right ascension and convert to longitude/latitude. For more, see the Glossary, under Observations.

[5]*TBOO* 12 p. 219. Again, this is a raw observation, the reference star being α Ceti.

It was observed December 10 at 7^h 20^m at 4° 45' Aries, parallax corrected.[6] *Its motion over two days was 1° 8'. Therefore, on December 8 at 7^h 20^m it was seen at 3° 37' Aries, while at our time of 6^h 12^m it was at 3° 35$\frac{1}{2}$' Aries. Hence, εκα is 42° 21' 30" and κεα is 96° 22' 14", and αε is 67,794, again longer, for it is yet closer to the sun's perigee.*

Fourth, in 1595 at our moment the equated longitude was 1^s 15° 58' 30" and the [equated] commutation was 5^s 28° 21' 55", which makes the angle καζ 1° 38' 5".

It was observed on October 27 at 12^h 20^m at 18° 52' 15" Taurus, retrograde.[7] *Its diurnal motion was 23'. And so on the 26th at 12^h 20^m it was at 19° 15' 15" Taurus, while at our time it was at 19° 21' 35" Taurus. Therefore, ακζ is 3° 23' 5", and the supplement of αζκ is 5° 1' 10", and αζ is 67,478.* But this last operation is untrustworthy, owing to the small angles of the triangle. For if an error of one or two minutes is made either in observing or in computing Mars's eccentric position using Tycho's hypothesis, the ratio of the angles is easily changed perceptibly. But for now I shall present all four lines for inspection.

Sun's mean position	22°	59'	Pisces	αθ	66,774
	10	6	Aquarius	αη	67,467
	27	13	Sagittarius	αε	67,794
	14	20	Scorpio	αζ	67,478

So the longest is αε, which is also the closest to the sun's perigee; the shortest is αθ, which is also the farthest from the sun's perigee; and αζ and αη are about equal, because they are also nearly equally removed from perigee.

Moreover, even if αζ is a little longer than αη which is nearer the perigee, this should be attributed to the smallness of the angles at ζ, through which such a small error as this could easily be introduced. Therefore, the circle δγ, which Copernicus described about the point α of uniformity of the earth's motion, is not the earth's path. There is instead some other circle θηεζ on which the earth is found, whose center lies in the same direction as the sun—that is, at β.

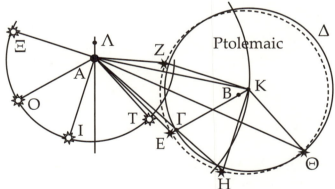

In the Ptolemaic form, *let the earth be at* A, *the sun's sphere be* ΞOIT, K *the putative center of the epicycle; that is, the center about which [is described] the epicycle* ΔΓ, *itself putative, which is equal to the [circle of the] theory of the sun.* The total equivalence between the hypotheses of Copernicus and Brahe requires this to be done, although for the present demonstration it doesn't matter what

---

[6]*TBOO* 12 p. 297. Brahe gives the longitude.

[7]*TBOO* 12 p. 450–451. Another raw observation; reference stars are α Ceti and Aldebaran.

ratio the sun's orb and the planet's epicycle have, provided that they have equal periods. *And let AΛ be Mars's line of apsides. Let AK, AΛ be parallel to ακ, αλ in the preceding Copernican form. From the center of the earth A, let the lines AΘ, AH, AE, AZ be drawn parallel to the previous lines κθ, κη, κε, κζ, and equal to them, so that Mars is at Θ in 1590, at H in 1592, at E in 1593, and at Z in 1595; and at the same time the sun's mean position at those times is AT, AI, AO, AX, respectively, so that KΘ and AT are parallel (as is known from the Ptolemaic hypothesis), and so*

134   *on for the rest. With Θ, H, E, and Z connected with K, it will be demonstrated as before, with exactly the same numbers, lines, and angles, that these lines are unequal, contrary to common opinion, and consequently that Mars does not traverse the circle ΓΔ, whose center is at the center of uniform motion K, but the circle ZHEΘ instead, whose center lies in the direction of B from K, very near to the line KB, which should be parallel to the line drawn from the earth A through the sun's perigee.*

Therefore, the epicycle's apogee lies in the direction of the sun's perigee. And because the epicycle, owing to the total equivalence just mentioned, is to be supposed equal to the sun's circuit, with ZK parallel to XA, EK to OA, HK to IA, and ΘK to TA, it is also likely that XA, OA, IA, and TA are unequal, and that the point of the sun's mean position (the center of the sun's epicycle, in the Brahean conception) does not stand at the same distance from the point of uniform motion throughout its circuit. This I remark only in passing: it has no effect upon the present demonstration, but serves as an extension to it.

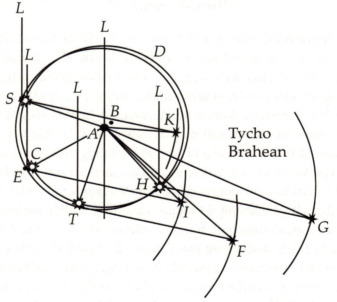

In the Tychonic form, *let A be the earth, and about it let the sun's concentric CD be described, and let this be considered to be the deferent of the system of the planets, since A is the point of uniform motion of the sun's concentric. Therefore the sun itself will be on another eccentric circle. Let its center be in the direction of B from A. Now let AL be the reference for Mars's line of apsides, so that the line of apsides, in the circling and translating of its eccentric, remains ever parallel to AL. And let the lines of the sun's mean motion at our four moments be AH, AT, AE, AS, and from A let the lines along which Mars is observed be extended in the direction of one or another degree of the zodiac in accordance with the description above. And since at all four*

135   *times Mars is supposed to be at the same place on the eccentric, its distances from the points of the sun's mean position will all be equal and parallel. Let them be GH, FT,*

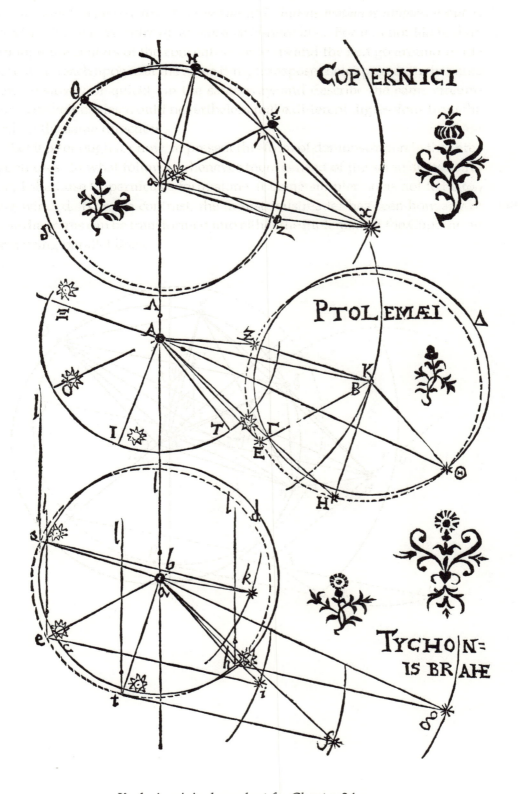

Kepler's original woodcut for Chapter 24

IE, and KS, all equal, and let the angles LHG, LTF, LEI, LSK, all be equal to the previous angle ΛAK *or* λακ, *so that at our moments Mars might be at G, F, I, K.* I would note in passing that these four points *G, F, I, K,* in actual fact will make an arc exactly equal in length and placement to the previous arc ΘHEZ in the Ptolemaic form. This is because there is no difference between the two other than that Ptolemy has an epicycle, equal to the [circle of the] theory of the sun, carried around on an eccentric, while Tycho has the eccentric carried around on the [circle of the] theory of the sun or on a circle equal to the Ptolemaic epicycle.

Once again, then, with the angles and numbers remaining the same, it will be demonstrated, contrary to common opinion, that the lines AH, AT, AE, AS are un-equal. And so that point on the eccentric whence originates the eccentricity of Mars and all the planets (which is here considered to be on the line of the sun's mean motion, following earlier practitioners) does not go around on that circle *DC* about whose center *A* it makes equal angles in equal times, but on the circle *HTES* whose center lies in the direction opposite the center of the sun's eccentric *B*, as has thus far become roughly clear from the lines themselves.

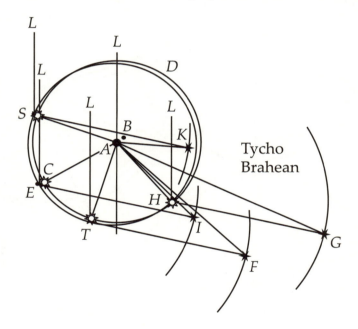

CHAPTER 25

From three distances of the sun from the center of the world, with known zodiacal positions, to find the apogee and eccentricity of the sun or earth.

I shall now once again test the quantity of the eccentricity and the position of the apogee in a single circle adapted to all three forms. For it is easily seen that they are simply opposites: for example, in the Copernican form the longest line is towards Gemini, while in the other forms it is towards Sagittarius. This is because Copernicus's observer is looking towards the center, and the others are looking away from it. Thus Copernicus too looks across the center at the same parts of the zodiac as the others.

Let the circle $\theta\eta\epsilon\zeta$, with center β, be set out, in which from the given point α there are the given lines $\alpha\theta$, $\alpha\eta$, $\alpha\epsilon$, and $\alpha\zeta$, as before, with the angles about α also given, for each of them is $42°\ 52'\ 47''$. What is sought is both the magnitude $\alpha\beta$ and the direction of that line with respect to the fixed stars or to the other lines. Let θ, η, and ϵ be selected and joined with one another, since three points suffice for investigating this.

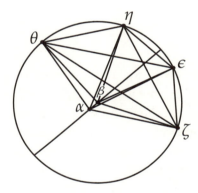

136

First, in the triangle $\theta\alpha\eta$ the sides and the included angle are given, $\theta\eta$ is sought, and is shown by trigonometry to be $49,169^1$ in the same units in which the sides $\alpha\theta$ and $\alpha\eta$ were expressed.

Second, in triangle $\alpha\epsilon\eta$ the angle $\alpha\epsilon\eta$ is sought, and found to be $68°\ 12'\ 26''$.

Third, in the triangle $\theta\alpha\epsilon$ the angle $\alpha\epsilon\theta$ is sought, and is found to be $46°\ 39'\ 10''$, which, subtracted from $\alpha\epsilon\eta$ leaves $21°\ 33'\ 16''$, which is the angle at the circumference $\theta\epsilon\eta$. Therefore, twice this amount, $43°\ 6'\ 32''$, will be $\theta\beta\eta$, the angle at the center, because β is by supposition the center of the circle. So in the isosceles triangle $\theta\beta\eta$ the angles are given, as well as the side $\theta\eta$ found previously. The size of $\theta\beta$, the radius of the circle, is sought, and found to be $66,923$. And since $\beta\theta\eta$ is $68°\ 26'\ 44''$, while before, when $\theta\eta$ was being found, $\alpha\theta\eta$ was $69°\ 18'\ 46''$, $\beta\theta\alpha$ is therefore $0°\ 52'\ 2''$. Next, in the triangle $\beta\theta\alpha$, from the sides and the included angle, $\theta\alpha\beta$ and $\alpha\beta$ are sought. And the angle $\theta\alpha\beta$ is found to be $97°\ 50'\ 30''$, so that $\alpha\beta$ is at $15°\ 8'\ 30''$ Gemini, because $\alpha\theta$ is at $22°\ 59'$ Virgo. Tycho, however, places the sun's apogee at $5\frac{1}{2}°$ Cancer. So you see that this very free enquiry has brought us within $20°$ of the correct Tychonic position. Also, $\alpha\beta$ is found to be 1023, and if $\theta\beta$ be taken as $100,000$, $\alpha\beta$ will become 1530. But the whole solar eccentricity is 3592, and its half

---

[1] This figure is in error: the correct value would be $49,073$. However, this does not affect the value of $\alpha\beta$ much, and brings the line of apsides some six degrees nearer the Tychonic position.

is 1796 or 1800. So here somewhat less than half of the solar eccentricity is claimed for the eccentricity of our circle. But you should bear in mind that observations near minimum values may be somewhat in error, and that use was made of a questionable mean longitude and equation from Tycho. This will become quite clear if you will carry out the same operation with the angle $\theta\eta\zeta$, and then with $\eta\epsilon\zeta$ and $\theta\epsilon\zeta$. For each time, $\alpha\beta$ has a somewhat different magnitude, and its position beneath the fixed stars will fall one side or the other of $5\frac{1}{2}^{\circ}$ Capricorn and Cancer.

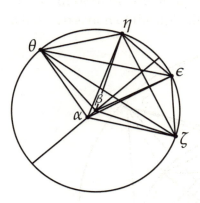

Consequently, we shall look into this more carefully below, where in many instances, through a fine demonstration, the eccentricity will be found to be half the solar eccentricity, and the apogee very near the Tychonic one.

It has thus been demonstrated in the Copernican form that the center of the earth's circuit is halfway between the sun's body and the point of uniform motion of that circuit. That is, that the earth proceeds nonuniformly on its orbit, slowing when it recedes far from the sun and speeding up when it approaches. This is in conformity with physical principles and with the analogy with other planets.

In the same way it has been demonstrated in the Ptolemaic form that the epicycle is eccentric with respect to the point about which its motion is uniform, and that its eccentricity is half of the sun's eccentricity as it is commonly determined, and in the opposite direction.

Finally, in the Tychonic form it has been demonstrated that the point whence originate the planets' eccentricities does not move on the sun's concentric, but throughout its course is unequally distant from the earth, about which it moves regularly and uniformly; and that it is farther away near the sun's perigee, and nearer at apogee, again by half of the sun's eccentricity. And so, since this Ptolemaic epicycle and this Brahean deferent have so great an analogy with the theory of the sun, it is likely that they also have a greater analogy: that is, the sun's true eccentricity, too, will only be half of that computed from the maximum equation; or, what is the same thing, the sun will make use of an equant, whose eccentricity is twice the eccentric's eccentricity.

I admit that this line of argument is a little weaker when applied to the Ptolemaic and Tychonic form, insofar as we are using the sun's mean motion, following the authorities. The argument consequently will become clearer when, moved by those reasons adduced in Chapter 6 above, I measure out the planet's motion by the sun's apparent motion.

Chapter 26

Demonstration from the same observations that the epicycle is eccentric with respect to the point of attachment or axis, and that the annual orb (and so also the earth's path around the sun, or the sun's around the earth) is eccentric with respect to the body of the sun or earth, with an eccentricity just half that which Tycho Brahe found through equations of the sun's motion.

We shall now go through the observations[1] again, carefully: On 1590 March 4 at 7^h 10^m, Mars was found by careful observation and calculation to be at 24° 22′ 56″ Aries with latitude 0° 3′ 20″ S. At that time, 8° Aries was setting, so Mars was rather low. Therefore, it was raised up towards the east by refraction, so it seems right that without refraction it would have been seen at 24° 20′ Aries. Its parallax can only be very small, particularly at this distance, for Mars was near the sun and therefore had receded very far from the center of the earth.

On 1592 January 23 at 7^h 20^m, using only one stellar distance from Mars, without the confirmation of another, Mars was found at 11° 32′ 44″ Aries with latitude 0° 1′ 36″ S. And so we shall make no changes for horizontal variations, although we suspect an uncertainty of one or two minutes.

On 1593 December 7 at 8^h 0^m Mars was found at 3° 6′ 50″ Aries with no danger of horizontal variation, with latitude 7′ 9″ S. However, the right ascension found using three stars showed a discrepancy of 4′, and the value taken as true was the mean between the extremes.

On 1595 October 25 at 8^h 10^m, the planet's distance from three fixed stars was observed, and by unanimous consensus the planet was found to be at 19° 39′ 25″ [Taurus] with latitude 0° 12′ 41″ S.   138

Now we shall reduce the three subsequent times to the first. Accordingly,

to its sidereal position on the eccentric in	1590	4	March	7^h	10^m
Mars will return to the same sidereal pos. in the years	1592	20	January	6	45
	1593	7	December	6	15
	1595	25	October	5	45

The motion over three days and 35 minutes of time in 1592 was 2° 9′ 4″, according to Magini.[2] Therefore, at our time Mars was seen at 9° 23′ 40″ [Aries].

---

[1] That is, the observations presented in Ch. 24.

[2] *Ephemerides coelestium motuum Io. Antonii Magini Patauini, ad annos XL. Ab anno Domini 1581. Vsque ad annum 1620.* Venice, 1582.

In 1593 the motion over 1 hour 45 min., from the diurnal motion of 33', is 2' 25''. And so at our time Mars's position comes out to be 3° 4' 27'' Aries. Likewise, in 1595 the motion over 2 hours 25 min., from a diurnal motion of 22' 11'', is 2' 14''. Therefore at our time the position of Mars comes out to be 19° 41' 39'' Taurus.

FROM THIS THERE FOLLOWS THE TABLE OF POSITIONS

of Mars, from observation;				of the sun, from Tycho's calculation.			
1590	24°	20'	Aries	24°	0'	25''	Pisces
1592	9	24	Aries	10	17	8	Aquarius
1593	3	4½	Aries	25	53	24	Sagittarius
1595	19	42	Taurus	11	41	34	Scorpio

Now because we have proposed to enquire how far the earth is from the center of the sun, we will need first of all to use the hypothesis constructed above in Ch. 16, from oppositions to the sun's apparent position, to find the position of the line drawn from the center of the sun through the body of Mars to the zodiac. And on 1595 October 25 at 5^h 45^m that line is found at 14° 19' 52'' Taurus. Therefore, for the three other times it is set back each time by 1' 36''. Thus, in 1593 it was at 14° 18' 16'' Taurus; in 1592, at 14° 16' 40'' Taurus; and in 1590, at 14° 15' 4'' Taurus.

LET THE FIRST DIAGRAM BE IN COPERNICUS'S FORM

And let α be the center of the sun; β the center of Mars's eccentric, drawn through o; χ the center of uniform motion for Mars's eccentric; γ the center of the earth's eccentric; δ, ε, ζ, η four positions of the earth, opposite the sun's apparent positions; θ the position of Mars on its eccentric. Let all points be connected with one another.

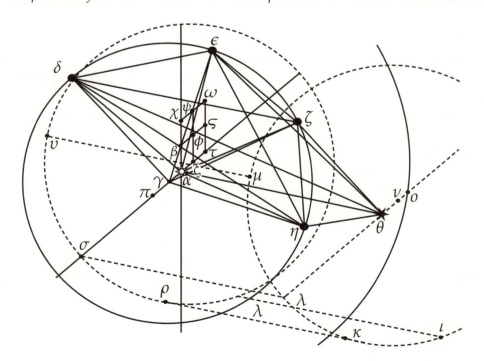

Now, in triangle δαθ							*Likewise, in triangle εαθ*					
because	δα	*is*	24°	0′	25″	*Pisces*	*because*	εα	*is*	10°	17′	8″ *Aquarius*
and	δθ		24	20	0	*Aries*	*and*	εθ		9	24	0 *Aries*
Angle	αδθ	*is*	30	19	35		*Angle*	αεθ	*is*	59	6	52
And because	δθ	*is*	24	20	0	*Aries*	*And because*	εθ	*is*	9	24	0 *Aries*
and	αθ		14	15	4	*Taurus*	*and*	αθ		14	16	40 *Taurus*

Therefore angle δθα *is* 19 55 4

Let αθ *be taken as* 100,000. αδ *is sought, which, by trigonometry, comes out to be* 67,467.

Therefore angle εθα *is* 34 52 40

So εα *comes out to be* 66,632.

In triangle ζαθ							*Finally, in triangle ηθα*					
because	ζα	*is*	25°	53′	24″	*Sagit.*	*because*	ηα	*is*	11°	41′	34″ *Scorpio*
and	ζθ		3	4	30	*Aries*	*and*	ηθ		19	42	0 *Taurus*
The suppl. of	αζθ	*is*	82	48	54		*The suppl. of*	αηθ	*is*	8	0	26
And because	ζθ	*is*	3	4	30	*Aries*	*And because*	ηθ	*is*	19	42	0 *Taurus*
and	αθ	*is*	14	18	16	*Taurus*	*and*	αθ	*is*	14	19	52 *Taurus*

Therefore ζθα *is* 41 13 46

So ζα *comes out to be* 66,429

Therefore ηθα *is* 5 22 8

So ηα *comes out to be* 67,220[3]

Here are the distances of the center of the sun from the earth, gathered together for you:

δα	67,467
εα	66,632
ζα	66,429
ηα	67,220

We shall now test the magnitude of the eccentricity that may be deduced from these distances. If the sun's theory lacks an equant, the eccentricity of its circle will turn out to be about 3600. This is because we used the true or apparent positions of the sun, whose point of uniform motion has to be that far (namely, 3600) from the center of the world, as Brahe has proved from solar observations. But if, on the other hand, the eccentricity turns out to be less, and is about half the Brahean value, we have won, and vindicated our contention, that the point of uniform motion found by Brahe is not the center of the sun's eccentric.

You can see at a glance (I would note in passing) that αζ *is the shortest, it being near the sun's perigee; next, that* αε *is longer, it being in Aquarius, 34 degrees from perigee; then* αη, *it being 54 degrees from perigee; and lastly, that* αδ *is longest, because it is 80 degrees from perigee. And since* αζ *is almost at perigee, it will be hardly any longer than the shortest. Similarly, since* αδ *is near the middle elongation, it will be only a little less than the mean distance. Therefore, the eccentricity will come out to be a little greater than 1038, which is the difference between* δα *and* ζα. *And if* δα *is assigned the measure of 100,000, 1038 will have the value 1539; and this is about what the eccentricity will amount to (it is actually a little greater). And this is much closer to 1800, half the Tychonic value, than to the full value of 3600.*   140

---

[3]An erroneous figure: this should be 67,171. The translator completed the calculation using this corrected value, and found the eccentricity αγ to be 2382. This is nearly the same as the value 2401 given by Delambre (*Histoire de l'Astronomie Moderne*, vol. I p. 433, cited in *KGW* 3 p. 466). Both Kepler's value and Delambre's emendation are considerably greater than the correct eccentricity of about 1660 parts, an error that vitiated many of Kepler's subsequent computations and tables. See Curtis Wilson, "The Error in Kepler's Acronychal Data for Mars," *Centaurus* 13 (1969), 263–268, also in Wilson, *Astronomy from Kepler to Newton* (Variorum, 1989).

The same is to be said of the sun's apogee. For because ζα is shortest, the perigee is about 25° 53′ Sagittarius. And because εα is shorter than ηα, the perigee is closer to 10° 17′ Aquarius than to 11° 42′ Scorpio. But the mean is 25° 57′ Sagittarius. Therefore, the perigee is beyond 25° 57′ Sagittarius and before 10° 17′ Aquarius; that is, in Capricorn.

This I wanted to give as a foretaste to make up for the coming labors. For now I shall follow the geometric path to investigate the position of the apogee and the eccentricity. And since three points determine a circle, I shall at first use the points δ, ζ, and η.

I proceed as in Ch. 25 above. Since the points δ, ζ, η are placed on a single circumference with center γ, the angle δηζ will be half of the angle δγζ, and its measure will be the arc δζ. Therefore, the ratio of δζ to the radius δγ, and to the eccentricity γα, will be given, together with the angle δαγ, because αγ lies in the direction of the apsides. But for the knowledge of the angle δηζ and of the line δζ, we need to solve the three triangles.

First, in δαζ, because αδ is in	24°	0′	25″	*Pisces*			
and αζ	25	53	24	*Sagit.*			
Therefore, δαζ is	88	7	1		*From this and αδ,*		67,467
Add 3′ 12″ for precession	88	10	13		*and αζ,*		66,429
The two remaining angles δ and ζ	91	49	47		*is found angle αδζ 45° 27′ 22″,*		
Half	45	54	54		*and its sine*		71271;
Tangent of this	103246.				*from which, with side αζ,*		
					δζ is found:		93,159.

Second, in δαη, since αδ is	24°	0′	25″	*Pisces*			
and αη is	11	41	34	*Scorpio*			
Therefore δαη is	132	18	51				
Add for precession		4	48				
	132	23	39				
The two remaining angles δ and η	47	36	21		*From this and αδ,*		67,467
Half	23	48	11		*and αη,*		67,220
Tangent	44110.				*is found angle αηδ 23° 51′ 0″*		

Third, in ζαη, since αζ is	25°	53′	24″	*Sagittarius*			
and αη is	11	41	34	*Scorpio*			
Therefore ζαη is	44	11	50		*From this and αζ,*		66,429
Add for precession		1	36		*and αη,*		67,220
	44	13	26		*is found angle αηζ 67° 3′ 12″.*		
The two remaining angles ζ, η	135	46	34				
Half	67	53	17				
Tangent	246120.						

So, since αηδ is	23°	51′	0″	*And since αδζ is*	45°	27′	22″
and αηζ is	67	3	12	*and γδζ is*		46	47	48
Therefore δηζ is	43	12	12	*Therefore γδα is*	1	20	26
Hence, δγζ is	86	24	24				

The other two γ, α,	178	39	34			
The other two δ, ζ,	93	35	36	*Half*	89	19	47
Half, γδζ,	46	47	48	*Tangent*	8540000.
Whose sine is	72893			*Let γδ be taken as*	100000
From this and δζ,					
δγ is found to be	68,141			*Of these units, αδ will be*	99,011

Of these units, αδ will be   99,011
Hence δγα is found: 68° 26′ 7″
So that αγ is at   15 34 18 *Capricorn*[4]
But the sine of δαγ,   93000
And the sine of γδα,   2340
Show that the eccentricity αγ is   2516.

And yet it was said before that using δ and ζ the eccentricity comes out to be a little greater than 1539, supposing that ζ is nearest to perigee. But since here (letting η be received into the company in place of ζ) the eccentricity comes out much greater, this hints, although erroneously, that there is some line at perigee which is still shorter than αζ. In order that this distance at perigee might be able to be shorter than αζ, this line of argument has moved the perigee to 16 Capricorn, that is, farther from αζ.

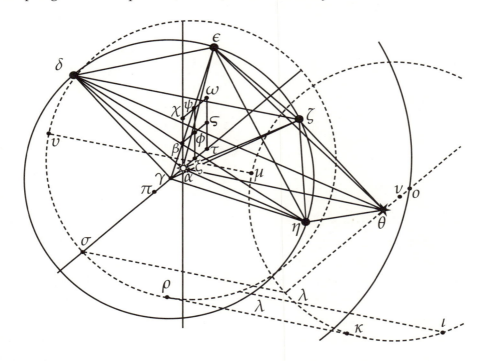

But since we already know that the sun's perigee is not in 16 Capricorn but in 6 Capricorn, it is appropriate to assign the cause of the slight error to

---

[4]Here Kepler seems to have made a simple mistake: he apparently subtracted the angle δγα from the position of the line δα to obtain the position of the line αγ. He needed to use the position of the line δγ instead, whose longitude is 1° 20′ 26″ less. However, the value he gives for the angle δγα is inconsistent with his values for the sides and included angle of the triangle δγα. According to the translator's calculations, δγα should be 66° 18′ 53″, and δαγ should be 112° 20′ 41″, whose supplement is 67° 39′ 19″, 1° 20′ 26″ greater than δγα. Perhaps the number he gives was intended to be the supplement of δαγ.

Using correct values throughout, the translator has found the resulting perigee to be at 11° 57′ 44″ Capricorn.

the point η and to the excessive length of the line $\alpha\eta$.[5] For the result of this is that the circle $\delta\epsilon\eta$ would be too large, and its radius $\delta\gamma$ too long; and consequently $\gamma\alpha$ would be too long, and γ would move perpendicularly away from the line $\delta\eta$, but obliquely from the point ζ. Thus, the line $\gamma\alpha$ would now be placed too far to the east. Therefore, letting δ and γ remain, let $\alpha\eta$ be supposed shorter. Then the center γ moves perpendicularly toward the line $\delta\eta$, and $\delta\gamma$ thus becomes shorter. And because γ approaches $\delta\eta$ perpendicularly, it recedes obliquely from the present $\gamma\alpha$. Hence, if a straight line be drawn from α through the new position of γ, it will be inclined back towards δ.

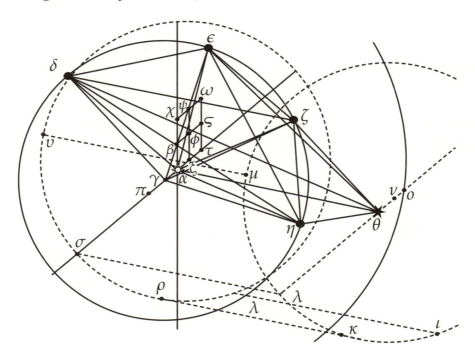

You thus see how shortening $\alpha\eta$ helps us in two ways. But to shorten $\alpha\eta$ is a very easy, small change, because the angles are small: it can be done by saying that the planet was seen in a slightly prior position along a line drawn from η below θ.[6] *For example, let the observed position of Mars be 19° 40′ Taurus, and the supplement of $\alpha\eta\theta$ be 7° 58′ 26″, and $\eta\theta\alpha$ 5° 20′ 8″: then $\alpha\eta$ will be 67,030. The second and third triangles are then changed, $\alpha\eta\delta$ becoming 23° 53′ 0″, and $\alpha\eta\zeta$ 67° 15′ 32″. Therefore, $\delta\eta\zeta$ is 43° 22′ 26″, and $\delta\gamma\zeta$ 86° 44′ 52″. The remaining angles are 93° 15′ 8″, whose half, $\gamma\delta\zeta$, is 46° 37′ 34″, and $\gamma\delta\alpha$ is 1° 10′ 12″. Hence*

142

---

[5]Here Kepler displays his remarkable insight into the interrelationship of the variables in the orbit. Although he did not find the error in the computed length of $\alpha\eta$, he sensed it was too long, and adjusted the data to shorten it, in what is surely the first deliberate published example of what has come to be called the 'fudge factor'.

[6]The letters η and θ have been exchanged to preserve the sense of the argument. It is clear that Kepler means that point θ will appear as if farther down in the diagram. However, it may be clearer to imagine the point η moving directly toward α (to η'), θ meanwhile remaining fixed. The angle $\alpha\eta\theta$ thus increases, which is in fact what happens in the numerical example that follows.

$\delta\gamma$ is 67,892. *And where this is* 100000, $\alpha\delta$ *will be* 99,416, *and* $\delta\gamma\alpha$ 73° 24′ 39″. *Consequently the perigee is in* 10° 36′ *Capricorn, and the eccentricity is still about* 2100.

Thus, as the true perigee was aproached the eccentricity decreased, and so when we get exactly to the correct perigee we shall also get exactly to the halving of the eccentricity.

But it is nonetheless worthwhile to find out, in addition, how much we gain by changing the line $\alpha\theta$, adding one minute to the computed eccentric position of Mars while keeping fixed the point of observation (that is, the point η) for the year 1595. *With $\alpha\theta$ accordingly moved forward, if these lines of sight $\eta\theta$, $\zeta\theta$, and the rest were to stay the same, it would come about that $\alpha\theta$ would be intersected by $\eta\theta$ at a place higher than θ; and on the other hand it would be intersected by $\zeta\theta$ and its counterparts at a place lower than θ.[7] So $\alpha\theta$ would not keep the same length. But since we are supposing that Mars is at the same eccentric position all four times, the length of $\alpha\theta$ will also be the same for all four times. Therefore, in order that the point of intersection θ be the same, and nonetheless the lines of sight go towards their original zodiacal positions, it will be necessary to draw a line parallel to $\eta\theta$ somewhat lower, thus making $\alpha\eta$ shorter; and also a line outside $\zeta\theta$ and parallel to it, by which $\alpha\zeta$ would be lengthened; and so on for the rest. Next, the whole labor is to be repeated from the beginning. For $de\theta\alpha$ will now be 19° 56′ 4″, $\epsilon\theta\alpha$ 34° 53′ 40″, $\zeta\theta\alpha$ 41° 14′ 46″, and $\eta\theta\alpha$ 5° 21′ 8″. Therefore $\delta\alpha$ will be 67,522, $\epsilon\alpha$ 66,660, $\zeta\alpha$ 66,451, $\eta\alpha$ 66,963. Hence, $\alpha\delta\zeta$ will be 45° 26′ 37″, $\alpha\eta\delta$ 23° 54′ 30″, $\alpha\eta\zeta$ 67° 20′ 48″. And $\delta\eta\zeta$ will be 43° 26′ 18″, and $\delta\gamma\zeta$ 86° 52′ 36″, $\gamma\delta\zeta$ 46° 33′ 42″, and $\gamma\delta\alpha$ 1° 7′ 5″ —a different angle from different beginnings. Now, when $\alpha\zeta$ is divided by the sine of $\alpha\delta\zeta$, and the quotient multiplied by the sine of $\delta\alpha\zeta$, the product is $\delta\zeta$, 93,252. Again, when this is divided into the sine of $\delta\gamma\zeta$, and the quotient is multiplied by the sine of $\delta\zeta\gamma$, the product will be $\delta\gamma$, 67,823. Hence the angle $\delta\gamma\alpha$ is 76° 37′ 30″, and the perigee is at 7° 23′ Capricorn, with an eccentricity about 1880,* which would clearly be 1800 if the perigee were brought back to $5\frac{1}{2}$ Capricorn, as could happen through the combined effect of both causes.

143

For if you now subtract only half a minute from the apparent position for 1595, we will be within range. And there could easily be a one minute error in the equations of the eccentric found by the hypothesis of Chapter 16.

---

[7]"Higher" and "lower" here (and in the third sentence following) mean "farther from α" and "nearer α," respectively.

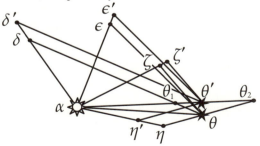

Kepler is saying that if the only line to change position is $\alpha\theta$, then the old lines from δ, ϵ, ζ, and η will intersect it at different points (for instance, at θ_1 and θ_2 in the adjacent diagram). Kepler's solution is to take a point θ' on the new line, with $\alpha\theta' = \alpha\theta$, and to draw lines from θ' parallel to the old lines from δ, ϵ, ζ, and η, intersecting the lines (or their extensions) from α to those points at δ', ϵ', ζ', and η'.

So, since the result is easily vitiated by the data from 1595, let us now omit them and use the remaining three points δ, ϵ, ζ, keeping the most recent correction of eccentric position, and forming the new triangles $\delta\alpha\epsilon$ and $\epsilon\alpha\zeta$.

Now because $\alpha\delta$ is	24°	0′	25″	*Pisces*	*From this and $\alpha\delta$,*		67,522	
and $\alpha\epsilon$	10	17	8	*Aquarius*	*and $\alpha\epsilon$,*		66,660	
Therefore angle $\delta\alpha\epsilon$ is	43	43	17		*$\alpha\delta\epsilon$ is found to be*	67°	12′	35″
For precession of equinoxes add		1	36		*But $\alpha\delta\zeta$ was, and remains,*	45	26	37
	43	44	53		*Therefore $\epsilon\delta\zeta$ is*	21	45	58
					and $\epsilon\gamma\zeta$ is	43	31	56
Similarly, because $\alpha\epsilon$ is	10°	17′	8″	*Aquarius*	*From this and $\alpha\epsilon$,*		66,660	
and $\alpha\zeta$	25	53	23	*Sagittarius*	*and $\alpha\zeta$,*		66,451	
Therefore angle $\epsilon\alpha\zeta$ is	44	23	44		*$\alpha\zeta\epsilon$ is found to be*	68°	0′	34″.
Precession of equinoxes		1	36		*To $\alpha\delta\zeta$,*	45	26	37
	44	25	20		*Add angle $\delta\alpha\zeta$,*	88	10	13
						133	36	50

$\alpha\zeta\delta$ will be	46	23	10
Therefore $\epsilon\zeta\delta$ will be	21	37	24
and $\epsilon\gamma\delta$	43	14	48
Hence $\delta\gamma\zeta$	86	46	44[8]
and $\gamma\delta\zeta$	46	36	38
While $\alpha\delta\zeta$ remains	45	26	37
Therefore $\gamma\delta\alpha$ will be	1	10	1

And because $\delta\zeta$ remains 93,252, as before, dividing the sine of $\gamma\delta\zeta$ by the sine of $\delta\gamma\zeta$ and multiplying the quotient by $\delta\zeta$ gives $\gamma\delta$ as 67,873.

But $\alpha\delta$ is 67,522.

From this and $\gamma\delta$,

$\delta\gamma\alpha$ is found to be	75°	8′	40″
and the perigee in	8	51	45 *Capricorn,*

about what it was before.

The eccentricity, somewhat more than 2000, is to be decreased to 1800 (as before), if the perigee were to be moved back to $5\frac{1}{2}°$ Capricorn. This is accomplished by lengthening $\alpha\epsilon$. And $\alpha\epsilon$ is lengthened if we say that the planet appeared one or two minutes before 9° 24′ Aries. For then, from the point θ set up by the other lines of observation, some line would be drawn outside $\theta\epsilon$ towards $\theta\zeta$.

One might, however, hold suspect such license in making small changes in the data, thinking that by taking the same liberty in changing whatever we don't like in the observations, the full Tychonic eccentricity might also at last be obtained. Anyone who thinks this way should make a try at it, and, comparing his changes with ours, he should judge whether the changes remain within the limits of observational precision. He also needs to beware lest, elated by a trust in one such iteration, he render himself all the more guilty afterwards, because of the very divergent apogees found for the sun.

I, to be sure, have laid all my prejudices and preferences out in the open here, so that I am more afraid of appearing to the reader to be importunate than I am of seeming untrustworthy.

And by the way, a remark for future use: if $\gamma\delta$ is made 100,000, $\alpha\theta$ will come out to be 147,443, or even greater when the figures that are still missing are obtained correctly.

144 Finally, to avoid prolixity, if $\alpha\theta$ be 147,700, and the eccentric position of

---

[8]This is obtained by adding $\epsilon\gamma\zeta$ and $\epsilon\gamma\delta$. Then $\gamma\delta\zeta$ is half the supplement of this sum.

Mars in 1595 be 14° 21′ 7″ Taurus, and the eccentricity of the earth be 1800, and the earth's path oval, as will be said in Chapters 30 and 44, the appearances will come out to be

					Should be		
24°	21′	13″	Aries;			24	20
9	23	20	Aries			9	24
3	2	30	Aries			3	$4\frac{1}{2}$
19	42	40	Taurus			19	42

From this figure, I conclude that $\alpha\theta$ is about 147,750.

And thus it has been demonstrated that $\alpha\gamma$ is about 1800, although it ought to have been 3600 if Tycho's discoveries were accommodated to the Copernican form and the sun's apparent motions. Consequently the point π of the earth's uniformity of motion must be sought on the line $\alpha\pi$, so that $\gamma\pi$ and $\gamma\alpha$ are equal. For if the earth is moved uniformly about π, that is, if $\delta\pi\epsilon$, $\epsilon\pi\zeta$, $\zeta\pi\eta$ are equal, Tycho's observations of the sun will remain unaltered, and $\pi\iota\alpha$ will be 3600. And at the same time, since the earth will be at the same distance from point γ when at the points δ, ϵ, ζ, η, the observations of Mars will also remain unchanged.

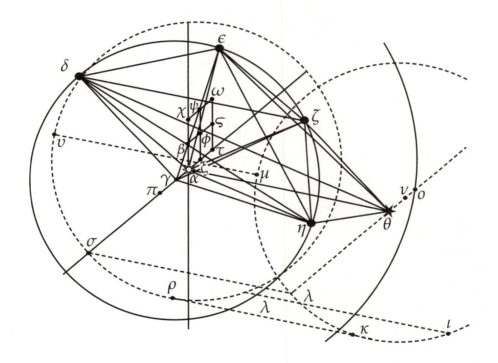

SECOND DIAGRAM: PTOLEMAIC FORM[9]

In the Ptolemaic form, the delineation can be done in two ways. In the first, let the earth replace the solar body at α, *and then lines of vision be drawn out from* α *parallel to* $\delta\theta$, $\epsilon\theta$, $\zeta\theta$, $\eta\theta$, *so that the Copernican positions of the earth* δ, ϵ, ζ, η *coalesce into one Ptolemaic position of the earth. Meanwhile, let the star Mars, which for Copernicus had stayed at one point* θ, *now be placed about* θ *at four locations:* ι, κ, λ, μ. *The description of this circle is as follows. Upward through* θ *let* $\theta\nu$ *be drawn*

Delineation of the theory of the epicycle.

---

[9]This heading is supplied to correspond to the first heading on p. 242. Kepler did not include the subsequent section titles.

Point of attachment.
See Part I.

equal and parallel to γα. *And about center* ν, *with radius* γε, *let the circle* ικλμ *be described. Thus the point* θ, *which we can call the point of attachment, moves around on the eccentric, which before, in Copernicus, the planet bodily traversed. While the epicycle is thus being borne around, the center* ν *is driven around* θ, *so that at one time it is inside* θα *and at another it is outside; however,* θν *is always parallel to itself and to the line* αγ. *And the epicycle will be moved uniformly, not about* θ *to which it is attached, nor about its center* ν, *but about a higher point,* o, *such that* θo *is twice* θν. *For the earth too will thus be moved*[10] *uniformly about* π, *not about the center of its orb* γ, *nor about the sun at* α.

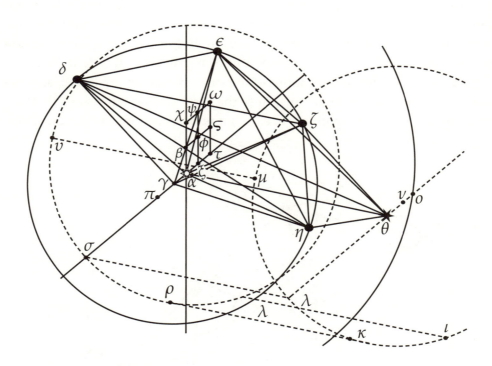

145

That these characteristics belong to the Ptolemaic epicycle, is properly demonstrated. But that they are carried over from the epicycle to the theory of the sun is shown by a probable argument only, pieced together from Ptolemaic opinions. *For, keeping everything as it was, let* ατ *be set up equal to* απ *and in the same line but in the opposite direction, so that* τα *may be the center of the sun's uniform motion, which the theorists had thought to be the center of the sun's orbit. Therefore, the line* θνo *will always be parallel to the line of the sun's apogee* ατ. *Now, if you decide that the diurnal parallax of Mars should be kept in the same ratio to the sun's parallax as that given out by Tycho,* ικλμ *will also be equal to the [circle of the] sun's theory, and consequently* θo *will also be equal to the eccentricity of the point* τ, *about which the sun moves uniformly. But* ικλμ *also moves in the same direction in which the sun moves on its circle, according to Ptolemy, and at the same times both the sun on its eccentric and the planet on its epicycle are found in the same, or at least corresponding, places, so that lines from* τ *through the sun and from* o *through the planet are ever parallel, again as taught by Ptolemy. So, since all other things are in agreement, why not this, too: that, just as* ικλμ *is moved uniformly, not about the*

---

[10]Reading *movebitur* instead of *movebatur*.

center ν but about the point o above it, as is demonstrated here by the transposition of the eccentric of the earth into the epicycle, where we took θ in place of α, ν for γ, and o for π: so likewise, these points are distinct in the sun, so that the eccentricity ατ, which is found from solar observations, is to be bisected at ξ, with ξ the center of the sun's eccentric λρσν? For Ptolemy made use of such a procedure to make it appear that if the sun's apparent positions were used, exactly the same eccentricity would be used on the planet's epicycle as was found in the sun. So since the observations give evidence of the double eccentricity of the Ptolemaic epicycle (because, as was said, the parallel relationships of the lines leave the triangles the same as in the Copernican form), the spirit of Ptolemy urges us to bisect the sun's eccentricity as well, so that the lines λι, ρκ, σλ, νμ remain parallel.

So by this reasoning even Ptolemy will be persuaded that ατ, the eccentricity of the sun's motion found by Tycho, should be bisected at ξ, so that the center of the sun's orbit is at ξ, and the uniformity of motion at τ.

Now this argument in the Ptolemaic form (as I just now began to say) is no firmer than the Ptolemaic world system itself. For anyone who believes Ptolemy, thinking that for the three superior planets there are three theories of epicycles, exactly equal to the theory of the sun, in quantity and quality, in lines as well as motions, in absolutely all respects—this same person will not admit this one inconsistency, but will also gladly derive this bisection for the solar theory from the epicycle, as if from an image in a mirror to the face itself.

Ptolemy's hypothesis refuted in passing.

And finally, when a comparison of hypotheses has been made, and it has appeared that four theories of the sun (or rather, six, as will be said elsewhere) can be generated from a single theory of the earth, like many images from one substantial face, the sun itself, the clearest of truth, will melt all this Ptolemaic apparatus like butter, and will disperse the followers of Ptolemy, some to Copernicus's camp, and some to Brahe's.

146

Here, one might raise a question. The Ptolemaic epicycle has three notable points: the center ν, the point θ which we have called the point of attachment, and the point o about which its motion is uniform. Now since it is said that the line θo remains parallel to ατ throughout the entire circuit, what are the properties of the circuits described by the other two points ν and o? *In order to show this, let lines be drawn from ξ and τα parallel to αβ, and from β and χ parallel to ατ, and let them be extended so as to intersect one another; and let the intersection of the lines from ξ and β be φ, from ξ and χ be ψ, from τ and β be ς, and from τ and χ be ω. Now, just as the point θ is moved regularly around χ, traversing an eccentric described about β, so also ν is moved regularly around ψ, traversing an eccentric described about φ. Also, o is moved regularly about ω, traversing a third eccentric likewise equal to the others, described about ς. For all three of these eccentrics the zodiacal position of the apogee is the same, owing to the lines' αχ, ξψ, τω being parallel. But the word "apogee" cannot be applied properly to any of these, apart from the first, that belongs to the point θ, since its line of apsides αβχ is drawn through the earth itself, which was placed at α, but not at ξ or τ.*

It is indeed true that straight lines can be drawn from the earth α through the centers of the remaining two eccentrics φ and ς, which may properly be called "lines of the apogee". These will fall to the west of the apogee αχ; that is, αφ will be at 24° Leo, and ας at 19° Leo, approximately. But then these lines will not pass through the pertinent point of uniform motion of each eccentric. Thus, if any of Ptolemy's followers does not wish to attach the epicycle to the eccentric at the point θ, *but prefers to relate it to the center* ν, *he will be driven to use two lines of apsides: one,* αφ, *for the eccentric, and the other,* αψ, *for the equant; and also two eccentricities,* αφ *and* αψ. *How intricate and inconvenient this is (for of its absurdity enough has been said in Chapter 6), let anyone so inclined judge.*

147   *The same will happen if anyone wishes to attach the epicycle to the eccentric at the point* ο, *about which the epicycle revolves uniformly. For then the eccentric bearing the point* ο *will have two apogees and eccentricities, one for the center on the line* αϛ, *and the other for the point of uniformity of motion on the line* αω. *The remaining possibilities are either to attach the epicycle at* θ, *or improperly to take apogees for the eccentrics that bear the points* ν *and* ο, *and to compute the eccentricities from the points* ζ *and* τ, *not from the reference point, the earth* α.

What has been given so far is the first delineation in the Ptolemaic form. The other can be constructed as follows. Let the Copernican positions of the earth δ, ε, ζ, η, merge, not in α, but in γ, so that in this diagram, γ, not α, denotes the earth, the center of the world. Here, the epicycle too, as well as its three eccentrics belonging to the points θ, ν, ο, will be shifted from their positions by the amount αγ, and a perfect equivalence will result. I shall forego further explanation, lest the reader become too confused, for this has really only been mentioned for smatterers or curiosity seekers.

Tychonic Form[11]

In the Tychonic form, there is no need of any new delineation. A very brief sketch is sufficient. *The eccentric's point of attachment is set at the four different positions* λ, ρ, σ, υ, *so the planet is at* ι, κ, λ, μ, *and* ιλ, καρ, λσ, μυ, *and* θα *parallel. Now Tycho himself made the center of the circle of Mars carry the double epicycle, and said that it goes around* α *uniformly on a circle concentric with the sun: for this idea he was indebted to Ptolemy. And in this matter, he, along with Ptolemy and Copernicus, was strongly urged by me (in Ch. 6 of Part I) to seek that point of attachment, whether it is the center of the concentric or of an eccentric, in the center of the solar body, this being supported by physical arguments and by the demonstration of its geometrical possibility. Additional support is provided by the valid argument of Ch. 22 and 23, that unless this were done, even if the observations were referred to the sun's mean position, the Ptolemaic epicycle and the Brahean deferent would be made eccentric, in directions exactly opposite to the sun's eccentricity. I have also promised stronger arguments, deduced from Brahe's own observations, for abandoning the sun's concentric, and in Ch. 52 and 57 below I will produce them.[12] But it has already been proved here in Chapter 26 that this center of Mars's concentric (or the point from which Mars's eccentricity originates) is found, not on an equal eccentric described about the center of the sun's point of uniform motion* τ, *as Brahe along with the other authorities had believed, but on an eccentric described about* ζ, *which is in the middle position between* α *and* τ.

---

[11]This heading is supplied to correspond to the first heading on p. 242. Kepler did not include the subsequent section titles.

[12]See all of Ch. 52, especially p. 394. Ch. 57 does not refer to Brahe's observations, but does argue that the planet's reciprocation takes place along the line from the planet through the sun: see especially p. 418.

*Therefore, if the center of Mars's concentric goes around with the sun, it never-
theless goes around on an eccentric described about ξ, and consequently the sun itself
will go around on an eccentric described about ξ. But its motion is uniform about τ.
Therefore, the sun's eccentricity ατ must be bisected at ξ.* For it is not likely that,
although the centers of the concentrics of Mars and the sun go around in the
same way, reach apogee at the same time, transpose their apogees in the same
way, go slowly or quickly in the same way, and describe the same circumferences, their circles would nevertheless make different digressions from the
earth in the same direction.

Let it be enough for now, to present this form of demonstration in the three
hypotheses. In what follows, whenever there is need of the same demonstration, I shall use Copernicus's form alone, it being simpler, so as not to be too
long-winded. Now, in contrast, the industrious reader has seen how any of   148
these diagrams can be transformed into either the Ptolemaic or the Copernican
form using parallel lines.

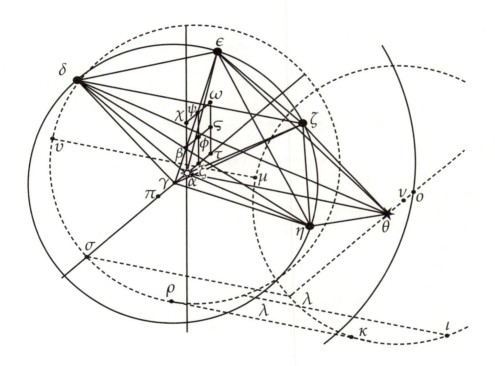

CHAPTER 27

From four other observations of the star Mars outside the acronychal situation but still in the same eccentric position, to demonstrate the eccentricity of the earth's orb, with its aphelion and the ratio of the orbs at that place, together with the eccentric position of Mars on the zodiac.

Hitherto we have almost exclusively used the aphelion of Mars, along with the correction of the mean motion and the hypothesis of the equations found above. If these should err by a single minute in defining the planet's zodiacal longitude, as can easily happen, this creates considerable difficulty for us in the present undertaking.

So now, at this point, we shall take for granted nothing whatever except Mars's periodic time, concerning which there can be no doubt, and the sun's zodiacal positions from Tycho's calculation. We shall also, I grant, assume an eccentric position, as is the common procedure in demonstrations leading to an impossibility, but we shall test that very position using repeated suppositions.

These are the observations.

			H	M							
1585	May	7	11	26	at 25°	55′	Leo	Lat. 1°	33′ N.
	May	12	10	8	at 28	$3\frac{1}{2}$	Leo	Lat. 1	$24\frac{1}{2}$ N.
1587	March	27	9	40	at 18	$21\frac{3}{4}$	Virgo	Lat. 2	$55\frac{2}{3}$ N.
	April	1	9	30	at 17	11	Virgo	Lat. 2	$43\frac{1}{2}$ N.
1589	Febr.	12	5	13 am	at 8	48	Scorpio	Lat. 2	9 N.
1590	Dec.	28	7	8 am	at 8	6	Scorpio	Lat. 1	14 N.
1591	Jan.	5	6	50 am	at 12	$44\frac{2}{5}$	Scorpio	Lat. 1	$23\frac{1}{4}$ N.

Since in 1589 there is but a single day that can be related to the others, and nothing else was observed for a long time before and after, let us refer the other times to this one. The catalog of them, along with the apparent positions of the sun and Mars, and with the eccentric position of Mars, is this:

Time	(am)	Sun	Mars	For the first trial, let the position on the eccentric be
1585 May 10	6:11	28° 55¾′ Taurus	26° 54½′ Leo	5° 22′ 2″ ⎫
1587 March 28	5:42	16 50⅖ Aries	18 12 Virgo	5 23 38 ⎪ Libra
1589 Febr. 12	5:13	3 41⅔ Pisces	8 48¼ Scorpio	5 25 14 ⎬
1590 Dec. 31	4:44	19 6⅘ Capr.[1]	9° 47⅙ Scorpio	5 26 50 ⎭

Let the diagram be made as before, with α the sun, β the center of the earth's eccentric, ζ, δ, ε, γ the four positions of the earth, η the position of Mars on its eccentric,

---

[1]This should be 19° 34′ Capricorn, a large error which escaped notice because it has little effect upon the size of angle $\epsilon\gamma\zeta$, used as a test below. Although $\alpha\gamma$ is shortened to 60,113, $\epsilon\gamma\zeta$ becomes 21° 21′ 37″.

and let each point be connected with all the others. Now, from the data,

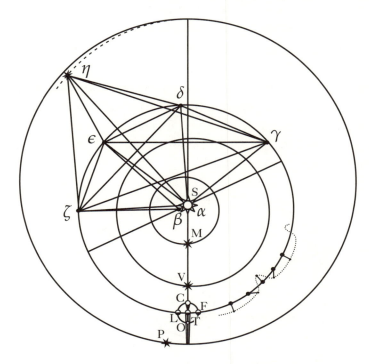

149

the known angles will be		hence are given		
αζη	87° 58′ 45″	αηζ	38° 27′ 32″	αζ	62227½
αεη	151 21 36	αηε	17 11 38	αε	61675
αδη	114 53 25	αηδ	33 23 1	αδ	60658
αγη	69 19 38	αηγ	34 20 20	αγ	60291

by the method of
Chapter 26,
preceding.

Now because there stand two angles upon the arc ζε at the circumference of the circle, namely ζδε, ζγε, by Euclid III. 21 these must be equal. And in order that they come out equal, αη must be moved forward and backward above α beneath the zodiac as long as [is necessary].[2] *And since in this first trial the zodiacal position for αη is given, let it therefore be tried whether ζδε, ζγε might be equal, for then it will be established that the position of αη is correct.*

In the four triangles	ζαδ,	δαε,	εαγ,	ζαγ
the same number of angles is sought, namely	ζδα,	εδα,	εγα,	ζγα
in order to get	εδζ,		εγζ	

Now in any of these triangles the angles at α are given by the position of the sun from Tycho, and the correction for the precession of the equinoxes. But the sides comprehending those angles have just been found. Therefore, the angles, too, will be given.

And   found angles

ζαδ	*is*	85° 17′ 7″	ζδα	48° 8′ 59″	*hence* εδζ *is* 21° 28′ 1″	*they differ by 9′.*
εαδ	*is*	43 10 20	εδα	69 37 0		
εαγ	*is*	87 46 48	εγα	46 47 36	*hence* εγζ *is* 21 19 6	
ζαγ	*is*	129 53 45	ζγα	25 28 30		

150

Since these angles didn't quite come out equal, I made a second trial with αη's

---

[2]Kepler's use of the Latin *tantisper*, "so long as," seems to be missing an object clause, which is supplied by the bracketed words.

sidereal position moved forward 2′. And I found $\epsilon\delta\zeta$ *to be* 21° 40′ 9″, $\epsilon\gamma\zeta$ 21° 22′ 14″, *differing by* 18′, *which is twice the previous discrepancy. Whence it is understood that* $\alpha\eta$ *should have been moved backwards to a lesser longitude, rather than forwards.*

For a third trial, then, supposing Mars's eccentric position in 1585 was 5° 20′ 2″ *Libra,* $\epsilon\delta\zeta$ *came out to be* 21° 15′ 54″, *and* $\epsilon\gamma\zeta$ 21° 13′ 54″. *There remains a difference of* 2′, *which we may safely ignore.* Nevertheless, by extrapolating we realize that at this place Mars's eccentric position has to be moved back[3] through $2\frac{1}{2}′$, just as previously in Chapter 22 it was moved forward 1′ on the opposite semicircle. Both of these are brought about by an increase of the eccentricity and a slight retraction of the aphelion.

Limit of accuracy of the hypothesis of Chapter 16 for longitudinal positions.

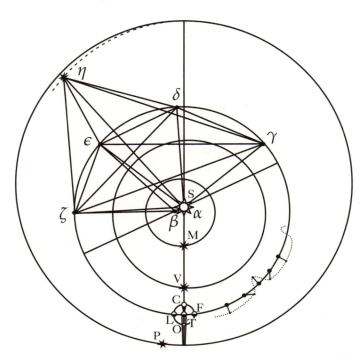

Let us now proceed to the investigation of the rest. *And because each of the angles in question has decreased, they will decrease further when* $\alpha\eta$ *is moved back. Therefore, let each be* 21° 13′, *and* $\zeta\beta\epsilon$, *the double angle at the center, be* 42° 26′. *Therefore,* $\zeta\epsilon\beta$ *is* 68° 47′.

In triangle $\zeta\alpha\epsilon$ *angle* $\zeta\alpha\epsilon$ *is* 42° 6′ 57″, *and the sides are given by a new correction, so that* $\alpha\zeta$ *is* 62, 177 *and* $\alpha\epsilon$ 61, 525, *approximately. Hence,* $\zeta\epsilon\alpha$ *is given as* 69° 43′ 31″, *and* $\zeta\epsilon$ 44, 518. *But this same* $\zeta\epsilon$, *from angle* $\zeta\beta\epsilon$ *(which* $\zeta\epsilon$ *subtends), is* 72, 379 *where* $\epsilon\beta$ *is* 100, 000. *Therefore, where* $\epsilon\beta$ *is* 100, 000, $\alpha\eta$ *is* 162, 818, *and thus* $\alpha\epsilon$ *is* 100, 174. *But when* $\zeta\epsilon\beta$ *is subtracted from* $\zeta\epsilon\alpha$, *the remainder* $\beta\epsilon\alpha$ *is* 0° 56′ 31″, *and* $\beta\alpha\epsilon$ *is* 83° 30′. *Therefore, the aphelion is at* 10° 19′ *Capricorn, while the eccentricity* $\alpha\beta$ *is* 1653.

151

Again we have come rather close to half of 3600, and would doubtless obtain exactly that if we had the apogee perfectly.

Still, it should be noted that if we suppose that the earth's path is not a perfect circle, but is narrower at the sides, $\alpha\eta$ comes out here to be a little

---

[3] *Anticipandum*, which should mean "anticipated" or "set ahead." However, the context shows that the position must be moved back in longitude.

less than 163,100. And then, with $1\frac{1}{2}'$ subtracted from the eccentric position, and taking the value 1800 for the earth's eccentricity and $5\frac{1}{2}°$ Capricorn as the aphelion, the following apparent positions are produced:

	26° 55′ Leo	18° 11$\frac{2}{3}$′ Virgo	8° 49′ Scorpio	9° 44$\frac{1}{3}$′ Scorpio
Should have been:	26 54$\frac{1}{2}$	18 12	8 48$\frac{1}{4}$	9 47$\frac{1}{6}$

This supposition also agrees with my observations of 1604 February 29/ March 10, for on the night following that day with my instruments I found Mars culminating at 26° 18$\frac{4}{5}'$ Libra. And calculation based upon these assumptions puts it at 26° 17$\frac{1}{2}'$ Libra. Moreover, at $8\frac{2}{3}^h$, a few hours before the observation, it was again in the same eccentric position.

Besides, since Mars has some latitude here, the value for $\alpha\eta$ just found is the distance in the plane of the ecliptic of the point η from the center of the sun, to which point a perpendicular is drawn from the body of Mars, as was noted in Chapter 20 above. So the true distance of the planet's body itself from the center of the sun is made a little longer by 37 units.

CHAPTER 28

Assuming not only the zodiacal positions of the sun, but also the sun's distances from the earth found using an eccentricity of 1800, through a number of observations of Mars at the same eccentric position, to see whether by unanimous consent the same distance of Mars from the sun, and the same eccentric position, are elicited. By which argument, it will be confirmed that the solar eccentricity of 1800 is correct, and was properly assumed.

The reader should not be surprised that in this third turn I am now not pre-supposing the eccentric position of Mars as it is given by the hypothesis of acronychal observations found above. For I have said that that hypothesis was only vicarious, not natural, and thus possesses only as much trustworthiness as is derived from the observations; and it could deviate somewhat in the intermediate positions between observations. Besides, it helps us to have at hand various methods of demonstration by which to explore carefully the distances of Mars at all places throughout the entire circle. Accordingly, a new form of demonstration will follow here.

152

THE OBSERVATIONS ARE THESE.

		D	H		it was at				Lat.		
1583	April	22	$9\frac{2}{3}$		it was at	1°	17′	Leo,	1°	$50\frac{2}{3}′$	N.
1585	March	9	$9\frac{1}{6}$		it was at	11	$49\frac{1}{10}$	Leo,	3	$29\frac{1}{10}$	N.
	March	11	5		it was at	11	$45\frac{1}{2}$	Leo,	3	$24\frac{1}{8}$	N.
	March	12	5		it was at	11	$45\frac{3}{4}$	Leo,	3	$21\frac{2}{3}$	N.
1587	January	26	5	am	it was at	4	$41\frac{3}{4}$	Libra,	3	26	N.
	January	28	5	am	it was at	4	41	Libra,	3	27	N.
1588	December	5	$6\frac{1}{2}$	am	it was at	9	23	Libra,	1	$44\frac{3}{4}$	N.
	December	15	$6\frac{1}{6}$	am	it was at	14	$35\frac{2}{3}$	Libra,	1	54	N.
1590	October	31	$6\frac{1}{4}$	am	it was at	2	$57\frac{1}{3}$	Libra,	1	$15\frac{1}{2}$	N.

When the times of the other observations are adjusted so as to return Mars to the same eccentric position which it had at the last time, we are given the following times. Along with these are added the requisite positions of the sun, and the distances of the sun and the earth computed from the hypothesis so far established. And indeed, we have taken up this labor in order to test these very things. A little later, in Ch. 30, a technique for computing these distances will follow.[1]

---

[1]Kepler is somewhat reticent here about how he obtained his figures for the distances. And understandably so, since the numbers he gives are derived from an elliptical orbit, not from the circle he is ostensibly considering here. So instead of raising that question here, he refers

	D. H. (am)		Mars at			Sun at				Distance Sun to earth
1583 April	23	$8\frac{1}{10}$	$1°\ 29\frac{1}{2}'$	Leo	$12°\ 10'$	$3''$	Taurus			101,049
1585 March	10	$7\frac{2}{3}$	$11\ 48\frac{1}{3}$	Leo	29 41	4	Pisces			99,770
1587 January	26	$7\frac{1}{6}$	$4\ 41\frac{3}{4}$	Libra	16 5	55	Aquarius			98,613
1588 December	13	$6\frac{3}{4}$	$13\ 35\frac{2}{3}$	Libra	1 44	53	Capricorn			98,203
1590 October	31	$6\frac{1}{4}$	$2\ 57\frac{1}{3}$	Libra	17 28	33	Scorpio			98,770

Now for the deduction of the observations from the days of the observations to the times we have chosen. For the first time, the diurnal motion was taken from Magini, since over the space of a few hours there is no danger of error. The other times are supported by observations before and after. However, for the penultimate time I also looked up the sequence of diurnal motions in Magini. Around December 15 the diurnal motion was 30′, and around December 5 it was 32′. For the last time, although Mars, being at an altitude of 23°, is affected by refraction, so that 2′ might easily be wanting in the latitude, this refraction nevertheless hardly affects the longitude of Mars. (Tycho claimed that the refraction of the fixed stars, also applicable to the planets, ceases at this altitude, although the solar refraction reaches higher, and at this altitude is about 4′. This distinction was discussed and demolished in my *Optical astronomy* p. 137,[2] and would be rendered even more dubious if any changes are to be made in the parallax of the sun.)

Let α be the sun's body, αβ the eccentricity of the earth's orb (1800), and the line of apsides be at $5\frac{1}{2}°$ Cancer, positions of the earth ζ, ε, δ, γ, θ, and the body of the planet at the same eccentric position η all five times, since the intervals span complete periods of Mars. And let all points be connected with one another. It is desired to find αη, and its zodiacal position, that is, angle ηαθ,

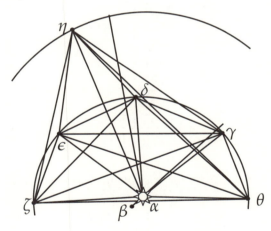

153

ηαγ, or some other angle at α. We shall do this from two earth positions in the following manner. *For a start, let these be ε, δ. And in triangle εαδ, given the sides εα 99,770, αδ 98,613, and angle εαδ, the rest are sought, namely, the angles δ and ε and the side δε.*[3]

---

the reader ahead to Ch. 30, where he presents his table of distances and notes his use of the ellipse.

What this shows is that, in this chapter at least, Kepler is not reproducing the arguments that led him to introduce an equant into the theory of the sun or earth. Rather, he is presenting an additional argument, worked out after his discovery of the elliptical orbit, to confirm the sun/earth equant.

[2]Kepler, *Optics*, Ch. 4 sect. 8, esp. p. 150 (in the English translation).

[3]In the computations below, bracketed material is provided by the translator. The computation is Kepler's usual procedure for applying the law of tangents. For more on this, see the sample computation in the translator's Introduction, p. xxv.

αε 99770		29° 41' 4" *Pisces* [sun's position in 1585]	
αδ 98613		16 5 55 *Aquarius* [sun's position in 1587]	
[diff.] 1157		43 35 9	
[sum] 198383		*Precession* 1 36	
991915 |5	εαδ 43 36 45	[sin εαδ] 68977	99770 [αε]
165085		136 23 15 [suppl.]	[sin αδε] 93376	73870 [from left]
158706	8	68 11 38 [half suppl.]	653632	7 664830
6379 |3	*Tang.* 249813	36138	66483
[583 = diff./sum]		583 [from left]	28013 |3 5171
		12491	8125	517
68° 11' 38"		1998	7470 |8 73700 δε
50 3		75	655	7
αδε 69 1 41		1456 [product]	654 |0 [sin εαδ]/[sin αδε]
αεδ 67 21 35			1|

With these matters investigated, one goes on up to triangle εηδ.

For since	εα is	29° 41' 4"	*Pisces*	δα 16° 5' 55"	*Aquarius*	
and	εη	11 48 20	*Leo*	δη 4 41 45	*Libra*	
	αεη will be	132 7 16		αδη 131 24 10		
But	αεδ was just	67 21 35		αδε 69 1 41		
Remainder	ηεδ	64 45 41		ηδε 62 22 29		

154 (margin)

The remaining angle εηδ to make up two right angles with these is 52° 51' 49".

So, angles ε, η, δ and one side εδ being given, side εη will also be given.

Sin. εηδ 79718		εδ 73700	
Sin. ηδε 88600	8	89972	εη
8838		719776	8
7974	9	17224	
864		8997	1
797	9	8226	
67		8097	9
62	7	129	
5	1	90	1
		39	5

Finally, let the triangle ηεα also be solved, in which now are given:

εη	81915		*And αεη, as before, is* 132° 7' 16"		Sin. εαη 36556		
εα	99770		*Supplement* 47 52 44		Sin. αεη 74173	2	
Diff.	17855		*Half* 23 56 22		73112	0	
Sum	181685		*Tangent* 44396		10610		
1635165	9*		*9823		7311	2	
150335			39956		3299	9	
145348	8		3552		3290	0	
4987			89		9	3	
3834	2		[from above] 23° 56' 22"	13			
1153			[tan⁻¹ .04361] 2 29 50	4361	εη	81915	
1150	3		εαη 21 26 32			202903	
But in 1585, αε was in			29° 41' 4" *Virgo*			1638300	
Therefore αη in 1585 was in			8 14 32 *Virgo*			16383	
						7372	
						25	

The magnitude sought, αη, comes out to be 166208

If the other three observations at ζ, γ, θ, will allow this same position and length for $\alpha\eta$, we shall have excellent confirmation of them.

So, in the same manner in which we have hitherto worked with ϵ and δ, we shall now work with ζ, γ, seeking the same $\alpha\eta$.

155

For the angles, and line, γ, ζ

$\alpha\zeta$ 101049		12° 10′ 3″ *Taurus*		
$\alpha\gamma$ 98203		1 44 53 *Capricorn*		
2846		130 25 10		
19925|1		*Precession* 4 48		
8535		130 29 58		
7970|4		49 30 2	76041|	179055
565		24 45 1	42468|1	101049
399|3		46101	335730	1790550
166		1438	297276|7	17905
159|8		4610	38454	716
		1844	38221|9	162
		138	233|0	180933 $\zeta\gamma$
24° 45′ 1″		37	212|5	
22 48	– – – – – – –	663	21|5	
$\alpha\gamma\zeta$ 25 7 49				
$\alpha\zeta\gamma$ 24 22 13				

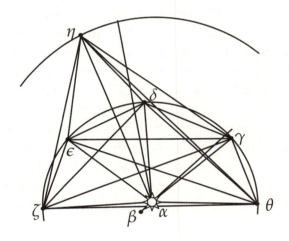

And now in $\zeta\gamma\eta$.

Because	$\zeta\eta$	*is*	1° 29½′	*Leo*	$\gamma\eta$ 13° 35′ 40″ *Libra*			
and	$\zeta\alpha$	*is*	12 10 3	*Taurus*	$\gamma\alpha$ 1 44 53 *Capricorn*			
Therefore	$\eta\zeta\alpha$	*is*	79 19 27		$\eta\gamma\alpha$ 78 9 13			
But	$\gamma\zeta\alpha$	*is*	24 22 13		$\zeta\gamma\alpha$ 25 7 49			
Therefore	$\eta\zeta\gamma$	*is*	54 57 14		$\eta\gamma\zeta$ 53 1 24 ,			

and the remain- $\gamma\eta\zeta$, is 72° 1′ 22″
ing angle to
make up two
right angles
with these,

The same is also proved from this: $\zeta\eta$ *is at*   1°   29½′   *Leo*
       and $\gamma\eta$ *is at*   13   35   40   *Libra,*
or, with the precession over the interval subtracted, at   13   35   52   *Libra*
            Therefore, $\gamma\eta\zeta$ *is*   72   1   22

Next, given the angles of the triangle ζηγ, and side γζ, the side ζη is sought.

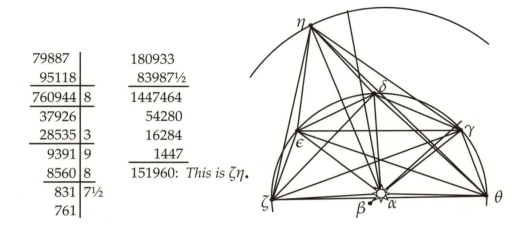

79887		180933		
95118		83987½		
760944	8	1447464		
37926		54280		
28535	3	16284		
9391	9	1447		
8560	8	151960: *This is ζη.*		
831	7½			
761				

156

Finally, in triangle ηζα the sides and the comprehended angle are given.

ζη	151960					98269	
ζα	101049		79° 19′ 27″			89861	10
	50911	2	100 40 33			84080	
	253009		50 20 16			80875	9
	506018	0	120612			3205	
	3096		20122			2696	3
	2530	1	241224			509	5
	562	2	1206			449	
	506		50° 20′ 16″	241		60	7
	56	2	13 38 39	24			
				24270			

ζαη comes out to be	63	58				151960
But αζ is at	12	10	3	*Scorpio in 1583*		109357
Therefore αη is at	8	11	31	*Virgo in 1583*		1519600
Precession		1	36			136764
So it would be at	8	13	8	*Virgo in 1585*		4559
Previous position	8	14	32	*Virgo in 1585*		760
Difference		1	24			106

αη comes out to be 166179
Previous value 166208
Difference 29

And so it appears that with two other observations, at ζ and γ, we have come upon the same thing, within the limits of observational accuracy. For an error of a minute and a half in observing, or in deducing the observed position to a day when it was not observed, can be committed.

But let us also see the evidence of the fifth position, θ, that is, of the observation at θ.

We know that θα *is at*	17° 28′ 33″ *Scorpio, and suppose that* θα *is* 98,770
and θη *was observed at*	2 57 20 *Libra*
Therefore the angle αθη *is*	44° 31′ 13″.

If I lengthen αη subtending this angle, I will thereby move αη farther forward in longitude, and vice versa.

157

So let αη be 166,208, as initially determined.

Now as αη is to [sin] αθη so is αθ to [sin] αηθ.

98770				
166208				
831040	5			
156660				
149587	9			
7073				
6648	4			
425				
332	2			
93	6			

70116			
59426			
415982			
594			
59			
41665 *Produces αηθ*	24° 37′ 28″		
But θη points toward	2	57 20	*Libra in year 90*
Therefore αη is at	8	19 52	*Virgo in year 90*
Precession		4 48	
So it is at	8	15 4	*in year 85*
Which was at first	8	14 32	
Difference		0′ 32″	

And so through the very slightest shortening of αη, it will fall exactly in the same place with the first two observations.

And so it appears from this that we rightly assumed and posited the distances αζ, αε, αδ, αγ, αθ, and the eccentricity αβ as well. For it is impossible to take other distances than these, which nonetheless also fit as nearly as possible on a circle and have the appropriate zodiacal positions, and still obtain the same magnitude for αη, and its zodiacal position, from all five observations.

But concerning the length of αη, we shall put our trust mostly in observations ζ, γ, θ. For also in the common method of measuring distances of things on earth, the farther the standing points are from one another, the the more accurately distance of the mark is obtained.

For the zodiacal position, however, we shall put our trust in the observations at ε, δ, instead. For if there is some slight error in the length of αη, it is presented to the observer at ε, δ quite obliquely, and the angle does not change perceptibly.

Nor is this to be forgotten: that there is no perceptible lengthening of αη over the seven years from 1583 to 1590, because the progression of the aphelion is very slow.

Summary: On 1590 October 31 at $6\frac{1}{4}^h$ in the morning, Mars's eccentric position was 8° 19′ 20″ Virgo, while the hypothesis constructed using acronychal observations places it at 8° 19′ 29″ Virgo. Its distance was 166,180, which must be lengthened because of the latitude, so that from this distance, the distance from the actual body of Mars to the center of the sun comes out to be about 166,228.

CHAPTER 29

A Method of Deducing the Distances of the Sun and Earth from the Known Eccentricity

I think it has been well enough confirmed that the distances of the sun and the earth are to be deduced by halving the eccentricity obtained by Tycho. This is also abundantly confirmed by observation of the sun's summer and winter diameter, as I have shown in the *Astronomiae pars optica* chapter 11.[1] But it is also wonderfully confirmed in the *Mysterium cosmographicum*, in the table in ch. 15 p. 53,[2] where the equations of the center for Mars, Venus, and Mercury were deficient when the lunar orb was interposed, but excessive when it was omitted. Now, the orb of the moon being retained while the eccentricity of the sun is bisected, they come out about right.

Furthermore, the same thing will again be confirmed more frequently and much more clearly when we use the distances forthcoming from the bisection (as we have just begun to do in the last chapter) and shall see the phenomena follow from them. Therefore, in order that these distances be ready at hand for our future use, I shall show how they may easily be computed, using a geometrical demonstration.

On the line αδ let α be the body of the sun (or the earth for Tycho, or the center of attachment of the epicycle for Ptolemy); β the center of the eccentric ζδη of the earth (or of the sun and of the annual orb for Tycho, or of the epicycle for Ptolemy); and αβ being extended, let it intersect the eccentric at δ, ε, so that δ is the aphelion or apogee and ε the perihelion or perigee; and let βγ be equal to αβ. Also, let γ be the center of motion or of uniformity, at which the earth (for Ptolemy, the center of the epicycle, for Tycho the sun and the point of attachment of all the eccentrics) sets out equal angles in equal times. And let αγ be 3600, from the observations of Tycho and the Landgrave,[3] but αβ, according to my recently introduced adjustment, be 1800. Now let ζη be drawn through α perpendicular to δε, intersecting the circle at ζ, η, and also through α let the straight line θι be drawn, at any inclination whatever, intersecting the circumference at θ, ι; and let the four points θ, ι, ζ, η be connected with the center β. And let this also be posited at the start: that although the earth (sun, or planet) is moved uniformly around γ and thus nonuniformly around β, it nevertheless remains on the circumference of the circle described about β. Now, by the equivalence shown in the second chapter (which, for the sake of avoiding confusion, I shall not apply to the general Ptolemaic hypothesis), this is exactly as if one were to say that the earth

---

[1] Kepler, *Optics* Chapter 11, Problem 2, English trans., p. 353.

[2] *Mysterium Cosmographicum: The Secret of the Universe*, A. M. Duncan, trans., New York 1981, pp. 162–3 (second table).

[3] The observations of William IV, Landgrave of Hesse, which were known to Kepler through Tycho, were published several years later by W. Snel under the title *Coeli et siderum in eo errantium observationes Hassiacae...*, Leiden 1618 (cited in *KGW* 3 p. 466)

(or sun) is moved nonuniformly on a concentric with an epicycle[4] about center α, the semidiameter of the epicycle being equal to αβ; and the arcs described on the concentric by the center of the epicycle being similar to the arcs of the epicycle described by the earth (or sun), so that both the earth (or sun) and the center of the epicycle are moved unequally in equal times, so as to become slow, or again speed up, simultaneously. I am going to postpone for a little while the physical explanation of this hypothesis.

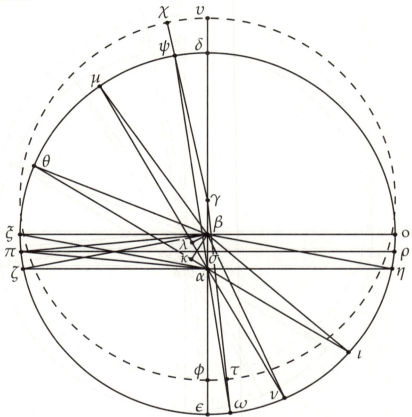

Now, with these things supposed, I shall proceed to the work of finding the distances. *And because βδ is 100,000 and βα 1800, and αβδ is a straight line, by addition of the two the aphelial distance αδ is obtained; and because βε too is 100,000, when αβ is subtracted the perihelial distance αε remains.*

Longest and shortest distance.

And because βαζ is right, and ζβ is 100,000 (that is, the whole sine), therefore αβ is the sine of the angle αζβ. Consequently αζβ is 1° 1′ 53″, which is the optical part of the equation of the sun or earth. Now the maximum equation at the middle elongations, which is composed of the optical and physical parts, has the whole eccentricity 3600 (or 3592) as its sine. Therefore, when the sun or earth goes from δ to ζ, it will in fact take two days longer than one fourth of the periodic time, but it nevertheless does only one day's journey beyond one fourth of its total circuit. So in this distance, or quarter of the periodic time,[5] it will take one day longer than it should, owing to the physical weakening.

Distances of intermediate positions.

159

But to proceed to the distance αζ. *In the right triangle ζαβ, since one of the acute angles is given, the other, ζβα, will be the difference between the first and one*

---

[4]ὁμοκεντρεπικύκλῳ.

[5]Kepler surely means "quarter of the circumference" here.

right angle, that is, 88° 58′ 7″. And therefore αζ will be the sine of this angle, 99984. And the opposite line αη is the same size.

For finding the intermediate distances of two opposite degrees of equated anomaly, let θι be inspected, passing through the body α whence the eccentricity is computed. *Now δαθ and δαι are equated anomalies, and are opposite, in that α is between them and is collinear with them. Now let a line βκ fall from β perpendicular to θι, so as to make θκ, κι equal. In the right triangle βκα the base βα is given, as well as the angles καβ (from the integral number of degrees of equated anomaly chosen) and its complement κβα. Therefore, the sides κα, κβ will not be unknown. And κβ is the sine of the angle κθβ or κιβ. This being given, its complement, θβκ or ιβκ, will also be known, and its sine, which is the line θκ or κι. And when κα is added to κθ, αθ is obtained; and when the same is subtracted from κι, αι is obtained. The former distance corresponds to the equated anomaly δαθ, and the latter to the equated anomaly δαι, which line has a line equal to itself in the preceding semicircle, standing as far from aphelion in semicircle δθ as αι itself does in semicircle δη.*

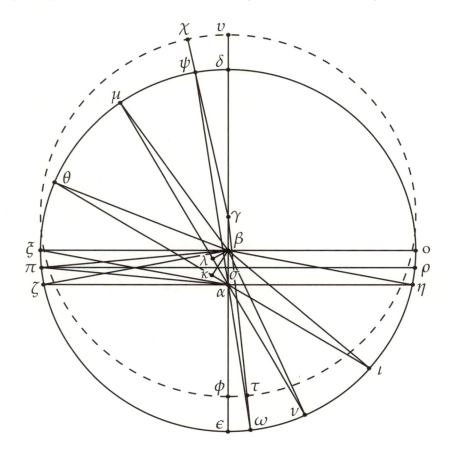

Now through α let the straight line μν be drawn intersecting the circle in μ, ν, and making the angle μαδ equal to the angle κβα. And from β let βλ fall perpendicular to μν bisecting μν at λ. And let μ, ν be connected to β. Now, since καβ is an angle of an integral number of degrees, the remainder κβα, and μαδ equal to it, are also an integral number of degrees, and in the similar triangles βκα, βλα side κα is equal to side λβ, and κβ to λα. But λβ is the sine of angle λμβ, λνβ, and the complement of λμβ is λβμ, λβν. And the sine of this is the lines λμ, λν, and the difference between these and αμ, αν is λα. But the magnitudes λα, λβ have just now been found in triangle αβκ. Therefore, by the use of one triangle, four distances can be found

making equal angles about α with the line of apsides and its perpendicular $\zeta\eta$ drawn through α. For $\mu\alpha\zeta$ is equal to $\theta\alpha\delta$, and $\nu\alpha\eta$ to $\iota\alpha\epsilon$.[6]

So the greatest distance is at δ, the shortest at ϵ, but the mean, which is equal to $\beta\zeta$, is not at $\zeta\eta$. Neither is it on a line through β parallel to $\zeta\alpha$ which shall be called ξo. For $\alpha\zeta$ is less than $\beta\zeta$, since $\zeta\beta\alpha$ is subtended by a smaller line than $\zeta\alpha\beta$, which is right, and, $\alpha\xi$ being drawn, it is longer than $\beta\xi$, since it subtends a greater angle $\xi\beta\alpha$ (which is right) while $\xi\beta$ subtends a lesser angle $\xi\alpha\beta$.

But in order that the place of the mean distances be defined geometrically, *let $\alpha\beta$ be bisected at σ, and through this let $\pi\rho$ be drawn perpendicular to $\alpha\beta$, intersecting the circle at π, ρ. I say that these are the points that are equally distant from α and β.*

For let one of the points π be connected with α and β. The lines $\pi\alpha$ and $\pi\beta$ will subtend equal angles $\pi\sigma\alpha$ and $\pi\sigma\beta$ (since they are right), and $\alpha\sigma$, $\sigma\beta$ will be equal, and $\pi\sigma$ common. Therefore $\pi\alpha$, $\pi\beta$ are equal. And thus the demonstration taken from Reinhold[7] concerning the whole $\alpha\gamma$ and its midpoint β remains true for the point σ and its half $\alpha\beta$.

One might think that since at π the distance $\alpha\pi$ becomes equal to the semidiameter $\beta\pi$, the angle $\beta\pi\alpha$ is also greater than $\beta\zeta\alpha$, and thus the greatest equation occurs at π, *on the argument that the straight line $\beta\alpha$ is presented more directly from π than from ζ. However, this proposed line of reasoning is not true. For to the same extent that $\beta\alpha$ is more oblique with respect to ζ, π in turn is more distant than ζ, since $\pi\sigma$ is longer than $\zeta\alpha$. For $\pi\beta\sigma$ is greater than $\zeta\beta\alpha$, which $\zeta\alpha$ subtends.* So Ptolemy, and after him Reinhold in the *Theoricae*, demonstrated correctly that the greatest equation (eccentric equation alone, or optical part) ocurs at ζ.[8] I shall, however, set up this demonstration in another, simpler form. *Let any point be taken above ζ, such as θ, and any below η or ζ, such as ι. Let them be connected with α, and from β let perpendiculars $\beta\kappa$ fall to $\theta\alpha$ or $\iota\alpha$ extended. Now since $\delta\alpha\zeta$ and $\beta\kappa\alpha$ are equal, they being right angles, and $\kappa\beta\alpha$, $\kappa\alpha\beta$ together are equal to one right angle, when the same angle $\delta\alpha\theta$ or $\beta\alpha\kappa$ is subtracted from the equals, the equals $\theta\alpha\zeta$, $\kappa\beta\alpha$ will remain. And first, let a line be drawn through α above ζ, as $\theta\alpha$ was just drawn, whether θ is next to ζ or remote. At the same time let its perpendicular $\beta\kappa$ be inclined to $\beta\alpha$. Now $\beta\alpha$ is greater than any of the perpendiculars $\beta\kappa$, since $\beta\alpha$ is subtended by the right angle $\beta\kappa\alpha$, while $\beta\kappa$ is subtended by the smaller, acute angle $\beta\alpha\kappa$. Now since $\beta\zeta$, $\beta\theta$, $\beta\iota$ are equal, and $\beta\alpha\zeta$, $\beta\kappa\theta$, $\beta\kappa\iota$ are right, they fit onto the same semicircle, whose diameter is equal to $\beta\zeta$, $\beta\theta$, $\beta\iota$. And $\beta\alpha$, being longer, subtends a greater part of the circumference of any such semicircle than does $\beta\kappa$ or any of the perpendiculars; and for that reason, its angle $\beta\zeta\alpha$ will be greater than $\beta\theta\kappa$, or the angle of the equation for any other point above ζ, such as π or ξ. Which was to*

Where the mean between the longest and shortest distances is.

160

Where is the equation greatest?

---

[6]Stated algebraically, the problem here is to find the distance from the sun to the earth ($\alpha\theta$, say) as a function of the equated anomaly $\delta\alpha\theta$. Where $\beta\delta = a_T$, the major semi-axis of the earth's orbit,

$$\frac{\alpha\theta}{a_T} = \cos[\sin^{-1}(\alpha\beta\sin\delta\alpha\theta)] + \alpha\beta\cos\delta\alpha\theta.$$

[7]Erasmus Reinhold published the *Theoricae planetarum* of Georg Peurbach with detailed scholia in 1542 in Wittenberg, and the work was later reprinted. The passage cited is in the scholium to the theory of the sun, Part II Prop. iv. (citation from KGW 3 p. 467).

[8]Ptolemy, *Almagest* II.3, ed. Heiberg vol. 1 pp. 221–223.

be demonstrated.

Everything said in this chapter about the computing of distances of the sun and earth will also be valid for Mars, as long as one of the suppositions is that the orbits of planets are perfect circles. When this is seen as false, another method of computing them will be given.

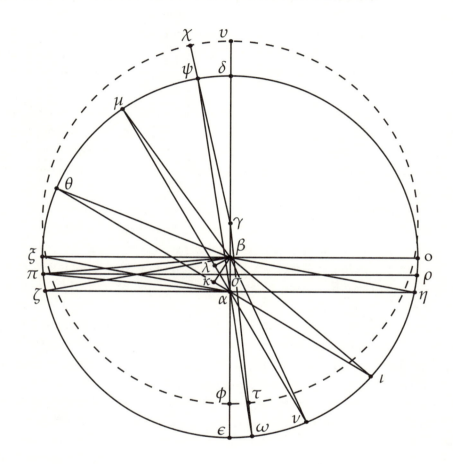

Chapter 30

Table of the distance of the sun from the earth and its use

We have gathered together here into a three-columned table the distances of the sun accumulated in this way [that is, in the way described in ch. 29], as if they were done for integral degrees of equated anomaly for the whole semicircle (for those in the other semicircle at equal distances from apogee are also equal to these). In the first column, which we have called "mean anomaly", are the angles $\delta\beta\mu$, $\delta\beta\theta$, $\delta\beta\xi$, $\delta\beta\iota$, $\delta\beta\nu$, composed of the integral angles $\delta\alpha\mu$, $\delta\alpha\theta$, $\delta\alpha\zeta$, $\delta\alpha\iota$, $\delta\alpha\nu$, and their optical or eccentric equations, namely, $\beta\mu\alpha$, $\beta\theta\alpha$, $\beta\xi\alpha$, $\beta\iota\alpha$, $\beta\nu\alpha$. In the second, the distances themselves, $\alpha\mu$, $\alpha\theta$, $\alpha\xi$, $\alpha\iota$, $\alpha\nu$, are placed together opposite [the mean anomalies]. In the third, under the heading "equated anomaly," are tabulated angles not depicted here, but the procedure for whose generation will be revealed in part here and in part in Chapters 31 and 40. For they are constituted by subtraction of the optical equations $\alpha\mu\beta$ and so forth from $\delta\alpha\mu$ and so forth. Thus we have given no column to the integral angles like $\delta\alpha\mu$, because they are the arithmetic mean between the angles of the columns at the sides, and thus are easily found in themselves, and are not of any use, as we shall hear. 161

Therefore, beginning with either the mean or the equated anomaly, either one going into its proper respective column as use will dictate; or, where it exceeds a semicircle, beginning with the full-circle complement of either of these; you find the requisite distance of the sun from the earth, in units of which the radius of the orb is 100,000 and the eccentricity is 1800. *Instructions for use of the table.*

It is true that in this way (that is, in associating the distance $\alpha\zeta$ of the angle $\delta\alpha\zeta$ with an angle which is as much smaller than $\delta\alpha\zeta$ as $\delta\alpha\zeta$ is smaller than $\delta\beta\zeta$) a path is attached to the circuit of the earth (or sun) about α which is oval rather than exactly circular. For the distance $\alpha\zeta$ (for example) was determined by the angle $\delta\alpha\zeta$, an integral 90°, and it was assumed in the operation that this angle $\delta\alpha\zeta$ was the equated anomaly. Now, however, you are told to get the distances using the angles of the anomaly that is called "equated" in our table, which have been diminished by the equation $\beta\zeta\alpha$. It thus happens that at 90° you do not get 99,984, although you would previously have determined it to be 99,984. For here, opposite 99,984, you find an equated anomaly of 88° 58′ 7″, which is not your value. For it was 90° that was proposed, which, standing farther down, shows [a distance of] 99,953, while, according to the law of the circle, $\alpha\zeta$ or $\alpha\eta$ should be 99,984. So all the distances are diminished at the sides, most greatly about ζ, η, none at δ, ϵ. Clearly, an oval is thus substituted for the circular path. You will obtain the same result if you begin with a mean anomaly obtained from whatever source. For when the diagram was set out above, the mean anomaly denoted angles about γ. But now you will begin with angles about β, smaller than the optical equation about γ. And 91° 1′ 53″ of mean anomaly shows you a distance of 99,984. But that was the *In this table I have supposed the path of the sun or earth to be oval.*

269

magnitude of $\delta\beta\zeta$ above. Nor was the mean anomaly there, for it was $\delta\gamma\zeta$, which is still greater. So the former mean anomaly of 91° 1′53″ had generated a longer distance there than a mean anomaly of the same magnitude, 91° 1′ 53″, shows here. All this, I say, is true. But there is no reason why you should be brought up short. For since we are considering differences of one degree, you see that the distances within one degree vary no more than 31 parts in one hundred thousand. Therefore, there would be no perceptible error, even if what has been done were to be out of the proper order. Below, in Ch. 44 and the following, by analogy with the rest of the planets, you will find the reason for adapting this arrangement to the theory of the sun as well. Therefore, this is not out of order, but perfectly correct, because it concerns the qualitative nature of the figure that the planet describes, which has been substituted.

As regards the quantity of the figure, the cure is excessive. The equated anomaly of 88° 58′ 7″, to which corresponds a mean anomaly of 91° 1′ 53″, ought not to show a value of 99, 984, but 100, 000, which is the mean between the distances of the figure and of the table. The reason for this assertion must be postponed until Ch. 55 and the following.

It was just said, however, that we shall not err perceptibly if we err by 31 units. It will therefore hurt us much less perceptibly if we err by only half that, or 16 units. So for the time being we safely admit this small error, in order to accommodate ourselves to an understanding of what has been advanced in the reading so far, and to avoid appearing to presuppose what was to be demonstrated.

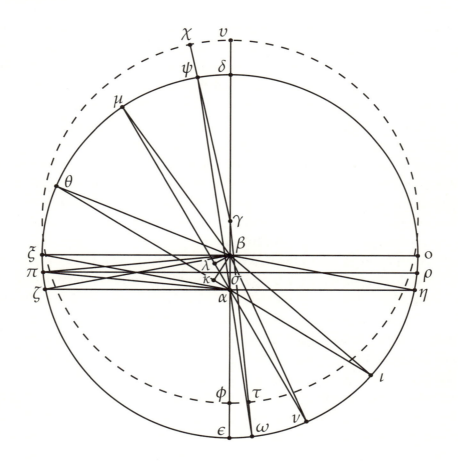

Mean Anomaly			Distance	Equated Anomaly			Mean Anomaly			Distance	Equated Anomaly		
D.	M.	S.		D.	M.	S.	D.	M.	S.		D.	M.	S.
0	0	0	101800	0	0	0	45	43	45	101265	44	16	15
1	1	5	101800	0	58	55	46	44	30	101242	45	15	30
2	2	10	101799	1	57	50	47	45	15	101219	46	14	45
3	3	14	101797	2	56	46	48	45	59	101195	47	14	1
4	4	18	101795	3	55	42	49	46	42	101172	48	13	18
5	5	23	101793	4	54	37	50	47	24	101147	49	12	36
6	6	27	101790	5	53	33	51	48	5	101123	50	11	55
7	7	31	101786	6	52	29	52	48	46	101098	51	11	14
8	8	36	101782	7	51	24	53	49	25	101073	52	10	35
9	9	40	101777	8	50	20	54	50	4	101047	53	9	56
10	10	44	101772	9	49	16	55	50	41	101022	54	9	19
11	11	48	101766	10	48	12	56	51	18	100995	55	8	42
12	12	52	101760	11	47	8	57	51	54	100969	56	8	6
13	13	55	101753	12	46	5	58	52	29	100942	57	7	31
14	14	58	101746	13	45	2	59	53	3	100915	58	6	57
15	16	1	101738	14	43	59	60	53	35	100888	59	6	25
16	17	3	101729	15	42	57	61	54	7	100860	60	5	53
17	18	6	101720	16	41	54	62	54	38	100832	61	5	22
18	19	8	101710	17	40	52	63	55	8	100804	62	4	52
19	20	9	101700	18	39	51	64	55	37	100775	63	4	23
20	21	10	101689	19	38	50	65	56	5	100747	64	3	55
21	22	11	101678	20	37	49	66	56	32	100719	65	3	28
22	23	11	101667	21	36	49	67	56	58	100690	66	3	2
23	24	11	101654	22	35	49	68	57	22	100660	67	2	38
24	25	10	101642	23	34	50	69	57	46	100631	68	2	14
25	26	9	101628	24	33	51	70	58	9	100601	69	1	51
26	27	8	101615	25	32	52	71	58	30	100571	70	1	30
27	28	6	101600	26	31	54	72	58	51	100542	71	1	9
28	29	3	101586	27	30	57	73	59	11	100511	72	0	49
29	30	0	101570	28	30	0	74	59	29	100481	73	0	31
30	30	56	101555	29	29	4	75	59	46	100451	74	0	14
31	31	52	101539	30	28	8	77	0	2	100420	74	59	58
32	32	47	101522	31	27	13	78	0	18	100389	75	59	42
33	33	42	101505	32	26	18	79	0	32	100359	76	59	28
34	34	36	101487	33	25	24	80	0	45	100328	77	59	15
35	35	29	101469	34	24	31	81	0	57	100297	78	59	3
36	36	22	101451	35	23	43	82	1	7	100266	79	58	53
37	37	14	101432	36	22	46	83	1	16	100235	80	58	44
38	38	6	101412	37	21	54	84	1	25	100203	81	58	36
39	38	57	101392	38	21	3	85	1	32	100172	82	58	28
40	39	47	101372	39	20	13	86	1	38	100141	83	58	22
41	40	36	101351	40	19	24	87	1	43	100109	84	58	17
42	41	24	101330	41	18	36	88	1	46	100078	85	58	14
43	42	12	101309	42	17	48	89	1	49	100047	86	58	11
44	42	59	101287	43	17	1	90	1	51	100015	87	58	9
45	43	45	101265	44	16	15	91	1	53	99984	88	58	7

163

Mean Anomaly			Distance	Equated Anomaly			Mean Anomaly			Distance	Equated Anomaly		
D.	M.	S.		D.	M.	S.	D.	M.	S.		D.	M.	S.
91	1	53	99984	88	58	7	135	43	45	98719	134	16	15
92	1	51	99952	89	58	9	136	42	59	98698	135	17	1
93	1	49	99921	90	58	11	137	42	12	98676	136	17	48
94	1	46	99890	91	58	14	138	41	24	98655	137	18	36
95	1	43	99858	92	58	17	139	40	36	98634	138	19	24
96	1	38	99827	93	58	22	140	39	47	98614	139	20	13
97	1	32	99796	94	58	28	141	38	57	98595	140	21	3
98	1	25	99765	95	58	35	142	38	6	98575	141	21	54
99	1	16	99734	96	58	44	143	37	14	98557	142	22	46
100	1	7	99703	97	58	53	144	36	22	98538	143	23	38
101	0	57	99672	98	59	3	145	35	30	98520	144	24	30
102	0	45	99641	99	59	15	146	34	36	98503	145	25	24
103	0	31	99610	100	59	29	147	33	42	98486	146	26	18
104	0	18	99580	101	59	42	148	32	47	98469	147	27	13
105	0	2	99549	102	59	58	149	31	52	98453	148	28	8
105	59	46	99519	104	0	14	150	30	56	98437	149	29	4
106	59	29	99489	105	0	31	151	30	0	98422	150	30	0
107	59	11	99459	106	0	49	152	29	3	98407	151	30	57
108	58	51	99429	107	1	9	153	28	6	98393	152	31	54
109	58	31	99399	108	1	29	154	27	8	98379	153	32	52
110	58	9	99370	109	1	51	155	26	9	98366	154	33	51
111	57	46	99341	110	2	14	156	25	10	98353	155	34	50
112	57	23	99312	111	2	37	157	24	11	98341	156	35	49
113	56	58	99283	112	3	2	158	23	11	98329	157	36	49
114	56	32	99254	113	3	28	159	22	11	98317	158	37	49
115	56	5	99226	114	3	55	160	21	10	98307	159	38	50
116	55	37	99198	115	4	23	161	20	9	98296	160	39	51
117	55	8	99170	116	4	52	162	19	8	98286	161	40	52
118	54	38	99142	117	5	22	163	18	6	98277	162	41	54
119	54	7	99115	118	5	53	164	17	3	98268	163	42	57
120	53	35	99088	119	6	25	165	16	1	98260	164	43	59
121	53	3	99061	120	6	57	166	14	58	98253	165	45	2
122	52	29	99034	121	7	31	167	13	55	98245	166	46	5
123	51	54	99008	122	8	6	168	12	52	98239	167	47	8
124	51	18	98982	123	8	42	169	11	48	98232	168	48	12
125	50	41	98957	124	9	19	170	10	44	98227	169	49	16
126	50	4	98931	125	9	56	171	9	40	98222	170	50	20
127	49	25	98906	126	10	35	172	8	36	98217	171	51	24
128	48	46	98882	127	11	14	173	7	31	98213	172	52	29
129	48	5	98857	128	11	55	174	6	27	98210	173	53	33
130	47	25	98833	129	12	35	175	5	23	98207	174	54	37
131	46	42	98810	130	13	18	176	4	18	98204	175	55	42
132	45	59	98787	131	14	1	177	3	14	98202	176	56	46
133	45	15	98764	132	14	45	178	2	10	98201	177	57	50
134	44	31	98741	133	15	29	179	1	5	98200	178	58	55
135	43	45	98719	134	16	15	180	0	0	98200	180	0	0

Chapter 31

That the bisection of the sun's eccentricity does not perceptibly alter the equations of the sun set out by Tycho; and concerning four ways of computing them.

But lest there remain any suspicions preventing our moving onwards, we shall investigate, in the usual Ptolemaic form of the first inequality,* whether there be any difference in the solar equations consequent upon the now bisected eccentricity.

First let there be an unbisected eccentricity of 3600 *on the line of apsides AF, with CE and CD accordingly radii of the orb; and let the anomaly FAE be* 45°, *and FAD* 135°. *Now it is obvious that, however great the discrepancy may be, it will reach its maximum around these positions of anomaly. For in the middle elongations the equations come out exactly the same, since [an eccentricity of]* 3600, *when investigated in both the sines and tangents,[1] produces the same arc. Therefore, as the radius CE is to the sine of the angle CAE or CAD, so is the eccentricity CA to the equation CEA or CDA, which are both* 1° 27′ 31″. And in this first way, Ptolemy computed the equations of the sun, and, following Ptolemy, Copernicus; and, following both, Brahe: each of them using only the eccentricity *AC*, whose magnitude they found through their observations.

There now follows a second way of computing the same equations, of which Ptolemy made use in the other planets, and of which I should make use. For I have demonstrated in this third part that the center of the eccentric is not at the point *C*, the center of uniform motion, but at *B*, the midpoint between the center of the world *A* and the point of uniformity *C*.

Therefore let CA be bisected at B, and let EB, BD, be the radius of the orb. By the same method, the part of the equation BDA, BEA, will be 0° 43′ 46″, *which, added to EAB, DAB, will result in an angle EBC of* 45° 43′ 46″, *and DBC, of* 135° 43′ 46″. *As a result, from the sides and the included angle, BEC comes out to be* 43′ 38″, *and BDC* 43′ 42″. *Thus the whole angle CEA is* 1° 27′ 24″, *and CDA is* 1° 27′ 28″, *within a hair's breadth of the previous value.* And so, in the appendix to Tycho Brahe's *Progymnasmata* p. 821, where the difference of the two calculations is given as $1\frac{1}{6}'$, you should read $\frac{1}{6}'$[2] And this is in accord with the conclusions of

* In the following chapters, the uncautious reader will become confused. The motion of the sun (for Brahe) or of the earth (for Copernicus) or of the epicycle (for Ptolemy), which is the cause of the second inequality for the other planets, also itself participates in the first inequality.

---

[1]That is, with a simple eccentric the angle of the equation (*AEC* or *ADC*) can be found using the law of sines alone, while when the equant is added, the angle is the sum of the optical part (*AEB* or *ADB*), found by the law of sines, and physical part (*BEC* or *BDC*), found by the law of tangents. Kepler is claiming that the two calculations produce identical results at the middle elongations, and series expansions of the angular relations show this to be true when the angle at *C* is 90°.

[2]Kepler made an error in computing *BEC* and *BDC*, which should have been 44′ 52″ and 42′ 39″, respectively. Thus *CEA* should have been 1° 28′ 38″, and *CDA*, 1° 26′ 25″. The differences

Ch. 4 applied to the form of the vicarious hypothesis.

And since you may observe that in this particular form of Ptolemaic hypothesis the parts of the equation are nearly equal (for the optical part was 43′ 46″ and the physical part was 43′ 38″ at *E* and 43′ 42″ at *D*), you see why, in the construction of the table in the preceding chapter, all I did was double the equation to establish the total equation. And this is the third way of computing the sun's equations. For at apogee and perigee both parts of the equation vanish, and in the middle elongations the parts are again equal, as was just now said. Therefore, since these three ways of computing the equations clearly coincide in eight places distributed about the entire circle, they will perceptibly coincide everywhere. This is a result of the smallness of the eccentricity: if it were greater, this coincidence would not take place everywhere.

165   Now I shall prepare myself for finding yet a fourth way to the equation, to be computed not through an arbitrary hypothesis but from the very nature of things, in eight chapters. Accordingly, this fourth way can finally follow in Chapter 40.

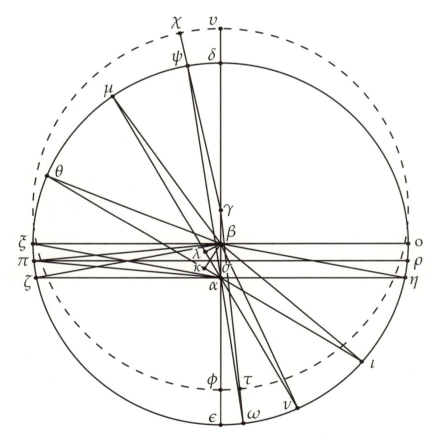

---

are 1′ 7″ and 1′ 6″, respectively, not far from the $1\frac{1}{6}′$ in the *Progymnasmata* (*TBOO* 3 p. 322 and the note on p. 407). But there is an interesting twist to the story: when Magini noticed the errors and pointed them out to Kepler (letters of 15 January and 23 February 1610, numbered 548 and 555 in *KGW* 16 pp. 270–274 and 285–7), Kepler replied (letter of 1 February 1610, number 551 in *KGW* 16 pp. 279–280) that he himself had written that appendix.

Caspar's analysis of the expanded equations for the orbits (mentioned in *KGW* 3 p. 467) shows that the maximum difference (which is also the difference between the physical and optical parts of the equation) amounts to 1′ 17″.

Chapter 32

The power that moves the planet in a circle diminishes with removal from its source.

I have said above that Ptolemy, well informed by the observations, bisected the eccentricities of the three superior planets, that Copernicus imitated this, and that Tycho's observations of Mars also urge the same conclusion, as has been seen in Chapters 19 and 20, and will appear with much greater certainty below in Chapter 42. In addition, Tycho closely imitated this in his lunar theory. And now the same thing has been demonstrated in the theory of the sun (for Tycho) or of the earth (for Copernicus). Further, there is nothing to prevent our believing the same of Venus and Mercury. Indeed, I now take it as proven that this is the origin of the belief that the centers of these planets' eccentrics move around on a small annual circle. Therefore all planets have this [double eccentricity]. Now in my *Mysterium cosmographicum*, published eight years ago,* I postponed arguing this case of the cause of the Ptolemaic equant for the sole reason that it could not be said on the basis of ordinary astronomy whether the sun or earth uses an equalizing point and has its eccentricity bisected. However, now that we have the confirmation of a sounder astronomy, it should be transparently clear that there is indeed an equant in the theory of the sun or earth. And, I say, now that this is demonstrated, it is proper to accept as true and legitimate the cause to which I assigned the Ptolemaic equant in the *Mysterium cosmographicum*, since it is universal and common to all the planets. So in this part of the work I shall make a further declaration of that cause.

* Now more than that.

And since the declaration will be general, I shall use the word, "planet". However, in this and the next few chapters, the reader may always understand by this, in particular, the earth for Copernicus or the sun for Tycho.

First, the reader should know that in every hypothesis constructed according to this Ptolemaic form, however great the eccentricity, the speed at perihelion and slowness at aphelion are very closely proportional to the lines drawn from the center of the world to the planet.

In the diagram of Chapter 29, in which α was the center of the world, β was the center of the eccentric δε, and γ was the point of the equant, let the equant circle νφ be described about center γ, with radius equal to βδ. And through the center of the world α, from which the eccentricity is reckoned (and in the business at hand, it is the sun for Copernicus and the earth for the others), let the straight line ψω be drawn, intersecting the eccentric at ψ and ω, so that the planet is at ψ and ω, having traversed the arcs of the eccentric δψ and εω, from apogee or aphelion and from perigee or perihelion, respectively. It is supposed that these arcs appear equal from α, since the straight line ψω makes the vertical angles ψαδ and ωαε, which are equal. But since δψ and εω are taken as minimal arcs, as if at the apsides δ and ε, they do not differ perceptibly from straight lines. And so, just as if δαψ and εαω were rectilinear

166

275

triangles, with right angles at δ and ε, and a common vertex α, δα will be to εα as arc δψ is to arc εω. But αδ is longer than αε. Therefore, arc δψ is longer than εω. These arcs, which are in fact unequal, appear equal from α. The question now is, how much time will the planet take to traverse each arc, according to Ptolemy's theory and hypothesis, when it has an equant? So let straight lines be drawn from the center γ through the points ψ and ω, intersecting the equant at χ, τ. Now Ptolemy will say, "since the whole circle of the equant υφ denotes the periodic time of the planet, then υχ is the measure of the time which the planet takes to traverse the arc of the eccentric ψδ, and φτ is the measure of the time which the planet takes to traverse the arc of the eccentric εω."

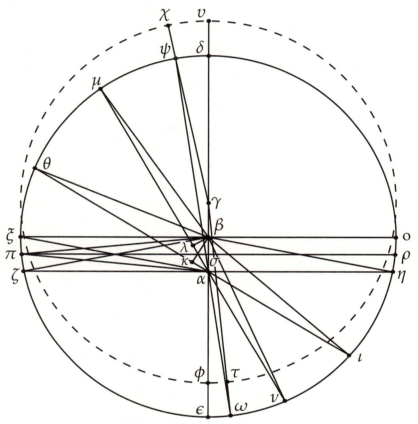

And now I myself say that υχ, thus designated as the arc of the time (as Ptolemy wished) is to the arc δψ which the planet traverses, very nearly as αδ, the distance of the arc δψ from the center of the world, is to δβ the mean distance of the points π and ρ from α. And likewise, the arc of the time φτ is to the arc of the planet's motion εω, approximately as αε, the distance of the arc εω from the center of the world α is to εβ and απ, the mean distance from the center of the world, which may be found at the points π and ρ. And now, as before, as γυ is to γδ so is υχ to δψ, and as γφ is to γε so is φτ to εω. But γυ is to γδ very nearly as βδ (or γυ) is to αδ, and this is shown by the fact that βδ is the arithmetic mean between γδ and αδ, since Ptolemy makes αβ, βγ equal. And further, the arithmetic mean between two terms whose ratio is near equality is only imperceptibly greater than the geometric mean. For example, the arithmetic mean between 10 and 12 is 11, and the geometric mean is about $10\frac{19}{20}$, so that there is less than the twentieth part of a unit between the two means. Nevertheless, these numbers are of the order of the eccentricity of Mars, which according to Ptolemy has the greatest eccentricity of all the planets.

And therefore, since the ratio γυ to γδ is imperceptibly greater than the ratio αδ to δβ, χυ will also have to ψδ a ratio imperceptibly greater than αδ to δβ. Likewise, as γε is to γφ, so is εω to φτ, but γε is to γφ approximately as εβ is to αε; that is, the former ratio is only imperceptibly less than the latter. Therefore, too, the ratio εω to φτ is only imperceptibly smaller than the ratio εβ to αε.

Now let us permute the ratios. For the ratio αδ to δβ is imperceptibly less than the ratio δβ or βε to εα, seeing that βδ or βε is the arithmetic mean between αδ and αε, as before. But it was proved that the ratio υχ to δψ is greater than the ratio αδ to δβ, from the smaller pair, and the ratio εω to φτ is less than the ratio εβ to αε, from the greater pair, so that in the two ratios αδ to δβ and εβ to αε, the amount by which the former terms will be greater and the latter smaller is, in the two ratios υχ to δψ and εω to φτ, the same as the amount by which the former is greater and the latter smaller. Consequently, there is some compensation even for that imperceptible difference, so that it is much more nearly true that the ratio of υχ to δψ is equal within a hair's breadth to the ratio of εω to φτ. 167

Therefore, the arcs δψ and εω being taken as equal (which hitherto were unequal), either δψ or εω will be a mean proportional between υχ, the increment of time at aphelion, and φτ, the increment of time at perihelion. Consequently, the ratio υχ to φτ (δψ and εω being equal) will be in the duplicate ratio of αδ to δβ or βε to εα, the former smaller and the latter greater by an imperceptible difference. But since the ratio αδ to αε is also the duplicate of either of these (for it is compounded of the two, which are nearly equal, taking the arithmetic mean δβ or βε), therefore, the arcs on the eccentric δψ and εω being equal, the ratio of the times υχ to φτ will be equal to the ratio αδ to αε. Or, more clearly, the planet takes a proportionally longer time to traverse some particular eccentric arc at δ than to traverse an equal eccentric arc at ε, according as αδ is greater than αε. And this follows from the way the Ptolemaic form* is ordered, and from its equalizing point, by means of a proof that is certain and valid so far as it concerns points near apogee and perigee. At other points there appears a very small discrepancy, which is all the smaller in effect the more obvious it is in the demonstration. This is because (for example) the ratio αμ to αν is smaller, and the ratio αθ to αι is much smaller, than αδ to αε, the greatest ratio, where the effect is also greatest.

* This refers to the particular form that accounts for the first inequality.

CHAPTER 33

The power that moves the planets resides in the body of the sun.

It was demonstrated in the previous chapter that the increments of time of a planet on equal parts of the eccentric circle (or on equal distances in the aethereal air) are in the same ratio to each other as the distances of those same spaces from the point whence the eccentricity is reckoned; or, more simply, to the extent that a planet is farther from the point that is taken as the center of the world, it is less strongly urged to move about that point. It is therefore necessary that the cause of this weakening is either in the very body of the planet, and in a motive force placed therein, or right at the supposed center of the world.

Now it is an axiom of the most common application in all of natural philosophy that of those things which occur at the same time and in the same manner, and which are always subject to like measurements, either one is the cause of the other or both are effects of the same cause. Just so, in this instance, the intension and remission of motion is always in the same ratio as the approach and recession from the center of the world. Thus, either that weakening will be the cause of the star's motion away from the center of the world, or the motion away will be the cause of the weakening, or both will have some cause in common. But no one can think up some third concurrent thing that would be the common cause of these two, and in the following chapters it will become clear that we have no need of feigning any such cause, since the two are sufficient in themselves.

Further, it is not in accord with nature that strength or weakness in longitudinal motion should be the cause of distance from the center. For distance from the center is prior both in thought and in nature to longitudinal motion. Indeed, longitudinal motion is never independent of distance from the center, since it requires a space in which to be performed, while distance from the center can be conceived without motion. Therefore, distance will be the cause of intensity of motion, and a greater or lesser distance will result in a greater or lesser amount of time.

And since distance belongs to the class of related things, whose being depends upon end points, while relation itself, without respect to end points, has no efficacy, it therefore follows (as has been said) that the cause of the variation of intensity of motion inheres in one or the other of the end points.

Now the body of a planet is never by itself made heavier in receding, nor lighter in approaching.

Moreover, that an animate force, which is seated in the mobile body of the planet and imparts a motion to the heavenly body, undergoes intension and remission so many times without ever becoming tired or growing old,—this will surely be absurd to say. Also, it is impossible to understand how this animate force could carry its body through the spaces of the world, since there

278

are no solid orbs, as Tycho Brahe has proved. And on the other hand, a round body lacks such aids as wings or feet, by the moving of which the soul might carry its body through the aethereal air as birds do in the atmosphere, by some kind of pressure upon, and counter-pressure from, that air.[1]

Therefore, the only remaining possibility is that the cause of this weakening and intension resides in the other endpoint, namely, in that point that is taken to be the center of the world, from which the distances are measured.

So now, if the distance of the center of the world from the body of a planet governs its slowness, and approach governs its speeding up, it is a necessary consequence that the source of motive power is at that supposed center of the world. And with this laid down, the manner in which the cause operates will also be clear. For it gives us to understand that the planets are moved rather in the manner of the steelyard or lever. For if the planet is moved with greater difficulty (and hence more slowly) by the power at the center when it is farther from the center, it is just as if I were to say that where the weight is farther from the fulcrum, it is thereby rendered heavier, not of itself, but by the power of the arm supporting it at that distance. And this is true, both of the steelyard or lever, and of the motion of the planets: that the weakening of power is in the ratio of the distances.

> The motive power is in the center of the system.

But which body is it that is at the center? Is there none, as for Copernicus when he is computing, and for Tycho in part? Is it the earth, as for Ptolemy and for Tycho in part? Or finally, is it the sun itself, as I, and Copernicus when he is speculating, would have it? This question I began to discuss in physical terms in Part 1. I there supposed as one of the principles what has now been expressly and geometrically proved in Chapter 32: that a planet is moved less vigorously when it recedes from the point whence the eccentricity is computed.

> The sun is in the center of the planetary system.

169

From this principle I presented a probable argument that the sun, rather than being at some other point occupied by no body, is at that point and at the center of the world (or the earth for Ptolemy). Allow me, then, to repeat that same probable argument, our foundation [for it] now demonstrated, in the present chapter. Then, you may remember, I demonstrated in the second part, that the phenomena at either end of the night follow beautifully if we reckon the oppositions of Mars according to the sun's apparent position. If this is done, then at the same time we set up the eccentricity and the distances from the very center of the sun's body, with the result that the sun itself again comes to be at the center of the world (for Copernicus), or at least at the center of the planetary system (for Tycho). But of these two arguments, one depends upon physical probability, and the other proceeds from possibility to actu-

---

[1] This clear adumbration of Newton's Third Law, couched as it is in terms of animal locomotion, lends support to the view (for which the translator is indebted to Robert Sacks of St. John's College, Santa Fe, New Mexico) that the intellectual ancestry of that law is to be found in Aristotle's *On the Motion of Animals*, Chapters 1 and 2. Aristotle's most pertinent statement is this: "For just as in the animal there must be something which is immovable if it is to have any motion, so *a fortiori* there must be something which is immovable outside the animal, supported upon which that which is moved moves. ...[T]here will be...no flying or swimming unless the air or sea were to offer resistance." (698b 13–18, translated by A. L. Peck, Loeb Classical Library, 1955.)

ality. And so in the third place I have postponed, because of its conceptual difficulty, demonstrating from the observations that we cannot avoid referring Mars to the apparent position of the sun, and drawing the line of apsides, which bisects the eccentric, directly through the sun's body, unless perhaps we wish to allow an eccentric such as will by no means be in accord with the parallax of the annual orb. Anyone who cannot tolerate the delay may read about this in chapter 52, and then may carry on here afterwards. For there nothing is assumed but the bare observations. You will find a similar proof in Part 5, from considerations of latitudes.

The motive power is in the sun. Therefore, with the sun belonging in the center of the system, the source of motive power, from what has now been demonstrated, belongs in the sun, since it too has now been found to be in the center of the world.

But indeed, if this very thing which I have just demonstrated *a posteriori* (from the observations) by a rather long deduction, if, I say, I had taken this as something to be demonstrated *a priori* (from the worthiness and eminence of the sun), so that the source of the world's life (which is visible in the motion of the heavens) is the same as the source of the light which forms the adornment of the entire machine, and which is also the source of the heat by which everything grows, I think I would have deserved an equal hearing.

The sun is in the center of the world, and does not move from place to place. Let Tycho Brahe himself, or anyone who prefers to follow his general hypothesis of the second inequality, consider by how close a likeness to the truth this physically elegant combination has for the most part been accepted (since for him, too, this substitution of the apparent position of the sun brings the sun back to the center of the planetary system) yet to some extent recoils from his hypothesis.

For it is obvious from what has been said that only one of the following can be true: either the power residing in the sun, which moves all the planets, by the same action moves the earth as well; or the sun, together with the planets linked to it through its motive force, is borne about the earth by some power which is seated in the earth.

Now Tycho himself destroyed the notion of real orbs, and I in turn have in this Third Part irrefutably demonstrated that there is an equant in the theory of the sun or earth. From this it follows that the motion of the sun itself (if it is moved) is intensified and remitted according as it is nearer or farther from the earth, and hence it would follow that the sun is moved by the earth. But if, on the other hand, the earth is in motion, it too will be moved by the sun 170 with greater or less velocity according as it is nearer or farther from it, while the power in the body of the sun remains perpetually constant. Between these two possibilities, therefore, there is no intermediate.

I myself agree with Copernicus, and allow that the earth is one of the planets.

The moon is driven around by the earth, but the sun and the other planets are not; the earth, on the other hand, is driven around by the sun. Now it is true that the same objection may be raised against Copernicus concerning the moon, that I raised against Tycho concerning the five planets; namely, that it appears absurd for the moon to be moved by the earth, and to be associated with it and bound to it as well, so that it too, as a secondary planet, is swept around the sun by the sun. Nevertheless, I prefer to allow one moon, akin to the earth in its corporeal disposition[2] (as I have shown in the *Optics*[3]) to be moved by a power seated in the earth but extended towards the sun, as will be described a little later in Ch. 37, than to ascribe to that same earth as well the motion of the sun and of all the planets bound to it.

But let us carry on in our consideration of this motive power residing in the sun, and let us now again observe its very close kinship with light.

The kinship of the solar motive power with light.

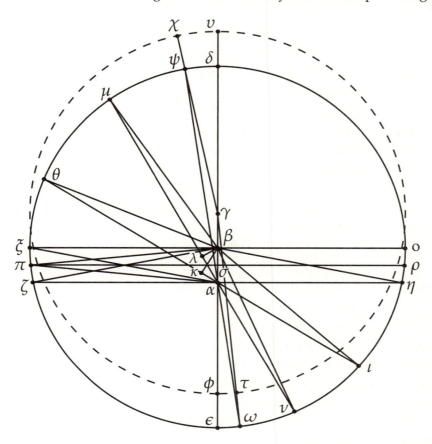

Since the perimeters of similar regular figures, even of circles, are to one another as their semidiameters, therefore as αδ is to αε, so is the circumference of the circle described about α through δ to the circumference of the circle described about the same point α through ε. But as αδ is to αε, so is the strength of the power at ε to the strength of the power at δ, inversely, by what was proved in Chapter 32. Therefore, as the circle at δ is to the smaller circle at ε, so, inversely, is the power at ε to the power at δ; that is, the power is weaker to the extent that it is more spread out, and stronger to the extent that it is more concentrated. Hence we understand that there is the same power in the whole circumference of the circle through δ as there is in the circumference of the smaller circle through ε. This was shown to be true of light in exactly the same way in the Astronomiae pars optica, *Chapter 1.[4] Therefore, in all respects and in all its attributes, the motive power from the sun coincides with light.*

---

[2]Changing the punctuation slightly so as to read *dispositione corporis* as modifying *cognatam* rather than *movendam*.

[3]Ch. 6, English trans. p. 243.

[4]Ch. 1 Proposition 9, English trans. p. 22.

Whether light is the
vehicle of the motive
power.

And although this light of the sun cannot be the moving power itself, I leave it to others to see whether light perhaps is equivalent to a kind of instrument or vehicle, of which the moving power makes use.

This seems gainsaid by the following: first, light is hindered by opaque things, and therefore if the moving power had light as a vehicle, rest of the moving bodies would follow upon darkness; again, light spreads spherically in straight lines, while the moving power, though spreading in straight lines, does so circularly; that is, it is exerted in but one region of the world, from east to west, and not the opposite, not at the poles, and so on. But we shall perhaps be able to reply to these objections in the chapters immediately following.

171

The moving power is an
immaterial *species* of the
solar body.

Finally, since there is just as much power in a larger and more distant circle as there is in a smaller and closer one, nothing of this power is lost in travelling from its source, nothing is scattered between the source and the movable body. The emission, then, in the same manner as light, is immaterial, unlike odours, which are accompanied by a diminution of substance, and unlike heat from a hot furnace, or anything similar which fills the intervening space. The remaining possibility, then, is that, just as light, which lights the whole earth, is an immaterial *species*[5] of that fire which is in the body of the sun, so this power which enfolds and bears the bodies of the planets, is an immaterial *species* residing in the sun itself, possessing inestimable strength, seeing that it is the primary activity of every motion in the universe.

This *species* belongs to the
second species of
continuous quantity, and
is a variety of surface.

Since, therefore, this *species* of the power, exactly as the *species* of light (for which see the *Astronomiae pars optica* Ch. 1[6]), cannot be considered as dispersed throughout the intermediate space between the source and the mobile body, but is seen as collected in the body in proportion to the amount of the circumference the mobile body occupies, this power (or *species*) will therefore not be any geometrical body, but is like a variety of surface, just as light is. To generalize this, the *species* of things that emanate immaterially are not by that emanation extended through the dimensions of a body, although they arise from a body (as this one does from the body of the sun). Instead, they are extended according to that very law of emission: they are not bounded by the emanation itself, but just as the surfaces of things to be illuminated cause light to be considered as a kind of surface, because they receive and terminate its emission, so the bodies of things that are moved appear to bring it about that this moving power be considered as if a sort of geometrical body, because their whole masses terminate or receive this emission of the motive *species,* so that the *species* can exist or subsist nowhere in the world but in the bodies of the mobile things themselves. And, exactly like light, between the source and the movable thing it does not exist, but will have a quasi-existence.

In what manner the
immaterial *species* of the
body of the sun may be
quantified.

Moreover, at the same time, a reply can be made here to a possible objection. For it was said above that this motive power is extended throughout the space of the world, in some places more spread out and in others more concentrated, and that the intensification and remission of the motions of the planets

---

[5]For the meaning of this technical term, for which there is no acceptable English equivalent, see the Glossary.

[6]*Optics,* English trans. p. 19.

are consequent upon this variation. Now, however, it has been said that this power is an immaterial *species* of its source, and never inheres in anything except a mobile subject, such as the body of a planet. But these appear to be contradictory: to lack matter and yet to be subject to geometrical dimensions; to be poured out throughout the whole world, and yet not to exist anywhere but where there is something movable.

The reply is this: although the motive power is not anything material, nevertheless, since it is designed for carrying matter (namely, the body of a planet), it is not free from geometrical laws, at least on account of this material action of carrying things about. Nor is there need for more, for we see that those motions are carried out in space and time, and that this power arises and emanates from the source through the space of the world, all of which are geometrical entities. So this power should indeed be subject to other geometrical necessities.

But lest I appear to philosophize with excessive insolence, I shall propose to the reader the clearly authentic example of light, since it also makes its nest in the sun, thence to break forth into the whole world as a companion to this motive power. Who, I ask, will say that light is something material? Nevertheless, it carries out its operations with respect to place, suffers alteration,[7] is reflected and refracted, and assumes quantities so as to be dense or rare, and to be capable of being taken as a surface wherever it falls upon something illuminable. Now just as it is said in optics, light too, and this motive power as well, does not exist in the intermediate space between the source and the illuminable, even though it passed through it, but quasi-existed there. Moreover, although light itself does indeed flow forth in no time, while this power creates motion in time, nonetheless the way in which both do so is the same, if you consider them correctly. Light manifests those things which are proper to it instantaneously, but so far as it is connected with matter, it too requires time. It illuminates a surface in a moment, because here matter need not undergo any alteration, for all illumination takes place according to surfaces, or at least as if a property of surfaces and not as a property of corporeality as such. On the other hand, light bleaches colors in time, since here it acts upon matter *qua* matter, making it hot and expelling the contrary cold which is embedded in the body's matter and is not on its surface. In precisely the same manner, this moving power perpetually and without any interval of time is present from the sun wherever there is a suitable movable body, for it receives nothing from the movable body to cause it to be there. On the other hand, it causes motion in time, since the movable body is material.

Or if it seems better, frame the comparison in this manner: light is constituted for illumination, and it is just as certain that power is constituted for motion. Light does everything that can be done to achieve the greatest illumination; nonetheless, it does not happen that color is most greatly illuminated. For color intermingles its own peculiar *species* with the illumination of light, thus forming some third entity. In like manner, there is no slowing in the moving power to prevent the planet's having as much speed as it has itself,

172
Comparison of light and motive power on the basis of quantification;

And on the basis of time.

Why the planets do not equal their mover, the immaterial *species* of the sun, in speed.

---

[7]Reading *mutationem* instead of *mutuum*.

but the planet's speed is not therefore that great, since something intervening prevents that, namely, some sort of matter possessed by the surrounding
aether, or the disposition of the movable body itself to rest (others might say,
"weight," but I do not entirely approve of that, except, indeed, where the earth
is concerned). It is the tempering effect of these, together with the weakening
of the motive power, that determines a planet's periodic time.

CHAPTER 34

The body of the sun is magnetic, and rotates in its space.

Concerning that power that is closely attached to the bodies of the planets and pulls them, we have already said how it is formed, how it is akin to light, and what it is in its metaphysical being. Next, we should contemplate the deeper nature of its source, shown by the outflowing *species* (acting as an archetype). 173 For it may appear that there lies hidden in the body of the sun something divine, which may be compared to our soul, from which flows that *species* driving the planets around, just as from the soul of someone throwing pebbles a *species* of motion comes to inhere in the pebbles thrown by him, even when he who threw them removes his hand from them. And to those who proceed soberly, other reflections will soon be provided.

Now because the power that is extended from the sun to the planets moves them in a circular course around the immovable body of the sun, this cannot happen, or be conceived in thought, in any other way than this, that the power traverses the same path along which it carries the other planets. This has been observed to some extent in catapults and all violent motions. Thus, Fracastoro and others, relying on a story told by the most ancient Egyptians, spoke with little probability when they said that perchance some of the planets, their orbits being deflected gradually beyond the poles of the world, would thus afterwards move in a path opposite to the rest and to their modern course.[1] For it is much more likely that the bodies of the planets are always borne in that direction in which the power emanating from the sun tends.

The power that moves the planets is whirled around.

But this *species* is immaterial, proceeding from its body out to this distance without the passing of any time, and is in all other respects like light. Therefore, it is not only required by the nature of the *species*, but likely in itself owing to this kinship with light, that along with the particles of its body or source it too is divided up, and when any particle of the solar body moves towards some part of the world, the particle of the immaterial *species* that from the beginning of creation corresponded to that particle of the body also always moves towards the same part. If this were not so, it would not be a *species*, and would come down from the body not in straight but in curved lines.

Since the *species* is moved in a circular course, in order thereby to confer motion upon the planets, the body of the sun, or source, must move with it, not, of course, from space to space in the world—for I have said, with Copernicus, that the body of the sun remains in the center of the world—but upon its center or axis, both immobile, its parts moving from place to place, while the whole body remains in the same place.

---

[1] Hieronymus Fracastorius, *Homocentrica*, Venice 1538, Sect. 3 Cap. 8. In this chapter, which bears the title, "Cur solis declinatio minuatur" (why the sun's declination may diminish), Fracastoro refers, in support of his remarkable opinion, to information received from the Egyptians by Herodotus and Pomponius Mela. (Citation from *KGW* 3 p. 468).

Instance in light.   In order that the force of the analogical argument may be that much more evident, I would like you to recall, reader, the demonstration in optics that vision occurs through the emanation of small sparks of light[2] toward the eye from the surfaces of the seen object. Now imagine that some orator in a great crowd of people, encircling him in an orb,[3] turns his face, or his whole body along with it, once around. Those of the audience to whom he turns his eyes directly also see his eyes, but those who stand behind him then lack the view of his eyes. But when he turns himself around, he turns his eyes around to everyone in the orb. Therefore, in a very short interval of time, all get a glimpse of his eyes. This they get by the arrival of a small spark of light or *species* of color descending from the eyes of the orator to the eyes of the spectators. Thus by turning his eyes around in the small space in which his head is located, he carries around along with it the rays of the spark of light in the very large

174   orb in which the eyes of the spectators all around are situated. For unless the spark of light went around, his spectators would not be recipients of his eyes' glance. Here you see clearly that the immaterial *species* of light either is moved around or stands still along with that of which it is the *species* either moving or standing still.[4]

The sun rotates.   Therefore, since the *species* of the source, or the power moving the planets, rotates about the center of the world, I conclude with good reason, following this example, that that of which it is the *species*, the sun, also rotates.

However, the same thing is also shown by the following argument. Motion that is local and subjected to time cannot inhere in a bare immaterial *species*, since such a *species* is incapable of receiving an effect of an applied motion unless the received motion is non-temporal, just as the power is immaterial. Also, although it has been proved that this moving power rotates, it cannot be allowed to have infinite speed (for then it would seem that the power will also impart infinite speed to the bodies), and therefore it completes its rotation in some period of time. Therefore, it cannot carry out this motion by itself, and it is as a consequence necessary that it is moved only because the body upon which it depends is moved.

By the same argument, it appears to be a correct conclusion that there does not exist within the boundaries of the solar body anything immaterial

---

[2]*Luculae.* Kepler uses this odd word in his *Optics,* Ch. 1 Propositions 15 and 32 (English trans. pp. 24 and 40). He says that in total darkness colored objects may still emit *luculae,* and that there probably is a *lucula* in the heart accompanying the heart's fire. Evidently, a *lucula* is a very small spark of light. It is not a "ray," for a ray "is nothing but the motion of light" (*Optics* Ch. 1 Prop. 8, English trans. p. 22.), nor is it a particle in the usual sense of the word, since it "lacks corporeal matter, but consists of its own sort of matter" (*Optics,* English trans. p. 19). The chief distinguishing feature of this matter is that it is "a kind of surface" (Prop. 8, p. 22). So it might be most nearly correct to imagine a *lucula* as a two dimensional "particle" or very small part of the lucid *species* of a body, whose motion (at right angles to its surface) constitutes a "ray."

[3]In Latin, as in English, the word *orbis* (orb) is ambiguous: it can denote either a circle or a sphere. Care has therefore been taken to preserve the ambiguity in the translation.

[4]In the *Dialogue on the Two Chief World Systems,* Galileo has Salviati say the following: "So the turning motion made by the fowling piece in following the flight of the bird with the sights, though slow, must be communicated to the ball also;..." (p. 172, Stillman Drake translation p. 178). Evidently, the notion of a sort of circular impetus or inertia had its attractions. However, Galileo's interlocutors reject the idea shortly after.

by whose rotation the *species* descending from that immaterial something also rotates. For again, local motion which takes time is not correctly attributed to anything immaterial. It therefore remains that the body of the sun itself rotates in the manner described above, indicating the poles of the zodiac by the poles of its rotation (by a line extended from the center of the body to the fixed stars through the poles), and indicating the ecliptic by the greatest circle of its body, thus furnishing a natural cause for these astronomical entities.

Natural cause of the zodiac.

Further, we see that the individual planets are not carried along with equal swiftness at every distance from the sun, nor is the speed of all of them at their various distances equal. For Saturn takes 30 years, Jupiter 12, Mars 23 months, earth 12, Venus $8\frac{1}{2}$, and Mercury 3. Nevertheless, it follows from what has been said that every orb of power emanating from the sun (in the space embraced by the lowest, Mercury, as well as that embraced by the highest, Saturn) is twisted around with a whirl equal to that which spins the solar body, with an equal period. (There is nothing absurd in this statement, for the emanating power is immaterial, and by its own nature would be capable of infinite speed if it were possible to impress a motion upon it from elsewhere, for then it could be impeded neither by weight, which it lacks, nor by the obstruction of the corporeal medium.) It is consequently clear that the planets are ill-suited to emulate the swiftness of the motive power. For Saturn is less receptive than Jupiter, since its returns are slower, while the orb of power at the path of Saturn returns with the same swiftness as the orb of power at the path of Jupiter, and so on in order, all the way to Mercury, which, by example of the superior planets, doubtless will move more slowly than the power that pulls it. It is therefore necessary that the nature of the planetary globes be material, from an inherent property, arising from the origin of things, to be inclined to rest or to the privation of motion.[5] When the tension between these things leads to a fight, that planet which is placed in a weaker power overcomes it more, and is moved more slowly by it; that which is closer to the sun overcomes it less.

In the bodies of the planets is a material inclination to rest in every place where they are put by themselves.

The motion of the planets is extrinsic.

175

This analogy shows that there is in all planets, even in the lowest, Mercury, a material force of disengaging itself somewhat from the orb of the sun's power.

From this it is concluded that the rotation of the solar body anticipates considerably the periodic times of all the planets; therefore, it must rotate in its space at least faster than once in a three-months' span.

The amount of time in which the rotation of the solar body traverses its space.

However, in my *Mysterium Cosmographicum* I pointed out that there is about the same ratio between the semidiameters of the sun's body and the orb of Mercury as there is between the semidiameters of the body of the earth and the orb of the moon. Hence, you may plausibly conclude that the period

---

[5]This property, which Kepler later called "inertia," (literally, "laziness") (*Epitomes Astronomiae Copernicanae*, Book 4 (Linz 1620) pp. 510–511, in *KGW* 7 p. 296), is significantly different from Aristotle's corporeal qualities. Aristotle argued that bodies have a natural and inherent tendency to move (see, for example, *On the Heavens* III.2). Kepler, in contrast, wrote, "For all corporeal stuff or *materia* of every thing in the whole world, there is this form, or rather this dead non-form: that it is heavy, and not befitted to travel by itself from one place to another, and must for that reason be pulled and driven by something living, or something else from outside." (from Kepler's German commentary on *On the Heavens*, in *KGW* 20.1 p. 164). Newton later adopted Kepler's term, though he used it to denote a resistance to a change in motion rather than to motion itself.

of the orb of Mercury would have the same ratio to the period of the body of the sun as the period of the orb of the moon has to the period of the body of the earth. And the semidiameter of the orb of the moon is sixty times the semidiameter of the body of the earth, while the period of the orb of the moon (or the month) is a little less than thirty times the period of the body of the earth (or day), and thus the ratio of the distances is double the ratio of the periodic times. Therefore, if the doubled ratio also holds for the sun and Mercury, since the diameter of the sun's body is about one sixtieth of the diameter of Mercury's orb, the time of rotation of the solar globe will be one thirtieth of 88 days, which is the period of Mercury's orb. Hence it is likely that the sun rotates in about three days.

Whether the earth's diurnal rotation comes from the rotation of the solar globe.

You may, on the other hand, prefer to prescribe the sun's diurnal period in such a way that the diurnal rotation of the earth is dispensed by the diurnal rotation of the sun, by some sort of magnetic force. I would certainly not object. Such a rapid rotation appears not to be alien to that body in which lies the first activation of every motion.

The monthly motion of the moon arises from the diurnal rotation of the earth.

This opinion (on the rotation of the solar body as the cause of the motion for the other planets) is beautifully confirmed by the example of the earth and the moon. For the chief, monthly motion of the moon, by the force of the demonstrations used in Chapters 32 and 33, is entirely derived from the earth as its source (for what the sun is for the rest of the planets there, the earth is for the moon in this demonstration). Consider, therefore, how our earth occasions the motion of the moon: while this our earth, and its immaterial *species* along with it, rotates twenty nine and one half times about its axis, at the moon this emitted *species* can drive it around only once in the same time, in (of course) the same direction in which the earth leads it.

Here, by the way, is a marvel: in any given time the center of the moon traverses twice as long a line about the center of the earth as any place on the surface of the earth beneath the great circle of the equator. For if equal spaces were measured out in equal times, the moon ought to return in sixty days, since the size of its orb is sixty times the size of the earth's globe.

This is surely because there is so much force in the immaterial *species* of the earth, while the lunar body is doubtless of great rarity and weak resistance. Thus, to remove your bewilderment, consider that on the principles we have supposed it would necessarily follow that if the moon were not to resist, by its material force, the motion impressed from outside by the earth, the moon

176   would be carried at exactly the same speed as the earth's immaterial *species*, that is, with the earth itself, and would complete its circuit in 24 hours, in which the earth also completes its circuit. For even if the tenuity of the earth's *species* is great at the distance of 60 semidiameters, the ratio of one to nothing is still the same as the ratio of sixty to nothing. Hence the immaterial *species* of the earth would prevail completely, if the moon did not resist.

What sort of body is the sun?

Here, one might inquire of me, what sort of body I consider the sun to be, from which this motive *species* descends. I would ask him to proceed under the guidance of a further analogy, and urge him to inspect more closely the example of the magnet brought up a little earlier, whose power resides in the entire body of the magnet, grows with its mass, and is itself also divided when the magnet is diminished. So in the sun the moving power appears so much stronger that it seems likely that its body is of all [those in the world] the most dense.

And the power of attracting iron is spread out in an orb from the magnet so that there exists a certain orb within which iron is attracted, but more strongly so as the iron comes nearer into the embrace of that orb. In exactly the same way the power moving the planets is propagated from the sun in an orb, and is weaker in the more remote parts of the orb.

Likeness of the sun's body to a magnet.

The magnet, however, does not attract with all its parts, but has filaments (so to speak) or straight fibers (seat of the motor power) extended throughout its length, so that if a little strip of iron is placed in a middle position between the heads of the magnet at the side, the magnet does not attract it, but only directs it parallel to its own fibers. Thus it is credible that there is in the sun no force whatever attracting the planets, as there is in the magnet, (for then they would approach the sun until they were quite joined with it), but only a directing force, and consequently that it has circular fibers all set up in the same direction, which are indicated by the zodiac circle.

The difference between the solar body and a magnet.

Therefore, as the sun forever turns itself, the motive force or the outflowing of the *species* from the sun's magnetic fibers, diffused through all the distances of the planets, also rotates in an orb, and does so in the same time as the sun, just as when a magnet is moved about, the magnetic power is also moved, and the iron along with it, following the magnetic force.

The principle of motion in the sun and in the magnet is the same.

The example of the magnet I have hit upon is a very pretty one, and entirely suited to the subject; indeed, it is little short of being the very truth. So why should I speak of the magnet as if it were an example? For, by the demonstration of the Englishman William Gilbert,[6] the earth itself is a big magnet, and is said by the same author, a defender of Copernicus, to rotate once a day, just as I conjecture about the sun. And because of this very thing, that it has magnetic fibers intersecting the line of its motion at right angles, those fibers therefore lie in various circles about the poles of the earth parallel to its motion. I am therefore absolutely within my rights to state that the moon is carried along by the rotation of the earth and the motion of its magnetic power, only thirty times slower.

By example of the earth, it is proved that there are magnets in the heavens.

I know that the earth's filaments and its motion indicate the equator, while the circuit of the moon is generally related to the zodiac—on this point there will be more in Chapter 37 and Part 5. With this one exception, everything fits: the earth is intimately related to the lunar period, just as the sun is to that of the other planets. And just as the planets are eccentric with respect to the sun, so is the moon with respect to the earth. So it is certain that the earth is looked upon by the moon's mover as a kind of pole star (so to speak), just as the sun is looked upon by the movers belonging to the rest of the planets, for which see Chapter 38. It is therefore plausible, since the earth moves the moon through its *species* and is a magnetic body, while the sun moves the planets similarly through an emitted *species*, that the sun is likewise a magnetic body.

Likeness of the earth and the sun, with respect to the motion impressed upon the planets.

177

---

[6]William Gilbert, *De magnete magneticisque corporibus et de magno Magnete Tellure physiologia nova*, London, 1600. Translated by P. F. Mottelay (John Wiley and Sons, 1893).

Chapter 35

Whether the motion from the sun, like its light, is subject to privation in the planets through occultations.[1]

This is a good time for me to take up the objections raised in Chapter 33, where to the kinship of light and motive power were opposed, first, the mutual occultations of the celestial bodies, and then, the different [manner of] emanation of the *species* of the two.

And concerning the first, it is worthy of consideration whether, just as one opaque body intercepts another's sunlight, mobile bodies similarly impede one another in motion when they lie in line with the sun. If so, light would clearly be the vehicle or instrument of the motive power.

It might appear that in order to avoid this as much as possible, God introduced the relative inclinations of all the eccentrics, deviations from the ecliptic, and transpositions of the nodes, as well as the proportions of the bodies and the attenuation of shadows in a cone. And since it would not be possible completely to prevent the stars' occasionally lining up with the sun, it is tempting to suppose that the very slow motions of the apsides and nodes (which are, as it were, a kind of aberration of the epicycles from their periodic times) derive their origin thence.

But it is answered, first, that the analogy between light and motive power is not to be disturbed by rashly confusing their properties. Light is impeded by the opaque, but is not impeded by a body, because light is light, and does not act upon the body but on the surface (or as if on the surface). Power acts upon the body without respect to its opacity. Therefore, since it is not correlated with the opaque, it will likewise not be impeded by the opaque.

On this account I would nearly separate light from moving power, unless I were to come upon examples in nature which leave to the rays of light, even when impeded, a certain efficacy in those locations where their entry is prohibited. But I am not chiefly concerned here with the association of light with the motive power.

Instead, in order to dissolve this suspicion that the motions are impeded, let us take another example from the magnet. Its power is not at all impeded by the interposition of matter (because, of course, it is immaterial), but passes through sheets of silver, copper, gold, glass, bone, wood, and attracts iron lying beyond these sheets exactly as if no sheets were there. Granted, it is impeded by the interposition of a magnetic plate. But the cause is ready at hand: the plate acts as a counterpart to the magnet. It therefore overcomes, by its strength, the more distant magnet lying beyond it. And although it is impeded by the interposition of the iron plate, this too belongs to the magnetic nature, and it immediately drinks up the magnet's power and uses it as if it

On the causes of latitudes,

And motions of the apsides.

178

---

[1] ἀντίφραξις.

290

were its own.

Therefore, in order that we may deny that the motion of the celestial bodies is impeded by central conjunctions of two of them, we must say that the nature of the sun differs from the nature of the rest of the celestial bodies more than the nature of the magnet differs from the nature of iron. Also, we must deny that the planets drink up the power all at once from the sun in the same way as the iron drinks it from the magnet. The question of whether they drink up any of it at all, I defer to Chapter 57.

In respect to the likelihood of the cause of the apogees' motion, it proves nothing concerning this common solar power being impeded by occultation.[2] For the motion of the apogees could be quite different; for instance, it could have an animate origin. You will find a certain obscure opinion on this point below in Chapter 57.[3]

On the motion of the apsides, once more.

Further, if the motion of the apogees were to arise because the motion of the planets around the sun is impeded by the occultation of the motive *species* emanating from the sun, the motion in longitude would be slowed, either by a progressive motion of latitude (by which the apogees would move back) or equally by slowing of the latitudinal motion. Thus, the apogees would stand still, although the observations testify that they move forward.

But the question whether, once the sun is preserved as the source of motion, the motions proper to the celestial bodies are impeded by occultation,[4] will also be discussed in Chapter 57.

---

[2]ἀντίφραξις.

[3]See p. 426.

[4]ἀντίφραξις.

CHAPTER 36

By what measure the motive power from the sun is attenuated as it spreads through the world.

There follows another, rather more difficult objection, arising from the second argument that was raised in Chapter 33 against the kinship of light and motive power, which seems irreconcilably at odds with our study of immaterial *species*. This objection wearied me for a long time without offering any prospect of solution.

It was demonstrated in Chapter 32 that the intension and remission of the planets' motion is in simple proportion to the distances. It appears, however, that the power emanating from the sun should be intensified and remitted in the duplicate or triplicate ratio of the distances or lines of efflux. Therefore, the intension and remission of the planets' motion will not be a result of the attenuation of the power emanating from the sun. The logical consequence appears to be proved in the following manner, for light as well as for moving power; however, discussions of light are clearer. The reader should read in "motive power". *Initially, let α be any point on the sun's body. It will therefore spread out its rays to every orb, and from a demonstration in optics, as the amplitude of the greater spherical surface γ, considered as an imaginary terminus for those rays, is to the amplitude of the smaller, β, so will the density of light at the smaller orb β be to its density at the greater γ.*

Next, let δε be any luminous great circle on the body of the sun. Thus its individual points, of which there are an infinity, spread out rays to the individual hemispheres β and γ in the same ratio. And the apparent magnitude of the diameter of the circle at the shorter distance (i. e., the angle δβε) is to the apparent magnitude at the greater distance (i. e., the angle δγε), conversely, as the longer distance αγ from such a circular line (which at a distance appears straight) is to the shorter δβ. So, since this diameter appears longer from the nearby point β than from the distant point γ, in the same ratio, while the radiation belonging to any given point is denser at a nearby point β than at a distant point γ, it is therefore apparent that the density of radiation of the circle at the nearby point β will have to the density of radiation at the distant point γ the [inverse of the] duplicate ratio of αβ to αγ .

Thirdly, let δαε be the apparent disc of the sun's body, and since similar surfaces (as are the apparent circular discs here) are in the duplicate ratio of their diameters, while the apparent diameters of the sun are in the simple inverse ratio of the distances αγ,αβ, the circular discs will therefore appear in the duplicate ratio of the distances αγ, αβ. But since the radiation of the circle δε at γ and β was just proved to be in the duplicate ratio of the distances αβ, αγ, while both are causes of the density [of radiation], it appears that the radiation of the disc, with respect to density or strength,

179

292

is in the triplicate ratio of the distances $\alpha\gamma$, $\alpha\beta$.

For example, if the distances $\alpha\gamma$ *to* $\alpha\beta$ *were as 2 to 1, the radiations of the point at* $\alpha\gamma$ *and* $\alpha\beta$ *would be as 1 to 2, with respect to the density of light, and the apparent diameters of circles would be 1 at* γ *and 2 at* β.

Therefore, the radiation of the diameter of the circle $\delta\epsilon$ *would be 1 at* γ *and 4 at* β. *But the discs are in the duplicate ratio of the diameters. Therefore the apparent magnitude of the disc at* γ *would be 1, and at* β, *4, as if you were to say that the disc* $\delta\alpha\epsilon$ *when seen from* β *appears to contain four times as many points as when seen from* γ. *Any of these points illuminates twice as densely at* β *than at* γ. *Therefore, when the ratios are compounded, the density of radiation of the whole disc* $\delta\alpha\epsilon$ *at* γ *would have to the density of radiation of the whole disc* $\delta\alpha\epsilon$ *at* β *the ratio 1 to 8.*

It does not trouble us here that we are reckoning from the sun's apparent disc, although it is a spherical surface. For the ratio of equally many things to each other mutually is the same. But the spherical surface was demonstrated by Archimedes to be four times the area of the greatest circle described in the sphere. Therefore, a body at γ, twice as far away as at β, should have appeared to be fully eight times as obscurely illuminated at γ as at β, not just twice. For it seems that the brightness of the rays ought to be intensified because of the fact that bodies appear to be magnified when they approach, as [for example] Venus, when at the perigee of the epicycle, defines a more evident shadow around bodies than at apogee. Therefore, by the force of the analogy which we have instituted between light and motive force, the same should be thought to apply to the motive force.

180

To this objection, I answer decisively, that in the initial supposition involving the point I made a false assumption. Even though I did indeed say something like that in the *Optics*, you should bear in mind that I was speaking of optics, whose points and lines are not quite indivisibles. For, as concerns the point, since it has no magnitude, while the rays are amplified with the magnitudes of bodies, it follows that the radiation of a point in itself is nothing, and hence what has no radiation has no greater or lesser density. Thus the initial assumption of the ratio of distances $\alpha\beta$ to $\alpha\gamma$ collapses.

Rather, because of this very thing, we say that any point shines more strongly or weakly to the extent that that point designates to us a greater or lesser quantity.

In the second supposition, concerning the circle, and the third, concerning the disc, there are two false assumptions. The first is that a mathematical circle, lacking breadth, is supposed to shine. For it can no more shine itself than can the point from whose motion the circle is supposed to be generated. You are no closer to having a surface when you posit a line of three *stadia*[1] than you are in positing a line of three feet.

Second, it is supposed that the optical magnification of the diameter or of the disc adds to the strength of the rays, although this is but a deception of the visual faculty, and belongs to the genus of theoretical entities, which lack any efficacy. The physical identity of the circle $\delta\epsilon$, the surface $\delta\alpha\epsilon$ (when light is in question) and the body $\delta\epsilon$ (when power is in question) remains the

---

[1]The *stadium* was 625 *pedes*, or one eighth of a Roman mile. It is thus roughly equivalent to a furlong, in the English system of measures.

same whether it is viewed from γ or from β, and will always act the same and have the same effect, spreading the same amount of power or light in the more diffuse orb γ as in the more compact orb β . Nothing is lost in the journey: the entire *species* carries through to any distance, however remote. It is attenuated only in the extensions of the spheres, so that in the individual points of the spheres, such as γ and β, it is rarer in the former, and denser in the latter, in the inverse ratio of the distances $\alpha\beta$ to $\alpha\gamma$. This is the sole cause of the weakening, not the diminishing of the source $\delta\epsilon$, which in fact does not happen, being but an optical illusion.

Indeed, if I might argue from Euclid's *Optics* here, less light arrives at the nearer, β, than at the more remote, for the reason that at β a smaller circle bounds the visible hemisphere of the luminous body $\delta\epsilon$ than at γ. Therefore, not so many particles of the sun $\delta\epsilon$ can be seen from β as from γ. But this is quite imperceptible, hard to express [even] in enormous numbers.

And I, after giving this answer to myself, am having a good laugh at my wretched alarm arising from this obscure business.

But the objection can rebound in the opposite direction, thus. If there is the same amount of light spread out in a large sphere as there is gathered together in a small sphere, there will not be the same amount of power in both places, for power is not considered orbically in a sphere as light is, but in the circle in which the planet proceeds. Thus, the magnetic filaments of the sun were supposed, above, to be set up only in longitude, not towards the poles or in other directions.

The answer is that the cause of light and motive power is exactly the same, and there is a deception in the reasoning. For in light, the rays do not flow out solely from the individual points and circles of the body to the corresponding points and circles of the sphere. Thus, the rays from γ do not come only from α (by which arrangement it would be impossible to ascribe a density to light in the spheres, since it would have no quantity in its origin, descending as it does from a point). Instead, the rays flow out from the whole hemisphere of the lucid body to the individual points of the imaginary spherical surface: thus, the rays from both δ and ϵ flow out to γ. The same thing takes place likewise with power. Even though the magnetic filaments of the solar body are ordered according to zodiacal longitude, and even though there is but a single great circle of the sun's body beneath the zodiac or ecliptic, and roughly beneath the orbit of the planet, and (finally) even though the other smaller circles (which are compacted to the size of points beneath the poles) are subordinated to their corresponding circles in the sphere of the planet, nevertheless, the rays from all the filaments of the solar body (standing up from one hemisphere of the body) flow down and converge on the individual points of the path of any of the planets as well as on those poles that are above the poles of the sun's body, and the body of the planet is transported according to the measure of the density of this entire *species* compounded of all the filaments.

It does not, however, follow that just as the sun shines equally in all directions, the planets, too, as you might fear, are moved indiscriminately in all

181

Why the planets always stay close to the zodiac.

directions. For the magnetic filaments of the sun, considered in themselves, do not move, but only inasmuch as the sun, rotating very rapidly in its space, carries the filaments around, and with them the moving *species* spreading abroad from them. Therefore the planet will not go backwards, because the sun always rotates forwards. The planet will not go to the poles (even though there might be some of the *species* from the sun's body at those points) because the filaments of the solar body are not extended in the direction of the poles, nor does the sun rotate in that direction, but rather in the direction in which its filaments urge it.

On these suppositions, it is so far from being the case that the planets are carried towards the poles that there is instead a single region of the zodiac, the mean between the poles, through which all planets of necessity would move in longitude without deflection, if they were to cease their own proper motions (of which see Chapter 38 below). *For the* species *of the solar hemisphere which takes up a post at some point of the zodiac, such as the point ζ in the present diagram, is the total of the filaments of the semicircle all tending in the same direction in concert, as, from [the region from] θ through κ, from λ through μ,*

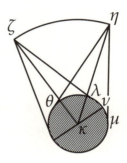

etc.[2] *But when you move off towards the poles of the world, as at η, then because one pole ν of the sun's body, and the complete circles λμ that surround the pole ν, are appropriated under the purview of ημ, the* species *is made up of filaments tending in opposite directions, for the opposite parts λ and μ of the circles move in opposite directions. Therefore that* species *θημ descending near the poles is less well adapted to the motion carrying the planets along.*

---

[2]Kepler's point here seems to be this: as the sun rotates, the filaments at θ tend around towards κ (which is to be taken as on the sun's surface), and the filaments at λ tend around towards μ, in the same direction.

CHAPTER 37

How the power moving the moon may be constituted.

182

Since in Chapter 34, in passing, I made mention of the motion of the moon, it would be appropriate to treat the matter in greater detail, so that no reservation introduced by the moon might deflect the reader into withholding his ready assent from me on the entire treatise; so that it might instead be marvelously confirmed, by the clearest consideration of the lunar motion; and finally, to ensure that the physical aspects of astronomy are treated fully in this book. For even though there are a few things in the theory of the moon that must be deferred, as they are to be treated differently or explained in more detail, they nevertheless will have their origin here.

After long and painstaking observations of the moon in every position in relation to the sun, Tycho Brahe expressed the opinion that in the moon, besides the anomaly of the epicycle, and besides that monthly anomaly which was also known to Ptolemy, the mean motion itself (so named in relation to these two inequalities) is not yet quite "mean." That is, it intensifies at the conjunctions and oppositions with the sun, and is remitted at the quadratures. Thus, even if it were undisturbed by epicycles, the moon itself, even though moving on a circle concentric with the earth, would move around nonuni-

183 formly.[1]

Let S be the body of the sun, M the orb of Mercury, V of Venus, T of the earth, P of Mars, and so on. And let all of them always move from right to left when in the upper part of the diagram. Now let CLOF be the orb of the moon, O the moon at opposition, C at conjunction, L, F at quadratures, and let CLOF remain for now a concentric described about the earth at T, and let it move in the direction OFCL. The question is, by what cause is the moon made to move more swiftly about T at C, O than at F, L, since we have just now mentally removed the eccentricity and the epicycles. Here, I know, the reader expects me to say that it is swifter at O because at that place its motion is in the same direction as the motion of all the planets. But this is not the true cause. For in that case, what would happen at C is what in fact happens, that the moon, in its compounded motion, is slowest, since its proper motion FCL resists considerably this common rightward motion.[2] For it should be noted that the moon, from point C on its orb, is borne less in the leftward direction L than the earth is borne to the right on its orb. Therefore, the moon, by a motion compounded of its proper motion and that which it has in common with the earth, also always moves to the left when above, as when the earth is at δ, but here, when the earth is down low

---

[1] What is being described here is the "variation", discovered by Brahe. The monthly inequality mentioned earlier, already known to Ptolemy, is the "evection". Brahe's discovery of the variation and his development of lunar theory is treated by Victor E. Thoren, *The Lord of Uraniborg* (Cambridge University Press, 1990), pp. 325–333. For his lunar theory, see *TBOO* 2 pp.101 sqq.

[2] Here Kepler clearly has gotten "left" and "right" confused in the text, which error is accordingly corrected in the translation.

at T, it is carried to the right, but slowly at C and swiftly at O. A motion of this sort is expressed approximately by the spiral lines set out here.

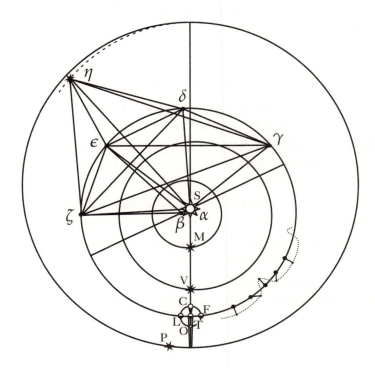

But perhaps you are expecting me to say instead that this phenomenon arises from the sun's motive power being weaker at O and stronger at C? Much less will I say this. For I would thus make it slow at O and C, and fast at F, L, which is contrary to what is desired. For if it is propelled weakly at O, it therefore is slow, and if it is held back more strongly at C, so that its tendency to move in the opposite direction from C to L is less, it will therefore again move slowly from C to L. Thus it was not correct for us to free the moon from the earth and commit it to the sun. This would cause it at length to wander away from the earth, just as the apogees wander from their places. It would be preferable to attribute to the earth a force that retains the moon, like a sort of chain, which would be there even if the moon did not circle the earth at all. On this supposition, the moon is, as it were, in the same boat with the earth, in the same power of the sun, and now, as if freed from this motion from the sun, it is rotated separately by the earth.

Therefore, I consider the cause of the swiftness at *O* and *C* to be none other than this, that the earth *T* derives its power of moving the moon from the sun *S*, and preserves it by the continuity of the line *TS*. Thus *SCTO* can properly be called the "diameter of power" since these two (*T* and *S*) are the sources of the entire motion.

On this supposition, that monthly inequality known to Ptolemy also will follow. *For if the power coming from the same source T is stronger at C, O than at F, L, then if the apogee is near C, O, there is a greater loss of speed than if the apogee were at F, L. Thus when going from the apogee at O or C the equations at F, L are greater than when going from the apogee at F, L to the conjunctions and oppositions O, C.*

You see, then, that these physical speculations are so devised that they can even account for the phenomena of the moon. The moon is not driven to circle the earth primarily by the sun, but by a power lying hidden in the earth itself, and casting forth its immaterial *species* to the body of the moon, but more strongly along the line that connects the centers of the sun (the primary source) and the earth.

It is, however, hard to explain more clearly how that diameter comes to be more powerful. For neither the sun's power nor the earth's, going out to the

184   moon, is any swifter when the moon falls on that diameter. That the rotations of these bodies (and therefore also of the *species*) are uniform and forever constant, is a highly reasonable presumption. There remains only the possibility described here, that the power is indeed not swifter, but more robust, when it emanates from the earth in the directions nearer to the line *ST*, for the reason that it was originally drawn off from the sun to the earth along that line.

Nevertheless, it is the sun which, either immediately or through that faculty which effects annual motion for the earth, is the principal director of the motion which the earth confers upon the moon. This is demonstrated chiefly by the moon's making its circuit beneath the zodiac, like the annual circuit of the centre of the earth, although the earth's diurnal motion, which confers the monthly motion upon the moon, takes place beneath the equator.

CHAPTER 38

*Besides the common motive force of the sun, the planets are endowed with an inherent force [*vis insita*], and the motion of each of them is compounded of the two causes.*

I have spoken of the origin of the motion which rotates the planets around the sun or the moon around the earth; that is, I have spoken of the natural causes of the circle which in the theories of the planets is called either eccentric or concentric, according to the various intentions of the authors. Now, something must be said of the natural cause of the eccentricity, or in the particular hypothesis of Copernicus, of the epicycle on a concentric. For the moving power from the sun has hitherto been uniform, having different degrees only at various amplitudes of the circles. Its innate quality is such that if a planet were to remain at the same distance from the sun it would be carried around perfectly uniformly, and would not experience any intension or remission of the solar motion. The modicum of nonuniformity perceived in the working of this power arises from the planet's having been transposed from one distance from the sun to another, so that it encounters one or another degree of strength of this power from the sun. The question therefore arises, if, as Brahe demonstrated, there are no solid orbs, how does the planet come to ascend from and descend towards the sun? Can this, too, really come from the sun? My answer is that it is to some extent from the sun, and to some extent not from the sun.

Examples of natural things, and this hitherto disparaged kinship of celestial things for terrestrial ones, cry out that in a simple body the operations which are more general are simpler, while the variables, if any (such as, in the motion of the planets, the varying distance from the sun, or the eccentricity), arise from the concurrence of extrinsic causes.

First argument, from the eccentricity.

Thus, in a river, the simple property of water is to descend towards the centre of the earth. But because its path is not direct, it flows in to those places where it finds a lower bed, stagnates where it meets with level ground, and is carried along with a roar where it comes upon steeper slopes; there is a place where it may be spun into whirlpools, if in a swift drop it should dash into projecting rocks. Where water itself, by its inherent force, endeavors to do nothing but descend towards the centre of the earth, a simple action by a simple property, nevertheless, flow and stagnation and waves and whirlpools and every variety [of phenomena] arise from the causes described, which are adventitious and accidental.

185

Particularly happy and better accommodated to our inquiry are the phenomena exhibited by the propulsion of boats. Imagine a cable or rope hanging high up across a river, suspended from both banks, and a pulley running along the rope, holding, by another rope, a skiff[1] floating in the river. If the ferryman in the skiff, otherwise at rest, fastens his rudder or oar in the right manner, the skiff, carried crosswise by the simple force of the downward-moving river, is transported from one bank to the other, as the pulley runs along the cable above. On broader rivers they make the skiffs go in circles, send them hither and thither, and play a thousand tricks, without touching the bottom or the banks, but by the use of the oar alone, directing the singular and most simple flow of the river to their own ends.

In very much the same manner, the power moving out from the sun into the world through the *species* is a kind of rapid torrent, which sweeps along all the planets, as well as, perhaps, the entire aethereal air, from west to east. It is not itself suited to attracting bodies to the sun or driving them further from it, which would be an infinitely troublesome task. It is therefore necessary that the planets themselves, rather like the skiffs, have their own motive powers, as if they had riders [*vectores*] or ferrymen, by whose forethought they accom- plish not only the approach to the sun and recession from the sun, but also (and this might be called the second argument) the declinations of latitudes; and as if from one bank to the other (that is, from north to south and back), travel across this river (which itself only follows the course of the ecliptic) from north to south and back.

Second argument, from latitudes.

It is certain from what has been said above that the power which comes from the sun is simple. But now, the eccentrics of the planets do not just decline from the ecliptic, but go in various directions, intersecting one another and the ecliptic. Therefore, other causes are conjoined with the motive power from the sun.

---

[1]It may be significant that *cymba* (here translated as "skiff") was often used by classical authors (Virgil, Horace, and others) to denote specifically the boat in which Charon ferried the dead across the river Styx. Further, *portitor*, here translated "ferryman", primarily means "customs officer" or "duty collector," and was for that reason a name for Charon himself. Because of this connection, it later came to denote any ferryman or carrier.

The verb used to describe the attachment of the oar or rudder is *religare*, cognate with the English word "religion."

CHAPTER 39

By what path and by what means the powers seated in the planets ought to move them, in order that the planet's circular orbit, such as is commonly thought to exist, be brought about in the aethereal air.

And so, let these points of great certainty in what has been demonstrated be axioms for us. First, that the body of a planet is inclined by nature to rest in every place where it is put by itself. Second, that it is transported from one longitudinal position to another by that power which originates in the sun. Third, if the distance of the planet from the sun were not altered, a circular path would result from this transport. Fourth, supposing the same planet to be in turn at two distances from the sun, remaining there for one whole circuit, the periodic times will be in the duplicate ratio of the distances or magnitude of the circles. Fifth, the bare and solitary power residing in the body of a planet itself is not sufficient for transporting its body from place to place, since it lacks feet, wings, and feathers by which it might press upon the aethereal air. And nevertheless, sixth, the approach and recession of a planet to and from the sun arises from that power which is proper to the planet. All these axioms are agreeable to nature in themselves, and have been demonstrated previously.

186

Axioms for theory of the celestial motions.

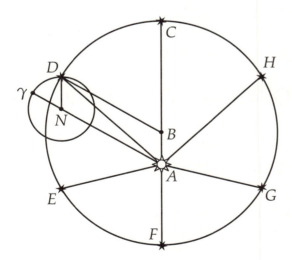
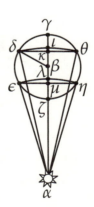

Now let us work with geometrical figures in order to see what laws will be required to represent any desired planetary orbit. Let the orbit of the planet be a circle, as has been believed until now, and let it be eccentric with respect to the sun, the source of power. *Let that eccentric CD be described about center B with radius BC, and on it let BC be the line of apsides, A the sun, and BA the eccentricity. Let the eccentric be divided into any number of equal parts, beginning from the line of apsides at C, and let the ends of these parts be connected with A. Therefore, CA, DA, EA, FA, GA, HA, will be the distances of the end points of the equal parts from the*

I.

What the planet shall do by the motion of its own body when its composite path is made to be a perfect circle. That is, what sort of distances from the sun will it produce?

source of power. Now with center β, radius βγ, equal to AB, let the epicycle γδ be described, divided into as many equal parts as the eccentric, beginning from γ, and let the line γβ be extended so as to make βα equal to BC, and let the point α be connected with the end points of the equal parts of the epicycle, by the lines γα, δα, εα, ζα, ηα, θα. These lines will be respectively equal to the distances drawn to the eccentric from A, this having been demonstrated above in Chapter 2. Next, with center α and radius δα, let the arc διθ be described, intersecting the diameter γζ at ι, and about the same center α, with radius αε, let the arc ελη be described, intersecting the diameter γζ at λ, and let the end points of the parts equidistant from the aphelion of the epicycle γ be connected by the lines δθ, εη, which will intersect the diameter at κ, μ, so that αδ or αι is longer than ακ, and αε or αλ longer than αμ.

First way: the planet itself moves on an epicycle. Chapter 49 depends primarily upon this.

187

If it were possible for the planet to move on a perfect epicycle by its inherent force,[1] and for its orbit at the same time to be a perfect circle, then we would have to consider similar arcs to be swept out in the same times, on the eccentric and on the epicycle. Consequently, it would now immediately become clear by what means, and by what measure, the distance αι would be made equal to AD. For since αι and αθ are equal, when the planet moves from γ to θ the distance αθ necessarily and without any special contrivance comes out right, equal to AD.

Absurdities of this account.

1.

But aside from one appearing to be in conflict with the fifth axiom, who claims that the planet moves locally from γ to θ by its inherent force, many additional absurdities are also involved.

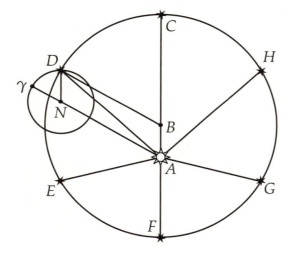
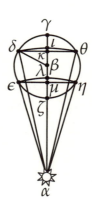

For let AN be drawn parallel to BD, and let AN be equal to BD, and about center N let an epicycle be described which shall go through D. Now since CD is a perfect circle, the same angles are swept out by the planet D about the center of the eccentric B as by the center of the epicycle N about the center of the sun A (through the equivalence demonstrated in Chapter 2), as long as the diameter of the epicycle ND with the planet at D always remains parallel to AB with respect to its position in the world. Therefore, the speed of the center of the epicycle N about the sun A would be made the same as the speed of the planet D about the center of the eccentric B, so that those motions would be intensified at the same time and remitted at the same time.

---

[1] *Vis insita.*

And since intension and remission depends upon greater or less distance of the body   2.
of the planet from the sun, therefore the center of the epicycle, remaining at the same
distance, would be contrived to move slowly or swiftly on account of the planet's being
farther from or nearer to the sun.

And although the power driving the planets is faster than any of them, as is shown
in Chapter 34, we are here to suppose in our imagination a single ray of power AN   3.
coming from the sun, as if there were a line upon which the center of the epicycle N
would always remain. And this line, with that same center N, would be now slow,   4.
now swift, again contrary to what was said above, that the power always produces the
same speed at the same distance. Moreover, we would be required to suppose that the   5.
planet has its own rotatory motion away from this imaginary ray AN in the opposite
direction, traversing unequal distances in equal times, according as this ray itself   This last is avoided below
were swifter or slower. In adopting this account, we would indeed approach   in Ch. 49, though the
other absurdities remain.
more closely the geometrical suppositions of the ancients, but we would stray
very far from physical theory, as is shown in Ch. 2. And my thoughts on the
matter have not sufficed to discover a way in which these things can happen
naturally.

Now all this would be conceived more simply were we to consider the   A second way in which
diameter of the epicycle *ND* as always remaining parallel to itself. For then   the planet might produce
the eccentric.
the planet would carry out its motion by a mental image, not of the epicycle,
but of the center of the eccentric *B*, and by keeping itself always at the same
distance from that center.

But at the beginning of this work, in Chapter 2, it was said to be most   Absurdities
absurd that a planet (even if you give it a mind) may imagine for itself a center   1.
and from it a distance, when in that center there is no particular body for the
planet to be aware of.

Now you might say that the planet observes the sun *A*, and already knows
beforehand, through memory, which ordered distances from the sun a perfect
eccentric would have to attain. But first, this is more indirect, and depends
upon something intermediate connecting the effect of the perfectly circular
path with the indication of the waxing and waning of the sun's diameter, and
this too in some mind. But that intermediate can be nothing but the position
of the center of the eccentric *B* at a certain distance from the sun, which, as has
just now been said, cannot be known by an unassisted mind.

I do not deny that a center can be conceived, and about it a circle. But I do   188
say this, that if the center is established by thought alone, without respect to
time and without any external indication, it is impossible to set up about it, in
reality, the perfectly circular path of some movable body.

Besides, if the planet were to derive its correct distances from the sun,   2.
ordained by the rule of the circle, from memory, it could also derive from
memory the equal arcs of the eccentric which are to be traversed in unequal
times, and which are to be traversed by an extrinsic force originating in the
sun, as if it were obtaining the values right from the Prutenic or Alphonsine tables. Thus it would know from memory beforehand what this extrinsic
and mindless power originating in the sun was going to do. All of these are
absurd:—

Chiefly since, as Aristotle affirms, there is no knowledge of the infinite,   3.
while the infinite is involved in this intension and remission.

But it is all right, because even the observations themselves will not allow *CD* to be a perfect circle, below in Chapter 44, so these theoretical arguments, weak though they are thought to be, do not stand alone, and are that much the less vulnerable to scornful rejection.

It is consequently more fitting that the planet itself require no assistance, whether of epicycle or eccentric, but rather that the task which it either performs by itself or has a part in performing is a reciprocating path along the diameter *γζ* directed towards the sun *α*.

The question now is, what is the measure by which the planet metes out the correct distances for any given time?

Now to us the measure is clear from geometry and the diagram. *For whenever the solar power moves the planet forward to the line DA, we then find the angle CBD and make γβδ equal to it; and thus we say that αδ, or αι which is its equal, is the correct distance from A to the planet at D.* But this measure which we have proposed for humans, we have just now denied the planet when we removed it from the circumference of the epicycle and restricted it to the straits of the diameter *γζ*.

Indeed, in this inquiry it is easier to say what is not the case than to say what is. *This is because at the moments when the sun has placed the planet on the lines drawn from A through C, D, E, F, G, H, the planet itself is presumed to have produced the distances γα, ια, λα, ζα, λα, ια, respectively.*Consequently, if the path of the planet is a perfect circle, then to equal parts of the eccentric *CD, DE, EF,* correspond unequal descents of the planet along the diameter, namely, *γι, ιλ, λζ.* Moreover, the order is perturbed: the highest is not the smallest, nor the lowest the greatest, but the middle parts *ιλ* are greatest, and the extremes *γι* and *λζ* are smaller, and the highest *γι* are a little smaller than the corresponding lowest, *λζ.* For *γκ* and *μζ* are equal, and *γι* is smaller than *γκ*, while *λζ* is greater than *μζ*.

189

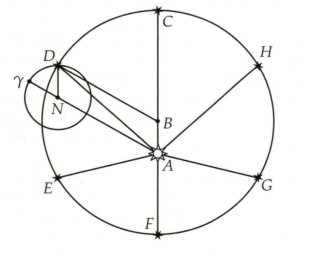
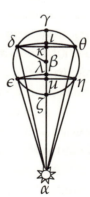

Furthermore, this same cause prevents *γι, ιλ, λζ,* being made proportional to either the times of the equal arcs traversed *CD, DE, EF,* or to the angles at the sun *CAD, DAE, EAF.* For the time or duration of the planet on equal parts of the eccentric *CD, DE, EF,* is continuously diminished from the highest to the lowest points, and the angles at the sun are continuously increased, but the reciprocations $\gamma\iota$ are increased in the middle regions, such as $\iota\lambda$.

Therefore, if the path of the planet is a perfect circle, the measure of the planet's descent on the diameter $\gamma\zeta$ is neither time, nor distance traversed on the eccentric, nor the angle at the sun.

And physical theories, too, also decisively repudiate these measures.

What, then, if we should say this: although the motion of the planet does not take place on an epicycle, this reciprocation is measured out in such a way that distances from the sun are produced that are similar to those which exist in an epicycle actually traversed? *Nor from an imaginary epicycle or eccentric.*

First, this attributes to the power belonging to the planet a knowledge of the imaginary epicycle and of its effects in setting out distances from the sun; and further, it attributes knowledge of the future speed or slowness which the common motion from the sun is going to cause. For it is necessary to suppose here an imaginary intension and remission of motion on the imaginary epicycle that is the same as that of the motion on the real eccentric. This is more incredible than the previous accounts, where the motion of the body was combined with knowledge of the epicycle or eccentric. Therefore, the objections raised there should be understood as applying here as well, the judgements being nearly identical.

Nevertheless, for want of a better opinion, we must at present put up with this one. And as for its involving many absurdities, in Chapter 57 below, a certain physicist will quite readily allow this: that on the testimony of the observations the path of the planet is not a circle. *In Chapter 57 below, the measure of this reciprocation will be made plain.*

So far, the discussion has concerned the measure relating to the form of this reciprocation. It now remains for us to find the measure of this measure; that is, the measure of the quantity or of motion through place. For it is not enough for the planet to know how far it should be from the sun: it also has to know what to do in order to be at the correct distance. *II. By what means or measure may a planet grasp its distance from the sun?*

Now anyone who is so attracted to the supposition of a perfectly circular orbit as to locate in the planet a mind which could preside over the reciprocation, can say only this: that this planetary mind observes the increasing and decreasing size of the solar diameter, and understands, using this as an indication, what distances from the sun it should cause to occur at any given time. For example, sailors cannot know from the sea itself how far they have travelled over the waters, since the course, viewed in that way, has no distinct limits. Instead, they find this either from the amount of time they have sailed, if wind and sea remain constant and the ship does not stop, or from the direction of the wind and the changing elevation of the pole, or from all or several of these in conjunction, or, may it please the gods, by a contrivance of a number of wheels, with paddles lowered into the water (for certain conceited mechanics are proposing an instrument of this sort, who ascribe the calm of the continents to the waves of the Ocean).[2] In just the same way, the mind of the planet cannot by itself measure its position, or the distance between it- *To what extent is a perception of the size of the solar body to be attributed to the planets?*

190

---

[2]The device that Kepler is describing is an early instance of what was known to yachtsmen of the last century as a "taffrail log," which keeps track of distance travelled through the water by registering turns of a paddlewheel (or in recent times, a screw). Kepler points out that this only gives motion relative to the water, which may itself be in motion.

Thus the planets would
become surveyors,
measuring their distances
from the sun from a
single reference point,
through the apparent size
of the solar body.

self and the sun, since between them there is pure aethereal air, devoid of any means of indication. So it makes use either of the increment of time, and an equal exertion of forces through that time (which has just been denied above), or of a physical machine, which is ridiculous (for by the example of the sun and moon we suppose the celestial bodies to be round, and it is therefore also probable that the entire field of the aetheral air moves around with the planets),[3] or finally, of some suitable means of indication that are variable with the altered distance of the planet from the sun. And other than the sun's apparent diameter alone, nothing else presents itself. *Thus we humans know that the sun is 229 of its own semidiameters distant from us when its diameter subtends 30′, and 222 semidiameters when it subtends 31′.*[4]

If it were indeed certain that this proper motion of the planet along the diameter of the epicycle could not be carried out by any material and corporeal or magnetic power of the planet, nor by an unassisted animate power, but that it is governed by a planetary mind, nothing absurd would be stated.[5] For that the sun is observed by the planets in other respects as well, the latitudes bear witness. For by these latitudes the planets would depart from the middle and royal road of this solar power, as from the mainstream of a river, and move to the sides, as is said in Chapter 38, unless they meanwhile paid attention to the sun, and carried out their approach and recession along a line drawn to its center. They would then describe circles which, seen from the earth or the center of the world, would appear smaller, parallel to some great circle. But all planets describe great circles that intersect the ecliptic at points that are opposite with respect to the sun, as was demonstrated for Mars from observations above in Chapters 12, 13, and 14. Therefore, the diameter of reciprocation $\gamma\zeta$ is also directed towards the sun, and the latitudes heed the sun in every respect. However, in Part 5 below I will also transfer this characteristic of the latitude from the elements of mind to the elements of nature and magnetic faculties.

There exists in the planets
something like a mind,
which pays attention to
the body of the sun.

See the marginal note to
Ch. 63.

Objections which can be
raised against perception
of the solar body.

1.
Small angle subtended.

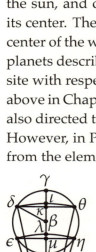

Now one cannot say in reply to me that the solar diameter and its variation is far too small to be used as a standard. For it is certain that there is no planet for which it entirely vanishes. Since on earth it is thirty minutes, on Mars it will be twenty, on Jupiter seven, and on Saturn three, while on Venus it will be forty, and on Mercury at least eighty and sometimes as high as one hundred and twenty. Do not grumble about the smallness of the body, but rather the unapt coarseness of human senses, which cannot follow such small things.

One should on the contrary note that this body, however small, is nonetheless capable of moving such distant bodies in a circle, as is

---

[3]That is, it would be of no use to know the relative velocity of a planet through the aethereal air.

[4]Substituting "semidiameters" for "diameters" where the sense requires.

[5]This was the standard view of the day, included in virtually all the introductory university textbooks in natural philosophy. The planetary mind was usually identified with the biblical angels, thus bringing Aristotle into harmony with scripture. However, some theorists believed that the mind is united with the planet as soul is with body. This view was widely regarded as heretical, as it suggested that the Prime Mover is the soul of the world (*anima mundi*), and thus that the universe is somehow God's body.

demonstrated in the preceding chapters. The illumination of the world by such a tiny corpuscle is known to all. And so it is credible that if the movers are endowed with some faculty of observing its diameter, this faculty is as much more acute than our eyes as its work, and the perpetual motion, is more constant than our own troubled and confused schemes.

So then, Kepler, would you give each of the planets a pair of eyes? By no means, nor is this necessary, no more than that they need feet or wings in order to move. But Brahe has recently eliminated solid orbs. Now our theorizing has not emptied nature's treasure house: we still cannot establish, through our own knowledge, how many senses there ought to be. There are even examples at hand worthy of our admiration. For tell us, in physical terms: with what eyes shall the animate faculties of sublunar bodies look upon the positions of the stars in the zodiac, so that when a harmonic arrangement (which we call "aspects") is found among them, the bodies leap up and become inflamed for their task? Was it with her eyes that my mother noted the positions of the stars in order to know that she was born with Saturn, Jupiter, Mars, Venus, and Mercury, all in sextiles and trines? And could it have been by the same means that she gave birth to her children, and especially to me her first born, chiefly on those days when as many as possible of the same aspects, especially of Saturn and Jupiter, recurred, or when as many pristine positions as possible were occupied by squares, oppositions, and conjunctions? I have observed those things in all cases whatever that have occurred to this very day. But why do I say these things that are just as absurd as the previous ones, but for those who have exercised themselves in natural matters more diligently than is usual nowadays?

So our hypothetical person who says that the planet's path is a perfect circle will say this: that the planet performs its reciprocation so as to make the diameters of the sun, at the end points of equal arcs of the eccentric, appear very nearly* inversely proportional to the lines $\delta\alpha$, $\epsilon\alpha$, $\zeta\alpha$, or to $\iota\alpha$, $\lambda\alpha$, $\zeta\alpha$, which are equal to them, taken with respect to the longest line $\gamma\alpha$; and that through this consideration of the diameters of the sun at the chosen moments of time, come the proximities of ι, λ, ζ to γ.

It should be known, however, that the increases of the diameter of the sun and the arcs of the epicycle do not square with each other well, and so the motive mind will have to have a very good memory in order to adjust the unequal versed sines of the arcs on the epicycle to the equal increases of the solar diameter. For this, see Ch. 56 and 57 below.

Let that be enough concerning the means of indicating distances. There remains a third point about the animate faculty that carries the planet around that I briefly draw attention to: anyone who says that the body of a planet is moved by an inherent force is just plain wrong. This we rejected at the beginning. But it is also impossible simply to ascribe this force to the sun instead. For the same force that attracts the planet also repels it in turn, and this is inconsistent with the simplicity of the solar body. But anyone who by some unique argument reduces this motion to the bodies of the sun and the planet in concerted action gives a new cast to the material of this entire chapter, and under this rubric Chapter 57 below is specifically assigned for this topic.

You see, my thoughtful and intelligent reader, that the opinion of a per-

191
2.
Defect in the means of perception.

* For in Ch. 57 the ratio will be somewhat different.

III.
By what animate faculty the mind might obtain the means of determining the distance of its body from the sun.

fect eccentric circle for the path of a planet drags many incredible things into physical theories. This is not, indeed, because it makes the solar diameter an indicator for the planetary mind, for this opinion will perhaps turn out to be closest to the truth, but because it ascribes incredible faculties to the mover, both mental and animate.

192   But we, who are close to the truth, should find out how to cast these theories (which, though not everywhere perfect, are nonetheless suitable for the sun's motions) in numerical form. In the end, it will be helpful for a more exact discovery of the truth, reserved for Chapter 57, that we had previously worked on them here.

CHAPTER 40

An imperfect method for computing the equations from the physical hypothesis, which nonetheless suffices for the theory of the sun or earth.

Such a long-winded discussion was necessary to prepare a way for a natural form for the equations, on which I am going to be very busy in Part 4. Now we must return to the equations of the sun's eccentric in particular, which is the main subject of this third part, and for the sake of which the general discussion of the last eight chapters has been presented.

My first error was to suppose that the path of the planet is a perfect circle, a supposition that was all the more noxious a thief of time the more it was endowed with the authority of all philosophers, and the more convenient it was for metaphysics in particular. Accordingly, let the path of the planet be a perfect eccentric, for in the theory of the sun the amount by which it differs from the oval path is imperceptible. Those things that are going to be needed for the other planets, on account of this deviation, follow below in ch. 59 and 60.

Since, therefore, the time increments of a planet over equal parts of the eccentric are to one another as the distances of those parts, and since the individual points of the entire semicircle of the eccentric are all at different distances, it was no easy task I set myself when I sought to find how the sums of the individual distances may be obtained. For unless we can find the sum of all of them (and they are infinite in number) we cannot say how great the time increment is for any one of them. Thus the equation will not be known. For the whole sum of the distances is to the whole periodic time as any partial sum of the distances is to its corresponding time.

Through elongations of the planet from the center of the sun, to find the physical part of the equation.

I consequently began by dividing the eccentric into 360 parts, as if these were least particles, and supposed that within one such part the distance does not change. I then found the distances at the beginnings of the parts or degrees by the method of chapter 29, and added them all up. Next, I assigned an artificial round number to the periodic time; although it is in fact 365 days and 6 hours, I set it equal to 360 degrees, or a full circle, which for the astronomers is the mean anomaly. As a result, I have so arranged it that as the sum of the distances is to the sum of the time, so is any given distance to its time. Finally, I added the times over the individual degrees and compared these times, or degrees of mean anomaly, with the degrees of the eccentric anomaly, or the number of parts whose distance was sought. This furnished the physical equation, to which the optical equation, found by the method of Chapter 29 with those same distances, was to be added in order to have the whole.

Term:
What is the mean anomaly?

193

However, since this procedure is mechanical and tedious, and since it is impossible to compute the equation given the ratio for one individual degree without the others, I looked around for other means. And since I knew that

Through the areas to find the physical part of the equation.

the points of the eccentric are infinite, and their distances are infinite, it struck me that all these distances are contained in the plane of the eccentric. For I had remembered that Archimedes, in seeking the ratio of the circumference to the diameter, once thus divided a circle into an infinity of triangles – this being the hidden force of his *reductio ad absurdum*. Accordingly, instead of dividing the circumference, as before, I now cut the plane of the eccentric into 360 parts by lines drawn from the point whence the eccentricity is reckoned.

Let AB be the line of apsides, A the sun (or earth, for Ptolemy); B the center of the eccentric CD, whose semicircle CD shall be divided into any number of equal parts CG, GH, HE, EI, IK, KD, and let the points A and B be connected with the points of division. Therefore, AC will be the greatest distance, AD the least, and the others, in order, are AG, AH, AE, AI, AK. And since triangles under equal altitudes are as their bases, and the sectors, or triangles, CBG, GBH, and so on (standing upon least parts of the circumference and therefore not differing from straight lines) all have the same altitude, the equal sides BC, BG, BH, they are therefore all equal. But all the triangles are contained in the area CDE, and all the arcs or bases are contained in the circumference CED. Therefore, by composition, as the area CDE is to the arc CED so is the area CBG to the arc CG, and alternately, as arc CED is to CG, CH, and the rest in order, so is the area CDE to the areas CBG, CBH, and the rest in order. Therefore, no error is introduced if the areas be taken for the arcs in this way, and substituting the areas CGB, CHB for the angles of eccentric anomaly CBG, CBH.

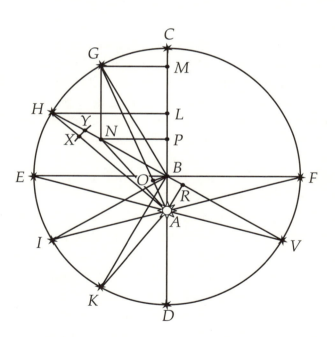

Further, just as the straight lines from B to the infinite parts of the circumference are all contained in the area of the semicircle CDE, and the straight lines from B to the infinite parts of the arc CH are all contained in the area CBH, therefore also the straight lines from A to the same infinite parts of the circumference or arc make up the same thing. And finally, since those drawn from A and B both fill up one and the same semicircle CDE, while those from A are the very distances whose sum is sought, it therefore seemed to me I could conclude that by computing the area CAH or CAE I would have the sum of the infinite distances in CH or CE, not because the infinite can be traversed, but because I thought that the measure of the faculty by which the collected distances mete out the times is contained in this area, so that we would be able to obtain it by knowing the area without an enumeration of least parts.

194   *Therefore, from the above, as the area CDE is to half the periodic time, which we have proclaimed to be 180°, so are the areas CAG, CAH to the times on CG and CH.*

Thus the area CGA becomes a measure of the time or mean anomaly corresponding to the arc of the eccentric CG, since the mean anomaly measures the time.[1]

Earlier, however, the part CGB of this area CAG was the measure of the eccentric anomaly, whose optical equation is the angle BGA. Therefore, the remaining area, that of the triangle BGA, is the excess (for this place) of the mean anomaly over the eccentric anomaly, and the angle BGA of that triangle is the excess of the eccentric anomaly CBG over the equated anomaly CAG. Thus the knowledge of this one triangle provides both parts of the equation corresponding to the equated anomaly GAC.

And hence it is manifest why in Chapters 30 and 31 above I said that in the theory of the sun the parts of the equation are very nearly equal. *That is, any given arc, and the angle at the center which it subtends (as CG and CBG in the figure), are measured by its area, which is called the "sector", as area CBG. Therefore, with one arm of the compass at G, with radius GB, let an arc of the circumference be described intersecting GA at O. Hence, as the area GBC is to the angle GBC, so is the area BGO to the angle BGO. But the angle BGO is the optical part of the equation. Thus the area GOB, through the doubling of the part of the equation, measures the optical part of the equation, for in our account presented earlier the whole area GBA was to be consulted for the physical part of the equation.*

Now clearly the genuine measure of the physical part of the equation AGB exceeds the proposed measure of the optical part OGB by the small space or area OAB (while near the perigee the latter in turn exceeds the former by a small area). Nevertheless, if the eccentricity is small, as is that of the sun or earth, with which we are concerned in this third part, this is not perceptible. For the nearer it approaches to the line of apsides, the narrower becomes the whole triangle AGB and consequently also its little part AOB, however much its altitude AO increases at the same time. On the other hand, in the middle elongations, the angle BEA with its sector is at a certain point directly measured by the area BEA, and the excess begins to turn into a defect.

Therefore, the greatest difference that can occur is that accumulated at the octants, or locations intermediate between the apsides and the quadrants. How great that difference is, will now be shown.

Since for some time now I have used the same form of computation by means of areas in the theory of Mars, I could not ignore this difference on account of the planet's great eccentricity, nor did the doubling of the optical equation avoid all perceptible error. It was therefore necessary to investigate the area of the triangle of the equation [*triangulum aequatorium*]. This can be done in various ways, but I shall go on to state the easiest.

The reason why, in the third method in Ch. 31 above, part of the equation was simply doubled to give the whole equation.

195

Short cut investigation of the area of the triangle of the equation.

---

[1]This is the first published statement of what we now call "Kepler's Second Law." It is clear from Kepler's account that it is a computational contrivance that gets around the difficulty of computing eccentric anomalies using the distance law enunciated in the third paragraph of the present chapter. Kepler was also concerned that the area formulation and the distance formulation were not quite equivalent (the "paralogism" mentioned later in this chapter).

It is remarkable that Kepler states the area relationship as an improper proportion: an area is said to have a ratio to a time—a statement that makes no sense in Euclid's theory of proportions. Of course Kepler could have set area to area in the same ratio as time to time, which would have been perfectly Euclidean, but for some reason he chose not to. Inadvertence, or a deliberate spurning of nearly two millennia of mathematical reasoning?

It is well known that triangles with equal altitudes are proportional to their bases. I say also, that triangles on equal bases are proportional to their altitudes.

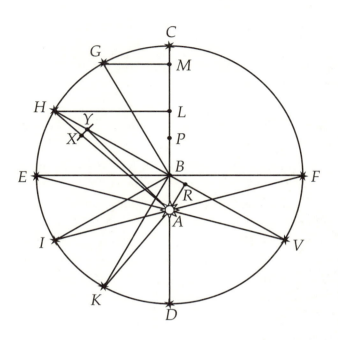

Let AGB, AHB stand upon the same base AB extended to C. From G let the line GN be drawn parallel to the common base AB and intersecting HB at N, and let N be connected with A. From the three vertices G, H, N, of the triangles let GM, HL, MP be drawn perpendicular to the base, determining the altitudes of the triangles. Therefore, since GN and MP are parallel, and GM, NP are perpendicular [to them], GM and NP will be equal. But GM is the altitude of triangle AGB, and NP is the altitude of triangle ANB. Therefore triangles ANB and AGB have equal altitudes, and since they are both on the same base AB, they are equal. And since ANB is part of AHB, and the base line HB is common, together with the vertex A, the triangles NAB and HAB have equal altitudes. Therefore, as the base NB is to BH, so is NAB to HAB. But NAB and GAB were proved equal. Therefore, as NB is to BH, so is GAB to HAB. But as BN is to BH, so is NP to HL, because NBP and HBL are similar triangles. Therefore also, as NP is to HL, so is GAB to HAB. But NP and GM are equal. Therefore, as GM is to HL, altitude to altitude, so is area GAB to area HAB. Q.E.D.

Value of the triangle at 90° of eccentric anomaly.

Now let BE be perpendicular to CD, and let the triangle BEA have a right angle at B. BE will be the altitude, and BA the base. Therefore, by Euclid I.42, 900, or half the base BA (which for the sun is 1800), multiplied by the altitude BE, 100,000, which is the radius of the circle, gives the area of the triangle BEA, that is, 90,000,000. But the area of a circle of radius 100,000 (from the most recent investigations of Adrian Romanus, a most expert geometer)[2] *is 31,415,926,536, with no error in even the last digit. And as this the area of the circle is to the 360° of mean anomaly or time (that is, 21,600′ or 1,296,000″), so is the area of the triangle, 90,000,000, to 3713″; that is, 1° 1′ 53″. So the area BEA has a value of 1° 1′ 53″. But in Chapters 29 and 30, the angle BEA was also 1° 1′ 53″. Therefore, both parts of the equation are equal at this place, that is, near 90°.*

Adrian Romanus

At other degrees of eccentric anomaly, we proceed as follows. *Since BEA is 3713″, as its altitude EB is to HL or GM, the altitudes of the other triangles—that*

---

[2]Adrian Romanus, or Romain, or van Roomen (1561–1615), Flemish physician, ecclesiastic, and mathematician, author of numerous mathematical and astronomical works. The computation of π appears in his *Idea mathematicae pars prima…*, (Louvain 1593): he divided the circle into 2^{30} triangles to attain an effective accuracy of fifteen decimal places.

is, as the whole sine is to the sines of the eccentric anomaly HBC, GBC—so is 3713″
to the areas of the remaining triangles. So 3713″ will be multiplied by the sines of the
angles at B, and, with the last five digits struck, the remainder will be the physical part
of the equation expressed in seconds, corresponding to the angle at B. For example, let
HBC be 45° 43′ 46″, as it was above in Ch. 31. Therefore, its sine, 71605, multiplied
by 3713″, with the last five digits struck, gives 2659″, that is, 44′ 19″. In the table
above, we assumed this to be equal to the optical part of the equation, 43′ 46″.

Therefore, this little area *ABO* at its greatest does not exceed 33″.

And this is that fourth procedure for computing the eccentric equations, of which I began to speak above near the end of Chapter 34, which closely expresses the very nature of things and the foregoing theories of Chapters 32 and 33.

196

Defect of this operation with the area of the triangle, arising from the supposition of a circular orbit.

Nevertheless, my argument contains a paralogism, not, indeed, of great moment. It arises from this: that while Archimedes did indeed divide the circle into an infinity of triangles, they stood upon the circumference at right angles, so that their vertices were at the center of the circle *B*. But one cannot proceed in the same way with triangles standing upon the circumference with their vertices at *A*, because the circumference is intersected obliquely by the straight lines from *A* in all places other than *C* and *D*.

You could have found this error empirically, as I myself did, by taking all the distances *AC*, *AG*, *AH*, at the individual whole degrees of the angle *CBG*, *GBH*, and adding them all up. (These distances, though they are presented in the table in Chapter 30, correspond in position to individual whole degrees of the angle at *A*, and consequently to angles at *B* cut minutewise. Nevertheless, one could easily find, by interpolation, the distance from *A* corresponding to any angle of an integral number of degrees about *B*.) Now the sum comes out to be greater than 36,000,000, although 360 of the distances from *B* add up to exactly 36,000,000. But, on the contrary, if both sums were measured by the same area of the circle, the sums ought to have been equal.

By "angles cut minutewise" I mean angles expressed in degrees and some odd minutes.

A demonstration of the error, on the other hand, is as follows. *Through B let any straight line other than CD be drawn, intersecting the circumference, and let it be EF; and let the points of intersection E and F be connected with A. Now since the point A does not lie on the line EF, EAF is a figure, a triangle. Therefore, EA, AF together are longer than EF, by Euclid I.22. But the area of the circle contains the sum of all lines EF, and therefore it contains a sum which is less than all the lines EA, AF, since any two opposite points on the eccentric, together with A, determine such a triangle, with the exception of C, D, A, where instead of a triangle there is a straight line.*

I would add, in passing, that it is also proved, in the same way, that the distances from *A* corresponding to all of the 360 integral degrees of the angle at *A* (which are in the table in Chapter 30 above), added up in one sum, are less than 36,000,000. *For through the point A let any straight line other than CD be drawn (let it be EV), and let E and V be connected with B. In the triangle EBV, the straight lines EB, BV together will be longer than EA, AV, two opposite distances. But all 360 of EB, BV taken together are 36,000,000. Therefore, all 360 of EA, AV taken together are less than 36,000,000.*

To return to what I was saying, this method of finding the equations is not only very easy indeed, and based upon the natural causes of the motion explained above, but also agrees most precisely with the observations in the theory of the sun or earth. Nevertheless, it errs in two respects. First, it supposes that the orbit of the planet is a perfect circle, which, as will be demonstrated below in Ch. 44, is not true. Second, it uses a plane which does not exactly measure the distances of all points from the sun. Nevertheless, as if by a miracle, each of these exactly cancels the effect of the other, as is demonstrated below in Chapter 59.[3]

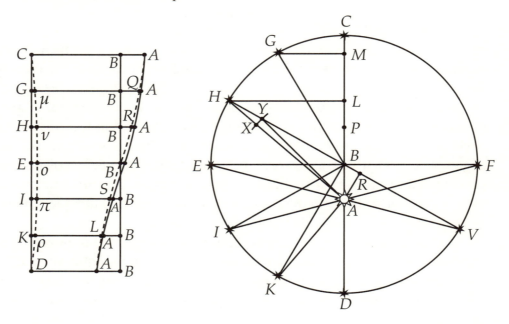

And because at the present time there are first-rate geometers who on occasion labor endlessly on matters whose usefulness is not so evident, I call upon one and all to help me here to find some plane figure equal to the sum of the distances. I have indeed myself found it geometrically – in a broad sense – but let them show me how to express numerically what I have delineated geometrically; that is, let them show how to square the figure I have found. *Let the circumference CED be unfolded into a straight line and divided into as many parts as before at the points G, H, E, I, K, and let perpendiculars be set up at the points of division equal to the radius CB, and let the parallelogram be completed. This will be double the Archimedean triangle, by which he measured the area of the semicircle. Now if you were to make individual parallelograms in this manner in the individual sectors, then the whole parallelogram divided into parts will be equivalent to the whole area of the semicircle; that is, the ratio 2:1 holds everywhere.*

Now, in the same manner, let the distances CA, GA, and so on, be set up, and the points A be connected by the conchoid* A, A, A, A, drawn through the individual points (which are potentially infinite): the figure AACD will be equivalent to all the

---

[3]Readers who hope to find a clear explanation in Ch. 59 will be disappointed: Propositions 13–15 of that chapter, where he raises this question again, are no more than qualitative gropings towards a solution. A clear solution required separation of the radial and circumsolar components of the planet's motion, and it was not until the *Epitome* of 1621 that this was done. For a good treatment of this perplexing matter, see Eric Aiton, "Infinitesimals and the Area Law'", in F. Kraft, K. Meyer, and B. Sticker, eds., *Internationales Kepler-Symposium, Weil der Stadt, 1971* (Hildesheim, 1973), pp. 285–305.

distances from A. For similarly the individual lines AG, AH have approximately made up the one parallelogram, except that the conchoid is not parallel to CD, but inclined to the radii GA, HA, EA, exactly as the distances are inclined to the circumference in the circle itself. Hence, there is nothing wrong with the conchoid's being made longer than the semicircle CD.

But EA is longer than EB, so that, if CA, GQ, HR, EB, IS, KL, DA, be taken, equal to lines determined by perpendiculars drawn from A to the lines between the points and B (as, in the circular diagram, the perpendicular AR is dropped to HB extended, defining HR which is shorter than HA), the figure between the conchoid AQRBSLA and CD would be quite equal to the figure CBBD. For the conchoid would intersect BB at the line EA, and because BA at the top and bottom are equal, and BQ is equal to LB, BR to SB, and so on, the figures BBRQA and BBALS would therefore be congruent. One of these is added to, and the other subtracted from, the equal figures CBBE and EBBD, and therefore, the whole figure between AQRBSLA and CD is equal to that between BB and CD. Hence, the small area between the two conchoids AQRBSLA and AAAAAA is the measure of the excesses of the distances from A over the distances from B—and the standard of measure is the same as that by which the parallelogram is set equal to all the distances from B.

It should also be noted that this area is not of the same breadth at places equally removed from the line *EA*, but wider below. *For in the circular diagram let HBR be extended to V, so that AH, AV correspond respectively to the upper angle HBE and the lower angle FBV, which are equal and are equally removed from the middle points E and F. And about center A with radius AV let the arc XY be drawn through AH and BH. Now if you connect A and Y,*[4] *AYR will be exactly congruent to the triangle AVR, for AV and AY and AX are equal, by construction, and are the longer sides, while VR, RY are equal and are the shorter. But from the point H outside the circumference XY two lines are drawn: HX through the center A, and HY not through the center. Therefore, HY is longer than HX, and therefore the greater AV or AX is increased by the shorter XH, and the lesser VR or RY is increased by the longer YH—and nevertheless, the whole RH remains shorter than the whole AH. Therefore, the difference between RH and AH is less than the difference between RY and AX, that is, between VR and VA. And consequently in the conchoid SA is greater, and RA less, although IE, EH are equal.* Therefore, the area between the two conchoids is not bisected by *EA*. However, it appears to be bisected by *BB*, which some geometer should investigate, who should at the same time show how to square the area between the conchoids, so that it may be expressible in numbers. In Chapter 43 below you will find a rough estimate of this area.

Granted, these general considerations of the computation of the physical equation are not yet well enough supported by the geometrical apparatus. Nevertheless, I wished to given them a preliminary treatment here, so that when all the planetary inequalities are determined (as, in particular, we presupposed that the course of the sun or earth is a perfect eccentric, which will be denied concerning Mars below in Ch. 44 and 53), this operation will not be so sharply divorced from its basis in physical theory. For, touching the theory of the sun, with which we have been concerned to this point, we in-

The area between the conchoids is of unequal breadth at places equally removed from the center.

198

---

[4]Kepler does not provide a diagram showing this connection. The adjacent figure is supplied by the translator.

troduce no discrepancy either by misjudging the area of the conchoid, which we have taken to be less than it really is, or the assumption of a perfect eccentric, in which we appear to be erring in excess—to what extent cannot yet be said, since all has not been presented. But the very things which have been rejected in this chapter as paralogisms will be taken up again below, when we shall have come to a perfectly correct way of expressing the equations, when the thing that gave rise to the paralogism will have been eliminated from that hypothesis.

I have described within a hair's breadth, through most certain observations and proofs, the cause and measure of the second inequality, which makes the planets appear stationary, direct, and retrograde. It has been shown that this second inequality itself shares something in common with the first inequality, and that the theory of the sun or earth (for Copernicus) or of the epicycle (for Ptolemy) is like the theory of the other planets. Also, the physical causes of this inequality have been found, and have been adapted to the calculation in the theory of the sun. Therefore, it is fitting that I now bring to a close this third part, as a morning task followed by lunch, with the master of the *Metamorphoses* adding his voice to mine:

> PART REMAINS OF WHAT HAS BEEN BEGUN, PART OF THE WORK IS FINISHED:
> THE ANCHOR IS CAST; HERE LET THE CRAFT LIE. [5]

---

[5]Ovid, *Ars amatoria*, I. 771–2

Part 4

of the

COMMENTARIES

ON

THE MOTIONS OF THE STAR

MARS

INVESTIGATION

OF THE TRUE MEASURE

OF THE FIRST INEQUALITY

FROM

PHYSICAL CAUSES

AND

THE AUTHOR'S OWN IDEAS

Those things demonstrated in the third part pertain to all the planets, whence they can be called, not unjustly, the "key to a deeper astronomy." We should rejoice all the more at this discovery as it becomes clearer that there could have been no way of investigating them other than through observations of the star Mars. Ptolemy did take notice of this bisection of the sun's eccentricity in Venus as well as Mercury, and for that reason introduced the eccentrics on the eccentrics, or, what is the same thing, the gyrations of the center of the epicycle: this demonstration is reserved for treatises proper to these planets. Nevertheless, the circumstances of the observations themselves, and Venus's small departures from the sun which allow it only to be observed when low down at night, would have introduced the greatest of impediments to a methodical investigation of this subject were it based upon anything other than Mars. With Mercury, the attempt was even more absurd because it very rarely emerges from the sun's rays, while it is farther from earth than Mars and Venus when they are observed at their nearest. The truth therefore would have come to us, as with Ptolemy, through a search of the broadest plains, and, as it were, by feeling about with our hands in the deepest shadows.

But now, using the star Mars as an example, we shall state how much of the first inequality, which is occasioned by the eccentric, and which is different for each planet, we owe to this common second inequality, which has been found in Part III.

CHAPTER 41

A trial examination of the apsides and eccentricity, and of the ratio of the orbs, using the observations recently employed, made at locations other than opposition with the sun, with, however, a false assumption.

In the second part, above, I tried to find the aphelion and eccentricity, as well as the distances of the star Mars from the sun on the entire circle, using acronychal observations in imitation of the ancients. And indeed, the eccentric equations corresponded closely to other observations made elsewhere than at opposition with the sun. However, the eccentricity and the distances from the sun were first repudiated by the annual parallaxes of longitude and latitude. Therefore, in order that the distances of the star from the center of the sun could be found throughout the circumference of the eccentric, the second inequality (epicyclic, for Ptolemy, or belonging to the annual orb for Tycho and Copernicus) had to be explored first, in Part 3. But now, if the planet's path were a perfect circle, the planet's first inequality, which exists by reason of the eccentric, could be investigated immediately. For in Chapter 25 above we presented a method by which, given the distances of three points of the circumference from some point within the circumference, and the angles at that point, to find the position and size of the circle with respect to that point, the center and eccentricity, along with the apsides.

Now, in Chapter 26, the distance of Mars from the center of the sun was found to be 147,750 in 14° 21′ 7″ Taurus, at the node, on 1595 October 25. Again, in Chapter 27, the distance of Mars was found to be somewhat less than 163,100 at 5° 25′ 20″ Libra, and that was on 1590 December 31. And because Mars is 41 degrees from the node, multiplying the sine of 41° by the sine of the greatest inclination, found in Ch. 13, yields an inclination at that point of 1° 12′ 40″. The secant of this exceeds the radius by 22 parts in one hundred thousand, which, in our dimensions, is 34 units. Therefore, the corrected distance at this place would be somewhat less than 163,134. Let it remain 163,100. But the secant of this inclination divided into the secant of 41° gives the secant of an arc 50″ longer. Therefore, 50″ must be subtracted from the position of Mars, so as to make it 5° 24′ 30″ Libra.

Thirdly, in Chapter 28 the distance of Mars was found to be 166,180 at 8° 19′ 20″ Virgo, on 1590 October 31, 68 degrees from the node. So the inclination at that place is 1° 42′ 40″. The secant of this is 45 units larger [than the radius], or 75 in our dimensions. Therefore, the corrected distance is 166,255. The subtraction from Mars's position, to reduce it to the ecliptic, is 16″.

These three positions, referred to the same year, 1590, and the month of October, through corrections for the precession of the equinoxes, are:

147,750	14°	16′	52″	Taurus
163,100	5	24	21	Libra
166,255	8	19	4	Virgo[1]

202

It is clear that the aphelion is nearer to the eighth degree of Virgo than to the others, because its distance is longer. *So, following [the pattern of] the demonstration in Chapter 25, let α be the center of the solar body. From it let αθ, αη, ακ be drawn in the same ratio as the distances are produced numerically here, and let all the points be joined. And let the angle καθ be 114° 2′ 12″, which is the angle from 14° Taurus to 8° Virgo. Similarly, let καη be 27° 5′ 17″, which is the angle from 8° Virgo to 5° Libra. And ηαθ is the sum of the two. For the sun is assumed to be the center of the zodiac.*

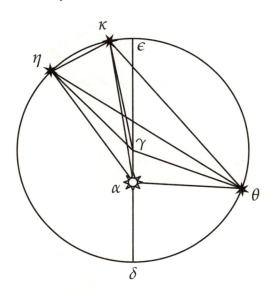

We need now to investigate the circle which passes through ηκθ, so that η, κ, θ may be three positions of the planet.

In the Ptolemaic form, α will be the earth, the center of the zodiac, and η, κ, θ three positions of the point of attachment of the epicycle. Everything else stays the same.

So, in triangle ηαθ, the angle with its sides being given, the angle αθη is found to be 20° 26′ 13″. Likewise, in καθ, angle αθκ is given as 35° 10′ 17″. Subtracting αθη from this leaves ηθκ, 14° 44′ 4″. Let γ be the center of the circle in question. Let αγ be drawn, and extended to the aphelion ε and perihelion δ; and let η and κ be joined to γ.

Now, since ηθκ stands on the circumference, and ηγκ on the center, subtending the same arc ηκ, ηγκ will therefore be twice the angle ηθκ, or 29° 28′ 8″. And where ηγ is 100,000, κη will be 50,868, which is double the sine of half ηγκ.

Now, in triangle ηακ, the angle with its sides again being given, κηα is found to be 78° 44′ 1″, and through this, κη [is found to be] 77,187 where ηα is 163,100. Therefore, in the units of which κη formerly was 50,868 and ηγ was 100,000, ηα becomes 107,486. And since ηγκ is 29° 28′ 8″, κηγ will be half the supplement, because ηγ, κγ are equal. Therefore, κηγ is 75° 15′ 56″. Subtract κηα from this. The remainder is γηα.

Thus, in triangle γηα, the angle with its sides is given. Therefore, ηαγ is known to be 38° 15′ 45″. And consequently (since αη is in 5° 24′ 21″ Libra) the line of apsides αγ will be in 27° 8′ 36″ Leo. But through the angle ηαγ the eccentricity αγ, 9768, is also found, in units of which ηγ is 100,000. Finally, in the units of which αη is 163,100, ηγ will be 151,740. But in the same units, the semidiameter of the annual orb was also 100,000. Therefore, the ratio of the orbs is that which 100,000 has to 151,740.

---

[1]There are several minor errors in the computations leading to these figures; none of them, however, effects an error of more than 20″ in the results.

How erroneous all of this is, you can gather from this: that however many times you take, instead of one or more of the distances $\alpha\theta$, $\alpha\eta$, $\alpha\kappa$ that were used, some other distance, corresponding to another place on the eccentric,[2] and found by an equally certain irrefutable line of argument, each time you do this all of those things come out differently.

And in the following chapter, [the ratio] will be found with greatest certainty to be that which $100,000$ has to $152,640$, approximately; eccentricity 9264, where the radius is $100,000$. The aphelion for 1590 October 31 was found in Ch. 16 above to be at 28° 53′ Leo, which will be confirmed, in the next chapter, to be within 11′ of the truth.

---

[2]The Latin reads, "competentem alii loco eccentrici," which means, "corresponding to a place on another eccentric." Since Kepler is showing that choosing different heliocentric longitudes and distances will produce different eccentric circles, this reading has some plausibility. However, his point seems to be that a different set of positions will result in a different eccentric circle, and to presuppose that the new position lies on a different eccentric appears to prejudge the issue unnecessarily. It therefore seems best to read "alio" instead of "alii."

CHAPTER 42

Through several observations at places other than the acronychal position, with Mars near aphelion, and again several others with Mars near perihelion, to find the exact location of the aphelion, the correction of the mean motion, the true eccentricity, and the ratio of the orbs.

You have just seen, reader, that we have to start anew. For you perceive that three eccentric positions of Mars and the same number of distances from the sun, when the law of the circle is applied to them, reject the aphelion found above (with little uncertainty). This is the source of our suspicion that the planet's path is not a circle. On this supposition, one could not use three distances to learn the others. Therefore, the distance at any particular place has to be deduced from its own observations, and especially those at aphelion and perihelion, through the comparison of which we learn the true eccentricity.

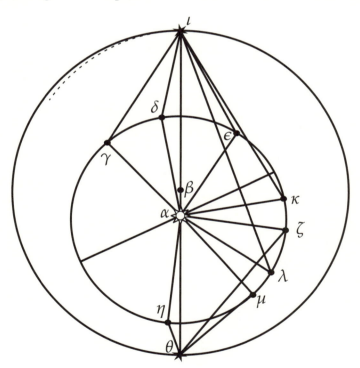

Let α be the center of the world, αβ the line of apsides, and ιθ the eccentric upon center β, with ι the aphelion and θ the perihelion. From Chapter 41, or better, from Chapter 16, we understand that the observations in which Mars is nearest to ι are these.

I. On 1585 February 17 at 10^h, the planet was seen at 15° $12\frac{1}{2}'$ Leo, with latitude 4° 16′ North.[1]

---

[1]*TBOO* 10 p. 392. The latitude is given as $4°13\frac{1}{4}'$.

II. On 1586 December 27 at 4^h in the morning, at 29° $42\frac{2}{3}'$ Virgo, lat. 2° $46\frac{3}{5}'$ N.[2]

[IIb.][3] And on 1587 January 1 at 7^h 8^m in the morning, at 1° 4' 36" Libra, Lat. 2° 54" N. And on January 9 in the morning, at 2° $51\frac{1}{2}'$ Libra, lat. 3° 6' N.[4]

[III.] On 1588 November 10, at 6^h 30^m in the morning there was 31° 27' between Mars and Cor Leonis. Mars's declination was 3° $16\frac{1}{4}'$ North. Therefore, Mars was at 25° 31' Virgo, latitude 1° 36' 45" N.[5] [IIIb.] On December 5 at 6^h in the morning there were 45° 17' between Mars and Cor Leonis. The declination was 2° 5' south. Therefore Mars was at 9° $19\frac{2}{5}'$ Libra, latitude 1° $53\frac{1}{2}'$ north.[6] These observations were not, however, confirmed by fixed stars on the other side of Mars.   204

[IV.] On 1590 October 6, at 4^h 45^m in the morning, Mars was observed at an altitude of $12\frac{1}{2}$ degrees, [and distances taken] from the Tail of Leo[7] and the Heart of Hydra,[8] with its declination. But since neither of the fixed stars was extended straight from Mars in the direction of the longitude, it happened that the two right ascensions, constructed through the given declination, disagreed by 6'.[9] This can easily happen if some very small amount is wanting in the

---

[2]*TBOO* 11 p. 66. The longitude given there, 39° Virgo, is obviously wrong.

[3]Kepler neglected to continue the numbering of observations that he had begun immediately above. Conjectural numbers have been inserted to identify the remaining observations.

[4]*TBOO* 11 p. 177 (raw observations only).

[5]*TBOO* 11 p. 276. It has been remarked that Kepler's computational errors have a way of cancelling each other or otherwise vanishing, and this is a good example. The longitude given here is 9' greater than the data show it to be. However, the longitude for December 5 is 14' less than that required by the data, and since the position given in the table below is the result of an interpolation between the two, the longitude given there is only about $3\frac{1}{2}'$ off. From there onward, the nature of the computations tends to reduce the error even further, so that in the computation of the position of the line $\alpha\iota$ it is less than one minute. For the record, $\kappa\iota$ should be at 2° 39' 10" Libra, angle $\alpha\kappa\iota$ should be 68° 16', and $\alpha\iota$ as computed from this position should be at 29° 21' 35" Leo.

[6]*TBOO* 11 p. 276.

[7]Denebola.

[8]Alphard, α Hydrae.

[9]Kepler's meaning here may be somewhat obscure. The following diagram illustrates the positions of Mars and the two reference stars, and the measured angles, which are from Brahe's observations book (*TBOO* 12 p. 44) and his fixed-star tables (*TBOO* vol. 3).

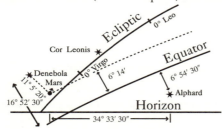

A determination of Mars's position made by measuring "straight from Mars in the direction of the longitude," as, for example, a measurement of the angle between Mars and Cor Leonis, would determine Mars's longitude directly. However, Brahe chose to ignore Cor, and measure angles from Mars to Alphard and Denebola instead. Each distance, combined with the measured declination (probably measured with a separate instrument on an equatorial mount) could be used independently to determine Mars's right ascension (i. e., position along the equator). As Kepler says, these right ascensions turn out to disagree by 6', the ascension computed

declination. Indeed, they appear not to have had much confidence in this, with the result that they measured Mars from the Tail of Leo, which is at the same longitude, all the distance being latitudinal, with the aim of knowing Mars's latitude with greater certainty from this rather than from the declination. But let the declination of 6° 14′ stand, as well as the distance from the Heart of Hydra of 34° 33$\frac{1}{2}$′. Its right ascension would thus be 168° 56$\frac{1}{4}$′.[10] Therefore, its position would be 17° 16$\frac{3}{4}$′ Virgo, latitude 1° 16$\frac{2}{3}$′ north.[11] The table of refraction for the fixed stars shows 4 minutes at this altitude, while the table for the sun shows more. Also, Virgo is rising steeply. Therefore Mars has to be put forward (eastward) about 3 minutes or (using the solar refractions) a little more, whence it was subtracted by refraction. The parallax was quite small, so it hardly removes anything from the refraction. Mars would have been at 17° 20′ Virgo.[12]

[V.] On 1600 March 5/15 at 8$\frac{1}{2}$^{h} pm Mars was at 29° 12$\frac{1}{2}$′ Cancer, latitude 3° 23′ N. And on March 6/16 at 8$\frac{1}{2}$^{h} at 29° 18′ Cancer, lat. 3° 19$\frac{3}{4}$′ N.[13]

Now the times that return Mars to the same place on the eccentric correspond to one another as follows:[14]

				observed position of Mars				And of the sun				And distances of the sun from earth from Ch. 30
1585	17	Feb.	10.	0 pm	15	12	30	Leo	9	22	37	Pisces	99,170
1587	5	Jan.	9.	31 pm	2	8	30	Libra	25	21	16	Capr.	98,300
1588	22	Nov.	9.	2$\frac{1}{2}$pm	2	35	40	Libra	10	55	8	Sagit.	98,355
1590	10	Oct.	8.	35 pm	20	13	30	Virgo	26	58	46	Libra	99,300
1600	6	Mar.	6.	17$\frac{1}{2}$pm	29	18	30	Cancer	26	31	36	Pisces	99,667

The procedure for referring the observations to the appropriate times is this. Since in 1587 the diurnal motions of Mars are decreasing, as is apparent both in Magini and in the observations on the three days, I have assumed the following diurnal motions: 17, 16, 16, 16, 15, 15, 14, 14, 13, 13, 13, 12, 12.

On 1588 November 10 the observation is 39 minutes less than the midday position of Magini. On December 5 it is 33 minutes less. And our moment is between these. Therefore, we have also taken the intermediate difference of 36′.

In 1590 the observation is solitary, and, as was seen, was itself not well made. Nevertheless, the diurnal motion in Magini is a constant 37′ over many

---

from Alphard (about 168°56′) being the greater.

[10]Recomputation shows this to be about a minute too high, an insignificant difference.

[11]Oddly, this longitude and latitude appear to be more nearly in agreement with the distance measured from Denebola. Recomputation from the given R. A. and declination gives a longitude of 17°23$\frac{1}{2}$′ Virgo, latitude 1°20$\frac{1}{2}$′ North.

[12]17°26′ Virgo.

[13]*TBOO* 12 p.224.

[14]Kepler has made adjustments upon all observations other than the first, so that the times may be 686 days, 23 hours, and 31 minutes apart.

days[15]

Now to the point: and while I have so far presented many methods of finding or testing the eccentric positions and distances, I nevertheless here follow yet another one, it being the easiest. *Let δ, ε, κ, λ, γ be positions of the earth, with δ, γ on the left and ε, κ, λ on the right side of the eccentric. And since the lines αδ, αε, ακ, αλ, αγ are given, and also the angles αδι, αει, ακι, αλι, αγι, I shall take a third element common to all the triangles, namely the side αλι, which is one of the magnitudes sought, and using this side I shall find the angles at ι and see whether they place the line αι at the same zodiacal position (except to the extent that it is moved forward in the later times by the precession of the equinoxes). From this I am going to know whether the value assumed for αι was any good.*   205

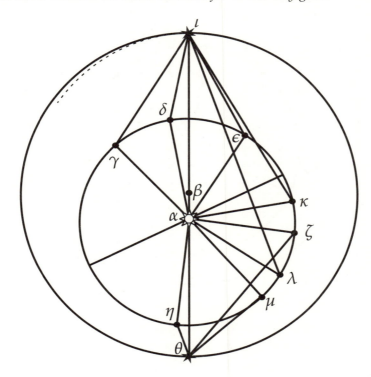

The basis of the method is this: that as αι is to [the sines of] the angles δ, ε, κ, λ, γ, so are αδ, αε, ακ, αλ, αγ to [the sines of] the angles at ι.

γα	26°	31'	36"	Pisces	δα	9°	22'	37"	Pisces	εα	25°	21'	16"	Capricorn
γι	29	18	30	Cancer	δι	15	12	30	Leo	ει	2	8	30	Libra
αγι	122	46	54		αδι	155	49	53		αει	113	12	46	

κα	10°	55'	8"	Sagittarius	λα	26°	58'	46"	Libra
κι	2	35	40	Libra	λι	20	13	30	Virgo
ακι	68	19	28		αλι	36	45	16	

The sines of these, multiplied by the earth-sun distance, and divided by the magnitude assumed for αι, 166,700, yields the sines of the angles which, added to the observed positions of Mars at γ, δ, and subtracted at ε, κ, λ, put the line αι at the following positions:

---

[15]From October 6 at 4^h 45^m to October 10 at 8^h 35^m is 4^d 15^h 50^m, which, at 37' per day, amounts to 2° 52' 25". Kepler's figures show that he added 2° 53' 30", an additional minute that is not of great consequence.

γ				δ				ε				κ				λ			
29°	28′	44″	Leo	29°	18′	19″	Leo	29°	19′	21″	Leo	29°	20′	40″	Leo	29°	20′	30″	Leo
								Should have been at											
29	30	51		29	18	0		29	19	36		29	21	12		29	22	48	
								or at											
29	29	51		29	17	0		29	18	36		29	20	12		29	21	48	

That is, the five positions ought to have differed by no more than the amount occasioned by the precession of the equinoxes.

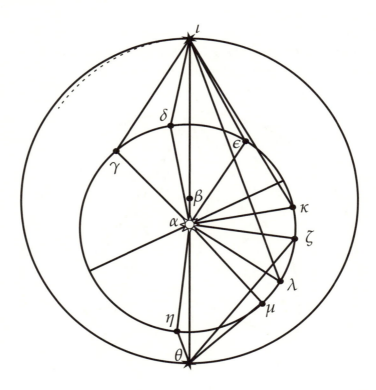

You see from the diagram that if, other things remaining the same, you will take αι to be shorter, it is going to be moved forward at γ, δ and back at ε, κ, λ, but not by an equal distance for all of them. And as soon as you do this, you will make matters worse at δ, κ, λ, and better at γ, ε. The opposite will happen if you will lengthen αι. But it is fitting to have these small errors distributed among all the positions. Therefore, the distance αι is not to be changed at all, and the planet, at the prescribed times, is at the positions last mentioned.

206   If you wish to seek confirmation using the method of Chapter 28, to test the consensus, *the points δ, ε being joined, you will find δε to be 74,058, δεα 68° 36′ 0″, εδα 67° 21′ 3″. Therefore εδι is 88° 28′ 50″, and δει is 44° 36′ 46″, and ειδ is 46° 54′ 24″. Therefore, ιε is 101,380, and εαι is 33° 58′ 33″. Therefore, in 1587 αι was at 29° 19′ 49″ Leo (we just now chose 29° 18′ 36″, the difference of one minute keeping it in agreement with other positions). Finally, αι is 166,725, and the position of κ is in agreement [with the former one].*

Since $166,666\frac{2}{3}$ is the sesquialter of the radius, $100,000$,[16] it is credible that this is the ratio of the mean distance of the earth from the sun to the greatest distance of Mars from the sun. But at present I shall base nothing upon conjecture.

---

[16]"Sesquialter" means "in the ratio of 3 to 2," which these numbers clearly are not.

And since the plane of the eccentric is inclined to the ecliptic here at an angle of 1° 48′, whose secant is 49 units above [the radius], or 82 of our present units, the most correct distance of Mars and the sun will be 166,780, as far as can be told from these observations, which, you will recall, were deduced from ones that were rather distant instead of being optimally obtained on the very days in question.

Let us now proceed to the perigee, where the catalog of observations, and a middling knowledge of the mean motion, show the following to be the nearest observations:

I. On 1589 Nov. 1 at $6\frac{1}{6}^h$ in the evening, Mars was at 20° $59\frac{1}{4}′$ Capricorn, with latitude 1° 36′ south.[17]

II. On 1591 Sept. 26 at $7^h 10^m$ at 18° 36′ Capricorn, lat. 2° $49\frac{1}{5}′$ south.[18]

III. On 1593 July 31 at $1\frac{3}{4}^h$ am at 17° $39\frac{1}{2}′$ Pisces, lat. 6° $6\frac{1}{4}′$ south,[19] and August 11 at $1\frac{3}{4}^h$ am at 16° $7\frac{1}{2}′$ Pisces, Lat. 6° $18\frac{5}{6}′$ south.[20]

The times correspond thus:

			Mars				Sun			Sun-Earth Distance
1589	1	Nov.	$6\frac{1}{6}^h$ pm	20°	$59\frac{1}{4}′$	Capr.	19°	13′	56″ Scorpio	98,730
1591	19	Sept.	$5^{\,h} 42^m$	14°	$18\frac{1}{2}′$	Capr.	5°	47′	3″ Libra	99,946
1593	6	Aug.	$5^{\,h} 14^m$	16°	56′	Pisces	23°	26′	13″ Leo	101,183

For 1591 we need to take it on faith that the diurnal motions are the same as those of Magini, since the observation is solitary. And since in Magini it moves 4° 16′ in 7 days, on September 19 at $7\frac{1}{6}^h$ Mars will be at 14° 20′ Capricorn, and at $6\frac{1}{6}^h$ it will be at 14° $18\frac{1}{2}′$ Capricorn. About the station on July 16 or 17, Mars was about 1° 16′ farther forward in the calculation than in Magini. Now, on September 26, it is still 0° 53′ farther forward. Therefore, over 70 days the difference has been diminished by about 23 minutes. So if we interpolate, this difference will be about 2 minutes greater on September 19. We shall therefore believe that at our given moment Mars is at 14° 20′ Capricorn.

In 1593 Mars left its station. And on midnight of July 30 the position of Mars disagrees with Magini's midday position by 3° $25\frac{1}{2}′$, and on August 10 by 3° $59\frac{1}{2}′$, so that the difference is increased, but gradually less so. Therefore, I have assumed a difference of 3° 46′ for August 6, so that at $1\frac{3}{4}$ hours after midnight it would be at 16° 52′ Pisces. And the diurnal motion was 10′. This is 8 hours 30 minutes past our time, which would account for about 4′ of Mars's retrograde motion. Therefore, at our time it was at 16° 56′ Pisces. It is certain that (on this point at least) we are no more than one minute high or low.

It was not observed more frequently at perigee. For in 1595 its arrival at perigee fell in the middle of summer, when twilight lasts all night in Denmark. In 1597 Tycho Brahe was travelling. And when it is near the sun in its winter semicircle it is long hidden, since its speed is not much less than the sun's.

207

---

[17]*TBOO* 11 p. 335 (raw observations).

[18]*TBOO* 12 pp. 144–5 (raw observations).

[19]*TBOO* 12 p. 284.

[20]*TBOO* 12 p. 285.

In the diagram, let Mars's eccentric position be θ, the positions of the earth, ζ, μ, η; and let

ζα	be	19°	13′	56″	Scorpio	μα	5°	47′	3″	Libra	ηα	23°	26′	13″	Leo
ζθ		20°	59′	15″	Capricorn	μθ	14°	18′	30″	Capricorn	ηθ	16°	56′	0″	Pisces
							or	20′							

Therefore, αζθ *is* 61   45   19   αμθ 98   31   27   αηθ 156   30   13

                                                    or   32   57

When the length of the common side αθ is assumed to be 138,400, its position comes out thus:

Through ζ   29° 55′ 20″ *Aquarius*   μ 29° 53′ 6″ *Aquarius*   η 29° 59′ 10″ *Aquarius*

                                        or   54′ 36″

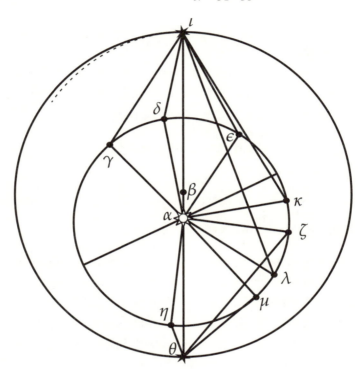

But if it was 55′ 20″ at ζ, it should have been 56′ 56″ at μ, and 58′ 32″ at η, for that is the amount of the precession of the equinoxes. It can thus be seen from the diagram that the line αθ determined through η goes too far forward, and through μ, ζ, too far back, in relation to that through η. Other things remaining unchanged, this happened because I assumed too small a value for αθ. Therefore, if I make it a hundred parts longer, namely, 138,500, the following positions come out:

From ζ 29° 57′ 10″ *Aquarius; from* μ 29° 55′ 36″   *from* η 29° 58′ 17″ *Aquarius*[21]

                        or 29° 57′ 6″ *Aquarius*

So now the positions of αθ have been made to be too close to one another,[22] *and the*

---

[21] Kepler's reasoning here is faulty. Consider, for example, the triangle αζθ. The line αζ is fixed in position and magnitude, and the angle αζθ is given. Therefore, ζθ is given in position. Kepler found that θ as determined from point ζ was too far forward; that is, that the point θ was placed to the left of its true position. This means that αθ was too long, for when θ is moved to the right along ζθ, αθ becomes shorter. Recomputation of Kepler's figures confirms this conclusion: they are way off. Correct values are: from ζ, 29° 53′ 18″ Aquarius; from μ, 29° 50′ 31″ (or 29° 51′ 47″) Aquarius; and from η, 29° 59′ 58″ Aquarius. Figures in agreement with Kepler's are obtained by decreasing αθ by 100.

[22] Here may be seen the probable cause of Kepler's error (see the preceding footnote): he seems to have thought of αθ as fixed in position, while the points ζ, μ, and η are moved around as the length of αθ is changed. Actually, ζ, μ, and η are fixed, and the three different positions of θ determined by them are moved around as the length of αθ is changed.

error is now more so in closeness than it was before in remoteness. Therefore, the most correct length of αθ will be about 138,430.

At this point the plane is inclined 1° 48′ (as it was before at the opposite position), and the secant is 49 units greater than the radius. But as 100,000 is to 138,430, so is this 49 to 68. Therefore, the correct length of the radius is approximately 138,500, at least from these observations involving long interpolations.

Investigation of the apsides, from the above

With all three observations taken into account, let the position of the line αθ on 1589 November 1 at $6\frac{1}{6}^h$ pm be taken as 29° 54′ 53″ Aquarius, so that in 1591 it would be 29° 56′ 30″, and in 1593, 29° 58′ 6″ Aquarius. The vicarious hypotheses of Chapter 16 shows it to be at 29° 52′ 55″ for the first of the times.   208

But previously and in like manner we took αι on 1588 November 22 at 9^h $2\frac{1}{2}^m$ to be 29° 20′ 12″ Leo.

From 1588 November 22 at 9^h $2\frac{1}{2}^m$ to 1589 November 1 at 6^h 10^m are 344 days diminished by 2^h $52\frac{1}{2}^m$, while a whole revolution to the same fixed star has 687 days diminished by 0^h 28 min. Therefore, our interval appears to exceed half the periodic time by a few hours.

Consider:

	Days		Hours		Min.	
343	Days	11	Hours	46	Min.	Half the period
343		21		$52\frac{1}{2}$		Our interval
Excess		10		$6\frac{1}{2}$		

And from the position at the earlier time, 29° 20′ 12″ Leo, to the position which Mars held at the later time, 29° 54′ 53″ Aquarius, is 180° 34′ 41″, or 180° 33′ 53″ with the precession of 48″ subtracted. Therefore, if the excess of 33′ 53″ beyond the semicircle is is sufficient for the 10 hours $6\frac{1}{2}$ minutes from Mars's diurnal motion on the eccentric, the aphelion would consequently be understood to be at 29° 20′ 12″ Leo.

But we know the diurnal motions of Mars on the eccentric near apogee and perigee from the distances just found and from the demonstrations of Chapter 32. For the diurnal motions are approximately in the [inverse] duplicate ratio of the distances. At apogee the diurnal motion is about 26′ 13″, at perigee 38′ 2″, since the mean diurnal motion is 31′ 27″.

Consider, then: if Mars, in moving from its apogee point, expends half its periodic time, at the end of this time, having traversed exactly 180 degrees, it is going to be at the perigee point. But now if it begins this space of time one day after it was at apogee, it will begin its course 26′ 13″ beyond apogee and will end it at 180° 38′ 2″. Therefore, in half the time it will traverse 11′ 49″ more than half the path. The opposite will happen if it were to begin one day before apogee.   209

Therefore, since our time, too, shows an arc greater [than a semicircle], our aphelion also should be moved forward. First, we shall credit half of our hours to the time before aphelion, and half after perihelion. The planet will then begin from 5′ 16″ before aphelion, which is thus put at 29° 25′ 28″ Leo, and it will come to 8′ 1″ after perihelion, the amount of travel being 13′ 17″ beyond 180°. But its path was seen to be 33′ 53″ beyond 180°. Therefore, it is still faster by 20′ 36″. Therefore, since to increase the path by 11′ 49″, one day, or the promotion of the planet to 26′ 13″ beyond aphelion, is needed, how much will the planet be promoted from aphelion to increase the path by 20′ 36″?

The rule of proportions shows it to be 1 day 17^h 54^m, or a distance from aphelion of 45′ 42″. Therefore, the aphelion is to be moved forward 45′ 42″ from the position we just gave it, 29° 25′ 28″ Leo.

So it will fall at	28°	39′	46″	Leo
On 1588 November 22, above	28	50	44	Leo
Difference		10′	58″	.

To which of the investigations of the aphelion one ought to give more trust, is uncertain. For it can easily happen that in the positioning and assuming of the lines $\alpha\iota$, $\alpha\theta$ we have erred by 4 minutes, two for the one and two for the other, owing to difficulties in the observations. And this is all that needs to be accumulated, through the compounding of errors, to change the aphelion by 11 minutes. Here, however, it is reasonable for us to trust the present operation.

Correction of the mean motion

When the aphelion is changed, the mean motion is changed as well. For at the same time at which in the previous investigation of the aphelion Mars is thought to fall at aphelion, with no equation, it has now passed the aphelion by 11 minutes. Therefore, it has an equation of 4 minutes, subtractive. Thus in its mean motion it has passed that original position by 4′.[23]

Investigation of the eccentricity

First, the distances found previously should be corrected, if necessary, to the extent that they are some small amount distant from the apsides just found, the aphelia by 40 minutes, perihelia by 75 minutes. But there is no perceptible change so close to the apsides.

Therefore, the	Aphelial distance	is	166, 780	($\alpha\iota$)
	Perihelial distance		138, 500	($\alpha\theta$)
	Sum		305, 280	($\iota\theta$)
	Half		152, 640	(semidiameter $\iota\beta$)
	Eccentricity		14, 140	($\alpha\beta$)

And as 152, 640 is to 100, 000, so is 14, 140 to the eccentricity 9264. But half the eccentricity of the equating point was 9282. The difference of 18 is clearly of no importance. You see how precisely the eccentricity of the equating point is to be bisected in Mars in order to establish the distance between the centers

---

[23] A diagram may be helpful here.

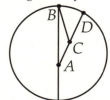

Suppose the aphelion is originally at B, but it is moved back 11′ to D. The corresponding equation, the angle ABC, is 4′. So the line CB, the mean motion, is now 4′ farther forward than its original position along the line AB, at the same time as before.

of the eccentric and the world. Above, in Chapter 32, I took this to be fundamental, and in the following chapters postponed its demonstration. Now, however, that obligation is discharged.

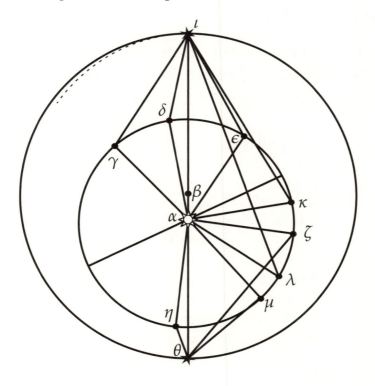

Chapter 43

On the defect in the equations accumulated by bisection of the eccentricity and the use of triangular areas, on the supposition that the planet's orbit is perfectly circular.

210   What was proved in Part III concerning the bisection of the eccentricity in the theory of the sun has now likewise been demonstrated with perfect certainty for Mars. And now that our evidence of this is complete, it would at last be time to proceed to the physical theories of Chapters 32 and the following, seeing that they are going to apply to all planets in common, had I not seen fit to present them earlier. I did so because there, in the theory of the sun or earth, the procedure for computing the equations from physical causes had to be completed with full perfection, and because I knew that where that method of constructing the equations is to be applied to the theory of Mars, much more difficult physical theories were to follow.

Now when the true configuration of the orbits is found, the eccentric equations, upon which alone the vicarious hypothesis found in Chapter 16 has hitherto depended, must necessarily follow by the same means. We shall therefore explore it in turn here.

Therefore, following what was demonstrated in Chapter 40 (all of which, in every detail, is to be understood as holding here), let the orbit of the planet, in accord with the well-worn opinion, be circular, even though Ch. 41 has just urged us to doubt it. *Therefore, at the eccentric anomaly of 90° the eccentricity 9264 found in Ch. 42 will be the tangent, which will give the optical part of the equation, 5° 17′ 34″. And since at the eccentric anomaly of 90° the area of the triangle is right-angled, the radius multiplied by half the eccentricity, 4632, gives the area of the triangle, 463,200,000. Now as the area of the circle, 31,415,926,536, is to 360 degrees or 1,296,000 seconds, so is this area just found, 463,200,000, to 19,108″, or 5° 18′ 28″, the physical part of the equation. Consequently, the whole equation is 10° 36′ 2″, so that to the mean anomaly of 95° 18′ 28″ corresponds the equated anomaly of 84° 42′ 26″.*

But according to the method of Chapter 18, the vicarious hypothesis, accurate enough for the longitudes, shows us that to the mean anomaly of 95° 18′ 28″ there ought to correspond the equated anomaly of 84° 42′ 2″. The difference is 24″.

Now let our eccentric anomaly be taken as 45° and 135°. *And as the whole sine is to the sine of these angles, so is 19,108″, the area of the greatest triangle of the equation, to the area at this position, 13,512″, or 3° 45′ 12″, so that by addition of this the physical part of the equation to the eccentric anomaly the mean anomalies of 48° 45′ 12″ and 138° 45′ 12″ are constructed. But from the given sides of the given angles, the angles of equated anomaly corresponding to these mean anomalies come out to be 41° 28′ 54″ and 130° 59′ 25″.* But by the vicarious hypothesis, as in Chapter 18 of this work, the same simple anomalies of 48° 45′ 12″ and 138°

45′ 12″ being taken, the equated anomaly for the former comes out to be 41°
20′ 33″, less than by the area of the triangle, the excess being 8′ 21″; and for
the latter, 131° 7′ 26″, more than by the area of the triangle, the defect being
8′. So, since it is certain that an error of this magnitude cannot be attributed
to our vicarious hypothesis, I had to accept that this procedure for finding the   211
equations was still imperfect.

Indeed, in Chapter 19 as well, when I tried out the bisection on Mars and
computed the equations using a motionless point of the equant in the Ptolemaic manner, a difference was found at about 45° of eccentric anomaly of
nearly the same amount, but in the opposite direction. For in the upper quadrant, the planet was closer to the aphelion, and in the lower to the perihelion,
than it should have been; while here in the upper quadrant it was farther from
the aphelion, and in the lower from the perihelion, than it should be. And so
in the upper quadrant it is moving too swiftly away from the aphelion, and
the same from the perihelion below. Therefore, it is slower than it should be
in the middle elongations.

I believe it has just occured to the reader that the cause of these errors might
perhaps lie in the flaw to which this operation with areas is subject, mentioned
in Chapter 40: that the areas are not equivalent to the distances that modify
the swiftness and slowness. But the present error cannot arise thence. For first
of all, the excess of the sum of the distances over the area of the circle is small:
just about as small, that is, as the little space between the conchoids. Then,
too, the area makes all the distances a little smaller than they should be, and
most of all those that are at the middle elongations. So if any error flows from
this, it lies in our not having made the planet take enough time in the middle
elongations. But the errors we are now seeing are in the opposite direction,
for we have made the planet take too much time in the middle elongations.

Refutation of false causes of this imperfection

The same can be raised in objection to anyone who might conceive a suspicion that the error arose because we rejected the double epicycle of Copernicus and Tycho, which makes the
orbit of the planet oval, and took up the Ptolemaic perfect circle in the present account. For it
was said at the end of Chapter 4 that the Copernican orbit moves outwards from the center by
246 parts, which would only increase the error,
rather than making an incursion towards the
center, as would suit our purposes, since we are
now following the idea that the time increments
are proportional to the distances.

But to make it clear to the eye that the area of
the conchoid of Ch. 40 is made very small, con-
sider that the secant of the angle 5° 19′ (the maximum optical equation) is
100, 432, which is the line *EA*. So from this excess of 432, which is the small
line *BA*, part of the line *EA*, we will be able to get an approximate idea of the
sum of all these excesses (such as, *QA, RA, BA, SA, LA*) in this way.

Estimate of the area between the two conchoids.

The secant of 89°, and its tangent, taken together, are as great as the sines
of all degrees of the whole semicircle, as Cardano helps us see in the part of

A short cut for finding the sum of all the sines at once.

De subtilitate in which he explains the properties of the circle.[1] A proof of this is given by Justus Byrgius.[2]

212   *Therefore, if all our remaining excesses (other then the greatest, 432) were [to the greatest][3] as [all] the sines in one semicircle are to the semidiameter, then as $100,000$ is to the sum of the secant and the tangent of 89° (that is, $11,458,869$), so, approximately, would 432 be to $49,934$, the approximate sum of all the excesses at integral degrees of the semicircle.[4] For the excesses of the distances in the upper quadrant are longer than those excesses of the secants, to about the same extent that they are shorter in the lower quadrant.*

By what ratio the excesses of the distances of the points on a circle from an eccentric point, or the breadth of the space between the conchoids, may grow.

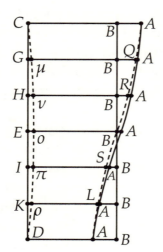

But nevertheless, it is not true that the excesses QA, RA, SA, and so on, are to one another as the sines of the corresponding number of degrees.[5] Instead, they are approximately in the duplicate ratio of the sines. *As for example, the sine of 90° is twice the sine of 30°. Now the optical equation of 90° is 5° 19′, and half of its sine gives an arc which is likewise about its half, that is, 2° 39′ 15″, for the optical equation at 30° of eccentric anomaly, whose secant is $100,107$. And here 107, the excess of the secant over the radius, is about one fourth of the former, 432, while the sine of 30° would be half the sine of 90°.* Some geometer should see whether this argument be demonstrable. For me it suffices at present to answer

---

[1] Girolamo Cardano (1501–1576), physician, mathematician, and philosopher. Although perhaps best known today for his pioneering work on probability theory, his reputation among his contemporaries rested chiefly upon his unorthodox physical theories, presented in *De subtilitate rerum*, (Paris and Nuremberg, 1550, and many other editions), and *De rerum varietate*, (Basel, 1557). Cardano was also famous as an author of astrological works. Both *De subtilitate* and *De varietate* are included in Vol. III of his *Opera omnia* (10 vols., Lyons, 1663). The passage cited is in *De subtilitate* Book XVI (*De scientiis*), ed. 1550 p. 303; *Opera* III p. 593.

[2] Justus Byrgius, or Jost Bürgi (1552–1632), Swiss instrument maker and mathematician, who, like Kepler, was at the time employed by Rudolph II. The proof was thus most likely given to Kepler personally: nothing is known of it other than this one mention. (See Rudolf Wolf, *Johannes Kepler und Jost Buergi, Ein Vortrag*, Zurich: Schulthess, 1872.) In the Kepler manuscripts in St. Petersburg, Vol. 5, there is a section titled, "Byrgii Arithmeticam," with Kepler's annotations and corrections. Of this, Frisch (*Joannis Kepleri Astronomi Opera Omnia* III p. 497) writes that the main aim of the book was trigonometrical, and that only part of it survives in Kepler's papers.

As Kepler implies, Cardano provided no proof of the theorem; however, Frisch (*loc. cit.* notes that it had been proven by Archimedes in *On the Sphere and the Cylinder* I.21. Kepler gave a further elaboration of the theorem in a letter to David Fabricius of 11 October 1605.

[3] Supplied to make sense of the proportion.

[4] Here Kepler assumes as a working hypothesis that the small segments QA, RA, SA, and so on, between the conchoids, are to one another as the sines of their eccentric anomalies, and then uses the Cardano/Bürgi rule to compute an approximate sum. The hypothesis is, however, false, as he points out in the next paragraph: from a single example, he concludes that the small segments are proportional to the *squares* of the sines of the eccentric anomalies. Caspar's analysis using modern techniques (KGW 3 p. 471) shows this to be very nearly correct.

[5] *Sinus aliquotorum graduum.* The presence of an adjectival ending in the usually uninflected *aliquot* (some whole number) suggests an unusual usage. The context suggests that Kepler intends the number of sines to be equal to the number of the "excesses."

those very small questions with which I am occupied.

Therefore, to arrive at 432, parts are accumulated that are not proportional to the sines, but are always smaller, and at the 45th degree or thereabouts are but their halves. Before that point they are less than the halves, so that about 30° they are only the fourths, and at length become imperceptible.

And so, of the sum of 49,934, we retain only one seventh, or about 7000. This is also shown empirically, by computing all the distances degree by degree and adding them up.[6]

And because one distance of 100,000 has the value of 60′, this little sum has a value of no more than $4\frac{1}{5}'$,[7] which is nonetheless spread all around the circumference, so that about 45° and 135°, where it is greatest, this tiny error turns out to be imperceptible even in Mars.

Consequently, we must seek another occasion for this discrepancy.

---

[6]It is hard to see how Kepler could have believed that the mysterious figure of 7000 (which he may have hastily supposed to be the square root of 49,000) is confirmed by experience, since he himself computed the sum (p. 359 of this translation), and found it to be 37,781.

[7]Should be $22\frac{2}{3}'$.

CHAPTER 44

The path of the planet through the ethereal air is not a circle, not even with respect to the first inequality alone, even if you mentally remove the Brahean and Ptolemaic complex of spirals resulting from the second inequality in these two authors.

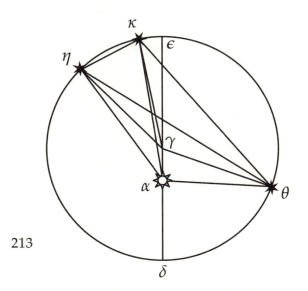

213

With the eccentricity and the ratio of the orbs established with the utmost certainty, it might appear strange to an astronomer that there remains yet another impediment in the way of astronomy's triumph. And me, Lord knows, I had triumphed for two full years. Nevertheless, by a comparison of the things which have been established in Chapters 41, 42, and 43, preceding, it is readily apparent what we are still lacking. The positions of the aphelia, eccentricity, and the ratio of the orbs, as constituted in the several places, differed greatly from one another. Nor were the computed physical equations in agreement with the observations (which the vicarious hypothesis represents). Let the diagram of Chapter 41 be brought back. *And because, in that diagram, in units of which* $\gamma\eta$ *was* 151,740, $\gamma\alpha$ *would have been* 14,822, *when* $\gamma\alpha$ *and* $\gamma\eta$ *or* $\gamma\epsilon$ *are added,* $\alpha\epsilon$ *would be* 166,562.[1] *In Chapter 42 this was found to be* 166,780. *Likewise, when* $\gamma\alpha$ *is subtracted from* $\gamma\delta$, *the remainder,* $\alpha\delta$, *would be* 136,918, *which in Chapter 42 was found to be, in all,* 138,500.

Again, the true length of the lines $\gamma\epsilon$, $\gamma\alpha$, $\alpha\epsilon$, *and* $\alpha\delta$ *was found in Chapter 42.*[2] *If, therefore, what was supposed and used in Chapter 41 is true, that the path of the planet is a circle, it is not difficult to say how much* $\alpha\kappa$, $\alpha\eta$, $\alpha\theta$ *ought to be. Since in Oct. 1590* $\alpha\epsilon$ *was at* 28° 41' 40" *Leo, and* κ, η, θ *are as given in Chapter 41,*[3] *the angles* $\kappa\alpha\gamma$, $\eta\alpha\gamma$, $\theta\alpha\gamma$ *will be given. Therefore, the optical equation will also be given:*

---

[1]These numbers are derived from those at the bottom of p. 320. Keep in mind that Kepler is using them as an example of bad numbers. The radius $\eta\gamma$ is given as 151,740. The number 14,822 derives from multiplying the given (incorrect) eccentricity, 9768, by the (incorrect) ratio of the circles of Mars and earth, 151,740 : 100,000.

[2]The Latin for "length" is "longitudo," which can also mean "longitude." However, it is the length of these lines (the radius and eccentricity of the hypothetically circular orbit) that is given at the end of Chapter 42 (p. 330). Kepler is confident that these lengths are correct at aphelion and perihelion; the present investigation will show that the distances of η, κ, and θ from α, as defined by the circle, are greater than those derived from the observations.

[3]See the table at the top of p. 320

$\alpha\kappa\gamma$ 0° 53′ 13″, $\alpha\eta\gamma$ 3° 10′ 24″, $\alpha\theta\gamma$ 5° 8′ 47″. *And as the sine of these angles is to the truest eccentricity* $\alpha\gamma$, 14, 140,[4] *so are the sines of* $\kappa\gamma\epsilon$, $\eta\gamma\epsilon$, $\theta\gamma\alpha$ *to* $\alpha\kappa$, $\alpha\eta$, $\alpha\theta$.

They come out thus:	$\alpha\kappa$	166, 605	$\alpha\eta$	163, 883	$\alpha\theta$	148, 539
But in the observations they were found to be: [5]		166, 255		163, 100		147, 750
Difference:		350		783		789

If anyone wishes to attribute this difference to the slippery luck of observing, he must surely not have felt nor paid attention to the force of the demonstrations used hitherto, and will be shamelessly imputing to me the vilest fraud in corrupting the observations of Brahe. I therefore appeal to the observations of subsequent years, at least those that experienced observers made. For if in any respect I have given free rein to my inclinations in one direction, it will only go so much the farther into error on the other side. But there is no need of this. I am addressing this to you who are experienced in matters astronomical, who know that in astronomy there is no tolerance for the sophistical loopholes that beset other disciplines. To you I appeal. You see at κ a small defect from the circle, at η, θ on both sides, a rather large one, enough so that we cannot excuse it by uncertainties in observing (for in Chapter 42 I reckon an uncertainty of perhaps 200, or at most 300 units).

What, then, is to be said? Could this actually be the situation described in Chapter 6 above, in which by transposition of the reference point from the sun's mean motion to its apparent motion I set up another eccentric that makes an excursion towards the side of the sun's apogee? By no means. For there, it would approach from the one side by the same amount as it moves away on the other. Here, however, you see that the planet approaches the center from the circular orbit on both sides. This is confirmed by many other observations, some of which follow below in Chapters 51 and 53.

Clearly, then, [what is to be said] is this: the orbit of the planet is not a circle, but comes in gradually on both sides and returns again to the circle's distance at perigee. They are accustomed to call the shape of this sort of path "oval." 214

This same thing is also proved from Chapter 43 preceding. There it was *Second argument.* supposed that the area of a perfect eccentric is very closely equivalent to all the distances of the equal parts of the circumference of that eccentric from the source of the motive power, however many they are. Thus, the parts of the area measure the amounts of time which the planet spends on the parts of the corresponding eccentric circumference. But if that area about which the planet marks a boundary is not a perfect circle, but is diminished at the sides from the amplitude it has at the apsides, and if nevertheless this area circumscribed by an irregular orbit still measures the times which the planet takes to traverse the whole and its equal parts, then this diminished area measures a time equal to that measured by the previous undiminished area. So the parts of the diminished area nearest aphelion and perihelion will measure a greater time, because in those regions the diminution is narrowest, but the parts at the

---

[4]See "Investigation of the Eccentricity" on p. 330.

[5]These distances are taken from Chapter 41, p. 320, but were originally derived in Chapters 28, 27, and 26, respectively.

middle elongations measure less time than before, because the greatest diminution in the whole area occurs there. So if we now use the diminished area in adjusting the equations, the planet will become slower near aphelion and perihelion than it was in the previous faulty form of equation, and swifter near the middle elongations, because here the distances are lessened. Therefore, the times, when they are abstracted from the area and adjusted upward and downward, will be accumulated at aphelion and perihelion in much the same manner as, if one were to squeeze a fat-bellied sausage at its middle, he would squeeze and squash the ground meat, with which it is stuffed, outwards from the belly towards the two ends, emerging above and below his hand.[6]

And indeed, if contraries remedy one another, this is plainly the aptest medicine for purging the faults under which, in Chapter 43 above, our physical hypothesis was perceived to be laboring. For the planet is going to be swifter at the middle elongations, where previously it was perceived to be going slower than it should, and it will be slowed down above and below, near the apsides, where previously it did violence to the equations belonging to the eighths of the period through its excessive fleetness.

This, then, is the other argument by which it is proved that the orbit of the planet really is deflected from the established circle, making ingress towards the sides and the centre of the eccentric.

But for all that, this argument still did not have enough effect upon me to let me go beyond it and think about the planet's departure from the orbit. When I had sweated for the longest time trying to reconcile equations of this sort, I was finally discouraged by the absurdity of the measurements, and abandoned the whole enquiry until I was informed by the distances (found in the way shown in Chapter 41) about the departure from the [circular] orbit, and once more took up this problem of the equations.

And from this, what I promised I would prove, in Chapters 20 and 23 above, is now done: that the orbit of the planet is not a circle but of an oval shape.

---

[6]This realization that the physical theory required a reshaping of the orbit was the decisive insight that convinced Kepler to abandon perfect circularity. It may be worth noting that in the Mars Notebook, in a passage written some time before his acceptance of what he at first called the "circulus Martis ovalis," he noted that the distances at the sides of the orbit appeared to be somewhat less, but rejected this as implausible because it would make the orbit oval! (*KGW* 20.2 p. 547 l. 15–16). Kepler's first realization of the implications of the physical theory appears in *KGW* 20.2 p. 576 ff. A transcript and translation of this section of the Mars Notebook is available in W. H. Donahue, "Kepler's First Thoughts on Oval Orbits: Text, Translation, and Commentary," *Journal for the History of Astronomy* 24 (1993) pp. 71–100.

CHAPTER 45

On the natural causes of this deflection of the planet from the circle: first opinion examined.

When I was first informed in this manner by Brahe's most certain observa- tions that the orbit of the planet is not exactly circular but is deficient at the sides, I judged that I also knew the natural cause of the deflection from its footprints. For I had worked very hard on that subject in Chapter 39. And I suggest to the reader that he reread that entire chapter carefully before going on. For in that chapter I assigned the cause of the eccentricity to a certain power that is in the body of the planet. It therefore followed that the cause of this deflecting from the eccentric circle should also be ascribed to the same body of the planet. But then what they say in the proverb, "A hasty dog bears blind pups," happened to me. For in Chapter 39, I worked very energetically on the question of why I could not state a probable enough cause for a perfect circle's resulting from the orbit of the planet, as some absurdities always had to be attributed to the power that has its seat in the planet's body. Now, having seen from the observations that the planet's orbit is not perfectly circular, I immediately succumbed to this great persuasive impetus, believing that from those things which were called absurd in fabricating the circle in Chapter 39, now transmuted into a more probable form, an orbit of the planet that would be both correct and in agreement with the observations would be effected. If I had embarked upon this path a little more thoughtfully, I might have immediately arrived at the truth of the matter. But since I was blind from desire, and did not pay attention to each and every part of Chapter 39, staying instead with the first thought to offer itself—a wonderfully probable one, owing to the uniformity of the epicyclic motion—I entered into new labyrinths, from which we will have to extract ourselves in this Chapter 45 and the following ones all the way to 50.

Let the diagram from Chapter 39 be repeated. The weaker opinion in that chapter was that, in order to describe a perfect circle, the planet effects an epicycle by its inherent force, thus disengaging its body from the ray of power from the sun. As, for example, if the ray of power from the sun be AC, and move forward at an unequal pace from AC to Aγ, while the planet be initially at C, and from that time forth, by its inherent force, it disengage itself from [the ray] AC or Aγ. Thus, at the time when AC comes to Aγ, the planet from C or γ would come to D,

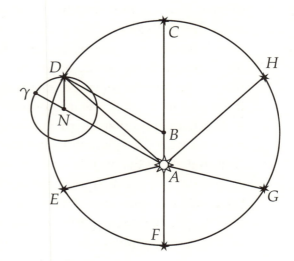

and would also do this at a nonuniform pace, more swiftly or slowly in the same proportion as [the length of] AC. For in this fashion the line ND through the center of the epicycle and the planet remains ever parallel to the line AB. However, I said in Chapter 39 that it appeared absurd to me that the planet [in moving] from γ to D at a nonuniform pace disengages itself from the ray of the solar power, and thus accommodates itself by its own force to the extrinsic force from the sun, and has foreknowledge of its speed and the decrease thereof. Therefore, to avoid this absurdity, let AC still go nonuniformly, but let the planet go uniformly from γ to D. Let us see whether what follows is anything like what we have proved in the preceding chapter from the observations.

216

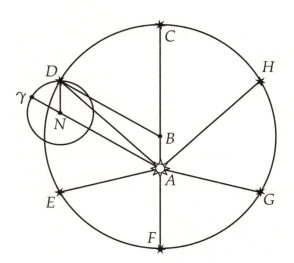

While the center of the epicycle N and its aphelion [in moving] from the line AC to Aγ will be slow from C to γ, it being near the eccentric's aphelion C, let the planet [in moving] from γ to D be supposed not to be slow but to proceed with its mean motion. Consequently, the angle γND will be greater than the angle γAC. So ND will not be parallel to AB but will be inclined towards AC. Thus the planet D will not stay on the circle which it began to describe from C, the one, that is, which goes through CF, but will encroach from the circumference D and the parallel ND towards CA. And this same thing was also affirmed in the previous chapter by the distances AD computed from the observations, namely, that they do not reach all the way to the circumference of the circle CF. This same thing was also affirmed by the physical equations constructed through the summation of the distances AC, AD, namely, that the planet ought to be faster at the sides of the eccentric; that is, that its distances from the sun ought to be supposed smaller. Therefore, since this consensus brought to bear a considerable force of persuasion, I concluded forthwith that the planet's incursion at the sides is the result of this: that the power moving the planet and administering the distances according to the law of the circle supersedes the power of the sun, in that the former made equal progress in equal times thus sending the planet down uniformly towards the sun according to the law of the epicycle, while the latter, in its varying degrees [of speed obtained] through the varying distances, moved the planet in its care forward nonuniformly, and more slowly when it is high. It thus happened that the distances of equal arcs on the epicycle were accumulated near the aphelion C and the perihelion F, and were more sparsely scattered about the middle elongations. In this way, all the shorter ones were drawn back upwards from the correct [circular] distances from perihelion to the place of longer ones. That error therefore began to become rooted in me which I had happily begun to refute in Chapter 39 above, that it is a property of the planetary power to lead the body of the planet around in the path of an epicycle. Had the diameter of the epicycle ND remained equidistant from

AB, I could have shed my erroneous opinion, and could have ascribed (as is perfectly correct) all promotion in zodiacal longitude to the sun, leaving to the planet only the reciprocation on the diameter $\gamma\zeta$, as in part of Chapter 39. But because the observations testified that the diameter of the epicycle is inclined in the middle elongations, this error concerning the motion of the planet on the circumference of the epicycle, whose motion would be regular measured with respect to the line $AN\gamma$ going from the sun A through the center of   217 the epicycle N, was admirably confirmed in me. Think yourself, reader, and you will feel the force of the argument. For I did not think it possible for the planet's orbit to be rendered oval in any other way.

When, therefore, these things occurred to me, quite certain of the quantity of the incursion at the sides (that is, that the numbers would be in agreement), I celebrated another triumph over Mars. Nor did it appear to me any difficulty, if there were some discord among the numbers, to dissipate it through some slight adjustment in the equations of the center[1] all around, so as to make it imperceptible.

And we, good reader, can fairly indulge in so splendid a triumph for a brief day's respite (for the following five chapters, that is), meanwhile repressing the rumors of renewed rebellion, lest its splendor die before we enjoy it. If anything will be left of it afterwards, we shall go through it in the proper time and order. We are merry indeed now, but [will be] active and energetic then.

---

[1] τῷ προσθαφαιρεῖν

CHAPTER 46

How the line of the planet's motion can be described from the opinion of Chapter 45, and what its properties are.

In the preceding chapter a cause was stated by which it could happen that the planet depart from a circular orbit. However, the geometrical description of the path cannot be carried out using this model. For the epicycle is inclined according to the length of the distances, while the multitude and length of the distances is in turn dependent upon the rotation of the epicycle. *And because the sum of the distances is contained in the plane of the eccentric, as was demonstrated in Chapter 40, that sum cannot be found unless the epicycle be transformed into an eccentric. But it was demonstrated in Chapter 2, and repeated in Chapter 39, and used in Chapter 40, that if a concentric be described about center α with semidiameter equal to βδ, and on it an epicycle with semidiameter αβ; and then about center β an eccentric δλ with eccentricity αβ; and afterwards the circumferences of both the epicycle and the eccentric δλ be divided into similar parts; the distances of the points of division, both of the epicycle and the eccentric, from the chosen point α are respectively equal to one another in length. On this premise, since in Chapter 40 we posited an eccentric to present a plain and easy demonstration, and a method of computing the distances, here, too, we can examine the distances on the eccentric, even though we are supposing them to be meted out by the uniform motion of the planet's epicycle. This procedure seems to open a way to us to a geometrical description of the planetary path that follows from the hypothesis of Chapter 45. Let us therefore say, for the sake*
218 *of understanding, that in the circuit of the epicycle the planet makes digressions from the sun α of the same magnitude as if it were on the circumference of a perfect eccentric δλ (and let this be a semicircle defined by the straight line λαβδ) describing equal arcs in equal times, such as δε, εζ, ζθ, θι, ικ, κλ. It does this in such a manner that the angles at β are equal, and β is the point of uniform motion, at least for this position for which the distances are being sought. Let the points of division be connected to α and β. Now this semicircle is purely fictitious: it should be drawn only for computing the sum of a number of distances. If the planet were moved forward with the same degree of power from the sun at both δ and λ, in the same manner as the epicyclical rotation is supposed to be always uniformly set in motion, then it really would traverse these equal parts of the eccentric, from which we have taken the distances, in equal times; and also, the distances corresponding to the times denoted by the points of division would be these very ones, αδ, αε, αζ, αθ, αι, ακ, αλ, not only in quantity, but also in their identical position. In a word, the path of the planet would be the circle δθλ.*

The planet does in fact represent quantitatively the reported distances resulting from the uniform rotation of the epicycle, but is itself moved forward unequally in equal times by the sun, less at δ, more at λ. Thus in the time signified and measured

342

by δβε*, it does not traverse the space δε, although it does attain the distance αε. And in [that same] time (measured by the angle λβκ, equal to εβδ) it traverses more space than κλ, although it attains the length of the distance ακ.[1] Therefore, the planet has a length of distance αε before it is actually moved forward to ε, and a length of distance ακ before it is moved forward to κ; and inversely, when it is moved forward to ε or κ, it has already been at the distances αε and ακ, and for that reason it will now be somewhat nearer. Thus the planet, when at ε, κ, and all the other points of this sort, is nearer to the point α than are the points ε, κ on the circumference. So the planet moves inward from the established distance of the circle δλ towards the point α which is near the center β, never coinciding with this circle at any points other than δ, λ. For the manner of the incursion is the same in the opposite semicircle.

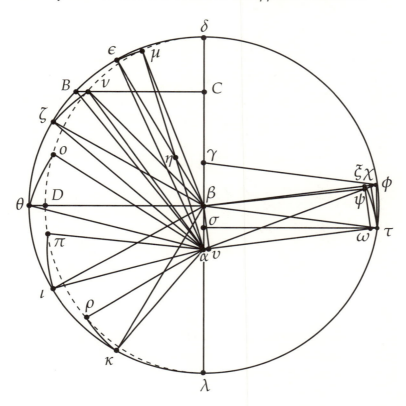

Also, the plane δαε, δαζ, and so on, contains in itself the sum of the distances of all the points on the arc of the epicycle, which, by Chapter 40, is similar to the arc δε. And yet the planet, in equal times (which are now being measured by δε, εζ), describes unequal arcs on its real path, short when it is far from the sun α, long when it moves near to the sun, in such a way that the arcs of the planetary path which are traversed in equal times are in the inverse ratio of the distances, by Ch. 32. It thus happens that the arc εδ (which is here the measure of the time) exceeds the arc of the path traversed, which let be μδ, to about the same extent that the area εαδ exceeds the sector εβδ, whose measure is the angle εβδ or the arc εδ.

If you declare the entire plane area to be 360° in number, just as the circumference

---

[1]*longitudinem distantiae.* This has been translated awkwardly because it is a strange phrase that illustrates Kepler's idiosyncratic use of the word *longitudo.* For more on this term, see the Glossary.

---

* In this place, when we are computing only the distance αε (that is, αμ), the angle δβε measures the time, the genuine and physical measure of which is the plane surface δαμ, as will be made clear below.

First attempt at a description of the oval.

of the circle, and the periodic time as well, then the number of the time, or δε (at this position) is approximately the mean, either arithmetic or geometric (for they hardly differ) between the number of the sum of the distances or the area εαδ, and the number of the planetary path or μδ. There occurs here a multiple obstacle to calculation.[2] First, that the plane of the circle is not perfectly equivalent to the sum of the distances, as was demonstrated in Chapter 40, even though it was said at the end of Chapter 43 that the defect is quite small.

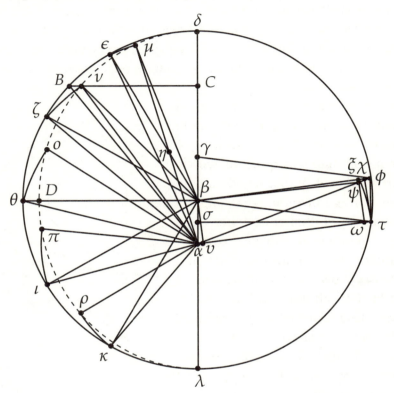

Chapter 48 is devoted to removing these difficulties.

219   Second, that the proportion just described is not exactly geometrical. True, the individual distances are to the individual mean distances in the inverse ratio of the individual arcs of the planetary path to the mean arcs. But the sum of a certain number of distances does not maintain the same ratio to the sum of the same number of mean distances, as the sum of the same number of arcs to the sum of the mean arcs, inversely. As you will see from an example. *Let the two distances be* 12 *and* 11, *the mean* 10, *and let the mean arc be the same. And let it be that, as the distance* 12 *is to the mean distance* 10, *so is the mean arc* 10 *to the arc* $8\frac{1}{3}$ *belonging to the distance* 12. *Let it also be that as the distance* 11 *is to* 10, *so is the arc* 10 *to* $9\frac{1}{11}$. *Compound the distances* 12 *and* 11 *into one sum, which will be* 23. *The sum of the two means is* 20, *the sum of the two arcs is* $17\frac{14}{33}$. *Here,* 10 *was indeed the mean proportional between* 12 *and* $8\frac{1}{3}$, *and between* 11 *and* $9\frac{1}{11}$, *but now the sum* 20 *is not the mean proportional between* 23 *and* $17\frac{14}{33}$, *but between* 23 *and* $17\frac{19}{23}$, *which is greater.*[3]

However, this ratio is valid for the arithmetic mean. *For example, let* 10 *be the arithmetic mean between* 12 *and* 8; *likewise, between* 11 *and* 9. *Compound* 12 *and*

---

[2] ἀμηχανία.

[3] This last number should be $17\frac{9}{23}$, which is less than $17\frac{14}{33}$, not greater, as Kepler states. However, it remains true that they are unequal, which is what Kepler is undertaking to prove.

11: *they make* 23. *Compound 8 and 9: they make* 17. *Therefore,* 20 *is again the arithmetic mean between* 17 *and* 23.   And since it was demonstrated in Chapter 32 that there is hardly any difference for the present undertaking between the arithmetic and geometric means, what is here denied to be true for all cases will therefore be only slightly different from the truth.

Third, even if the area $\epsilon\beta\delta$ were the exact geometric mean between $\epsilon\alpha\delta$ and $\mu\beta\delta$, nonetheless, it cannot be constructed geometrically. *For the triangle $\alpha\epsilon\beta$ ought to be equal to the sector $\epsilon\beta\mu$. But geometers have yet to devise a method by which a given angle can be cut in a given ratio.*

Fourth, if none of the above deter us, *the sector $\mu\beta\delta$ of the circle is still not the same as the so-called "sector" $\mu\beta\delta$ of the oval plane. Therefore, even if the arc $\mu\delta$ were defined as if it were on the circumference of a circle, nevertheless, nothing would follow concerning $\mu\delta$ defined as if it were an arc on the path of the planet, which is not a circle.* Therefore, even though this is helpful to those who would like to make use of numbers, to know that $\epsilon\beta\delta$ is a mean between $\epsilon\alpha\delta$ and $\mu\beta\delta$; nevertheless, for us, who strive after a geometrical path, this passage does not lie open.

Term:
Properly, a sector is part of a circular plane cut off by two straight lines from the center. It is thus used improperly of a plane that is not perfectly circular.

We shall therefore try another way. *And on our fictitious eccentric $\delta\theta\lambda$ the measure of the time is $\delta\epsilon$, $\delta\zeta$ for finding out the distances $\alpha\epsilon$, $\alpha\zeta$, while the ratio of the sectors $\delta\beta\epsilon$, $\delta\beta\zeta$ to one another is the same as that of the arcs $\delta\epsilon$, $\delta\zeta$. However, on the true path of the planet, the plane [area] lying between the arc of the path and the sun α is likewise the true measure of the time during which the planet is found on the arc lying above it, by Chapter 40. Therefore, from the point α of the diameter let straight lines be projected enclosing spaces equal to $\epsilon\beta\delta$, $\zeta\beta\delta$, so that the space $\epsilon\eta\mu$, which is subtracted from the space $\epsilon\beta\delta$, is equal to the space $\eta\alpha\beta$, which is added to that same space $\epsilon\beta\delta$. And let these lines be $\alpha\mu$, $\alpha\nu$. And about center α, with radii $\alpha\epsilon$, $\alpha\zeta$, let arcs $\epsilon\mu$, $\zeta\nu$ be drawn intersecting these lines at μ, ν. Will the points μ, ν, o, π, and so on, constructed in this way, be thus obtained correctly, so that in the times $\delta\epsilon$, $\delta\zeta$, $\delta\theta$, $\delta\iota$, $\delta\kappa$ the planet will arrive at them?* This is indeed approximately true, but here, too, three things are wanting. First, as above, the plane is not exactly equivalent to the sum of the distances. Second, there is no geometrical way showing how to cut a given semicircle in a given ratio with a straight line drawn from a given point on the diameter. Third, it is not known whether the shortfall for any of the planes $\mu\alpha\delta$, $\nu\alpha\delta$, and so on, produced by the deflection of μ, ν, from the circumference, is in the same ratio as the rest. Nevertheless, these will still be useful to those who wish, contrary to the geometrical usage, to proceed using least parts, with the aid of numbers.

Second attempt at describing our oval.

220

Since geometry has left us destitute, in order that we may have a description of the line which has been born to us out of the theory of Chapter 45, let us go seek the assistance of a contrivance[4] by fetching our vicarious hypothesis from Chapter 16, which places the lines $\alpha\mu$, $\alpha\nu$, and so on, on which the planet stands, at the correct zodiacal places at the correct times, combining it with the present fictitious eccentric $\delta\theta\lambda$, from which the theory of Chapter 45 has persuaded us that we have derived the correct length of the lines $\alpha\epsilon$, $\alpha\zeta$; that is, $\alpha\mu$, $\alpha\nu$.

Third attempt and method of describing the oval born in Chapter 45.

---

[4] ἀτεχνία.

And besides, it is a good idea for the sake of shedding some light to compare the two hypotheses with one another, combined into one diagram. For although both are deceptive on certain points, each is useful for investigating particular truths (to the extent that they can be known at this point). And in this diagram, many things which have been said so far are brought together into a single view.

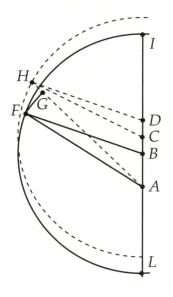

Let A be the center of the earth (or of the sun, for Copernicus), AI the line of apsides, AD the eccentricity of the point of the equant. And although it was denied in Chapter 19 that the point D could remain fixed, and AD the same, this is to be understood as true only if DA is bisected. But if we remain at liberty to divide DA as we please, as in Chapter 16, then this point can remain fixed. Therefore, let AD be divided in the ratio found in Chapter 16. Let the point of division be C, and let AC be 11,332, CD 7232. And about center C, with radius CH equal to 100,000, let the eccentric be described, as sketched out by the dotted line passing through H. This, then, will be the hypothesis of Chapter 16. For, taking any known angle of mean anomaly, let a dashed straight line, DH, be drawn from the center of the equant D bounded by a point on the circumference, containing between itself and the line of apsides the required angle,* which is the measure of the proposed time. And let the point H be connected to A. The angle IAH will thus be the equated anomaly and the true zodiacal position of AH, and the planet most certainly will be on the line AH at the given time and anomaly, by Ch. 16 and 18. But the distance AH will be false, and the planet will not be at the point H, because the division of AD at C and the eccentric H described about C are false by Ch. 19, 20, and 42. There it was shown that AD is to be bisected at B, so as to describe a more correct eccentric IL about B, but it will not be a perfect circle.   Now let the other hypothesis be delineated. And let AD be bisected at B, so that AB is 9282 (or, according to the numbers of Ch. 42, 9264), and about center B with radius CH, let another eccentric IL be described, which in this chapter I have also called a fiction,* for computing the correct distances. This is the one which in the diagram before the last was described as δθλ, about center β. And let the mean anomaly, which was previously proposed to us in the form of time, be transferred from D to B, and the straight line BF be drawn from B parallel to the former DH. And let the point of intersection F of the new eccentric be connected with A.

Therefore, by what was said in this Chapter 46, AF will be the distance of the planet at F from the center of the sun at A, which the hypothesis of Chapter 45 requires. But the angle BAF is false, and the zodiacal position of AF is false. For at the selected time and mean anomaly the planet is not found on AF. Before, however, the true line of the planet was AH, and the distance AH was false. So about center A with distance AF let the arc FG be described, intersecting AH at G. Thus the line AG, constituted by two manifestly false hypotheses, is nevertheless true in its position beneath the zodiac, and its length is consonant with the hypothesis of Chapter 45.

* In the vicarious hypothesis of Chapter 16, this is the proper measure of the time, for upon it is placed *D*, the point of the equant, in accord with the opinion of the ancients.

221

* It is true with respect to the shape that this was a fiction, since the path of the planet is not a circle. But with respect to position, and the center *B*, it is not a fiction, but true: thus, this fiction described about *B* is the opposite of the prior fiction about *C*.

Thus through the vicarious hypothesis of Chapter 16, which consists of the points *A*, *C*, *D*, and the eccentric *H*, we have made up for the defect of geometry, which was unable to show us the position of the line *AG* (onto which the correct distance *AF* is to be transferred) which we required of the hypothesis of Chapter 45.

One might ask, "Couldn't we, in the former diagram just as well as in the latter, *take as given the point* γ *of uniform motion, and from it draw* γμ, γν, γο, γπ, γρ *parallel to* βε, βζ, βθ, βι, βκ, *and draw the arcs* εμ, ζν, θο, ιπ, κρ *intersecting these parallels? And then understand the points of intersection to be the determinate places and positions of the distances?*

A fourth manner of description rejected.

The answer is no. For in so doing we shall err considerably in transferring the distances too high up, as is easily seen in the latter diagram. *For in it the line AH containing the true distances AF is lower than the line DH from the point of uniform motion D parallel to BF.*

Whichever of the described ways is used for delineating the line possessing the body of the planet, it now follows that this way indicated by the points δ, μ, ν, ο, π, ρ, λ, is truly oval, not the elliptical one to which the mechanics give that name from the egg (*ovum*), contrary to correct usage. For an egg is rotated about two vertices, one more blunt, the other sharper, and is visibly inclined at the sides. It is such a figure, I say, that we have created. *For the planet is fast at* λ, *slow at* δ, *and less fast at the former than it is slow at the latter. For there are more of the long distances exceeding the semidiameter than there are of the short ones (for they are longer up through* $92\frac{2}{3}^{\circ}$,[6] *and then shorter for* $87\frac{1}{3}^{\circ}$, *as can be demonstrated from the theory presented in Chapter 29). But in addition, that greater number of long [distances] is crowded into a narrower arc of the eccentric by being translated upwards, while these fewer [shorter] ones are spread out into a larger arc. So that to a mean anomaly of* $92\frac{2}{3}^{\circ *}$, *which contains* $92\frac{2}{3}^{\circ}$ *of distance,[7] there corresponds an eccentric anomaly of about* $87\frac{1}{3}^{\circ}$. *The remaining* $87\frac{1}{3}^{\circ}$ *of mean anomaly, with the same amount of distances shorter than the radius, is scattered over the remaining angle at the center of the eccentric,* $92\frac{2}{3}^{\circ}$. *Consequently, the short distances near perihelion are farther from one another than are the longer ones at aphelion. So even if the ratio between two neighboring perihelial distances remained constant, the part cut off from the circle would nevertheless be thinner about* ε, μ, δ *than about* ρ, κ, λ.[8] *For the*

What sort of oval is generated from these descriptions.

Dürer.[5]

* It amounts to this much in the erroneous opinion of chapter 45, on which we are wasting time here.

222

---

[5]Dürer, *Underweysung der Messung* (1525), p. 30, presents a way of accurately generating points on a true ellipse, which he nonetheless calls "eyer lini," or "egg line."

[6]Of mean anomaly.

[7]Because the sum of the distances is the area, which represents the time, which is the same as the mean anomaly.

[8]This appears not to follow from the premises. All that has been established is that the distances are more closely packed near aphelion than near perihelion. If "same ratio" here means a simple linear decrease, then the orbit's shape would be the opposite of what Kepler claims. For the changes of distance at aphelion and perihelion would be the same as they are everywhere else, in contrast with the changes under the epicyclic rule, which would be very small at aphelion and perihelion. This would create an outward-facing point at aphelion, and an inward-facing cusp at perihelion. If, on the other hand, "same ratio" means "the same as the ratio of decrease at aphelion" (as Kepler's next argument suggests), then we would only know that a given radial decrement occurs nearer to aphelion than to perihelion. And since the opposite occurs in the circle (as Kepler himself proved in Ch. 40), this establishes that the part cut off from the circle

short ones are transposed into the position of the longer ones in a shorter space at δ than at λ. But in addition, the distances themselves of the equal parts of the epicycle near to perihelion are in a greater ratio to one another than the distances of the parts near aphelion.[9] For it was demonstrated in Chapter 40 above that the conchoidal area is wider in its lower part than in its upper. Therefore, the conchoid must be thinned in greater steps over a shorter space at its lower point than at its upper, and in addition those greater intervals are compared to shorter lines. So on both counts the ratio is increased. With so many causes concurring, it appears that the part cut off from our eccentric circle is much wider below than above, at an equal distance from the apsides. Anyone can easily explore this using numbers, or by a mechanical delineation, by assuming some appreciable eccentricity.*

* A figure of this kind is to be found in the manuals of spherics and the commentaries of Reinhold on the theories of Peurbach, in the theory of Mercury.[10]

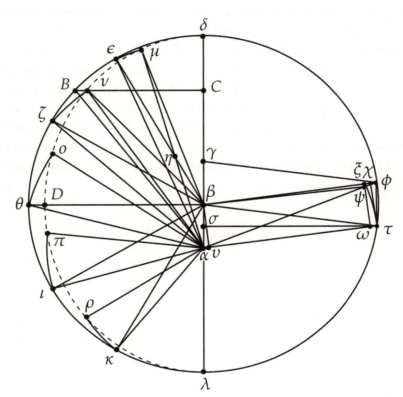

is *greater* at aphelion than at perihelion, which is the opposite of what Kepler wished to prove.

[9]This is true, but its effect is only to counteract the tendency described in the preceding footnote. In fact, the asymmetry of the oval is so slight as to be beyond Kepler's powers of analysis (and beyond any significance). See D. T. Whiteside, "Keplerian Planetary Eggs, Laid and Unlaid, 1600–1605", *Journal for the History of Astronomy* 5 (1974), pp. 1–21.

[10]Ptolemy's theory of Mercury (*Almagest* IX.7–11) has a "crank" mechanism on the deferent circle that creates a double perigee, as Ptolemy himself remarks. On p. 71 of *Novae theoricae planetarum* (Paris, 1550), Peurbach points out that this in effect makes the deferent have "a similarity with the outline of an oval plane."

CHAPTER 47

An attempt is made to find the quadrature of the oval-shaped plane which Chapter 45 produced, and which we have been busying ourselves to describe in Chapter 46; and through the quadrature, a method of finding the equations

We have accomplished nothing if, from the hypothesis we have taken, and the physical causes of Chapter 45, which we follow here as true, we shall not have constructed the correct equations, no less than the distances. But the equation is compounded from the parallaxes of the points on the eccentric and the elapsed time. The former of these I am accustomed to call the optical part of the equation, and the latter, the physical part. Now the elapsed time, even if it is really something different, is certainly measured most easily (if not most perfectly) by the the plane area circumscribed by the planet's path. We therefore turn to the measurement of the plane area of the eccentric ovoid, the rules for delineating which have been laid down already. Now there is going to be something lacking in our account that prevents our stating the true measure of this time. (For at the circumference of the ovoid the lines that join the parts of its circumference with the source of power are even more inclined than on the circle. This is even true as well of the lines that are drawn from the center of the eccentric to those same parts of the ovoid, although otherwise the radii from the center to the circumference of a perfect circle make perfect right angles.)[1] But the consequence of this is that the sum of the distances is not exactly measured by the plane surface, nor are the arcs of the ovoid exactly proportional to the distances. All these things will be clear from a rereading of Chapters 40 and 32. A guess as to how small this discrepancy is going to turn out to be, however, can be grasped from Ch. 43.

Terms:
What is the optical part of the equation, and what is the physical part.

223

And how else can we measure this plane surface, compare it to the plane surface of a circle, and divide it into prescribed parts, unless we find a square equal to the trimmed-off part, or the lunule cut off? Here we will have to summon up from tragedy a *deus*, or rather a sort of *ratio, ex machina*,[2] to teach us how to manufacture a quadrature of the ovoid, or of its border in the last diagram but one—that is, the lunule $\delta o\lambda\theta$—whose removal from the surface of the circle generates the ovoid $\delta o\lambda$. And just as I called upon the geometers before in Ch. 40 for the area of the conchoid, and begged their assistance, I do so again now for the ovoid (or, if you prefer, the "metopoid"[3]).

---

[1]Kepler's parentheses. In the Latin, these two parenthetical sentences are a subordinate clause in a rambling sentence that I have broken up for the sake of intelligibility. –trans.

[2]Kepler used the Greek terms θεὸς, λόγος τις, ἀπὸ μηχανῆς, which have been rendered in Latin here, since they are sufficiently familiar to readers of English.

[3]This is coined by Kepler from the Greek μέτωπον, meaning "forehead" or "face", and accordingly means, "face shaped figure". Kepler is trying to find a word that means "egg shaped"

349

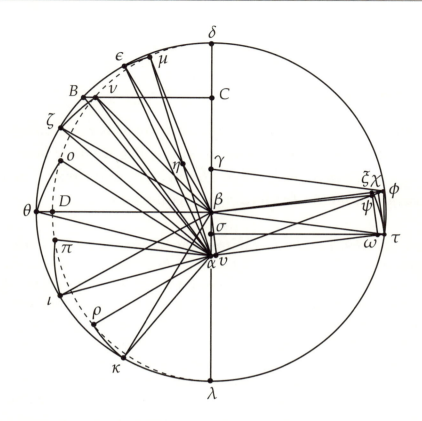

If our figure were a perfect ellipse, the job would have been done by Archimedes, who demonstrates in his book *On Spheroids*, prop. 6, 7, and 8, that the area of an ellipse is to the area of a circle sharing a common major diameter with the ellipse, as the rectangle contained by the diameters (or the "figure"[4] of the section) is to the square on the circle's diameter.

But let the figure be a perfect ellipse, for they hardly differ. Let us see what follows.

I say, therefore, that the lunule δολθ cut off from the semicircle will turn out to be imperceptibly greater than the small semicircle whose semidiameter is the eccentricity itself, αβ, or 9264. *For let αβ be bisected at σ (as in Ch. 29), and from σ let στ go out perpendicular to αβ. Let α and β be connected to τ. Now let γφ extend parallel to βτ, and let βφ, αφ be joined. And about center α with radius ατ let the arc τψ be drawn, intersecting αφ at ψ and βφ at ξ.*

Now since the point τ is equally remote from α and β, we are (following the Arabs in using the term most properly) at the middle elongation, that is, at the average distance of the planet τ from the sun α. And because γφ is parallel to βτ, the point ψ on the line αφ (in the diagram of the previous chapter) is the genuine and most true position of the translation of ατ to αψ. Therefore ψ is also the point of the planet's average

---

without suggesting an ellipse.

In the following treatment, Kepler approximated the oval with an ellipse having the same axes, supposing that the oval's excursions outside the ellipse are about the same as its incursions within it: thus, the areas of the oval and the ellipse would be very nearly equal. The trial ellipse, it should be noted, has an eccentricity greater than that of the true ellipse, and so the breadth of the lunule is greater than for the true ellipse. In fact, it is just twice the breadth of the true lunule, a circumstance that will play an important role in the further course of the investigation.

[4]This is the "area" mentioned in Apollonius's *Conics* I. 13 and elsewhere, which characterizes the ellipse.

distance. Hence, the little part of the line βψ between ψ and the circumference is the measure of the breadth of the lunule about the middle elongation, while the small line ξφ is greater than this breadth by some imperceptible magnitude.

Let a perpendicular be drawn from β to ατ, and let it be βυ. I say that ξφ, a part of the line βφ, is twice αυ.

For let τ and φ be joined, and from τ let τχ come out perpendicular to βφ. Similarly, from ξ let ξω come out perpendicular to ατ. Since the straight line αγ intersects the parallels γφ, βτ, [angles] βγφ and αβτ will be equal. And γβ is equal to αβ by construction. But also, βφ is equal to ατ, for both are equal to βτ by construction. Therefore, triangle γφβ is congruent with triangle βτα. Thus, γφ will be equal to βτ. But they are parallel by construction. Thus βγ and τφ, which connect the ends 224 *of equal parallels on the same side, will also be parallel and equal. But βγ is also equal to αβ. Therefore, αβ and τφ are also equal and parallel. Consequently, βφ and ατ will also be parallel. And because the angles at χ and υ are right, and the base τφ is equal to the base βα, and the angle βατ or βαυ is equal to the angle τφβ or τφχ, αυ and χφ will therefore be equal, as well as βυ and τχ perpendicular to them.*

Again, because the lines τχ and ξω are equal, being parallels between parallels,[5] *while βτ and αξ are equal, and the angles at χ and ω are right, the remaining sides of the triangles βχ and αω will also be equal. But βξ and υω are also equal, for they are parallels between the parallels βυ, ξω. Therefore, when the equals βξ and υω are subtracted [from βχ and αω], the remainders ξχ and αυ will be equal. But before, χφ and αυ were also equal. Therefore, ξφ is twice αυ.*

With these things demonstrated, we shall draw nearer to our proposition. *And to the diameter φβ of the circle (which should be understood to be extended to the other circumference), a straight line is drawn perpendicular from a point τ on the circumference, namely, τχ. Therefore, as φχ is to χτ, so is χτ to the remainder of the diameter. Therefore, the rectangle contained by χφ and the remaining part of the diameter is equal to the square on τχ.*

And because the square on τφ, that is, αβ, is equal to the [sum of the] squares on τχ and χφ, when equals are added, the rectangle contained by χφ and the entire diameter is equal to the square on αβ.

And because φξ is twice φχ, the rectangle contained by φξ (which is imperceptibly greater than the breadth of the lunule ψφ) and the semidiameter φβ is equal to the square on αβ.

But the rectangle contained by ξφ, φβ is the difference of the rectangle contained by ξβ, βφ and the square on βφ. And the lunules are also the difference between the areas of the ellipse and the circle. And as the rectangle contained by ξβ, βφ is to the square on βφ, so is the area of the ellipse to the area of the circle, approximately. * *Therefore also, as the square on βφ is to the rectangle ξφ, φβ, that is, the square on αβ, so is the area of the circle to the area of the two lunules, approximately. And by permutation, as the square on βφ is to the area of the circle, so is the square on αβ to the area of the lunules, approximately.*

But as the square on βφ is to the area of the circle of which βφ is the radius, so is the square on αβ to the circle of which αβ is the radius. Therefore, the area of the circle of which αβ is the radius imperceptibly exceeds the two lunules ψφ cut off. It

* Approximately, I say. For if βξ were the shorter semidiameter of the ellipse, and ξφ were the excess of the longer, then the ratio between the areas of the circle and the ellipse would be exactly the same. But βξ is not completely and exactly the shorter semidiameter.

---

[5]This is not immediately obvious, but is true because each of the lines is perpendicular to one of the pair of parallel lines ατ, βφ.

exactly equals the lunules γζφ *which are a little wider than they should be, because* ζφ *is imperceptibly longer than* ψφ, *as was said at the beginning.*

This demonstration also has its use in the true physical hypothesis.

So, granted what we have supposed, namely, that the area of the ellipse differs imperceptibly from the area of our ovoid, as a result of the compensation between the excess of the ovoid over the ellipse in the upper regions, and the defect in the lower regions,—these, as I said, being granted, we have squared our "new moon"[6] figures, and thus also the ovoidal one. Or, properly speaking, we have "circled" it. For Archimedes teaches us the ratio of the circle and the square.

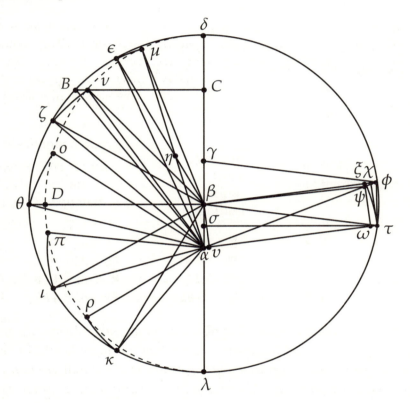

225   We shall now put this to use, thus. Since the area of the ovoid is less than the area of the circle by the area of the small circle described by the eccentricity, let the area of this small circle be computed next.*Now the ratio of the areas is the duplicate of the ratio of the diameters. And as* βφ, 100,000, *is to* βα, 9264, *so is* βα *to* ξφ, 858. *Therefore the ratio between* βφ *and* ξφ *is also the duplicate of the ratio between* βφ *and* βα. *Therefore, as* βφ, 100,000, *is to* ξφ, 858, *so is the area of the circle,* 31,415,900,000, *to the area of the small circle,* 269,500,000.

Therefore, when the area of the small circle is subtracted, the remainder is the area of the ovoid, 31,146,400,000, equivalent to 360 equal parts of the periodic time.

Those things which have been said so far are entirely consonant with the opinion of Chapter 45. Nevertheless, to use them, it is not enough to know the magnitude of the area of the ovoid. Indeed, we need also understand how to divide it, from the center β or the point α, in a given ratio. *For example, in the previous diagram, let the point* θ *be taken, and let the planet be observed on the line*

---

[6]*Menoides.* This is borrowed from the Greek μηνοειδές, which denotes something crescent shaped. However, it is derived from μείς, "month," a connection that was thought good to maintain in the translation.

αθ, but let it recede from the circumference θ towards the sun α. Therefore, given the eccentricity αβ and the angle θαβ, and supposing that the planet is at the point θ of the circumference, the angle θβδ will be given, and hence also the sector of the perfect circle, namely, θδβ, and the area of the triangle θβα, that is, the whole area θδα which (with the exceptions in Chapter 40 above) ought to have been the measure of the time elapsed while the planet moves from δ to θ, if the planet had gone in a perfect circle. But because it describes an oval inside the circle, not embracing the full area of the perfect circle, so, exactly as just a moment ago we needed knowledge of the area of the whole ovoid, we now need to know what portion of the ovoid is contained between the lines δα and αθ, that is, what portion the area of the part δθ of the lunule is of the area that measures the two lunules, namely, of the area of the small circle on the eccentricity. For when this is subtracted from the portion of the circle cut off by the lines αθ, αδ, the remainder will be the portion of the ovoid cut off by the same lines αθ, αδ. Thus, finally, the whole oviform will be correctly compared to its part δαθ in order to find the time, or the slowing of the planet, which occurs between the lines αδ and αθ.

Once again, now, where is the geometer who will show us how to do this? *Let the last diagram of Chapter 40 be presented again, in which CD is the semicircle stretched out into a straight line, divided into equal parts, and DE is a quadrant. And on the line EA from E let some line [Eo] be extended towards A which bears the ratio to the longest line BA (the one on the line CA) which that BA has to BC. And let the rest, Gμ, Hν, Iπ, Kρ, be set up similarly in the appropriate magnitudes, having the breadth of the lunule at any given position, so that Gμ is a little shorter than Kρ, and Hν shorter than Iπ (although they are the same distance from C and D), in accordance with what was demonstrated in Ch. 46. Thus the lunule, insofar as it shortens the distances, will be delineated and unfolded partwise onto a straight line.*

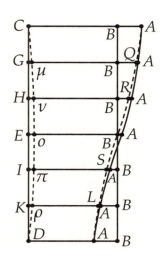

And because the whole space between CD and AA is twice the area of the stretched-out semicircle CD, the geometer should consider whether the small space between the curve CμνoπρD and the straight line CED is also going to be twice the lunule cut off from the area of the circle.

Nothing appears to contradict the possibility of this being so. For when the lunule really is a lunule, CD is then curved inwards while retaining its same length. But CμνoπρD, which was just constructed longer than CED, is then much shorter. Therefore, the lunule then contains a much smaller area than now.   But in fact, O geometers, this is not a demonstration. Therefore, you will assist me. And if this turns out to be true, you will teach me a method by which may be known the magnitude, not only of the whole small area between the straight line CED and the curve CoD, which I have so far said is equal to the small circle on the eccentricity (for two lunules are equal to the small circle, and this small area is now supposed to be twice one of the lunules), but also of any part of it, at any given length of the parts CG, CH; and by which this may be compared to the area between CD and BB.

So once again, as before in Ch. 46, since there is no way out through geom-

226

etry, we shall be content with a contrivance.[7] And no wonder, for the opinion born in Ch. 45, which threw us into these difficulties, is false.

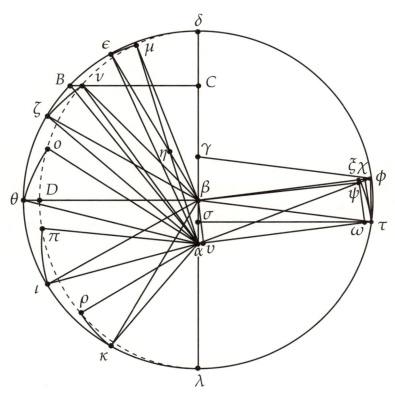

Therefore, let the previous diagram from Chapter 46 be considered again. If the area δoλ, which is an ovoid, were a perfect ellipse, when the ellipse δoλ and the circular area δθλ are described on the common longer diameter δλ, and the planes of the two figures are divided by lines BC applied ordinatewise (that is, perpendicularly to the longer diameter δλ) from one side of the longer diameter, the portions of the ellipse vδC would always remain in the same ratio to the portions of the circle BδC. This is demonstrated by the authors who wrote on conics, and Archimedes takes it over in On Spheroids *Prop. 5. If this were so, there would indeed be no need to know the oviform area. For then we would substitute the area of the circle for the area of the ellipse, and the parts of the circle for the similar parts of the ellipse.*

Let it be that δoλ is a perfect ellipse, for it is but slightly different from one. And from any point on the ellipse, v say, let a perpendicular be dropped to δλ, which let be vC, and let it be extended until it intersects the circle at B. And let B and v be joined to α. Now as βφ is to βξ so is CB to Cv, from the assumption of a perfect ellipse and Prop. 5 of On Spheroids, *and also as BC is to Cv so is the area BδC to the area vδC. But also as BC is to Cv, so is the area BαC to the area vαC. Therefore, as βφ is to βξ, so is the area αBδ to the area ανδ.*

Where the time is to be numbered for finding the distance of the planet from the sun.
First, let it be, at the planet's proposed time of departure from δ, that as the periodic time is to four right angles, so is the proposed time to the angle about β (δβζ, say), and let the distance αζ, to which αv is equal, be computed.

Where the time is to be numbered for finding the eccentric equation.
Again, let it be that as half the periodic time is to the known area of the semicircle δθλ, so is the proposed time (whose measure we have just now said is something else, δζ, when the distance αζ was computed) to the area αBδ. Thus the area is given. Now

227

---

[7]ἀτεχνίᾳ. See above, p. 345

a value for angle Bβδ must be found, such that its sine BC multiplied by half αβ (that is, the area of the triangle αBβ), together with the sector Bβδ, adds up to equal the area just found from the time. Here one has to proceed by trial and error.[8] *When you have obtained the angle Bβδ, then in triangle Bβα, from the angle β and the known sides αβ, βB, the angle Bαδ will become known. And because the ratio Bν to BC is known, Bαν will also be known, and when it is subtracted, there will remain ναδ, the correct equated angle for the time selected.*

For example, as in Ch. 43, let the mean anomaly, that is, the artificial or astronomical numbering of time, be 95° 18' 28''. And because 360° is equivalent to the area of the perfect circle, 31,415,926,536, 95° 18' 28'' will therefore be equivalent to the area 8,317,172,671. Let this be θαδ. Now, if the eccentric anomaly δθ were 90°, as I suppose conjecturally, its sector θβδ would be 7,853,981,670. And the sine θβ of 90° is 100,000. This multiplied by half the eccentricity αβ, 4632, gives 463,200,000 as the area θβα. The sum of the areas is 8,317,181,670, which is θαδ, and which exceeds what it should be by some small amount. We therefore guessed well that the angle or anomaly of the eccentric δβθ is 90°. And because its sine is 100,000, the lunule θD cut off at θ will be 858. Therefore, the shorter semidiameter Dβ will be 99,142, which is to 100,000 as 9264 is to 9344. This is the tangent of the angle αDβ, 5° 20' 18'', making the equated anomaly Dαδ 84° 39' 42''. The vicarious hypothesis shows this to be 84° 42' 2'', the difference being 2' 20''.

The investigation of the eccentricity in Ch. 42 depends upon aphelial and perihelial distances, and in these there can be some slight error which is increased tenfold in establishing the eccentricity. Therefore, it should be noted in passing that if a perfectly reliable way of equating through the physical causes were finally found, a perfectly true eccentricity could afterwards be established, and through it the distances of aphelion and perihelion could be entirely corrected. *For example, provided we can trust the vicarious hypothesis for the planet's zodiacal longitude, and suppose that everything we have assumed here and in Ch. 45 is true, the equation has been made 2' 20'' too large here, while the optical and physical effects upon the equation at the middle elongations are equal, as here. So the error being bisected, the half (1' 10'') should be subtracted from the angle last found, 5° 20' 18'', making it 5° 19' 8'': this shows a tangent of 9310.*[9] *Before, it was 9344. The difference of 34 subtracted from the eccentricity 9264 would leave a corrected eccentricity of 9230.* But we shall not pursue this now, since our assumptions are wrong on the very small quantities. Let it be enough to have noted it for future use in the chapters following next.

Let us also see what promise this form of computing equations holds at the eighths of the periodic time. Let the mean anomaly be 48° 45' 12'', as in Ch. 43. And since the numerical measure by which the areas are expressed is a matter of indifference, we shall retain the number 360° for the area of the circle, and 19,108'' for the area of the greatest triangle (for just now, in the other system of numbering, it was 463,200,000). *Let us guess that the anomaly of the eccentric, or Bβδ in the diagram, is 45°. Therefore, the sine, BC, is 70,711. This, multiplied by*

The method of equating used here should be noted. For in the end we are going to follow it when it is established that the path of the planet is a perfect ellipse, but nearer by half to the circle. The distance alone will be found by another method.

Term:
What the "mean anomaly" is.

A method of correcting the eccentricity, inserted in passing.

---

[8]Ubi conjectatione et regula falsi opus est.

[9]Kepler is using the law of tangents here, in reverse, in order to determine the eccentricity corresponding to a corrected equation.

the greatest triangle 19, 108″ *and with the zeroes dropped, gives the triangle* Bαβ *as*
13, 512″, *or* 3° 45′ 12″, *for this location. This, added to the sector* Bβδ, 45°, *gives* 48°
45′ 12″ *for the area* Bαδ, *which is also the mean anomaly we assumed. We therefore*

228 *guessed the angle* β *well. Now, as the radius* βϕ *is to* βξ, 99, 142,[10] *so is BC,* 70, 711,
to Cν, 70, 104. *And because BC is* 70, 711, Cβ, *the sine of its complement, will also*
be 70, 711 *at this location. Therefore, Cα is* 79, 975. *But as this is to* 100, 000, *so is Cν*
to the tangent of the required angle ναC, 41° 14′ 9″. *The vicarious hypothesis shows*
41° 20′ 33″.

The same things are also easily investigated in the lower octant. *Let the*
mean anomaly be 138° 45′ 12″, *and let the area, whose angle at* α *is sought, be ex-*
pressed in the same units. We will find that the sine of an angle of 135° *at* β, *which*
is 70, 711, *makes the sum of the sector and the area of the triangle this much. And be-*
cause, as before, the sine 70, 711 *is shortened in order to constitute a line of the ellipse*
that is applied ordinatewise,[11] becoming 70, 104, *this is now to be combined with the*
sine of the complement of 135°, *which is* 70, 711, *now not increased by the eccentricity*
αβ, *as before, but decreased by it, making it* 61, 447. *Thus, as this is to* 100, 000, *so is*
70, 104 *to the tangent of the required angle,* 48° 45′ 55″, *or its supplement* 131° 14′
5″. *The vicarious hypothesis shows* 131° 7′ 26″. Compare this with Ch. 43, and
with other methods, using this table.

Common mean anomalies	Through a simple eccentricity	Through bisection of the eccentricity and doubling of the upper part of the equation	Through bisection of the eccentricity and a stable equant point, in the Ptolemaic manner	The vicarious hypothesis, through a free division, nearly agreeing with the truth in its results	The physical hypothesis, through supposition of a perfect circle	The physical hypothesis, through supposition of the opinion of Chapter 45 and of a perfect ellipse

Correspond to different equated anomalies:

48° 45′ 12″	41° 40′ 14″	40° 45′ 52″	41° 15′ 31″	41° 20′ 33″	41° 28′ 54″	41° 14′ 9″
95° 18′ 28″	84° 40′ 44″	84° 37′ 48″	84° 41′ 22″	84° 42′ 2″	84° 42′ 26″	84° 39′ 42″
138° 45′ 12″	130° 40′ 46″	131° 45′ 0″	131° 15′ 31″	131° 7′ 26″	130° 59′ 25″	131° 14′ 5″
		Excess and defect verge in opposite directions, if the lower part be doubled.				
	Ch. 20 and 29	Ch. 29	Ch. 19	Ch. 16 and 29	Ch. 43 and 40 [12]	The present Ch. 47

By these indications we
are assured that we are on
the track that will carry
us through at last to the
natural and most true
causes of the equations,
as well as of the celestial
motions.
So, of the two physical hypotheses for computing the eccentric equations,
that one shows equations nearer the truth, which previously, in Ch. 45, also
gave truer distances, namely, the last one. And, what may seem strange,
by a slight increase in the eccentricity it becomes equivalent to the Ptolemaic
method, using a stable equant point and a bisected eccentricity.

And since we convicted this Ptolemaic method of error above, the physical

---

[10]Since βξ = βϕ − ξϕ, and ξϕ was found earlier to be 858.

[11]See Apollonius, *Conics* I Def. 4, trans. R. C. Taliaferro, p. 3.

[12]All the editions have "XXIX" here; however, the chapter referred to must be 40, where the
physical hypothesis is introduced and is first used to compute equations.

method, which is in effect the same as the Ptolemaic, must also be somewhat askew of the truth. And it does indeed make the planet slow near the apsides, and too swift about the middle longitudes. This is the first argument by which it is proved that either the opinion of Chapter 45 is erroneous, or it has been transposed into numbers by an erroneous method.

But because the plane area of the circle is not equivalent to the aggregate of all the distances, nor is the oval figure that, according to the opinion of Ch. 45, Mars describes, a perfect ellipse, as we have assumed, the causes for the departure from the truth are still hidden. For in addition to these two causes 229 of error in the calculation, the error can be consistent with a third cause, of the very foundation, the opinion expressed in Ch. 45. So we have not yet set up equations according to the law of the opinion of Ch. 45, and have not yet done justice to the hypothesis we took up there, because we have been abandoned by geometry. Therefore, we cannot yet charge it with being erroneous. For a calculation that is going to do this points to the rule of innocence for itself.[13]

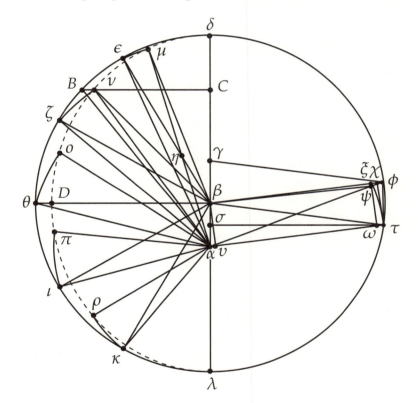

---

[13]"Hoc enim facturus calculus legem sibi ipsi indicit innocentiae." This is a somewhat cryptic sentence. I take Kepler to mean that if an algorithm could be found that allowed the physical model of Ch. 45 to produce observationally testable predictions, the mathematical soundness of the procedure would be adequate evidence that the errors, if any, are not created by the procedure itself. However, other readings may be possible.

CHAPTER 48

A method of computing the eccentric equations by a numerical measure and division of the circumference of the ovoid described in Ch. 46.

So, since the calculation taken up in the preceding chapter was abandoned by geometry on so many counts, so as to be suspected of culpability for the excesses and defects which we noted in the eccentric equations in that chapter, I finally sought refuge in the numberings of arithmetic, by which I attempted to avoid the obstacles which stood in the way of our describing the path of the planet in Chapter 46. For first, because the area was not the perfect measure of the sum of the distances, I dismissed the area and computed the distances of individual parts of the equally divided circumference. Second, since the ratio did not remain the same when any given number of terms of the geometrical proportion were added, I therefore investigated each individual proportion separately for each distance in relation to its minimal arc. Third, since the sum of any particular number of distances in Ch. 46 could not be established geometrically, I established it arithmetically. For there was no difficulty in that. Fourth, in carrying this out I had no dealings with sectors, whether circular or oval, and therefore it could obviously no longer be a hindrance to me that these sectors differ from one another.

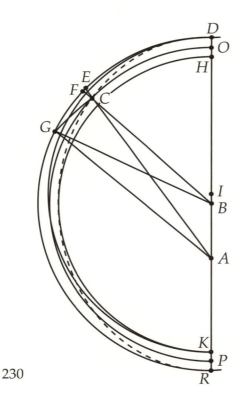

And so, with renewed effort, I brooded upon it, in order to know at least at the end whether the equations shown to us by the vicarious hypothesis also follow from the hypothesis under consideration, which gives the correct distances (that is, the one following the opinion of Chapter 45.)

I approached the matter thus. *About center B, with radius BD, let the circle DGR be described, in which the line of apsides is DR, and A is the source of power or the center of the sun. On the circle DG let the point G be chosen, and joined to B and A. And at first, let the angle GBD be the measure of the time, for computing the distance. Consequently, GA will be the true distance of the planet from A, although the planet has not moved all the way from D to G. For this method of computing or demonstrating the distance, from Ch. 45, has so far had the status of an assumption. But let DG be a small part of the circle, such as 1° out of the 360°. And since all the distances AG of this sort at the ends D and G of all the degrees DG can be computed in this way, by what was demonstrated in Chapter 29, I gathered all the 360 distances AG into one sum, in a very long addition. This was found to be 36,075,562[1] (with an eccentricity of 9165), corresponding to the entire oval path of Mars. Now about center A, with radius AG, let an arc, GC, be described towards D. And because the greater the distance the shorter the planet's path, when the distance of an arc of the circle DG is given (which arc is now, when we are computing the distance GA, nothing but the measure of the time), the length of the oval path DC will also be given, which the planet describes in the time DG under consideration (which is a simple anomaly of 1°). For as the length of the whole of the oval circumference is to the sum of all the distances, so is the distance of the arc DC (found by the arc DG) to the length of its oval arc DC. For it was proved in Ch. 33 above, and used in Chapter 46 (where the foundations for this operation were laid) that the arcs traversed are inversely as the distances. However, I applied this precaution: that AD and AG, the distances of the ends C and D of the arc from A, be added, and the mean of the sum be taken as the genuine distance of the whole arc DC. For let some eccentric circle DK described about center B, be divided into any number of parts, say, at D, G, L, K, M, N, and from the beginnings of the parts let arcs be drawn about the center of the world A until they intersect the lines drawn from A to the ends of the arcs, as DO, GP, LQ, KR, MS, NT. The areas on the left semicircle, ADO, AGP, ALQ, will be greater than they should be, while the arcs on the right, ANT, AMS, AKR, will be less than they should be. So, when least arcs are in question, the one is compensated by the other, so that TNA and ODA are very nearly equal to the area GDNA.*

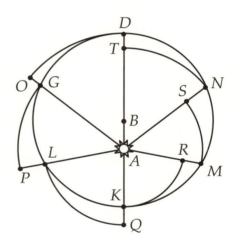

Next, the length *DC* in the previous diagram being thus given, to correspond to the given time *DG* and to the distance *GA* or *CA*, the angle of equated anomaly *CAD* now also should be found. *Let C be joined to B, and let AC be extended to E, where it intersects the circle, and BC to the intersection F. It was therefore not enough to know the length DC. The angle CBD ought to have been investigated as well. For because CD is shorter than FD, CD does not measure the angle FBD, or CBD. And on the other hand, even though CD is shorter than FD, if you place an observer at B it appears from B to be just as much a measure of the angle CBD as is FD. But according to the demonstrations of Chapter 32, it is true within all limits of sense perception that to the extent that FD is farther from B than is CD, FD is also longer than CD. Also, it is true within the same limits of sense perception (for the present purposes it does not matter how acute) that CE and CF are equal (thought in the truth of the matter CE is longer than CF, which is drawn through the center, by Euclid Book III prop. 7). I have therefore supposed, first, that CD and FD are equal, and both are a measure of the angle CBD or FBD, or even EBD, as if the arc EF were*   231

---

[1]This calculation of all the distances may be used to determine the "excesses" that Kepler tried to find through a mathematical short cut in Ch. 43. Subtracting 36,000,000 (sum of the radii of the circle) and dividing the remainder by 2 gives 37,781. See p. 335.

imperceptible. Thus, from knowing CD, the angle EBD was given. Therefore, in tri-angle EBA, from the angle EBA and the sides EB, BA, I sought the length of AE, whence I subtracted AC or AG computed earlier, and the remainder was CE or CF, the amount that the other end of CD approached the center B. So, when CE is bisected (for this can be done within the limits of perception) the approach of CD towards B was known, if it were to approach equally at all its points. But from the amount of approach, the optical parallax, or the apparent size of CD, was also given: that is, the angle CBD now corrected, which was previously assumed to be a little smaller, though in our numbers there is no error. Therefore, given the angle CBD, now corrected, which is the supplement of angle CBA, and the side CA, and the eccentricity BA, the equated anomaly CAD, which was sought, was given.

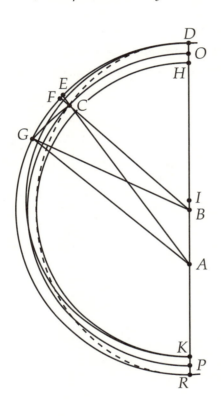

It was impossible to use this method to establish independently any equation other than the first, at a mean anomaly of 1°. All the rest, all the way to the 180th, always presupposed that the equation immediately preceding was known. I can't imagine anyone reading this not being overcome by the tedium of it even in the reading. So the reader may well judge how much vexation we (my calculator and I) derived hence, as we thrice followed this method through the 180° of anomaly, changing the eccentricity each time.

But the foundation of this calculation has not yet been laid out. For I said that I presupposed knowledge of the length of the whole oval. Whence, then, is this known? As for me, once I had descended to this clumsy numbering procedure, I did not manage to avoid clumsily presupposing the length, and then, when the whole thing was complete, seeing whether in the 180th operation I came out with an apparent position of more than 180°, or less. For if it had come out at exactly 180°, I knew that the length I had assumed for the oval was good, but if it was less, I had assumed it to be too small, and if more, greater.

Nevertheless, we are not left without a kind of geometrical helping hand for making a good guess as to the oval's length. *For as BD is to BA let BA be to DH, extended from D towards B. By what was demonstrated in Chapter 46, the rectangle contained by the breadth of the lunule and the semidiameter of the circle is nearly equal to the square on the eccentricity. Therefore, by Euclid VI. 17, the eccentricity is the mean proportional between the breadth of the lunule and the semidiameter. But this is how the diagram is set up. Therefore, DH is the breadth of the lunule.*

Let the half of DH also be taken, and extended from B towards D, and let it be BI. And about center I, with radius ID, let the circle DK be described, touching the eccentric at D. And also, about center B, with radius BH, let the circle HK be described, touching the previous circle at K. It is obvious that the circle HK is smaller than DK, and the circle DGR is greater than DK. And because circular circumferences are to one another as their semidiameters, as BD is to DI and BH, so is the greater circle DG to the smaller circles DK and KH. But DI is the arithmetic mean between DB and HB, because BI is half of HD. Therefore, the circle DK too, touching the smaller and larger circles described about the same center B, is the arithmetic mean between those circles that it touches.

If the oval path is continued, by hypothesis it will touch the greater circle at aphelion D and perihelion R, and to the smaller circle HK at the middle elongations. Thus, 232 *it is greater than the smaller circle HK, and smaller than the greater circle DR. It is therefore likely that the oval circumference is not much different from the length of the circular circumference DK.*

The following demonstration, however, makes one believe it to be a little larger.

Let the mean proportional between BH and BD be taken, and let it be BO, and about center B with radius BO let the circle OP be described. Thus, by Archimedes, On Spheroids 5, *the area of this circle OP will be equal to the area of an ellipse whose longer semidiameter is BD, and whose shorter is BH. And because the greatest of all figures of equal perimeter is the circle, conversely (in the common significance of the term), of figures with equal area, the one with the shortest perimeter will be the circle. Therefore, since the proposed figures, namely, the ellipse which has semidiameters DB, BH, and the circle OP, are equal in area, from the evidence just presented, the circumference of the ellipse will be longer than the circumference of the circle OP. But BO is imperceptibly less than ID, since BO is taken as the geometrical mean, and ID the arithmetic mean, between the same terms. For by the theory presented in Euclid's Book V,[2] since BO is the mean proportional between HB, BD, as the lesser, HB, is to the greater, BD, so is HO, the excess of the mean, to the defect, OD.[3] Therefore, since HB is less than BD, HO will also be less than OD. But BI is equal to the half of HD. Therefore, BI is greater than HO, and less than OD. Therefore, to the common semidiameter HB of the least circle HK unequals are added, namely, less than the half of DH to make BO, and half of DH to make DI. Therefore, DI is greater than BO. Thus the circle DK is greater than OP. But it is only imperceptibly greater, since DH is less than a hundredth of DB. Therefore, on the supposition that these circles are superabundantly equal, and on the supposition that the oval is a perfect ellipse, the circumference of the oval will be a little longer than the circle DK, and certainly longer than the circle OP. And because in Ch. 47 above DH was 858 where DB was 100,000, let half of DH, 429, be subtracted from DB, 100,000. The remainder will be 99,571. Next, as 100,000 is to 99,571, so almost exactly will the circumference of the circle be to the circumference of the oval, which is sought. And because the circumference of the circle has 360 degrees or 21,600', or 1,296,000'', a small part will be removed which has 5560'' or 92' 40''. And for the semi-circumference of the oval, 46' 20'' are to be subtracted, or even less, if the oval exceeds the circle DK at the place considered in the measuring.* As for me, I made no use of any demonstration, but by a most laborious and dogged calculation found the defect of the oval

---

[2]Euclid V.17.

[3]This is true of the *subduplicate* ratio of *HB* to *BD* rather than of the ratio itself. However, from *HB < BD* it follows correctly that *HO < OD*, and this inequality is what Kepler needs.

semicircle to be 45′ 45″, so that where the perfect semicircle is 180°, the oval would be 179° 14′ 15″.

Now this shortening of the oval circumference is necessarily equal to the opposite optical lengthening (for although this oval is shorter, it nonetheless appears contained within two right angles, or exactly 180 degrees, and is judged to be that long). Hence the reader may with good reason doubt whether in this process it is necessary first to shorten the entire oval, and afterwards to lengthen it again by parts optically. For from the diagram it seems apparent that the shortening is at its greatest where the approach towards the center *B* is greatest, and vice versa.

If in fact these variations did happen to be equal, the following method for computing the equations would arise.

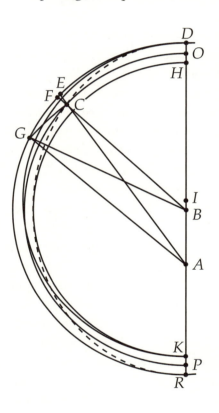

The first mean anomaly would be GBD, from which the distance GA would be computed, which, added to AD, the distance of the other end of the preceding from GD (which is always 1°), and the sum being halved, would give the uniform distance of the arc CD (the same for all its points). And we would then say that as the length of the semicircle is to the sum of all the distances on the semicircle, so is this distance of the arc GD to the length of FD, which is the apparent size of CD seen from B. Now from FD, as if it were a measure of the angle CBD, and from AC, AB, we would find the equated anomaly CAD by a shorter path than before.

But the reader should know that these two variations do not keep in step. For the optical amplification which arises from the approach of the path *DC* to the center *B*, happens chiefly about the middle elongations, and hardly at all at aphelion and perihelion. Contrariwise, the shortening of the oval path, which arises from the incursion of the planet towards the center, is about the same everywhere. For two opposite distances at the middle elongations of the eccentric add up to the sum of two near the line of apsides, one near aphelion and the other near perihelion. But the arcs of the oval circumference are in the inverse ratio of the distances. Therefore, two of these arcs, at the middle elongations, will be equal to two other arcs, one near aphelion and the other near perihelion. If these arcs of the oval path are equal, the diminution of the arcs at all four places will be approximately equal also. This is confirmed by experiment. For if the defect of the oval semicircle is 45′ 15″, the defect of the 180th part of the oval would be about 14″ about aphelion. And the amplification from the approach of the oval does not equal one second at aphelion.

So, as concerns the proposed ocular estimation of the diagram, it is not quite as simple as was said in the objection, that the shortening of the oval and its optical amplification compensate one another. This would indeed be so if all the arcs of the oval path were presented directly when seen from center *B*. But this happens only at the middle elongations. Near the apsides, on the other hand, these arcs are not at the same distance at both ends. Therefore, they are not made to be as much greater in their appearance, by approaching, as they are made smaller by being shortened.

So, following this method, I constructed equations for Mars at all degrees of the eccentric, and I did it three times. For the first time I took an eccentricity of 9165 that was not great enough, thinking that I had thus made this very certain through my treatment of the areas. And then I used more than 180° in the figuring, when I should have used less.

So, when this last operation showed more than 180°, which was absurd, for the second trial I assumed half the oval to be 179° 14′ 15″. At a mean anomaly of 45°, the result was:

equated anomaly	38°	5′	33″
while the vicarious hypothesis of Chapter 16 said . . .	38	4	54
Difference			39

At anomaly 90—equated anomaly	79°	31′	31″
The vicarious, index of truth	79	27	41
Difference		3	50

At anomaly 135. Equated	127	0	1
The veracious vicarious	126	51	9
Difference		8	52

And from this, chiefly from the anomaly of 90°, I realized that the eccentricity of 9165 was too small. This I corrected, using the method presented in passing in the preceding chapter. For seeing that at the middle elongations we have 3′ 50″ too much in the greatest equations, half of it, 1′ 55″, is given to the optical part, and the rest to the physical. And since 9165 subtends 5° 15′ 30″, you take 5° 17′ 25″, which yields 9227. And so, with a new eccentricity of 9230 (which is hardly different from the 9264 which I found in Ch. 42, nor is it much farther from 9282, which is half of the eccentricity of the equant in Ch. 16) I went through the whole job again. *First, the distances GA or CA were constructed at the individual whole numbers of degrees of equated distance anomaly GAD. Next, these were brought over to GD or GBD, the whole number of degrees of mean distance anomaly. Third, adjacent pairs, such as GA, AD, were added. Fourth, division by those divisors was carried out one hundred eighty times. The sum is 358° 28′ 30″, which is the length of the oval path. Fifth, the individual arcs of the oval path were added to one another in order. Sixth, the optical amplifications were borrowed from the previous unsuccessful operation, since I saw that over two computations they differed hardly at all. So these, too were added to the above sum, in order. Seventh, the sums of the arcs were increased by the sums of the optical amplifications.* Eighth, from the angle *CBD* thus found about the center of the eccentric *B*, and the distance *CA* as opposite side, and the eccentricity *AB* as the third side, I sought out the 180 angles of optical equation *ACB*, whence the total equations and the equated anomalies were derived. The resulting anomalies were:

234

Mean	Equated			in the Vicarious			Difference	
45	38°	2'	24″	38°	4'	54″	2'	30″
90	79	26	49	79	27	41	0	52
135	126	56	25	126	52	0	4	25

So the eccentricity still can be increased, and up above, [moving] from aphelion, the planet is made to be slightly slower than it should be, and the same near perihelion, and therefore it is swifter than it should be at the middle elongations, as was also found before in Chapter 47. So too many of the distances seem to be collected near the apsides, and not as many as required, or not as long as required, about the middle elongations. But a consideration of this follows in its proper place.

So, when I saw that the more skillfully and the more conveniently the physical causes, introduced in Ch. 45, are called upon for directing the principles of calculation, the closer I always have approached to the true equations provided by the vicarious hypothesis of Ch. 16, I greatly congratulated myself, and was confirmed in the opinion of Chapter 45.

On the other hand, since I was revulsed at the multiple contrivances[4] with which I contended in this chapter, I did not rest until I had established a more certain and direct way, and at the same time I began to suspect that what the opinion of Chapter 45 had required had in no way been achieved by the calculation.

---

[4] ἀτεχνίας.

CHAPTER 49

A critical examination of the previous method for the equations, and a more concise method, based upon the principles constituting the oval in the opinion of Chapter 45.

So, in order to see the cause of the contrivance[1] in this method, now fully presented, consider upon what foundations it rests. The planet is supposed to move uniformly on the epicycle, and to be swept around by the sun nonuniformly, according to the distance. From these two principles of motion, the oval path arises. But this method does not allow one to know what portion of the oval path corresponds to a given time, even if the distance of that portion be known, unless the length of the whole oval be known from the beginning. And the length of the oval cannot be known, except through the measure of the incursion of the planet from the circumference of the circle at the sides. But further, the measure of this incursion is not known before it is known what portion of the oval path is traversed in any given time. This, as you see, begs the question, and in our operation we presupposed what was being sought, namely, the length of the oval. This is not just a fault in our understanding, but is utterly alien to the primeval ordainer of the planetary courses: we have hitherto not found such an ungeometrical[2] preconception[3] in the rest of his works. Therefore a different approach must be taken for calling the opinion of Chapter 45 to the calculations, or if this cannot be done, the opinion itself will totter owing to its being suspected of circularity of argument.

From this consideration the implication has occurred to us that in using the uniform measure of time, we divided the composite oval path into unequal parts, and thus measured out the parts of this composite oval, unequal but equated again by the compensation of the distances, by the equal increments of time of the planet on all of them. And indeed, we had among our presuppositions that only the other power (that which is from the sun), intensifies according to distance, the power proper to the planet doing so not at all. But now, in this undertaking, we in a way encumber both forces with this relation to the distances, because we give the common work of the two, the oval, to the planet, according to the measure of the distances to be traversed.

Therefore, although we have approached rather close to the truth in the effect of this method, we have nothing in which to glory that the opinion of Chapter 45 has been expressed by it if we are abandoned by reason. We would therefore have appeared to be going about our business more correctly if, dis-

236

---

[1] ἀτεχνίας.

[2] ἀγεωμέτρητον. This is an allusion to the famous saying, ἀεὶ ὁ θεὸς γεωμετρεῖ (God eternally does geometry) attributed to Plato by Plutarch (*Quaestiones Convivales* 8.2.1).

[3] The "preconception" in question is presumably the prior knowledge of the length of the oval, which Kepler notes cannot be determined geometrically.

missing the composite oval and the quadrature of its area, the subject matter of Chapters 46, 47, and 48, we were to convert the calculation to the principles themselves of the oval path, assumed in Chapter 45. Let Chapter 45 be taken up again, *and about A, the center of the sun's body, with radius AD, let the circle DG of the center of the epicycle be described; and another, with center A, and radius AB, the circle of the aphelion; in which AGB is the line of apsides, and let the planet, when it is at aphelion,[4] be at B. Now let some time have passed from when the planet was at B, and let its measure be CDE, the angle on the epicycle, in order that as the aphelion of the epicycle B is translated to C, and the center of the epicycle G to D, the planet will have moved on the epicycle from C to E. Therefore, in order to know the angle DAB at the time CDE, consider that the planet has passed across from B to E by two powers. One made it move nearer the sun, and at the same time also drew it away from the line AC or AD, on which it had been previously, when AC was at AB.*

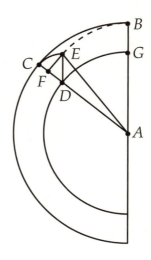

The other moved the planet forward along with the center of the epicycle, so that the center D of the epicycle was on the line AC, although it was formerly on AB. Now the power that drives the center of the epicycle around in a time designated by 360°, moves through 360°, or four right angles, about A, owing to the sum of 360 distances. Therefore, the sum of any number of distances being given from the time CDE, as before, the angle DAB will also be given. For the impression which the sun makes upon the body of the planet through the mediation of the distances AB, AE, is also presumed to make the same impression upon the center of the epicycle GD.

This is because if the planet had not disengaged itself from the ray of power AB or AC and move towards B, but only descended towards the sun, it would then still be

* Under a certain special condition, these things are true, namely, if the rays of power from the sun act as the planet's location, like a cart in which the planet is carried, as we suppose here. In itself, however, it is not true. On this point see the "first way" in Chapter 39. For among the five absurdities rejected there, we here omit only one, the last, and keep the other four.

237

at the point F on AC, on which line the center of the epicycle D also lies. And it has disengaged itself by the law of the epicycle, at the distance DE and the angle CDE (for this is prescribed by the opinion of Chapter 45, under which we are operating here). Thus by a kind of fiction for itself it places the center of its epicycle at D. For we have said in Ch. 39 how it is to be imagined that the power or fictitious radii of power AB, AC, and so on, serve as a position for the planet. Now, though, the ratio of the arcs BE of the oval path to the whole oval is not quite the same as the ratio of the corresponding arcs GD of the perfect circle to the whole circle. But neither is it true that as BC is to the whole perimeter of the circle BC so is the arc BF of the oval to the whole oval. But this should not stand in our way, for BE, and BF too, are composed of two powers, and if anything is disturbed in the proportion this comes from the planet's making its own descent on the circumference of the epicycle (following the opinion of Chapter 45). For if the planet had remained at the highest point of its epicycle, and had been subject to the same motive force from the sun, adumbrated by AB, AE, which is nonuniform (which indeed cannot happen simultaneously, for when the distance of the planet from the sun remains the same, the motive strength from the sun remains the same), then a*

---

[4] ἀφήλιος.

perfect arc of a greater circle BC would have been described, whose ratio to the whole BC is the same which the arc GD has to the whole GD.

I am indeed aware that if the planet is supposed to be on a smaller perimeter, that of the center of the epicycle DG, it will go much faster. But that is not a reason for assigning a greater speed to the center of the epicycle. For, by supposition, the center of the epicycle moves, not in its own right, since it is not a body, but because of the planet. It is thus presupposed that the planet moves its own body away from the solar rays according to the law of the epicycle, and makes use of certain rays of power from the sun for its position (ideas which were indeed rejected in Ch. 39, but taken up again in Ch. 45 in considerably altered form, and which are retained here in order to explain my attempts). On this basis, the foundation of the subsequent calculation is sound, whatever its result might be. For the oval is present here no less than before, since DE and AB do not remain parallel. For to the extent that the long distances AB, AE exceed the medium distances AG, AD, the arc DG, or the angle DAG, is made less than the angle CDE, the measure of the time. Thus DE inclines towards B, and E consequently makes an incursion from the circumference of the circle towards BA. For by Chapter 2, if DE had remained parallel to AB, then E would have been on the circumference.

This gives rise to the following method. Let the distances be found for each degree of mean anomaly. The method you have in Ch. 39 above, and I also used it above in Ch. 47 and 48. First, distances are found for non-integral degrees of mean anomaly, or *CE*. Then, by interpolation, they are carried over to integral degrees of *CE*. If you find this meandering route annoying, and if the greater labor of the direct way pleases you, and if, further, you want to have the whole thing presented in one overview, proceed thus.

Measure the time, or the artificial units expressing time, which is the astronomers' mean anomaly, on the epicycle CE, from its aphelion C, in a direction opposite the series of signs.[5] Thus the angle ADE, or its supplement CDE, is given as a whole number of degrees of mean anomaly. The radius AD is also given as 100,000, and the radius of the epicycle DE is 9264. Therefore, part of the equation, DAE, will be given, and the distance AE. Put both of these into a catalog, with the mean anomaly CE adjoined, for future use. In this manner, let all the distances AE be gathered and added, and the sum will be found to be about 36,075,562. For this sum was found using an eccentricity only slightly different from our present one, which is 9264. The 360th part of this has the value 100,210, and the same fraction of four right angles is one degree. Therefore, as each of the distances, in order, is to the distance 100,210, so is the arc of this distance 100,210 (60′) to the arcs belonging to the other distances. For, as was frequently announced in Ch. 39, 47, and 48, the ratio is inverse. Next, multiplying 60 minutes, or 3600 seconds, by 100,210, and dividing the product by each of the distances in the semicircle, 180 times, or better, by half the sum of adjacent pairs of distances (following the advice given in Ch. 48), yields the angles of the center of the epicycle, DAG. Next, beginning from the two least values of DAG, add them, and to the sum add the third, and again add the fourth to the sum of the three preceding, and so on, until you have accumulated all the 180 degrees. And if your final sum comes out to be exactly 180° it

238

---

[5]That is, in the direction of decreasing celestial longitude.

will prove to you that you did everything right, never departing from the instructions. And let these sums of yours, which are the angles DAG, again be inscribed in a catalog with the corresponding mean anomalies in the margin, for ready reference.

So, since an integral equation is to be computed, that is, the equated anomaly for a given mean anomaly, first, with the mean anomaly CDE measured on the epicycle, you extract the angle DAG or CAB from the latter catalog, the one with the sum of the angles. And with the same mean anomaly, you also extract the part of the equation CAE from the previous catalog. And when this is subtracted from the angle DAB, the remainder is the equated anomaly EAB. The variations in the other semicircle are known.

Let the mean anomaly be 45°.

The sum of its distances gives DAG as . . .	41°	26′	0″
By the same anomaly, the part CAE of the equation is given as	3°	30′	17″
Therefore, the equated anomaly EAB is	37°	56′	43″
Our vicarious hypothesis said	38°	4′	54″

Difference   8′

In this manner, at

Mean anomalies	We found an equated		While in the truer vicarious hypothesis		Difference
45	37	56 43	38	5	8 −
90	79	26 35	79	27	0
120	110	28 8	110	18½	9½ +
150	144	16 49	144⁶	8	9 +

Near the apsides the planet is made to be slower than it should be, and near the middle elongations swifter than it should be.

You will say that we have come out worse, since in Ch. 48 we came nearer the truth in our results. But, O good reader, if I were concerned with results, I could have avoided all this work, being content with the vicarious hypothesis. Be it known, therefore, that these errors are going to be our path to the truth. Meanwhile, let us be assured that at last we have brought the physical causes, which we supposed in Chapter 45, at least once to a calculation entirely free of error. At the same time, moreover, the calculation of Chapter 47, above, is confirmed, since this one is equivalent; and it is certain that what we held in suspicion there as ungeometrical,[7] have not obstructed us in any perceptible way. Thus if there remains any discrepancy between these equalities and the truth, it is to be attributed, not to the method of applying numbers, but to the opinion of Ch. 45, whence these numbers flow. This is to say, not that the opinion itself has immediately become totally false, but just that we have been hasty. For instead of waiting for the plenary judgement of the observations, when we understood the planet's path to be oval we immediately seized upon a certain quantity for that oval, solely on account of the elegance of the physical

239

---

[6]In *KGW* 3 p. 313, the number of degrees in the two equated anomalies is 114, while in the first edition they are 114 and 144, respectively. They should obviously both be 144°.

[7]ἀγεωμέτρητα. See the note to p. 365, above.

causes and the graceful uniformity of the epicyclic motion, which was falsely given credence.

Now the manner in which the ultimate and truest opinion is to be brought to a calculation, and made to conform most closely to these chapters, will be told in its place.* I am now going to complete the unfolding of my remaining trials.

* Ch. 56, 58, 59, 60.

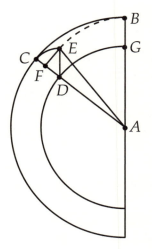

CHAPTER 50

On six other ways by which an attempt was made to construct the eccentric equations.

How small a heap of grain we have gathered from this threshing! But you also see what a huge cloud of husks there is now. They ought to have been hauled back to the beginning of Ch. 48, since before I investigated the arcs of the oval path I would have dealt with them. But for the sake of bringing light, it was appropriate to winnow them. Besides, we might end up finding a few useful grains.

IN THE FIRST AND SECOND METHOD, THE PROCEDURE WAS THIS

First, with eccentricity 9165, which is a little less than the correct value, I sought out all the distances, according to the procedure shown in Chapter 29.

Term:
What is the "anomaly of distance"

These corresponded to integral degrees of an anomaly occupying a middle position between the mean anomaly and the true equated anomaly. Although I am calling it "equated" for the time being, I nevertheless add a condition, that it be used only for the distances. I therefore name it "anomaly of distance".* *In the second diagram of Chapter 46 it is the angle FAB, and in the following one, CAD.*

* Although it is quantitatively a mean with respect to the others, you should nonetheless beware of calling it "mean," for in its proper usage, "mean" denotes the [anomaly of] time.

Second, I sought out the third proportional lines, each of which was to its distance as this distance was to the radius, 100, 000.

Third and fourth, I added the lines so found one by one,[1] and the sum of the distances was 35, 924, 252, less than 36, 000, 000. The cause of this you have in Chapter 40. But the sum of the proportional lines was found to be 36, 000, 000, which holds me in wonder. And because it is delightful, I wish some geometer would prove it to be necessary.[2] *About centers A, B let two equal circles IH and DC be described, and let the centers A, B be joined, and AB extended so as to intersect the circle about A at I and K and the circle about B at D and L. Then let the circle about A be divided into any number of equal parts, such as 360, beginning from I. And from A let lines AI, AH, AK, and so on, be drawn through the points of division I, H, K, and so on, intersecting the circle about B at the points D, C, L. Then let it be that as AI is*

A problem proposed to geometers. Since elsewhere there are three anomalies, of which 1. is called the mean, 2. the eccentric, and 3. the equated, in this diagram and in this particular attempt, in order to avoid confusion, let us understand the first to be in the arc CD or the angle CBD, the second to be in the angle CAD or the arc ED, and the third to be in the angle EAD.

---

[1]Reading "singillatim" for "sigillatim."

[2]The problem Kepler is proposing amounts to the straightforward (for modern mathematicians) evaluation of an integral, and the value does indeed turn out to be 360 times the radius. The procedure is sketched out by Caspar in *KGW* 3 p. 474.

to AD, so is AD to AG; and as AH is to AC, so is AC to AF; and finally, as AK is   240
to AL, so is AL to AM; and so on for all the rest. Let, I say, a geometer demonstrate
that the sum of the last 360, AG, AF, AM, joined together, is equal to the sum of the
first 360, AI, AH, AK, joined together.

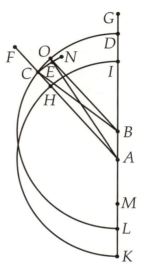

So, in this first method using the sums of the distances, I intended one thing (although erroneously and irrelevantly, namely, to add up the arcs *CD* or angles *CBD*, even though they were given at the beginning), but accomplished another, again erroneously. For I obtained, not the arcs, nor the angles, nor the path lengths, but the times on the unequal arcs of the planet's path, as if they were equal. And, following the rule of proportion, I said that as the sum of the means, *AD*, *AE*, *AL*, which is 35,924,252, is to the elapsed time of 360°, so is the sum of any number of distances to its elapsed time over the length of the path that includes these distances. *Let A be the sun, B the center of the eccentric CD, BC the semidiameter. Let B, A be joined to C. Here the distances CA were found corresponding to integral degrees of the angle CAD, and thus to unequal arcs on the circle CD, something that escaped my notice. So let CAD be 45°. From CB, BA, the angle CBD is given as 48° 42′ 59″. Therefore, if there were no physical cause of the equation, and CBD were the measure of the time or mean anomaly, then there would correspond to it this value for CAD, truly equated. But the planet is slower on CD, owing to its greater distance from A, and the distances are the measure of the elapsed time. Therefore, at the anomaly CAD of 45° I collected 45 distances at the beginning of the arcs, which are longer. Their sum was 4,869,307. I also collected 45 shorter ones, the ones at the ends of the arcs, by subtracting the longest, AD, 109,165, from the sum of the 46 distances, 4,975,577. The remainder was 4,866,412. The mean between the two sums, 4,867,852, I reduced to degrees, where 35,924,252 have the value 360°, or 99,790 have the value 1°. The result of this procedure was 48° 46′ 51″. And this ought to have been the time corresponding to the angle CAD. But the arc CD or the angle CBD was also found to be about that much, 48° 42′ 51″. This is absurd, and contrary to the hypothesis, which requires the planet to be slower at CD. The cause of this absurdity was immediately clear, namely, that in order to know the elapsed time for CD it would have been proper to take the distances corresponding to equal arcs of CD, while these distances just taken correspond to unequal arcs of CD, and are greater to the extent that the distances themselves are longer, by Ch. 32. Therefore, these distances had too small a numerical value.* Nevertheless, in order that I not lose all this labor, I subtracted the excess of this number of the elapsed time over the number of the angle *CAD*, from *CAD*, so as to leave as a remainder *EAD*, 41° 13′ 9″, and so that *AC*, *AE* might be equal. Here it was supposed that in the time *CBD* the planet traverses an angle *EBD* about the center of the eccentric *B* equal to *CAD*, and therefore, that as many distances from *A* were collected for the equal arcs *ED* of its eccentric as we found here on equal degrees of *CAD*. Thus, the same number of arcs which, in this calculation of ours, were spread out over *CD*, which were unequal and, in this locality, too great, are now understood to be compacted within the confines of arc *ED*, now divided into equal parts. Therefore, the angle *CBD* would here be the mean anomaly of distance, giving the angle *CAD*, for finding the distances *CA*, from which the angle *CAE*, the physical slowing and translation of *CA* to *EA*, is deduced.

I call it "mean," not from its being an intermediate quantity among three, but from the uniform and mean motion of time which this measures; that is, insofar as it is the distances that are being sought.

241

Although it cannot show much of a discrepancy from the prior method of Chapter 49, this procedure assumes without demonstration that *CAD* and *EBD* are equal, and consequently, that *CA* and *EB* are parallel, which was refuted above in the second diagram of Ch. 46. But now see how close this operation comes in its results. For

This differs slightly from the theory of Ch. 49 and the two in Ch. 47.

At a mean anomaly of			Was found an equated			Which in the vicarious is			Difference
48°	42′	59″	41°	13′	9″	41°	21′	0″	8−
95	15	31	84	44	18	84	39	18	5+
138	42	59	131	20	24	131	4	7	16+

The eccentricity was charged with being too small, and indeed, it really is greater, 9264 instead of 9165. Also, the planet was made to go too slowly near the apsides, and too fast near the middle elongations. But, dismissing this first method, which we seized upon by chance in thinking over the error committed at the beginning, let us turn to the implementation of the second method, born from a consideration of the same error.

In the second attempt, the third anomaly is *CAD*, the second is *CD* or *CBD*, and the first, the sum of the few lines *AG*, *AF*, whose measure is, by supposition, the area *CAD*, approximately as in Ch. 43.

The distances scattered over *CAD* had approximately the same numerical value as the sector *CBD*, and led the argument to an absurdity (for just as the area *CAD*, an approximate measure of the distances, is greater than the area of the sector *CBD*, the numerical value of the distances *CD* also had to have been greater than the sector *CBD*). The question therefore followed: did the lines *AF*, *AG* proportional to *AC*, *AD* correctly express the elapsed time over *CD*, thus allowing *CAD* to remain the true equated anomaly? The answer is, they did not. For if so, *AC* will remain in its place, which is the same at which its length was computed. Therefore, the orbit will be a perfect circle, which was refuted in Ch. 44. Therefore, the distances at the middle elongations, coming out longer than they should be, will make the planet slower than it should be there, and hence it will be faster at the apsides. But look at the result of the operation, which testifies to this itself.

For

Nearly coincides with the physical hypothesis of a perfect circle, Ch. 43.

At an equated anomaly of	There followed a mean anomaly of			But in the vicarious		Difference	
45°	52°	39′	40″	52°	53′	13′	−
90	100	29	12	100	34½	5	−
135	142	10	47	142	9	2	+

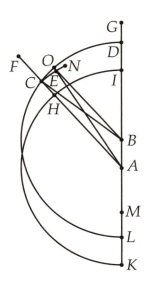

First, the eccentricity is charged with being too small, since the maximum equation comes out to be $10° \, 29\frac{1}{5}'$, which is $10° \, 34\frac{1}{2}'$ in the vicarious hypothesis. Second, at the time $52° \, 39\frac{2}{3}'$ the planet is found to have traversed as much of its path from the apsis as was traversed in the vicarious in the longer time of $52° \, 53'$. If the eccentricity were adjusted, all the values of the equated anomaly would be increased, so that, in the lower quadrant also, in the time of $37° \, 44'$ (the supplement of $142° \, 16'$, which has been adjusted by increasing the eccentricity), the planet will traverse as much of the path as it traverses in the vicarious in the longer time of $37° \, 51'$, which is the supplement of $142° \, 9'$: that is, both will traverse $45°$, which is the supplement of $135°$.

By the way, this false hypothesis has come very close to giving us true results: the difference, after correction, is not more than $8'$ or $7'$ at either place. You thus see that results must not be trusted. And you will note again what was observed in Ch. 47, that the truth lies at the midpoint between these two methods (the latter of which describes a perfect circle, and the former an oval, following the opinion of Ch. 45). Therefore, you can at least conclude now, as well as before in Ch. 47, that the lunules to be cut off from the perfect circle have only half the breadth of the one that follows from the opinion of Ch. 45. 242

THIRD AND FOURTH METHOD

So, since this second method did not accord with reason either, and in the first I learned that the distances are to be sought out corresponding to integral degrees of the angle CBD or to equal arcs of the eccentric CD, I also proceeded to the distances.

So, fifth (I am enumerating for you only those operations each of which is performed 180 times), I made use of interpolation to relate the distances found previously by dividing the mean anomalies CBD minutewise, or unequally, to the mean anomalies which are equal or are of an integral number of degrees. But now, CBD no longer remained the [mean] anomaly, as it formerly was in the first method, but was made the eccentric anomaly by this relating of the distances, as it also is in the second method.

Sixth, using the same distances as before, I sought out their proportionals, that is, the lines that are to the distances as the distances are to the radius, $100,000$. But this was unnecessary. Still, I wanted to be aware of all the possibilities.

Seventh and eighth, I again added the individual magnitudes, both of the distances AD, AC, and of their proportionals AG, AF. The sum of the distances came out to be $36,075,562$. The reason for its coming out greater than $36,000,000$ you have in Ch. 40. The sum of the proportionals came out to $36,384,621$.

I call an anomaly "minutewise" when it is not expressed as a whole number of degrees, but has some minutes added on.

In the third attempt, as in the second, CAD is again the third anomaly, CBD or CD the second, and the more densely crowded lines AD, AC, or the area measuring their sum (the area CAD), are the first anomaly, which is usually called the mean.

Now, in the previous diagram, we shall proceed demonstratively, using the equated anomaly *CAD* to obtain the eccentric anomaly *CBD*, and further, through this eccentric anomaly *CBD* learning the sum of the distances found on the arc *CD*. And by this sum of the distances, we shall obtain the elapsed time over the arc *CD*, or the mean anomaly. Or, in reverse order, for convenience's sake, if an angle *CBD* of an integral number of degrees (such as 45°) is used to find *CAB*, and 45 correct distances are obtained, these things, I say, follow demonstratively. But again, as before in the second method, CAD becomes the true equated anomaly, and thus *CA* remains in place, and the orbit *DC* will be a perfect circle. Since this is false, as was shown in Ch. 44, it necessarily follows that the distances at the middle elongations are taken to be too long here, and consequently that the times are made longer than they should be, and shorter at the apsides.

This method will in all respects be almost exactly equivalent to the previous method, using proportionals. For the number of distances we have gathered is now greater than before, to about the same extent that the proportionals, the same in number as the distances, were longer than the distances then. But, for safety's sake, witness the result of this calculation. For

at a simple anomaly			there results an equated			While in the vicarious			Difference	
48°	38′	31″	41°	31′	0″	41°	17′	6″	14′+	nearly co-
95	13	58	84	45	50	84	37	45	8+	incides with
138	45	41	131	1	52	131	7	13	5−	the preceding.

243

The eccentricity is again charged with being less than it should be. In other respects, the errors are the same as in the preceding one. The reason why the signs for excess are turned into signs for defect, is that here the difference shows errors in the equated anomaly, while there they are in the mean anomaly. And this is the third method.

In the fourth try, if any remedy were attempted, a monster would be produced, with CBD the third anomaly, and the area CAD the second anomaly. But the sum of the more densely packed lines FA, GA, is the first anomaly.

In substituting the proportionals *AG*, *AF* for the distances *AD*, *AC*, which is the fourth method, we are going to make the two parts of the equation into three. For the area *CAD* measures the sum of the distances *CA*, *DA*. It is therefore much less than the sum of the lines *FA*, *GA*. And even if we attempt a remedy like the one used on the first method, we shall still have doubled our errors. For since the distances themselves cannot be admitted, owing to their excessive length at the middle elongations, the proportionals will be even less tolerable, since they are longer. And if you want to test them with the results of a calculation, you will find that to a mean anomaly of 53° 23′ 56″ there corresponds an equated anomaly of 46° 0′, which in the vicarious hypothesis comes out to be about 45° 27′, a difference of 33′, clearly absurd.

Fifth and sixth method

So, since I accomplished nothing with these four methods, I then took the mean anomaly and the distances assigned to it (in the fifth operation) and went over to the table of the vicarious hypothesis of Chapter 16, and the true equated anomaly. Let the second diagram in Chapter 46 be taken up again.

Then, the distances *AF* belonging to integral degrees of mean anomaly *IBF* or *IDH*, also belong to the degrees and minutes of the equated anomaly *IAH*, which in the table mentioned corresponded to the mean anomaly *IDH* itself. Therefore,

Ninth, I related these distances obtained from the minutewise equated anomalies of the vicarious hypothesis of Ch. 16, that is, from the unequal angles *HAI*, to the individual integral degrees of the equated anomaly *HAI*, that is, to equal parts.

Tenth, with the same distances, thus set up, I sought the [third] proportionals, as in the second and sixth operations.

Eleventh and twelfth, I added each, according to its kind, and the sum of the distances was 35,770,014, and the sum of the proportionals, 35,692,048. In this translation of distances we moved all the long ones upwards, and made them fewer, establishing large arcs *IG* of the oval path above, at aphelion, and thus attributing the individual distances to the individual degrees, not of the anomaly *FAB*, as in the first method, but of HAB, which is the true equated anomaly. There are no more of these degrees in the upper semicircle than in the lower. Therefore, there are now more of the short distances than there are of the long ones, whence it not only happens that the sum of the 360 distances comes out smaller than the sum of the 360 diameters, but also the sum of the proportionals comes out smaller than was the sum of the distances themselves.

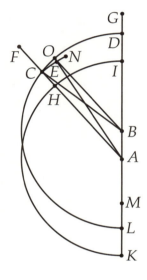

So, in the matter of the fifth method and the sums of the distances, reason again cries out against the method of basing equations upon it. Let the diagram belonging to this chapter be repeated, and let it be recalled to mind what was said about the first method. For in it the equated anomaly of distance *CAD* was divided into equal degrees. Hence it happened that *CD* was divided into unequal and [excessively] large parts, and had too few distances. From this, seizing upon a sort of accidental remedy for the error, we concluded from the sum of the distances on *CD* that to those distances belonged a shorter arc *ED*, so as to transfer *AC* over to *AE*, and thus *ED* could be obtained divided into equal parts with a distance established at any of its degrees. However, it was not from the sum of the distances found on *CD*, but from a mingling of the vicarious hypothesis with the hypothesis of the distances framed in Chapter 46, that the translation of *AC* to *AE* was now made and perfected, and the arc *ED* was attributed to the mean anomaly (which we have numbered on the arc *CD* for finding the distance *CA* or *EA*). And this was nonetheless done in such a way that *BE* and *AC* are not exactly parallel, as in the first method. And now that this has been done by a mingling of hypotheses, as I was saying, there is no need to go through the operation again, as in the first method. Instead, only one thing needs to be found: do the few distances *AC*, *AE* gathered into one sum in this fifth method, produce the same physical equation

244

Note the respect in which this is "mean." See the marginalia above.

that resulted artificially from the two mingled hypotheses?

Consider here how the distances are arranged in this last procedure. The angle *EAD*, whose terminus *E* is at distance *AC* from the sun, was divided into equal degrees in this last procedure, to each of which was given a distance. In this way the arc *ED* of the oval path standing upon the angle *EAD* ends up divided into unequal parts, and it receives too few distances. So the mean anomaly already given from the vicarious hypothesis cannot be had from the sum of the distances on *EAD*.

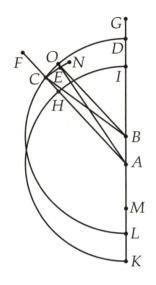

Now in the first method above, when *CD* received too few distances, with the angle *CAD* divided into equal degrees we substituted *ED* for *CD* as the arc suited to those distances. So similarly, in this fifth method, since *ED* has received too few distances, with the angle *EAD* divided into equal degrees, if it is permissible again to make use of a clumsy remedy, we would substitute *ND* for *ED*, as the arc to which those distances belong. *In order to find the distance CA let the mean anomaly CBD be 48° 44′. Given angle B, and CB, BA, CA is given as 105,784,[3] and CAB as 45°. The vicarious hypothesis requires AC to be transferred to AE. And we now divide ED, which the vicarious hypothesis declares to be 41° 22′, into equal degrees, and through them we collected no more than 41 distances and part of a 42nd. And these, gathered into a single sum, will constitute a mean anomaly which is by no means equal to the one first taken, DC, but is equal to another, DO, which shows the distance AO, to be transferred to AN.* "Work was begun on an amphora; why has a pitcher come forth from the whirling wheel?"[4] For the question was whether all the distances on the equal degrees of *ED*, gethered into a sum, would show the mean anomaly *DC*. But the operation gave me an answer concerning *ND*, and the anomaly *DO*.

245

Finally, let us turn to the sixth method, and the proportionals that are adapted to the demonstration of Ch. 32. For the true quantities of orbital arcs that appear equal from the center of the sun are in the ratio of the distances: thus to the extent that *AE* is longer, so is *ED*.

But the times of truly equal arcs, as measured on the orbit, are also in the ratio of the distances. For to the extent that *ED* is farther away from *A*, the planet also takes longer to traverse the arc *ED*.

Therefore, the times that the planet accumulates on those arcs that appear equal from the center of the sun, are in the duplicate ratio of the distances.

But *AF* is to *AH* likewise in the duplicate ratio of the distance *AC* or *AE* to the mean *AH*. And so the measures of the times that the planet accumulates over equal degrees of the angle *EAD* are the proportional lines *AG*, *AF*,

In this fifth method the third anomaly is indeed EAD, and its mean anomaly (first in order) is CD or CBD, and also the same anomaly of distance for the distance CA or EA. But the area EAD measures some sum of distances EA, DA, which is alien to this equated anomaly EAD, since it is the measure of time belonging to the arc DN, and the equated anomaly DAN. Again, therefore, a monster.

This sixth method, after a very simple correction of those things in which the opinion of Ch. 45 still errs, can also be of use in the truest physical hypothesis; and it is succinct and evident.

---

[3]This is the number that would be obtained if *AB* were 8478, *CBD* meanwhile remaining 48° 44′. Where *AB* is 9165, as it has been throughout this chapter, *CA* is 106,268 at this angle. If *CAB* is to be 45°, *CBD* should be 48° 43′, giving *CA* a value of 106,270.

[4]Horace, *Ars Poetica* 21–2.

belonging to the integral degrees or equal parts of the angle of true equated anomaly *EAD*.

Therefore, let the proportionals to the distances at equal degrees of equated anomaly be tested, just as other distances were also tested above in this chapter. As, since 35,692,048, the sum of all 360 distances, at all 360 equal parts of the angle at the sun, is equivalent to an elapsed time of 360°, what is the value of a just and correct sum at any given degree of equated anomaly?

In this manner is found

at equated anoma- lies of	Mean anoma- lies			Those yielded by the vicarious			Difference	
41	48°	24′	3″	48°	19′	2″	5 +	Coincides with those of Chapter 49
81	91	30	39	91	34	8	$3\frac{1}{2}$ −	
91	101	28	10	101	34	7	6 −	
131	138	28	5	138	39	28	11 −	

The eccentricity is again charged with being less than it should be. With this corrected, the difference above, at 41°, will be about 8′+, and below, about $7\frac{1}{2}'$ −. Thus here, too, the planet is not made to go fast enough at the apsides, and so there are too many distances near the apsides, and consequently less than there should be at the middle elongations. But it comes quite close to the truth, and clearly coincides with the method of Chapter 49. For if you consider well, the same thing is done here as was done in Ch. 49. There we computed the optical part of the equation by itself, and the physical part also by itself, while here we are computing them both together. There we had introduced fictitious rays of power in order to be able also to ascribe to the epicycle its own task of disengaging itself from those fictitious rays (for in the truth of the matter no rays go around as slowly as the center of the planetary epicycle goes, as was said in Ch. 39). And nevertheless, we left all the physical force carrying the planet around (as concerns the effect) to the sun, so that the epicycle would only serve to adjust the distances. Here, we made use of the same power of the sun for the physical translation, while we again computed the distances from the epicycle, and gave its equal parts to equal times, that is, to equal degrees of mean anomaly, as the opinion of Ch. 45 maintains. And if we ended up taking as many distances in any given part of the time as there are degrees of equated anomaly, they are still derived from the distances of the mean anomaly, and in length they are the same. And this form is so much easier that we can here put aside the other persuasion concerning the planet's epicyclic motion, and take one step closer to the truth of the physical cause, leaving to the epicyclic mode nothing but a reciprocation on the diameter—although this is still flawed, as has was clear from these equations, at least. For, as was noted just above in considering the second method, this preoccupation with epicyclic motion is excessive, showing distances at the middle elongations that are too small, from which it happens that at that place the planet exceeds its measure of speed, and at the apsides falls short of its measure. But it suffices us to express the opinion of Ch. 45 in our calculation. Therefore, although one might here raise

In this sixth method, the third anomaly is *EAD*, the second *ED*, while the first is the sum of the lines *AG*, *AF*, where *AF* or *AC* is understood to be translated to *AE*. Nonetheless, in computing the distance *AE*, that is, *AC* (from which *AF* is derived), *DC* or *DBC* is also the first anomaly. It is depicted twice here because two things are being investigated: time and distance.

246

the objection from Ch. 32 that this ratio of diurnal motions cannot be constant, since the parts of the eccentric near the apsides are presented directly to the sun, and the intermediate are presented obliquely, so that they thus appear differently from the way they would if they were presented directly— if any-one, I say, were to raise this objection, I shall answer as I did in Ch. 49, that this obliquity at the intermediate parts is added by the planet in its own right, and is produced through its descent. Thus, it is not to be imputed to the motive cause arising from the sun, nor is it affected by that cause.

Therefore, studious reader, from such a great number of chapters and methods, you have only two methods of equating that conform to the opinion of Ch. 45. One is by the physical hypothesis intermingled with an epicycle set up on the longitude, described in Ch. 49. The other is by this chapter, and its sixth method, in accordance with a more purely physical hypothesis, where the epicycle governs nothing but the descent towards the sun, or if anyone should wish to set it up to affect the latitude, [the epicycle should be] perpendicular to the plane of the ecliptic. And both of these two use different means to produce the same effect. You will thus be able to place confidence in them more safely when examining the opinion of Chapter 45.

And through a hitherto empty trust in the true physical causes that have been discovered, let a triumph over Mars once again be celebrated. Now, some rumor, I know not what, calls me to new tumults and new labors.[5]

---

[5]Cf. Horace, *Satires* II.126: "Saeviat, atque novos moveat fortuna tumultus;...". Although Kepler's concluding sentence has the ring of a literary allusion, no closer match has yet been found.

Chapter 51

Distances of Mars from the sun are explored and compared, at an equal distance from aphelion on either semicircle; and at the same time the trustworthiness of the vicarious hypothesis is explored.

While I am thus celebrating a triumph over the motions of Mars, and fetter him in the prison of tables and the leg-irons of eccentric equations, considering him utterly defeated, it is announced in various places that the victory is futile, and war is breaking out again with full force. For while the enemy was in the house as a captive, and hence lightly esteemed, he burst all the chains of the equations and broke out of the prison of the tables. That is, no method administered geometrically under the rule of the opinion of Ch. 45 was able to emulate in numerical accuracy the vicarious hypothesis of Chapter 16 (which has true equations derived from false causes). Outdoors, meanwhile, spies positioned throughout the whole circuit of the eccentric—I mean the true distances—have overthrown my troops of physical causes called forth from Ch. 45, and have shaken off their yoke, retaking liberty. And now there was not much to prevent the fugitive enemy's joining forces with his fellow rebels and reducing me to desperation, unless I had sent new reinforcements of physical reasoning in a hurry to the scattered, straggling veterans, and, informed with all diligence, had stuck to the trail without delay in the direction whither the captive had fled. In the following few chapters, I shall be telling of both these campaigns in the order in which they were waged.

247

And, to speak initially of the first of these, I shall begin by seeking out the distances of several places on the eccentric where the evidence is most trustworthy. Therefore, let it be our intention to explore the distances near the mean anomaly of 90° and 270°.

On 1589 May 6 at $11\frac{1}{3}^h$ Mars was observed at 27° $7\frac{1}{3}'$ Libra with latitude 0° $6\frac{2}{3}'$ north.[1] The true position of the sun at this time is calculated as 25° $48\frac{2}{3}'$ Taurus, and its distance from earth 101,361. The mean longitude of Mars was 7^s 26° 0′ 36″, and therefore its eccentric position was 15° 32′ 13″ Scorpio. But our vicarious hypothesis of Chapter 16 did not come nearer than $2\frac{1}{3}'$ to the true or observed position of Mars in an acronychal situation, and thus in so sensitive a procedure the computation of the equated anomaly cannot be trusted. Therefore, to the method of Chapters 27, 28, and 42 I shall add another observation, which nevertheless uses a freer method. Indeed, as I remarked in Ch. 12 above, Mars was not very often observed twice in this region. Therefore, we should be content with two observations. For with this one just now presented, there is associated another, from 1594 December 28.

> At a mean anomaly of 87.

> A short cut: given the distance of a planet with no latitude from a fixed star with a known latitude, to find the planet's longitude.

---

[1] *TBOO* 11 p. 334 (raw observation).

At $7\frac{1}{4}^{h}$ on the morning of that day, Mars's mean longitude is calculated to be 7^{s} 26° 13′ 39″,[2] a few minutes beyond the other. And at that time Mars, at an altitude of eight or nine degrees, was observed to be 50° 34′ from Spica Virginis. *So, since it stood very close to the ecliptic, in the right triangle between Spica, its ecliptic position, and Mars, the base is given as 50° 34′ and the side between Spica and the ecliptic is 1° 59′, which is Spica's latitude. Therefore, the remaining side is 50° 32′ 18″. Thus, since Spica was at 18° 11′ Libra, Mars fell at 8° 43′ 18″ Sagittarius. The declination of this position from the equator was 21° 50′ 20″ [S.].*

However, Mars was found to have a declination of 21° 41′ [S.]. Therefore, it displayed a small amount of north latitude, 9′ 20″. And on the following 1595 January 4 it still had 3′ of north latitude.[3] Our observation is hereby confirmed. But if you assume this to be the true latitude of Mars, its ecliptic position will not be changed perceptibly. So you may safely pronounce its position to be 8° 43′ Sagittarius. And because Mars was near the sun, it was very far from earth, and thus had a much smaller parallax than the sun, which we shall ignore. But we cannot similarly ignore the refraction, which I shall now remove. *For the sun's position was 16° 47′ 10″ Capricorn, distance from earth 98,232, and its right ascension was 288° 12′. Therefore, 306° 57′ on the equator was rising, and along with it 29° Sagittarius,[4] at which the angle between the ecliptic and the horizon is 26°, its complement 64°. And because the altitudinal refraction as shown by the table of refractions of the fixed stars is 6′ 30″, and from the table of the sun, 11′, when the star is at an altitude of $8\frac{1}{2}$°, 5′ 51″ or 9′ 53″ are to be subtracted from the latitude. The latitude from the former would be 3′ 29″ N., and from the latter 0′ 33″ S. And the longitudinal refraction is 2′ 39″ or 4′ 34″.[5]*

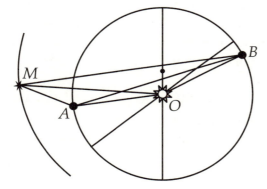

Of these two methods of determining the refraction, I shall follow the one which is confirmed by the latitudes, as follows. *In the earlier observation, the observed latitude was $6\frac{2}{3}$ North. And because Mars was near the earth, and the angle at the sun was 10° 17′, while at the earth it was 28° 41′, this latitude requires an inclination of 2′ 30″. Therefore, in our later observation the inclination will also be 2′ 30″, or a little less, since we are 8′ closer to the node. But with the inclination assumed to be 2′ 30″, since here the angle at the sun is 61°, and at earth 38°, a latitude of about 1′ 50″ N. must follow,*

---

[2]This should be 7^{s} 26° 17′ 40″. The mean longitude given by Kepler corresponds to a time of 4:15 am; perhaps he used the wrong hour. In any event, this has no effect upon the position, which he calculates with respect to Spica, Brahe's reference star for this observation. The observation itself appears in *TBOO* 12 p. 348.

[3]*TBOO* 12 p. 441.

[4]Recomputation indicates this to be incorrect. The correct rising degree would be 9° Capricorn, and the angle between the ecliptic and the horizon would be 38°, complement 52°. This change has little effect, however.

[5]The latitudinal refraction should be 5′ 7″ or 8′ 40″, making the corrected latitude 4′ 13″ N. or 0° 40″ N. The longitudinal refraction should be 3′ 28″ or 5′ 52″.

as indicated by our parallactic table.[6] *But by using the refraction of the fixed stars, we were left with a latitude of 3′ 29″ N., while by using the solar refraction we were moved down to 0° 33′ S. Thus in the second our refraction was greater than would be correct, and in the first, less. So the correct refraction is between the two, namely, 3′ 36″.*   That is, for us Mars will be put at $8° 46\frac{1}{3}′$ Sagittarius. *Let O be the sun, B and A points on the earth's orbit, A the position of the earth in the earlier observation, B in the later, M Mars. Let the lines be connected. And although Mars does not return to exactly the same place, let it nonetheless be represented in both instances by the line OM. Thus MAO is 28° 41′ 14″, and AO is 101,365. Let MO, the distance of Mars from the sun (which is being sought here), be taken as if known, and let it be 154,200. Thus OM will fall at 15° 31′ 3″ Scorpio. And if OM is assumed to be 154,200 in the earlier observation, it ought to be taken as shorter in the later one. Now one degree at this place on the eccentric changes the distance by 240 units, whatever form you use for constructing the distances.*[7] *Therefore, since the mean longitudes here differ by 13 minutes, and when a subtraction is made to account for precession, only eight, the proportional part of 240 is 32. Consequently, in the second observation, we have assumed OM to be 154,168. But OBM is also known, it being 38° 0′ 40″, and OB is 98,232. Therefore, OMB is given as 23° 6′ 11″. Consequently, on the second occasion OM was at 15° 40′ 9″ Scorpio, differing from the earlier eccentric position by 9 minutes. It should have differed by somewhat more. For the mean anomalies differed by 8′ 3″, to which corresponds 7′ 49″ in the equated anomaly of the eccentric at this place. Add to these the intervening precession of the equinoxes of 4′ 48″. Thus 12′ 37″ are accumulated. Mars therefore ought to fall at 15° 43′ 40″ Scorpio. Therefore, we should take a somewhat different value for the distance OM, and should change it so that the lines represented by OM move about another $2\frac{2}{3}′$ apart from each other. For when the earth is at A, OM should move back in longitude , and forward when the earth is at B. But this occurs if you increase OM. So let it be 154,400 in the first instance, and in the second, 154,368. For then OM falls first at 15° 29′ 34″ Scorpio, and second at 15° 42′ 18″ Scorpio.*

And for the first time the mean anomaly is 87° 9′ 24″, and 87° 16′ 30″[8] for the second. This will do for the mean longitude of the earlier one.

For the other mean longitude, an observation in the month of December 1595 will serve, being well supported by the consensus of a number of consecutive days, and at a place where the vicarious hypothesis also exactly represented the acronychal position of Mars in the preceding October. For the sake of consensus, we shall also add an observation of October 1597. In the other years it was not observed at this eccentric position. For the eccentric position falls at 10° Gemini. Thus Mars was observed last at this place in November 1580.[9] In 1582 its arrival at the place fell in October, when [Tycho's]

At a mean anomaly of the full-circle complement of 87.

249

---

[6]The parallactic table appears in Kepler's *Optics*. The table is bound into the 1604 edition at p. 275, and in *KGW* 2 p. 240. In the English translation, both parts of the table are in a pocket inside the back cover. Instructions for its use are to be found on fol. 320 of the 1604 edition, *KGW* 2 pp. 275-6; in the English translation on pp. 329–30; or in the translator's introduction to the present work.

[7]On the oval path, that is. The change on the circle would be greater.

[8]This should have been about four minutes greater, owing to the error in the mean longitude.

[9]*TBOO* 10 p. 84.

great interest in observing was not yet aroused. In 1584 it came in September, in 1586 in July, in 1588 in June, in 1590 in April, and in 1592 in March, at which times, being near the sun owing to the short, bright nights in Denmark, it was neglected, while, whenever there was an opportunity, they were intent upon the fixed stars, the moon, and the other planets. But at the end of 1593 and the beginning of 1594, when it was at quadrature with the sun, the observation was not continued beyond this aspect because astronomers are usually chiefly interested in the quadrature. So, in 1595 Dec. 17 at 7^h 6^m in the evening the planet was observed at 11° 31' 27'' Taurus, with latitude 1° 40' 44'' N.[10] The sun's position was 5° 39' 3'' Capricorn. Its distance from earth was 98, 200.

The mean longitude of Mars is concluded to be 2^s 2° 4' 22''.

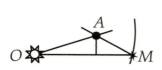

And since the aphelion is 4^s 28° 58' 10'', the distance of the position from aphelion is 86° 53' 48'' backwards. Previously, it was nearly the same as that, namely, 87° 9' 24''. Therefore, these two positions are nearly the same distance from aphelion. Now from our vicarious hypothesis, there corresponds to this simple anomaly an equated anomaly of 76° 25' 48''. Which, subtracted from the position of the aphelion, leaves Mars's eccentric position, 12° 32' 22'' Gemini. *Let A be the earth, O the sun, M Mars. AO is given as 98, 200. And because OM is at 12° 32' 22'' Gemini, while AM is at 11° 31' 27'' Taurus, therefore AMO is 31° 0' 55''. And because AO is at 5° 39' 3'' Capricorn, while AM is at 11° 31' 27'' Taurus, therefore, the supplement of OAM is 54° 7' 36''. Hence, since the sine of OAM is to OM as the sine of AMO is to AO, OM comes out to be* 154, 432. And since this position is 15 minutes closer to the apogee than the one from 1589, and at this position on the eccentric 1° causes a change of 240 units, therefore 60 units are to be subtracted for 15 minutes, since at positions farther removed from aphelion the distances are shorter. The distance thus comes out to be 154, 372. On the other hand, since the node is at about 16° 20' Taurus, and the eccentric position is 12° 32' Gemini, the planet is 26° 12' from the node. And the maximum inclination of the planes is 1° 50'. Therefore, the inclination at this place is 48' 32''. The secant of this exceeds the radius by 10 units, or in our dimensions, $15\frac{1}{2}$. So the distance of the actual point on Mars's orbit from the sun is 154, 387. And previously, at this distance from aphelion it was found to be approximately 154, 400 from the sun. Therefore, the distances of these two points on the eccentric from the sun is equal to within a hair's breadth. For the 13 units that are wanting in the latter distance are of no importance. I shall rejoice if I am able to come within an uncertainty of 100 units everywhere.

And now I shall add [the observation from] 1597, not so much to confirm the previous ones, which are certain in themselves, as to give the reader an opportunity to compare the observations of Tycho with the observations of others, by which means he will at length understand how much that man has benefited us. There do indeed exist observations of that author from the last

---

[10]*TBOO* 12 p. 458.

days of October of 1597,[11] but they were taken with a radius while travelling, and not brought to a calculation by the author himself, who knew how to correct distances taken with a radius, by applying a kind of parallactic table for the eye, as he informed us in the *Progymnasmata*. And so, since very different distances are ascribed to the same moment (possibly because they were supposed to be corrected immediately after the observations), they are to be dismissed. But at the same moment, I, though absent in Styria, made an observation, and (strange to say) did so with the eyes of Tycho Brahe, standing by the shore of the Baltic Sea. Here is the series of observations—can you hold your laughter, friends?[12]

On 1597 November 8, Saturday, or October 29, in the morning, Mars was   250 not yet on the line from the twelfth of Gemini to the fourth.[13] On the following day, it had already left it: it was nearer the ninth than the twelfth, and precisely on the line from 11 to 9, and also on the line from 1 to 5, or a little farther east. And the fifth was halfway between the first and Mars.

From these the position of Mars can be elicited, when certain stellar positions are assumed from Tycho Brahe's catalogue, which I had just now been professing to be my eyes. But because the Ninth is not described in Brahe's catalogue (since in its stead in the ninth place is another, distant by more than three degrees from the Ptolemaic, and less than all of them), we shall call upon the latitude of Mars as our counsel. For an approximate knowledge of it will suffice. *Now the mean longitude of Mars on the morning of October 29 at 5^h (which is an estimate, since I did not write down the time) is found to be 1^s 29° 10′ 43″. Therefore, its eccentric position was 9° 43′ Gemini, 23° 20′ from the node. Therefore, the inclination was 43′ 52″. But the sun was at 15° 40′ Scorpio, and the apparent position of Mars, by anticipation, was about $12\frac{1}{2}$° Cancer. Therefore, the latitude was 1° 36′ 24″.[14]* Let a computation be made of the longitude of a point on the line from the twelfth to the fourth, having a latitude of 1° $30\frac{1}{2}$′ North. *Since the*

---

[11] See for example *TBOO* 13 p. 11.

[12] Horace, *Ars Poetica* 5. Since the original is a question, I have replaced Kepler's period with a question mark. Kepler and his readers would also have been aware that the passage from Horace from which this quotation comes was used by Copernicus as one of the main arguments for his system, as contrasted with Ptolemy's jumble of disparate elements. See Copernicus, "Preface and Dedication to Pope Paul III," *De Revolutionibus*, Nuremberg 1543, fol. 3B.

[13] The numbers denote the order of the stars in Ptolemy's catalogue, under the constellation of Gemini. Tycho also followed the same order for the most part, except that some of his entries seem to denote different stars, as Kepler remarks below. Modern designations of the stars mentioned by Kepler are:

First	Castor
Fourth	Tau Geminorum
Fifth	Iota Geminorum
Ninth	(uncertain)
Eleventh	Zeta Geminorum
Twelfth	Delta Geminorum.

At the right is a diagram of the configuration.

[14] This should be about 1° 30′, and indeed, immediately below Kepler restates this as if it had been 1° 30′ 24″. Most likely Kepler's printer made the easy error of mistaking a "0" for a "6". Subsequent editions and translations have left this error uncorrected.

fourth is at 9° 54′ Cancer, latitude 7° 43′ N., and the twelfth is at 12° 56′ Cancer, lat.
0° 13½′ S., the longitude of our point, by interpolation, will be 12° 16′ 17″ Cancer.
But on 29 October, Mars was not yet here, and on the 30th it had already passed it.
The diurnal motion was no more than 5 minutes, half of which is 2½′, so that on the
morning of the 30th it was at 12° 18½′ Cancer. And so indeed it was at the end of
1600, but in 1597 it was at 12° 16′ Cancer. Three minutes of error in latitude
barely produce one in longitude. So the position is certain enough. *If you also*
explore it using the first and the fifth, and use the point on that line whose latitude
is 1° 30½′, it falls at 12° 9′ Cancer. And Mars was farther east, that is, forward in
longitude, at about 12° 16′ or a little before, also intermediate. Therefore, the latitude
computed by us is confirmed. For it should be approximately intermediate, and indeed
it is. That is, between the Martian latitude of 1° 30½′ and the 5° 42½′ of the fifth there
is 4° 12′, and between this and the 10° 2′ of the first there is an intermediate 4° 20′.

Therefore, let Mars be at 12° 16′ Cancer. On 1597 October 30 at 5^h am the
position of the sun is found to be 16° 38′ 8″ Scorpio, distance 98,820. [Mars's]
mean longitude was 1^s 29° 42′ 10″, aphelion 4^s 28° 57′ 10″, supplement of the
mean anomaly 89° 15′, of the equated anomaly 78° 43′ 23″, eccentric position
10° 13′ 47″ Gemini. Therefore, a distance of 153,753 is called forth from these.
And since we are are 2° 6′ lower from aphelion than before, we will add twice
240, the sum of the units corresponding to one degree:

	240
	240
And a tenth part:	24
Likewise another 15 parts, for the substitution of a line in the plane of Mars's orbit for the line in the plane of the ecliptic:	15
	153753
Produces	154272
Previous value	154400
Difference	128

251   If you subtract three minutes from the position of Mars, so that it would
be at 12° 13′ Cancer, which could be done in our observation, especially if the
time were different, this difference would be reconciled.

Second, I shall prove the same thing at parts closer to aphelion. On 1589
April 5 at 11^h 33^m Mars was observed at 7° 31′ 10″ Scorpio with latitude 1° 28′
13″ N.[15] It was very near the meridian, and consequently there were no horizontal variations. The mean longitude is concluded to be 7^s 9° 46′ 8″. And the
aphelion is at 4^s 28° 51′ 8″. Therefore, the mean anomaly is 70° 55′ 0″, to which

At a mean anomaly of 71°.

corresponds an equated anomaly of 61° 17′ 35″, by the vicarious hypothesis.
And so the eccentric position is 0° 8′ 43″ Scorpio; the sun's position, 25° 52′
43″ Aries; its distance from earth 100,560; the angle at the earth 11° 38′ 27″; at
the planet 7° 22′ 27″. Therefore, the distance of Mars from the sun is 158,090.[16]

---

[15]*TBOO* 11 p. 332.

[16]The translator's computations result in an eccentric position 20″ less than that given by
  Kepler—an insignificant difference, surely, yet this increases the distance to 158,208, owing
  to the small angles at the sun and Mars. Hence, this distance must be regarded as having a
  rather large uncertainty.

But again, so as not to trust the eccentric position, on account of the error of about two or three minutes which the vicarious hypothesis commits at this position on the eccentric, we shall appropriate a counterpart from 1591 Feb. 19, when, at $5\frac{1}{2}^h$ am Mars was observed to be 28° 11′ from the southern pan of Libra[17] (which in that year was at 9° $23\frac{1}{2}$′ Scorpio), with a latitude of 0° 26′ North.[18] So Mars fell at about 7° $34\frac{1}{2}$′ Sagittarius, approximately.[19] But since that eccentric position has a declination from the equator of 21° 39′ 10″, [while] the observed declination of Mars was 20° 50′ 30″, its latitude was therefore 48′ 40″. From this, the longitude is corrected, which becomes 7° $34\frac{1}{3}$′ Sagittarius. But the mean longitude is 7^s 8° 21′ 47″, to which corresponds an equated 59° 57′ 38″, and an eccentric position of 28° 51′ Libra. Therefore, the angle at the planet is 38° 43′ 20″; the sun's position 10° 14′ 25″ Pisces; therefore, the angle at the earth, 87° 20′ 0″; and the distance of the sun from the earth, 99, 210. Thus the distance of Mars from the sun comes out here to be 158, 428, longer than before, because here we are also nearer to aphelion by 1° 26′ 30″.[20] But at this place on the eccentric for one degree about 220 units are owed from the distance, or for the entire angular difference, 317, so that this place, if we carry it back to an anomaly similar to the preceding, has a distance of 158, 111 rather precisely. Whence it is proved that these two eccentric positions, treated by the method given above, will show exactly the same eccentric position as our vicarious hypothesis, except that on account of our nearness to 17° Scorpio we run the risk of being in error by one or two minutes. Moreover, in the latter of these, the distance from Aquila comes out to 54° 12′, which is not within 12′ of agreeing with the other observed data, and consequently this observation is not perfectly certain. Also, some small quantity should be added, owing to the latitude.

A suitable observation at a similar longitude on the other semicircle occurred on 1582 November 12 at $6\frac{3}{4}^h$ am, when the sun's position was 29° 35′ 17″ Scorpio.[21] Its distance was 98, 503, mean longitude of Mars 2^s 15° 10′ 20″, aphelion 4^s 28° 44′ 20″. Hence, the full-circle complement of the mean anomaly was 73° 34′, and of the equated anomaly, 63° 45′ 18″. Hence, the eccentric position was 24° 59′ 2″ Gemini. Then, I say, the planet was observed at 26° 35′ 30″ Cancer, making the angle of vision, the one at the earth, 57° 0′ 13″, and at the planet, 31° 36′ 28″. By these data it is determined that the planet's distance from the sun was 157, 631. And because the anomaly was previously 70° 55′, and is now 73° 34′, we are therefore lower by 2° 39′. For this, in the previously mentioned ratio, 586 units are owed. So from the analogy of this observation with the previous one, at a similar anomaly, a distance of 158, 217

At a mean anomaly of the full circle complement of 71°.

252

---

[17]This is α Librae.

[18]*TBOO* 12 p. 139.

[19]Substituting 7° $34\frac{1}{2}$′ for an obviously incorrect 7° $24\frac{1}{2}$′.

[20]As can easily be seen from the two equated anomalies, the second position is actually only 1° 20′ nearer aphelion than the first, which would result in a smaller adjustment and consequently a larger distance. However, since the distance for the first position was probably greater than that stated by Kepler, the two are very nearly in agreement (158, 200 for the first and 158, 150 for the second).

[21]*TBOO* 10 p. 174

is appropriate, where again about the same amount as before, or a little more, is to be added for the latitude. The difference is about 127, which is excused owing to the uncertainty of the prior observations. For it is very small, and may be neglected in our present undertaking, where we are considering magnitudes of 1800 or 3600 or still more.

At a mean anomaly of the full circle complement of 43.

But let us move yet higher towards aphelion, and explore those places where, by what was shown in Ch. 6, the dislocation of the eccentric occasioned by exchanging the sun's mean motion for its true motion can occur most clearly, namely, at the sun's apogee and the sign of Cancer.

On 1596 March 9 at $7^h 40^m$ pm, when the sun's position was 29° 31′ 24″ Pisces, the distance from earth 99,764, Mars's mean longitude 3^s 15° 35′ 0″, aphelion 4^s 28° 58′ 31″, the full-circle complement of the mean anomaly 43° 23′ 31″, equated anomaly 36° 40′ 2″, eccentric position from the vicarious hypothesis 22° 18′ 29″ Cancer—the planet was observed at 15° 49′ 12″ Gemini, latitude 1° 47′ 40″ N.[22] Therefore, the angle at the earth was 76° 17′ 48″, at the planet 36° 29′ 17″. Therefore, the distance of Mars from the sun was 162,994, or more correctly, this was the distance of the point in the plane of the ecliptic perpendicularly beneath the body of Mars.

But, for safety's sake, let another observation be added. And Mars was at precisely the same sidereal position on 1584 Nov. 25 at $10^h 20^m$, when the sun was at 14° 0′ 3″ Sagittarius, distance from earth 98,318. The mean anomaly was not perceptibly different from the previous one, because the motion of the aphelion is only very slightly faster than the motion of the fixed stars. Therefore, the eccentric position is the same, 22° 8′ 44″ Cancer, if you subtract the precession of 9′ 45″. But the planet was observed on Nov. 12 at $13^h 26^m$ at 23° 14′ 15″ Leo,[23] with latitude 2° 12′ 24″ N. On the 20th of Nov. following, at $18^h 30^m$ astronomically,[24] it appeared at 26° 0′ 30″ Leo.[25] Thus in 8 days 5 hours it was moved forward 2° 46′ 25″. In Magini, this is 2° 48′. Therefore, since our time follows by 4 days $15^h 49^m$, to which corresponds 1° 28′ of motion from Magini,[26] we shall add 1° 27′ according to the above ratio. So Mars could have been observed at 27° 27′ 30″ Leo, approximately. Therefore, the angle at the earth was 73° 27′ 27″, at the planet 35° 18′ 46″. Hence, the distance of Mars from the sun here was 163,051, exceeding the previous one by 57 units. This can be absorbed by a very slight change in the eccentric position, as, indeed, the vicarious hypothesis is not trustworthy to within one minute. Furthermore, I could easily have made some slight error in the application of the observation.

At a mean anomaly of 43°.

For a similar longitude in the other semicircle, we shall again take up the

---

[22]*TBOO* 13 p. 47.

[23]*TBOO* 10 p. 321. This was incorrectly copied as 23°14′5″, and stands thus in all editions.

[24]This indicates that the time is reckoned from noon; that is, that this observation was at 6:30 am on November 21.

[25]*TBOO* 10 p. 321 (raw observation).

[26]According to the figures from Magini, the diurnal motion was 20′ 28″, and the total for four days 15 hours 49 minutes would be 1° $35\frac{1}{3}$′. This results in a distance of 162,658, considerably less than that computed by Kepler.

observations of Chapter 27. There I derived a distance somewhat less than 163, 100 using the equation of the observations, but from the bare observations themselves I obtained 162, 818, in the plane of the ecliptic as before. Now for one of the times introduced in that chapter, 1589 Feb. 12[27] at 5^h 13^m am, the mean longitude was 6^s 12° 38′ 44″,[28] aphelion 4^s 28° 50′ 57″; and consequently, the mean anomaly was 43° 47′ 48″, lower than our previous one by 24 minutes. To this there corresponds [an adjustment of] about 64 units at this position on the eccentric. So the distance which was less than 163, 137[29] at an anomaly of   253 43° 48′ will again be increased at anomaly 43° 24′ according to this ratio, so as to make it almost exactly 163, 100 in this semicircle. In the previous one it was 163, 051 and 162, 996. Again, not an outstanding fit.

It should be noted, however, that in Chapter 27, to which I am referring here, the observations led us to subtract 1′ 30″ from the eccentric position computed from our vicarious hypothesis at $5\frac{1}{2}$° Libra, and this was through the observations of 1585, 1587, 1589, and 1590. Second, in Chapter 18 above, the acronychal observation of 1589, at 5° Scorpio, gave the same testimony, namely, that our vicarious hypothesis needs to be diminished by $2\frac{1}{5}$′. And in 1591, at 26° Sagittarius, there still was one minute to be subtracted. Third, in this very chapter, about 16° Scorpio, the observations of 1589 and 1594 required $3\frac{1}{2}$ minutes to be subtracted from the eccentric position computed from our vicarious hypothesis. So therefore this is constant about the middle elongation of this semicircle.

And likewise, near aphelion, we shall again take up the observations of   At a mean anomaly of 12°. Chapter 28, where at a mean anomaly of 11° 37′ a distance of 166, 180 or 166, 208 was found (without correction for latitude). This was in the descending semicircle. But it was at a similar anomaly on the ascending semicircle at   At a mean anomaly of the the following times.   full-circle complement of 12°.

On 1585 January 24 at 9^h, when the position of the sun was 15° 9′ 5″ Aquarius, its distance from earth 98, 590, the mean longitude of Mars 4^s 16° 50′ 10″, the aphelion 4^s 28° 46′ 41″, the remainder of the mean anomaly to complete the full circle 11° 56′ 31″, and the consequent eccentric position from the vicarious hypothesis 18° 49′ 0″ Leo: the planet was observed at 24° 9′ 30″ Leo, latitude 4° 31′ 0″ N.[30] The angle at earth was therefore 9° 0′ 25″, and at the planet 5° 20′ 30″. Therefore, the distance of Mars from the sun was 165, 792. But if

---

[27]The text has February 11, probably because the observation was on the morning following February 11. In Chapter 27 it is given as February 12, and this agrees with the eccentric position given there.

[28]This is the mean longitude for the same time one day later. The correct mean longitude for this time would be 6^s 12° 7′. The corresponding mean anomaly is 43° 16′, $7\frac{1}{2}$′ *higher* (that is, nearer aphelion) than before. So instead of adding 64 parts, Kepler should have subtracted 20, which would have resulted in a distance of about 163, 117, which is close to the distances in the other semicircle, as he stated them (but note that 163, 051 should have been 162, 658). Oddly enough, it is also close to what Kepler himself found: he would have noted more of a discrepancy, but in adding 64 to 163, 137 he found the sum to be 163,100!

[29]That is, 163, 100 increased by 37 to correct for the transfer from the plane of the ecliptic to the plane of Mars's orbit. See the end of Chapter 27.

[30]*TBOO* 10 p. 387, where the given longitude is 24° 4′ 30″ Leo. Neither latitude nor declination are given.

you subtract 1′ 30″ from the vicarious hypothesis here, as appeared necessary above in Ch. 18 in the computation of the acronychal opposition, the angle at the planet will be 5° 19′, and the distance of Mars from the sun 166,580. And the distance here is easily changed to this extent, because earth and Mars are close to one another. Therefore, for insurance, we shall bring in other positions.

On 1586 December 16 at $6\frac{1}{2}^h$ am, when the sun was at 4° 16′ 51″ Capricorn, 98,200 distant from earth, mean longitude of Mars 4^s 18′ 39′ 9″, remainder of mean anomaly 10° 9′ 41″, eccentric position from the vicarious hypothesis 20° 20′ 30″ Leo: the declination of Mars was found to be 3° 54′, right ascension from Arcturus and Spica 177° 27′.[31] Thus its longitude was 26° 6′ 24″ Virgo, latitude 2° 35′. Hence, the angle at earth was 81° 49′ 33″, at the planet 35° 45′ 54″. And the distance was 166,311, but by subtraction of 1′ 30″ from the eccentric position, 166,208. And at the previous distance from aphelion, 11° 37′, it would be about 70 units less. So it would be either 166,241 or 166,138.

On 1588 Nov. 6 at 6^h 50^m am, when the sun's position was 24° 3′ 43″ Scorpio, 98,630 distant from earth, mean longitude of Mars 4^s 20° 47′ 35″, remainder of the anomaly 8° 2′ 51″, eccentric position from the vicarious hypothesis 22° 7′ 48″ Leo: Mars was observed at 23° 16′ Virgo, lat. 1° 37′.[32] Hence, the angle at earth was 60° 47′ 43″, at the planet 31° 8′ 12″. And thus the planet's distance from the sun was 166,511. But by subtraction of 1′ 30″ from the position of the vicarious hypothesis, the distance becomes 166,396. And by this analogy, at the greater distance from aphelion of 11° 37′, where it is less by about 110, it is either 166,401 or 166,286. There is a discrepancy of 150 between this and the previous one. And if, keeping the correction of the eccentric position, we take a mean between the two, 166,230, as if saying that in the two observations of 1586 and 1588 there were some small observational errors in opposite senses [in the determination of the distance], we will hardly differ at all from the distance in the descending semicircle. Even this small difference will be able to be abolished by a slight retraction of the aphelion, of which more later. Thus, near aphelion also, as far as the senses can judge, we find the same distances from the sun at the same relationship to the aphelion in the two semicircles.

All three observations were made when Mars was in the east, and none with Mars in the west. For the rest of the observations [sc. of Mars at these mean longitudes] are lacking. Therefore, we will probably be safer to stay with the distance in the descending semicircle.

Third, let the same things we have explored above the middle elongations now be explored below, near perihelion.

In 1591 on the night following May 13, at 1 hour 40^m past midnight, when the sun was at 2° 8′ 43″ Gemini, distant from earth 101,487, while the mean longitude of Mars was 8^s 22° 18′ 4″, anomaly 113° 24′ 4″, equated 103° 15′ 48″, consequent eccentric position from the vicarious hypothesis 12° 9′ 48″

At a mean anomaly of 113 degrees.

---

[31] *TBOO* 11 p. 66. The right ascension is not given directly, but presumably was computed by Kepler using the distances from the reference stars.

[32] *TBOO* 11 p. 275 (raw observation).

254

Sagittarius (or, by analogy with 26 Sagittarius nearby, just now mentioned, 12°
8¾′ Sagittarius): Mars was observed at 2° 24½′ Capricorn, latitude 2° 15′ S.[33]
Therefore, the angle at the earth was 30° 15′ 44″, and at the planet either 20° 14′
39″ or 20° 15′ 42″. Hence, the distance of Mars (or of the point on the ecliptic)
from the sun was 147,802, or more correctly, 147,683. Here, you see that an
error of one minute in eccentric position causes the loss of 120 of our units at
this distance of Mars from earth, and at this distance of the sun in the opposite
position. So these slight discrepancies need no further attention. Besides, this
observation is well supported by others on many of the days nearby, right up
to the day of opposition with the sun. But since 12° 10′ Sagittarius is about $26\frac{1}{2}^{\circ}$
from the node, the secant of the inclination at this place exceeds the radius by
about 11 units, or 15 or 16 in our dimensions, so as to make the distance of
Mars from the sun almost exactly 147,820 or 147,700.

For a similar distance from aphelion in the other semicircle, we shall take
up again the observations of Chapter 26, where I derived a distance of Mars
from the sun of about 147,443 or 147,700 or 147,750. And at one of the times
noted therein, namely, 1590 March 4 at $7\frac{1}{5}^{h}$, the [mean] longitude of Mars was
1^s 4° 11′ 20″. Hence, the full-circle complement of the mean anomaly was 114°
41′. We are thus lower down from aphelion than we were before by one degree
and 17 minutes. And to one degree correspond 230 units at this position on
the eccentric. Therefore, the distance of 113° 24′ on the ascending semicircle
would be 147,743 or 148,000 or 148,050 (extrapolating from the observations
of Ch. 26). But on the descending semicircle, 147,820 or 147,700 was found
here. The difference is about 350 or 180 units, or none; it is rather uncertain.
For the observations with Mars at perigee are rather poorly obtained, on account of the low elevation of the zodiac and many other causes. And you see
in Chapter 26 that the true distance, hesitantly accepted, fluctuates between
147,443 and 147,750, a difference of 300 units which, in our present undertaking, are of no great importance, since Mars is so low and close to the sun or
the center of the world.

But let us descend here even farther towards perihelion, and explore the
same thing about 22 degrees before and after perihelion.

On 1589 Dec. 3 at 5^h 39^m, when the sun's position was 21° 44′ 56″ Sagittarius, and its distance from earth was 98,248, and the mean longitude of Mars
was 11^s 16° 27′ 53″, full-circle complement of the anomaly 162° 24′ 11″, and
the equated eccentric position 20° 4′ 32″ Pisces: Mars was observed at 15° 25′
33″ Aquarius, lat. 1° 11′ 47″ S.[34] But because it was found above in Ch. 42 that
our vicarious hypothesis errs somewhat near perihelion, we shall admit other
positions, as many as we can obtain, and inquire of them, using the method
of Chapter 42, the distance of Mars from the sun, and at the same time a more
correct eccentric position as well.

So, on 1591 Oct. 16 at 6^h 28^m, when the sun was at 2° 39′ 15″ Scorpio,
99,142 distant from earth, Mars's mean longitude 11^s 13° 53′ 57″, full-circle
complement of the anomaly 165° 0′ 9″, eccentric position from the vicarious

Margin notes:

At a mean anomaly of the full-circle complement of 113°.

255

At a mean anomaly of the full-circle complement of 162°.

---

[33]*TBOO* 12 p. 140.

[34]*TBOO* 11 p. 335 (raw observations).

hypothesis 16° 59′ 14″ Pisces: Mars was observed at 1° 27′ 18″ Aquarius, lat. 2° 10′ 52″ S.[35]

Also, on 1593 Sept. 8 at 10^h 38^m, when the sun was at 25° 41′ 0″ Virgo, 100,266 distant from earth, Mars's mean longitude 11^s 17° 10′ 17″, full-circle complement of the anomaly 161° 45′ 28″, and eccentric position from the vicarious hypothesis 20° 53′ 54″ Pisces: the planet was found at 8° 53′ 51″ Pisces with latitude 5° 14′ 30″ south.[36]

Finally, on 1595 July 22 at 2^h 40^m am, when the sun was at 7° 59′ 52″ Leo, 101,487 distant from earth, Mars's mean longitude 11^s 14° 9′ 5″, and anomaly 164° 48′ 55″, and consequent eccentric position from the vicarious hypothesis 17° 16′ 36″ Pisces: the apparent position of Mars, from the most select observations, was 4° 11′ 10″ Taurus, lat. 2° 30′ S.[37] Thus we twice have Mars in the most opportune position, in quadrature with the sun, while the positions of earth and Mars are also distant by a quadrant.

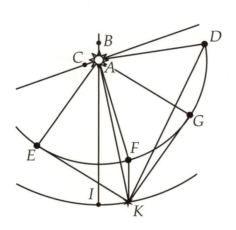

And so, following the method of Chapter 42, I shall make a test of the eccentric positions of the star, and to begin with I shall suppose that the distance of Mars at the first time was 139,212. Hence, the following ones were 139,033, 139,258, 139,045. For in such close proximity, the connection of the anomalies is easily known, as before. Let A be the sun, D, G, F, E, positions of the earth in 1589, 1591, 1593, 1595. Let K be the position of Mars, the same all four times (even though it is not quite the same in the observations). Let the points be connected. AD, AG, AF, AE are given in position and length. And the length of AK is introduced four times. Moreover, the lines of observation DK, GK, FK, EK have known positions. Therefore, ADK, AGK, AFK, AEK are given. So, through the opposition of the sides to the angles, DKA, GKA, FKA, EKA are also given. Hence, so is the position of KA, four times.

											Anomaly			AK
DA	21°	44′	56″	Sag.	98,248	DK	15°	25′	33″	Aqu.	162°	24′	11″	139,212
GA	2	39	15	Sco.	99,142	GK	1	27	18	Aqu.	165	0	9	139,033
FA	25	41	0	Vir.	100,266	FK	8	53	51	Pis.	161	45	28	139,258
EA	7	59	52	Leo	101,487	EK	4	11	10	Tau.	164	48	55	139,045

	Compl.				Yields				Thus AK is				Vicarious		
ADK	53°	40′	37″	DKA	34°	39′	23″	20°	5′	16″	Pisc.	20°	4′	32″	Pisc.
AGK	88	48	3	GKA	45	28	27	16	55	45	Pisc.	16	59	14	Pisc.
AFK	16	47	9	FKA	12	0	4	20	53	55	Pisc.	20	53	54	Pisc.
AEK	86	11	18	EKA	46	44	30	17	26	40	Pisc.	17	16	36	Pisc.

So, since the first and third position here agree rather closely, some less

---

[35]*TBOO* 12 p. 146.

[36]*TBOO* 12 p. 291 (raw observation).

[37]*TBOO* 12 p. 441.

thoughtful person will think that it should be established using these, the others being somehow reconciled. And I myself tried to do this for rather a long time. But since the second and fourth could not be reconciled, while the force of these observations was great, because in each the planet was observed in quadrature with the sun, and in the quadrilateral *AEKG* all the sides and angles are about equal, I therefore settled it as follows. From the vicarious hypothesis, you see that *AK* in the second observation ought to be distant from *AK* in the fourth by 17′ 22″. But by the assumption of this length, the two positions of *AK* are 30′ 55″ apart. So this is too much by 13′ 33″. And since all angles of the quadrilateral are about equal, I divided the excess in two, and added 6′ 46″ to the angles *EKA*, *GKA*. For in the observation at *E*, the line *AK* had moved forward too much, and not enough at *G*. So, with the two *AK*s moved back towards *E* and *G*, *EK* and *GK* staying fixed (for we are supposing the observations to be most certain), the angles at *K* will in all cases be increased. So now, given the angles *GKA*, 45° 35′ 13″, and *EKA*, 46° 51′ 16″, the other angles *G*, *E*, and the lines *EA*, *GA* remaining the same, *AK* comes out to be 138,765, and 138,787, differing from our assumed value by 258 units. So if we also subtract that much from the other two *AK*s, so as to make them 138,954 and 139,000, the resulting angles are *DKA* 34° 43′ 47″ and *AK* 20° 9′ 40″ [Pisces]; while *FKA* is 12° 1′ 24″ and *AK* 20° 55′ 15″ [Pisces]. But since I previously added 6′ 46″ at *G* and subtracted the same amount at *E*, I have therefore repositioned the eccentric positions at 17° 2′ 31″ Pisces at *G*, and 17° 19′ 54″ Pisces at *E*, increasing the position given by the vicarious hypothesis by 3′ 17″. Therefore, the same amount also ought to result at *D*,

namely,	20°	7′	49″	Pisces	At *F*	20°	57′	11″	Pisces
While I found	20	9	40			20	55	15	
Difference		1	51	more			1	56	less

And so I have also brought the other two positions near enough together. For their errors lie both past and short of the truth, which lends security. And to attribute an error of two minutes to observations at these positions, owing to the low elevation of the zodiac and horizontal variations, is not excessive.

At a similar anomaly on the descending semicircle, the observations at hand are no more than one, but it is certain enough. For on the night following June 29 in 1593, at 1ʰ 30ᵐ after midnight, when the sun was at 17° 25′ 42″ Cancer, 101,760 distant from earth, the longitude of Mars 10ˢ 10° 1′ 29″, anomaly 161° 5′ 29″, and the consequent position of Mars 6° 10′ 5″ Aquarius, it was observed at 13° 37′ 22″ Pisces, with latitude 4° 37′ S.[38] Hence, the supplement of the angle at earth was 56° 11′ 46″, at the planet, or the parallax of the annual orb, 37° 27′ 23″. From which the distance of Mars from the sun comes out to be 139,036. But above, at an anomaly of 161° 45′ 28″, where Mars was 40 minutes farther from aphelion than here, the distance was found and established as 139,000. And at this position on the eccentric these 40 minutes effect a change of 52 units. So here too, by extrapolation from our anomaly, there results a distance of 138,984 at an anomaly of 161° 45½′, an admirable consensus much to be suspected. For they can hardly all be so certain and neat. Furthermore, both distances must be increased somewhat owing to the

257

---

[38]*TBOO* 12 p. 282 (raw observation).

inclination of this position on the eccentric, which is at a maximum.

So, from this long induction, using a great many positions on the eccentric, it is clear that those distances of Mars from the sun are mutually equal whose points on the orbit are equally remote from aphelion, a question which we have investigated in Ch. 16 and 42. This is an evident way of showing that the aphelion we have obtained is correct, by Euclid III. 7.

At the same time, the distances of the sun from earth are confirmed, which were derived in Ch. 29 above and employed here in various ways. Nor is there any great discrepancy in the numbers that could testify to any flaw in them.

The implications of the observations presented in this chapter, and of the distances found through them, for the shaping of the planetary path, for which purpose we have produced them in this chapter, we shall postpone until Chapter 55. First, there is something that must be proved in Ch. 52 following, and in Ch. 53 many more observations are going to be called upon to testify.

CHAPTER 52

Demonstration from the observations of Chapter 51 that the planet's eccentric is set up, not about the center of the sun's epicycle, or the point of the sun's mean position, but about the actual body of the sun; and that the line of apsides goes through the latter rather than the former.

It is a happy accident that the distances found in Chapter 51 also inform us about this, which, though promised in Chapters 6, 26, and 33, I deliberately postponed until this point. For if I was correct in constructing the eccentric of Mars about the body of the sun, it is necessary that the planet really be at its greatest distance from the sun in the parts around 29° Leo, and that those parts which are at equal intervals from 29° Leo in either semicircle be at equal distances from the sun, and at unequal distances from the point that stands for the sun, which for Brahe is the center of the sun's epicycle. More specifically, the distances should be less in the descending semicircle. When this is proved, it will follow in addition that the parts around 24° Leo are neither the most distant from the sun's body, nor from the Copernican center of the world, which for Brahe is the center of the sun's epicycle, and also the center to which the planetary circle is attached; and the parts at an equal arc's removal from 24° Leo in either semicircle are at unequal distances from the sun and from the point that stands for it. *For let there be set out the sun's center A, Mars's line of apsides AC, eccentricity AC, and the eccentric ED with center C, and let the point F above AC be the point of uniform motion, G the aphelion, GFE and GFD equal angles, and let EA, DA be connected, which will be equal, as has now been proven. And through A let the line AB be drawn towards Capricorn, and let AB be extended from A towards Capricorn until its length be 1800 of the units of which AC was 14, 140 in Chapter 42, and AE, AD, 154, 400; and let B be the center of the earth's orb. Now because BA is directed towards 5½° Cancer, and AE towards 15½° Scorpio, the angle EAB is about 50° and acute, and EBA obtuse. Hence, EA is longer than EB. Likewise, since BA is directed towards 5½° Cancer, while AD is directed towards 12½° Gemini, BAD is therefore 157° and ABD is quite acute. Hence, AD, or AE which is equal to it, is shorter than BD. Therefore BE is much shorter than BD, and the difference is quite perceptible. For who are we to neglect AB, which is 1800 or even more, who could not tolerate observations with a mere 200 units' error? Hence, regions on opposite semicircles of the eccentric that are equidistant from G, such as E and D, are not*

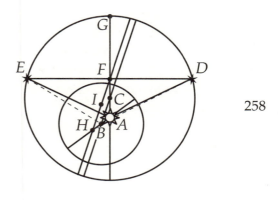

258

equidistant from any points not at the center other than those on the line CA that goes through the body of the sun.[1]

You may reply, however, that if BC is connected and extended, a new apsis is created where that line intersects the circle, and the point D is closer to that apsis than is the point E. So is it any wonder that BD is also longer? I answer that whatever lines are drawn, AE and AD always stay the same, since they are proven from the observations, in all three forms of hypotheses, and thus absolutely nothing in this derivation was assumed

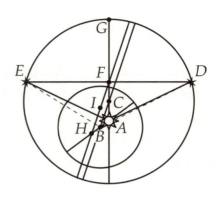

that could be subject to controversy. And so, with AE and AD remaining the same, let BC be drawn exactly as is objected against me. Nevertheless, that line BC by no means gives rise to a hypothesis that will fit the acronychal[2] *observations, as I proved in Chapter 6. Instead, to save the acronychal positions, one needs to substitute for BC a line FH through F parallel to CB, passing through F and H, the centers of uniform motion for Mars and the sun. But when this is done, the center of the eccentric is at the same time transferred from C to I, and there is more than a semicircle on the side of E, and less on the side of D. Nor are AE and AD left unchanged, but AE is lengthened, and AD is shortened. And since these lines are altered, the observations at positions other than acronychal will never be saved, since they give evidence that AE and AD are equal.* I don't think there is any need for computation. Never-

259   theless, if there is anyone who enjoys this labor (even though no astronomer should try anything with numbers whose foundations he has not previously seen in geometry, and geometry has just overturned the foundations of such an undertaking), he has an example of it in Ch. 24 above. There, I computed the distances of earth from *H*, the point of uniformity of the earth's motion, and the distance of Mars from the same point *H*, simultaneously in a single operation, using the same observations by which I afterwards computed the distances of earth and Mars from the center of the sun *A*, in Ch. 26.

For the peculiar cleverness of the method I have used is this: that it shows that whatever point in the plane of the earth's circle is chosen that has, with respect to the sun's body, a position described and determined through a number of observations both in zodiacal longitude and in distance from the sun, also shows the distance of earth and Mars from that chosen point; and it does these things without any knowledge of the equated anomaly on the eccentric corresponding to that point. In fact, the only reason why I used that knowledge in Chapter 26 was that it is a short cut.[3]

But in addition, there is another way to argue the point. It was proved in Chapter 44 above that the planet's orbit is not a circle but an oval, such that the diameter on it which is called the [line] of apsides is the longest. Just now,

---

[1]See Euclid I.12.

[2]ἀκρονυχίοις. Subsequent appearances of the word in this paragraph are in Latin.

[3]See p. 252.

in Ch. 51, it was proved that regions that are equally removed from the point of the aphelion G also make an equal incursion at the sides. There is thus a real oval situated about the line AC, and therefore, it is not situated about the line FH. And one who would compute the various distances of Mars from the point H by the method just recommended will find a great irregularity in the distances, incapable of being included by any means in a circle or in any other possible figure set up about FH.

So again the faith that was pledged in Chapter 6 and in many other places in this work, I have redeemed from all tincture of circular reasoning, and have shown that the eccentric of Mars cannot be referred to anything but the sun itself; and that, in consequence, it is not only reason that stands with me, but the observations themselves, in my releasing the observations of Mars from the sun's mean motion and measuring them out by the apparent motion of the sun.

CHAPTER 53

Another method of exploring the distances of Mars from the sun, using several contiguous observations before and after acronychal position: wherein the eccentric positions are also explored at the same time.

Since we are establishing new hypotheses here, in that we are enquiring into the natural cause of the eccentric equations, it is appropriate that we should explore everything as carefully as possible, lest in neglecting the foundations we build upon them a building doomed to ruin. And so it furthers us to explore this same thing (perfectly accurate distances of Mars from the sun) by many methods. *Let α be the sun, β the position of earth before opposition of Mars to the sun, and $\alpha\beta\delta$ the angle of vision, or the arc of the elongation of δ from the sun. Likewise, let γ be the position of the earth after opposition, and $\alpha\gamma\delta$ the angle of vision. Thus, at the first time, let the planet be on the line $\beta\delta$, and at the second, on the line $\gamma\delta$, and let it actually traverse $\theta\eta$. So, when the time of the two observations is given, the angle $\theta\alpha\eta$ will be given precisely enough by the vicarious hypothesis, for whatever eccentric position. If the pair of times are not far from each other, or if the planet is near the apsides or the middle elongations, the difference in length of the lines $\alpha\theta$, $\alpha\eta$ will also be known tolerably.*[1] *And we have included this much among our presuppositions only in order that there be no remaining difficulty here.*

260

And so, in dealing with the angles $\theta\beta\alpha$, $\eta\gamma\alpha$, given by observation, and $\beta\alpha$, $\gamma\alpha$, which are known from Part III, if we were to assume [a value for] $\theta\alpha$, and consequently also $\eta\alpha$, it is obvious that if this assumed value were longer than it should be, such as $\kappa\alpha$, $\iota\alpha$, then the angle $\iota\alpha\kappa$ would come out less than it should be; and if, on the other hand, it were shorter than it should be, such as $\zeta\alpha$, $\epsilon\alpha$, the angle $\epsilon\alpha\zeta$ would come out greater then it should be. So we must assume such distances as will make the angle of motion on the eccentric come out right.

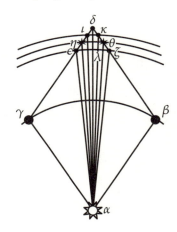

Any possible remaining error in the eccentric position will also occur in the same way. For let it be that $\theta\alpha$, $\eta\alpha$ are in their correct positions, and from that point let $\theta\alpha$ be carried forward, in error, through the angle $\theta\alpha\delta$. And let

---

[1]At aphelion and perihelion the rate of change in the distance vanishes; hence, near aphelion and perihelion the changes in the distance should be small, regardless of which theory is used to obtain the distances. At the middle elongations, although the rate of change in the distance is considerable in any theory, it would be nearly the same no matter what theory is used, since the orbital curves are nearly parallel. At other places, however, it makes considerable difference which theory is used to obtain the difference in distances, a problem which Kepler never addresses directly.

ηα likewise be carried forward, through the equal angle ηαε. You see that αδ, substituted for αθ, is going to be very much too long, and αε, following αη, will be quite short, contrary to what, by hypothesis, is known at the start. Furthermore, the angle γαβ ought not to be as small as possible, so that no error of observation, or at least no minimal one, in opposite directions in the sky (as can happen) could have any great effect. Now, with the help of this method, we must go through the years 1582 in Cancer, 1585 in Leo, 1587 in Virgo, 1589 in Scorpio, 1591 in Sagittarius, 1593 in Pisces, and 1595 in Taurus. For sufficient observations are at hand in all places.

If it seems a good idea to investigate demonstratively the elongation of the earth from the line through the sun and the planet at which any error in the distance of Mars from the sun would be most evidently perceived, let Ch. 6 be consulted. For, following that chapter, we shall define it as that angle at the sun whose sine has to the radius about the same ratio as the excess of Mars's distance from the sun over the sine of the complement of the angle has to the distance itself.

For let α be the sun, θ the planet, νζ the earth's orb. From θ let the straight line θμ be drawn perpendicular to θα. And on θμ let a number of centers be chosen, about which let circles through θ be described, until one of them be tangent to the earth's orb at ν.[2] The point ν will be where the defect of αθ at θ appears most evidently, that is, where it subtends the greatest angle.[3] From ν let νο be drawn parallel to μθ, intersecting αθ at ο. I say that ον is to να as οθ is to θα. For as νμ, or (which is the same thing) θμ, is to μα, so is ον to να. But νμ is to μα as οθ, and (very nearly) ζθ, is to θα.[4] Therefore, etc.

Let αθ be 161,000. Thus ζθ will be nearly 61,000. And as 161 is to 61, so is 100,000 to 37,888. This, taken as a sine, shows the angle ναθ to be 22° 15′, and greater, if instead of ζθ you take οθ.

So, many days, nearly 45, pass before the anomaly of relative motion is altered by $22\frac{1}{4}$ degrees. And before or after this time, αθ is much different. So at aphelion this angle of relative motion is about 28°, and at perihelion about $18\frac{1}{3}°$.   261

And now, having found the termini at which any error that may arise will be most evident, owing to an incorrect distance of Mars from the sun, it is easy for us to choose suitable observations, since many are available.

We shall begin from the opposition of 1582, from which year we shall choose the following observations.[5]

---

[2] Although Kepler proceeds by successive trials in this description, the points μ and ν can be determined by Euclidean geometry. A proof is supplied by A. E. L. Davis, "Grading the Eggs," *Centaurus* 35 (1992) p. 131.

[3] This is proved in Ch. 6, pp. 113–117, as Kepler says above.

[4] Euclid VI.2 and V.11.

[5] The sources of these observations are as follows. November 24: *TBOO* 10 p. 175; December 26: *TBOO* 10 p. 176; December 30: *TBOO* 10 p. 178. The provenance of the January observation is uncertain. There is an observation on that date in *TBOO* 10 p. 247, but only the meridian

	1582 November 24 4^h am			December 26 $8^h 30^m$			December 30 $8^h 10^m$			1583 January 26 $6^h 15^m$		
Observed at	26° 38' 30"	Cancer		17° 40' 30"	Cancer		16° 0' 30"	Cancer		8° 20' 30"	Cancer	
Observed latitude	2 49 10	N.		4 7 0	N		4 8 0	N.		2 52 12	N.[6]	
Sun at	11 40 40	Sagit.		15 4 12	Capr.		19 8 31	Capr.		16 33 20	Aquarius	
Sun - earth dist. $\alpha\beta$	98,345		$\alpha\beta$	98,226		$\alpha\gamma$	98,252		$\alpha\gamma$	98,624		
Mean anomaly[7]	67 28 13			49 39 10			47 51 35			34 8 15		
Eccentric position	0 43 34	Cancer		16 7 10	Cancer		17 57 32	Cancer		0° 9' 40"	Leo	
On the ecliptic: $\alpha\theta$	0 42 42	Cancer		16 6 23	Cancer	$\alpha\eta$	17 56 45	Cancer	$\alpha\eta$	0 9 30	Leo	
Resulting $\alpha\theta$	158,920			163,082		$\alpha\eta$	158,842		$\alpha\eta$		164,116	
Because of the lat.[8]	158,960			163,147			158,907			164,196		

The two intermediate ones differ by 4240. And indeed, the later one $\alpha\eta$, is shorter, although it should have been longer by 336.[9] So the sum of the two is 322,054. From this I subtract 336, and again add it. The halves of these are 160,859, which is $\alpha\theta$, and 161,363,[10] which is $\alpha\eta$. And $\alpha\theta$ will be at 16° 5' Cancer, and $\alpha\eta$ at 17° 55' Cancer.[11] So here, the vicarious hypothesis would lose $1\frac{1}{2}$ minutes.

But the distances themselves are not to be trusted, owing to the angle's being too small. For if the angle at δ be varied by one minute, through an error in observing, as easily happens, we shall be in error by a thousand units in either distance.

---

altitude is given. However, Brahe made observations on January 24, 25, and 27, and February 1 and 2, and Kepler's position may have been the result of a collation of several of these. For all four dates, Brahe gives only the raw observations, without calculating longitude and latitude.

[6]Comparison with the other data shows that this is much too low. In *TBOO* Brahe gives a meridian altitude for Mars of 61° 12' at $9^h 5^m 48^s$. Combining this with Kepler's given longitude and Brahe's equatorial elevation for Uraniburg (34° 5' 30") results in a latitude of approximately 3° 51'. Possibly Kepler miscopied the meridian altitude, writing a "2" instead of a "3."

[7]The mean anomalies given here are not consistent with those upon which the vicarious hypothesis was established (Chapter 18), nor with the data given in the penultimate paragraph of this chapter, nor, indeed, with each other. However, if one considers the mean anomalies and longitudes of all eight of the dates in these first two tables, one finds that most of them are consistent with an aphelion of 29° 10' 10" Leo at noon on 1601 January 1 old style (January 11 new style) ($9\frac{1}{2}'$ farther forward than the position given at the end of this chapter). If the mean longitude for that time is decreased by 20" to $10^s 7° 14' 14"$, computed mean anomalies and longitudes match Kepler's within a few seconds for five of the eight dates (the exceptions are those for December 1582 and March 1585). The different data may represent different stages in Kepler's attempts to "tune" the vicarious hypothesis to fit all the observations more accurately. The 15' Kepler adds to the mean anomalies in Chapter 54 may be a similar vestige.

[8]Not "By the latitude," as the Latin *Per latit.* would suggest. It is indeed possible to use the latitude to find the distance independently, but this procedure results in distances very different from Kepler's. Furthermore, there would not be much point in performing such a computation, as the resulting distances would not be very accurate. The numbers Kepler presents here make it clear that he was simply adding a small correction for the transfer from the plane of the ecliptic to the plane of Mars's orbit.

[9]This difference may be found by subtracting the corresponding distances given in the final table in this chapter. As Kepler says at the end of Chapter 56 (p. 409), these are the "diametral distances" (as defined in the Glossary). These distances are consistent with an as yet undetermined orbit intermediate between the oval hypothesis and the circle.

[10]This number is in error, in that in the computation Kepler added 336 to the half of 322,054. However, the computed longitude corresponding to this distance does not reflect this error; hence, Kepler apparently used the correct distance to find it.

[11]These positions are computed using the adjusted earth-Mars distance, the earth-sun distance, and the angle at earth, to find the angle at Mars, and using this together with Mars's observed position to find Mars's heliocentric longitude.

Therefore, let the two more remote ones be taken, which are found to differ by 5236. But we already know that they should differ by about 5570. So by an operation conducted as before, the more nearly correct values resulting are: $\alpha\theta$ 158,792, and $\alpha\eta$ 164,364, placing $\alpha\theta$ at 0° 41′ 0″ Cancer, and $\alpha\eta$ at 0° 8′ 30″ Leo. And it becomes certain, through observations on the four days at this position, that about $1\frac{1}{2}$ minutes must be subtracted from the eccentric positions derived from our vicarious hypothesis.

The distances found before are approximately confirmed as well, both before and beyond opposition, which turn out to have a magnitude between these. Unless, as the comparison indicates, they ought to be somewhat longer.[12]

But at the same time it is clear that if the angle $\theta\delta\eta$ had been off by one minute, both distances would have been off by about 50 units, no more. So in these distances there can barely be an error of the hundredth part of the uncertainty that there was in the previous ones.

Now, if a longitude that was taken up expresses satisfactorily the observed values for the distances for these four days, it will also express the observed values for the intervening days, namely, November 25, 26, and 27, and December 3, 17, 27, 28, and 29 of 1582, and January 16, 17, 18, 19, 21, and 22 of 1583.

Let us proceed to the opposition of 1585. For while the sun and Mars were 262 at opposition on January 31 of that year, the planet was observed at many closely-spaced positions over the two months preceding and the same number following. From among them we shall take these four observations.[13]

	1584 December 21 14^h	1585 Jan. 24 9^h	February 4 6^h 40^m	March 12 10^h 30^m
Mars was observed at	1° 13′ 30″ Virgo	24° 7′ 30″ Leo	19° 47′ 30″ Leo	11° 46′ 0″ Leo
Latitude	3° 31′ North	4° 31′ North	4° 28′ North	3° 22′ North
Sun at	10° 43′ 5″ Capr.	15° 9′ 5″ Aquarius	26° 10′ 31″ Aquarius	2° 16′ 42″ Aries
Distance from earth	98,210	98,595	98,840	99,850
Mean anomaly of Mars	29 46 53	12 4 21	6 21 31	12 47 15
Eccentric position	3° 54′ 34″ Leo	18° 49′ 0″ Leo	23° 34′ 47″ Leo	9° 23′ 28″ Virgo
On the ecliptic	3 53 56 Leo	18 49 3 Leo	23 35 0 Leo	9 24 7 Virgo
Resulting distance $\alpha\theta$	165,101	166,290	and $\alpha\eta$ 166,182	166,131
Because of the lat.	165,184	166,378	166,260	166,206

The two intermediate ones differ by 118. They should have differed by 187 in the opposite sense, so that $\alpha\theta$ would be 166,226 and $\alpha\eta$ 166,412. Therefore, $\alpha\theta$ falls at 18° 48′ 47″ Leo, and $\alpha\eta$ at 23° 34′ 48″ Leo. And the contemptibly small alteration of the eccentric position confirms the vicarious hypothesis for this place. But we learn from this that an error of one minute in observation at this place would vitiate the two distances by about 100 units.

When the more remote ones are consulted, their difference is found to be 1022. From what is known already approximately from the hypothesis, the difference should have been greater, namely, 1275. And, in fact, the fourth

---

[12]This comparison is necessary because Kepler's procedure will not show whether both distances are too short or too long. Therefore both distances need correction based on other nearby distances.

[13]Sources for these observations are as follows. December 21: *TBOO* 10 p. 322; January 24: *TBOO* 10 p. 387; February 4: *TBOO* 10 p. 389; March 12: *TBOO* 10 p. 396. All but the February observation are not worked out into latitude and longitude, and for the first and last observations the times are not stated.

[degree] of Leo is close to the eighteenth of Cancer, where previously something had to be subtracted from the eccentric position of the vicarious hypothesis. So, if you will subtract one minute at the fourth of Leo, you will now make $\alpha\theta$ a hundred units shorter, and if $2\frac{1}{2}'$, you will make it about $164,934$, which is short enough that $\alpha\eta$ can also keep the length $166,206$; and the last observation in the previous year 1583, which showed a length of $164,364$, can be reconciled with it. For they should have differed by 488, a certain enough value provided in advance by the hypothesis of the distances, while they do differ by 570.

Furthermore, it is possible to transfer half of this $2\frac{1}{2}'$ change in the eccentric position to the observations. For if either of them has erred by one minute, that will be able to effect an error of 50 units in either distance.

It would be tedious to repeat the same method, using the same words, for all the years of the oppositions. And so, in the following table, I have placed the observations themselves which I have consulted, and added what resulted from the computations. The hypotheses underlying the calculations are these. The sun's position is taken from Brahe. The sun-earth distance is from Ch. 30. The aphelion of Mars for the end of 1600[14] is $29°\ 0\frac{2}{3}'$ Leo.[15] The mean motion at the same time is $10^s\ 7°\ 14'\ 34''$.[16] The eccentricity and ratio of the orbs is as in Ch. 54.[17] To this I have added the distances of Mars from the sun as if previously known.[18] So if, using these distances, we match the proposed observations, these distances will be the correct ones, which is what I proposed to show in this chapter.

263

---

[14] From the mean motion given, this would appear to have been noon on 1601 January 1 (old style), or January 11 in the Gregorian calendar.

[15] If one compares this with the earlier figure of $28°\ 48'\ 55''$ Leo on 1587 March 6 old style (these numbers are from the table in Ch. 18), with $1'\ 4''$ per year added for the aphelial motion, one finds that Kepler has shifted the aphelion back $3'$.

[16] Comparison with the mean motion established through Kepler's previous data shows that he has subtracted about one minute from the mean longitudes. The combined effect of this change and the change in the aphelion is to add $2'$ to the mean anomalies.

[17] The computed longitudes in the table are consistent with the procedure given by Kepler in Chapter 60. The translator's recomputation using this procedure comes within a few arc seconds of most of Kepler's positions (notable exceptings being the positions for December 1582, off by $3'\ 15''$ and $2'\ 26''$, respectively). The vicarious hypothesis, in contrast, shows differences that gradually wax and wane, reaching a maximum of $2\frac{1}{2}'$ around the beginning of Pisces.

[18] As Kepler states in Chapter 56 below (409), these are the diametral distances, as defined at the beginning of Chapter 57.

Time	Sun's Position	Sun-Earth Distance	Sun-Mars Distance[1]	Mars's Eccentric Pos. on Ecliptic[2]	Computed Position	Observed Position	Difference	Latitude
1582 23 Nov. 16ʰ 0	11° 41′ Sagit.	98,345	158,852	0° 42′ 11″ Cancer	26 40 0 Cancer	26 38 30 Cancer	1′ 30″ +	North 2 49
26 Dec. 8 30	15 4 Capr.	98,226	162,104	16 7 18 Cancer	17 44 19 Cancer	17 40 30 Cancer	3 49 +	4 7
30 Dec. 8 10	19 9 Capr.	98,252	162,443	17 56 32 Cancer	16 6 20 Cancer	16 0 30 Cancer	5 50 +	4 8
1583 26 Jan. 6 15	16 33 Aquar.	98,624	164,421	0 6 24 Leo	8 17 57 Cancer	8 20 30 Cancer	2 33 −	2 52
1584 21 Dec. 14 0	10 16 Capr.	98,207	164,907	3 51 45 Leo	1 14 34 Virgo	1 13 30 Virgo	1 4 +	North 3 31
1585 24 Jan. 9 0	14 53 Aquar.	98,595	166,210	18 47 8 Leo	24 3 58 Leo	24 7 30 Leo	3 32 −	4 31
4 Feb. 6 40	26 10 Aquar.	98,830	166,400	23 33 41 Leo	19 43 52 Leo	19 47 0 Leo	3 8 −	4 28
12 Mar. 10 30	2 16 Aries	99,858	166,170	9 23 14 Virgo	11 43 31 Leo	11 46 0 Leo	2 29 −	3 22
1587 25 Jan. 17 0	16 1 Aquar.	98,611	166,232	8 13 40 Virgo	4 41 50 Libra	4 42 0 Libra	0 10 −	3 26
4 Mar. 13 24	24 0 Pisces	99,595	164,737	24 56 50 Virgo	26 24 41 Virgo	26 25 40 Virgo	0 59 −	3 38
10 Mar. 11 30	29 52 Pisces	99,780	164,382	27 35 54 Virgo	24 5 15 Virgo	24 5 15 Virgo	0 0	3 29
21 Apr. 9 30	10 48 Taurus	101,010	161,027	16 44 51 Libra	15 49 50 Virgo	15 48 20 Virgo	1 30 +	1 48
1589 8 Mar. 16 24	28 36 Pisces	99,736	161,000	16 55 14 Libra	12 14 7 Scorp.	12 16 50 Scorp.	2 43 −	2 4
13 Apr. 11 15	3 38 Taurus	100,810	157,141	4 1 50 Scorp.	4 45 0 Scorp.	4 43 20 Scorp.	1 40 +	1 10
15 Apr. 12 5	5 36 Taurus	100,866	156,900	5 1 41 Scorp.	3 58 57 Scorp.	3 58 20 Scorp.	0 37 +	1 4
6 May 11 20	25 49 Taurus	101,366	154,326	15 30 36 Scorp.	27 8 17 Libra	27 7 20 Libra	0 57 +	0 7
1591 13 May 14 0	2 10 Gemini	101,467	147,891	12 7 38 Sagit.	2 15 36 Capr.	2 20 0 Capr.	4 24 −	South 2 25
6 Jun. 12 20	24 59 Gemini	101,769	144,981	25 38 48 Sagit.	27 11 45 Sagit.	27 15 0 Sagit.	3 15 −	3 55
10 Jun. 11 50	28 47 Gemini	101,789	144,526	27 56 49 Sagit.	25 57 57 Sagit.	26 2 36 Sagit.	4 39 −	4 8
28 Jun. 10 24	15 51 Cancer	101,770	142,608	8 29 32 Capr.	21 4 21 Sagit.	21 10 0 Sagit.	5 39 −	4 45
1593 21 Jul. 14 0	8 26 Leo	101,498	138,376	20 1 38 Aquar.	17 43 14 Pisces	17 45 45 Pisces	2 31 −	5 46
22 Aug. 12 20	9 11 Virgo	100,761	138,463	10 15 25 Pisces	13 9 39 Pisces	13 10 15 Pisces	0 36 −	6 7
29 Aug. 10 20	15 54 Virgo[3]	100,562	138,682	14 37 15 Pisces	11 11 41 Pisces	11 14 0 Pisces	2 19 −	5 52
3 Oct. 8 0	20 15 Libra	99,500	140,697	6 19 39 Aries	7 49 54 Pisces	7 50 10 Pisces	0 16 −	3 17
1595 17 Sep. 16 45	4 18 Libra	99,990	143,222	22 49 19 Aries	26 5 45 Taurus	26 7 12 Taurus	1 27 −	1 42
27 Oct. 12 20	13 59 Scorp.	98,851	147,890	15 35 38 Taurus	18 50 46 Taurus	18 51 15 Taurus	0 29 −	0 6
3 Nov. 12 0	21 2 Scorp.	98,694	148,773	19 26 33 Taurus	16 18 33 Taurus	16 18 30 Taurus	0 3 +	North 0 17
18 Dec. 8 0	6 43 Capr.	98,200	154,539	13 2 29 Gemini	11 39 1 Taurus	11 40 0 Taurus	0 59 −	1 40

[1] Not the same as those found in the text: these are the "diametral distances" as determined in Ch. 56 (p. 409), as Kepler himself points out.

[2] Also not the same as those found in the text, nor are they the same as the adjusted positions given in the analysis of the data in each table. They are in fact positions computed from the ellipse and the area law.

[3] Corrected from 11° 54′ Virgo, on the basis of the other data.

These, then, are the distances that will result from an investigation using the method of this chapter from the observations set out here The apparent positions, on the other hand, when Mars's eccentric position is in Cancer, will come out about 4 minutes back from these, and in Sagittarius and Capricorn the same number of minutes forward. These small errors do not come from incorrect distances, for they would then be in opposite senses on opposite sides [of opposition], and not in the same sense.[19] I believe they can be reconciled by changing the sun's apogee by one degree, which is easily permitted by Brahe's observations. Nevertheless, I am not going to say anything definite at present. For the correction of both this apogee and the entire hypothesis is reserved for the book of Tables.[20]

---

[19]That is, for two adjacent observations on either side of opposition (such as those of December 1582), incorrect distances should have opposite effects upon the computed apparent positions, while in fact the two computed positions appear to be in error in the same direction. It should be noted, however, that for these two dates, the heliocentric longitudes are incorrect; thus, Kepler is trying to find a natural explanation for data that are actually erroneous.

[20]The "Tables" are the *Tabulae Rudolphinae*, which Kepler had hoped to produce soon after the publication of *Astronomia Nova*, but which were not published until 1627.

CHAPTER 54

A more accurate examination of the ratio of the orbs.

In Chapter 42, we did actually establish the ratio of the orbs from observations at positions other than acronychal, but they were not ones that were in agreement with one another entirely and to our full satisfaction.[1] Moreover, considered in itself, regardless of whether the most exact observations be available, the procedure is incapable of being brought to a certitude of 100 units. So it has to be done by polling and counting the votes. And in Chapter 28, at a mean anomaly of 11° 37′, which, after the correction of Chapter 53 preceding, becomes 11° 52′,[2] the distance of the point on the ecliptic to which a perpendicular dropped from the body of Mars descends, was found to be 166,180, or 166,208. And therefore, since this position is 23° from the northern limit, the inclination will be about 1° 43′, and excess of the secant will be 45 units, which will be about 70 in our dimensions. Therefore, the distance of Mars from the sun will be 166,250 or 166,278.

We shall now also compare the observations of Chapter 51, so as to be supported by an middling consensus. In 1586, with 10° 9′ 41″ of mean anomaly remaining, or 9° 54′ 41″ after correction, we found 166,311. But by subtracting $1\frac{1}{2}′$ from the position given by the vicarious hypothesis, we found 166,208. So for a subtraction of 3 minutes less than two degrees,[3] about 95 should be subtracted, making it 166,113. For the latitude, 80 must again be added, making it 166,193. Thus in 1588, when the remaining [mean] anomaly was 8° 2′ 51″, or 7° 47′ 51″ corrected, by a subtraction of $1\frac{1}{2}′$ from the position given by the vicarious hypothesis we found the distance to be 166,396. Thus, a position 4° 4′ lower will be shorter by about 102, making it 166,294. And, corrected for latitude, 166,284. This was previously found to be 166,193, from 1586. The mean is 166,238. In the descending semicircle, however, from 5 observations, we had found 166,250 or 166,278. So, although the difference is imperceptible, let us nevertheless take the mean, 166,260, giving more trust to the descending semicircle, as it is better confirmed by the observations.

Let it thus be [taken as] certain that at a mean anomaly of 11° 52′ the distance is 166,260. Hence, however great a hypothetical value you may conceive by some rough method, which is to be confirmed shortly thereafter, it follows that where the radius is 100,000, the distances at aphelion cannot increase more than 164 units, and even less if you use the hypothesis of a perfect circle.

---

[1] πληροφορίαν.

[2] The reader will no doubt recall that nowhere in Chapter 53 is there any mention of adding a full 15′ to the mean anomaly. Furthermore, the mean anomaly computed from the orbital parameters of the vicarious hypothesis was 11° 19′ 42″ at that position in Chapter 28. Perhaps this reference is to an earlier version of Chapter 53, and Kepler neglected to change it when the chapter was rewritten.

[3] That is, 1° 57′, which is the difference of the two mean anomalies.

But those units converted through a preconceived ratio of the orbs, as it is set up in Ch. 42, add about 250, and these added to 166,260 make 166,510. But above, in Ch. 42, we found 166,780, using weaker observations. The difference is 270 units.

We shall also treat likewise the perihelial distance which in Ch. 42 was found to be 138,500, from observations that were not solid enough.

265   Just now, in Ch. 51, at a remaining [mean] anomaly of $161° \ 45\frac{1}{2}'$, or $161° \ 30\frac{1}{2}'$ after correction, we found the distance, before correction for latitude, to be 139,000 or 138,984. So let 139,000 be at 21 Pisces. Since this position is 35 degrees from the limit, the inclination is therefore $1° \ 31\frac{1}{2}'$. The excess of the secant will be $35\frac{1}{2}$, which is equivalent to 49 of our units. And so the true distance of Mars from the sun is 139,049. But if the radius is 100,000, the perihelial distance is 575^4 units shorter than that at an anomaly of $161\frac{1}{2}°$, which becomes 876 of our units, or less, if you use a perfect circle. And when these are subtracted from 139,049, there remains the perihelial distance of 138,173. The difference from the value 138,500, found in Chapter 42, is 327.

So, according to this method, these distances are found:

Aphelial	166,510	And where 152,342 becomes 100,000,
Perihelial	138,173	14,169 becomes 9301.
Diameter	304,683	
Semidiameter	152,342	
Eccentricity	14,169	

Nevertheless, because our observations, especially at perigee, do not bear out that great a difference, and since it can happen that the vicarious hypothesis, since it is false, also might introduce some falsity into the eccentricity, let all the votes be tallied before the result is announced.

And so we shall adapt the aphelial distance found here, 166,510, to the eccentricity of Ch. 42, which was 9265. And as 109,265 is to 90,735, so is 166,510 to 138,274, where the radius is almost exactly 152,400.

Also, manifold experience has shown that the eccentricity that is most true and best fitted to the physical equations is between 9230 and 9300; that is, the eccentricity of Chapter 42, which is 9265.

Therefore, that we might not unduly abandon the perihelial distance found in this chapter, which is 138,173, nor unduly trust the aphelial distance of 166,510, let us conclude that the truest aphelial is 166,465, and the perihelial, 138,234, where the radius is 152,350.

---

[4]Kepler's estimate of the change in distance on the oval here is too large, although above, near aphelion, it was nearly correct. The figure here should have been 506 (using a provisional eccentricity of 9265 where the radius is 100,000). This would result in an eccentricity of 9261, very close to Kepler's "truest and best fitted" value of 9265.

CHAPTER 55

From the observations of Chapters 51 and 53, and the ratio of the orbs of Chapter 54, it is demonstrated that the hypothesis seized upon in Chapter 45 is in error, and makes the distances at the middle elongations shorter than they should be.

Indeed, I began to say this in Chapter 51. But since more observations, and more suitable ones, had to be provided to give evidence in Chapter 53, from which at the same time something else was also inferred in Chapter 52, the full demonstration was therefore postponed to this point.

There is no need for verbosity. At the mean anomalies of all the examples 266 appearing in Ch. 51 and 52 let the distances be computed according to the hypothesis of Chapter 45 and the ratio of orbs of Ch. 54, by the method I used from Chapter 46 through Chapter 50, and let them be compared to the distances of Ch. 51 and 53, found using infallible observations. It will be apparent that the more we descend from the apsides, the more the computed distances fall short of the observed distances, a result quite the opposite of what we saw in Ch. 44 above. For there, the distances computed according to the law of the circle were longer than the observed distances at the middle elongations, while here, the distances resulting from the hypothesis that makes the planet's orbit oval are shorter. It is therefore obvious that the planet's path is neither a circle nor such as to make as great an incursion from the circle at the sides as does the oval that arose from the opinion of Chapter 45 and was described in Chapter 46; but takes a middle course. And if, in turn, using the distances of Chapter 45, you compute the observed positions of Mars, especially those which, in Ch. 53, stood at some distance on either side of opposition, the planet before opposition will fall too far forward, and after opposition, too far back. This is most evident in the descending semicircle in 1589 and 1591, and in the ascending semicircle in 1582 and 1595. For in those places, the oval of Chapter 45 is 660 units[1] too small, while the perfect circle is too large by the same amount, and this can have an effect upon the appearances of 20 minutes and more. Thus, David Fabricius[2] was able to use his observations to charge my hypothesis of Chapter 45, which I had communicated to him as true, with this error of having distances that are too short at the middle elongations, writing at the very time when I was laboring to seek out the true hypothesis with re-

---

[1] Where the mean between the aphelial and perihelial distances is $152,350$. This would be about 433 units where this mean distance is $100,000$.

[2] For a biographical note on Fabricius, with many references, see J. R. Christianson, *On Tycho's Island* (Cambridge 2000) pp. 273–276. For Fabricius's role in the composition of *Astronomia Nova*, see J. R. Voelkel, *The Composition of Kepler's* Astronomia Nova (Princeton University Press 2001), esp. Ch. 8, pp. 170–210.

newed care. He was, in fact, quite close to arriving at the truth before me.[3] And since the perfect circle errs the same amount in the opposite direction, we argue rightly from this that the truth is in the middle, between the two.

Moreover, the equations computed from physical causes in Chapters 49 and 50 gave the same testimony, namely, that the lunule cut off from the perfect semicircle ought to have only half the breadth of the one which the opinion of Chapter 45 cuts off. Therefore, nothing prevents our saying that the matter is most certainly demonstrated: that the opinion of Chapter 45, in remedying the excess of the perfect circle, falls into the opposite defect.

So the physical causes of Ch. 45 go up in smoke.

---

[3]See Fabricius's letter of 27 October 1604 (Letter 297, in *KGW* 15 pp. 58–62), and Kepler's reply of 18 December 1604 (Letter 308, in *KGW* 15 pp. 78–81). Despite Kepler's generous acknowledgement of Fabricius's ostensible near-precedence, the latter's attachment to circularity and opposition to the introduction of physics into astronomy was such as to have eliminated any such possibility. See his letter of 20 January 1607 (Letter 408, in KGW 15 pp. 376-386), and Voelkel's account of the threat to Kepler's entire undertaking posed by Fabricius's alternative theory (Voelkel, *Composition,* (previous note), pp. 207–210).

CHAPTER 56

Demonstration from the observations already introduced, that the distances of Mars from the sun are to be chosen as if from the diameter of the epicycle.

The breadth of the lunule of Chapter 46 above, which the opinion of Chap- ter 45, which instructed us to cut it off from the semicircle, has produced for us—this breadth, I say, was found to be 858 units,[1] of which the semidiameter of the circle is 100,000. But then, by two arguments, which I have already presented in Chapters 49, 50, and 55, I concluded plainly that the breadth of the lunule is to be taken as only half that, namely, 429, or more correctly, 432, and in units of which the semidiameter of Mars is 152,350, nearly 660. I therefore began to think of the causes and the manner by which a lunule of such a breadth might be cut off.

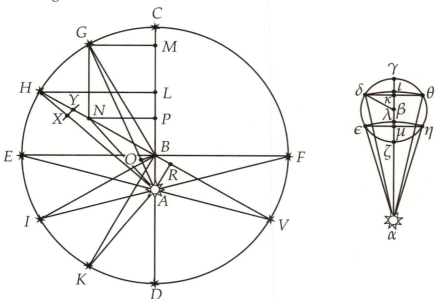

While anxiously turning this thought over in my mind, reflecting that absolutely nothing was articulated by Chapter 45, and consequently my triumph over Mars was futile, quite by chance I hit upon the secant of the angle 5° 18', which is the measure of the greatest optical equation. And when I saw that this was 100,429, it was as if I were awakened from sleep to see a new light, and I began to reason thus. At the middle elongations the optical part of the equation becomes a maximum. At the middle elongations the lunule or shortening of the distances is greatest, and has the same magnitude as the excess of the secant of the greatest optical equation 100,429 over the radius 100,000. Therefore, if the radius is substituted for the secant at the middle elongation, this accomplishes what the observations suggest. And, in the dia-

---

[1]See Ch. 47, p. 355.

gram in Chapter 40, I have concluded generally that if you use *HR* instead of
HA, *VR* instead of *VA*, and substitute *EB* for *EA*, and so on for all of them, the
effect on the rest of the eccentric positions will be the same as what was done
here at the middle elongations. And by equivalence, in the small diagram of
Chapter 39, $\alpha\kappa$ will be taken instead of the lines $\alpha\delta$ or $\alpha\iota$, and $\alpha\mu$ for $\alpha\epsilon$ or $\alpha\lambda$.

And so the reader should peruse Chapter 39 again. He will find that what
the observations additionally testify here was already urged there, from natural causes, namely, that it appears reasonable that the planet perform some
sort of reciprocation, moving on the diameter, as if of an epicycle, that is always directed towards the sun. He will also find that there is nothing more at
odds with this notion than this: that when we proposed to represent a perfect
circle, we were forced to make the highest parts $\gamma\iota$ of the reciprocation unequal to the lowest $\lambda\zeta$, which parts correspond to equal arcs on the eccentric,
the highest being short, and the lowest long. So, now that the planet's circular
path is denied, and $\kappa\alpha$, $\mu\alpha$ are taken instead of $\delta\alpha$, $\epsilon\alpha$, that is, instead of $\iota\alpha$, $\lambda\alpha$,
as was said, it follows further that those parts of the reciprocation, such as $\gamma\kappa$,
$\mu\zeta$, are equal. And that which had tormented us for a long time in Chapter 39
now surrenders to us in the face of the proof of the truth we have perceived.

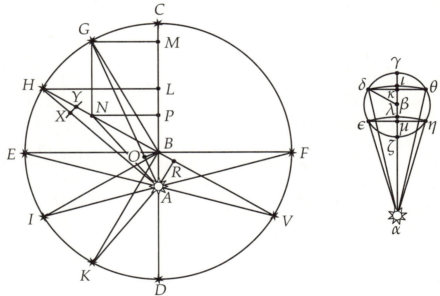

268   As for the middle parts $\kappa\mu$ still being larger than the extremes $\gamma\kappa$, $\mu\zeta$, it will
be said in Chapter 57 following that this is in accord with nature, contrary to
what we had been able to understand in Chapter 39.

But in addition, the difficulty that arose in Chapter 39 through supposing
that the increase of the sun's [apparent] diameter serve the planet as an index for its approaching and receding, now vanishes entirely, as will appear in
Chapter 57.

Thus, concerning the eccentric anomaly of 90°, I easily was able to see in
the manner just mentioned, that instead of the distance *EA* of the perfect circle,
EB is to be taken, corresponding to an equated anomaly *EAB*.

And although I have drawn a general conclusion concerning all the anomalies using a single one as an example, this was not a consequence just of that
one anomaly: there was need to strengthen it using closely spaced observations.

So now you understand the special capacity in which the observations of Chapters 51 and 53 are appointed to serve us, namely, to give this evidence.

Come, then: let the eccentric anomalies *CBG*, *CBH* be computed at the equated anomalies set out in those chapters, that is, at the angles *CAG*, *CAH*, and so on. Nor is there any need to strive after minute parts, nor be concerned about the imperfection of the eccentric equations that still remain in Ch. 19, 29, 43, 47, 48, 49, and 50. Use any of these methods, particularly the one in Ch. 43. You will not err in the equations by more than eight minutes.

When the angles are set up, seek out the lines, *HR* corresponding to the equated angle *HAC*, *RV* corresponding to the equated *VAC*, and so on for the others, and transpose them to the dimension of the orbs found in Ch. 54. You will find them as in the following table.

From the observations of Ch. 51		
On the descending semicircle	On the ascending semicircle	Computed from the reciprocation
166,180	166,401	166,228
166,208	166,296	
162,994	163,100	163,160
163,051		
158,091	158,217	158,074
158,111		
154,400	154,278	154,338
147,820	147,743	147,918
147,700	148,000	
	148,050	
139,000	138,984	139,093

In the observations of Chapter 53, there is no need to do the same thing. For I previously used this same method of reciprocation to find out the distances of Mars from the sun which I called upon in order to compute Mars's apparent positions. And since the observations were represented by these, they will therefore be correct.[2]

As you see, therefore, the distances measured on the diameter, found *a priori* in Ch. 39, are confirmed by closely spaced and very reliable observations throughout the entire perimeter of the eccentric.

---

[2] Because the relationship between the observations and the table in Ch. 53 is somewhat controversial, this short paragraph has been translated in a way that hews closely to the Latin, despite some awkwardness in the English.

CHAPTER 57

By what natural principles the planet may be made to reciprocate as if on the diameter of an epicycle.

269 It is clear, then, from the most reliable observations, that the course of the planet through the aethereal air is not a circle, but an oval figure, and that it reciprocates on the diameter of a small circle in the following manner. Suppose that, after describing equal arcs on the eccentric, the planet comes to be at the diametral distances $\gamma\alpha$, $\kappa\alpha$, $\mu\alpha$, $\zeta\alpha$, instead of the circumferential distances $\gamma\alpha$, $\delta\alpha$, $\epsilon\alpha$, $\zeta\alpha$ (that is, $\gamma\alpha$, $\iota\alpha$, $\lambda\alpha$, $\zeta\alpha$), upon which the perfect circle lies. It is clear from inspection that a lunule is cut off from the perfect semicircle of the eccentric, whose breadth at any point is equal to the differences between the two diverse distances, such as $\iota\kappa$, $\lambda\mu$. This is proposed not on the basis of arguments *a priori*, but of observations, as I have just said; so now the physical theories will proceed more correctly than they had hitherto.*** For it is not by any ratiocinative or mental process that a planetary mind assigns the equal parts of the reciprocation $\gamma\kappa$, $\kappa\mu$, $\mu\zeta$, to equal arcs CD, DE, EF, of the as yet untraversed eccentric, for the former are not equal. Instead, the reciprocation is coordinated with the space traversed on the eccentric in a natural way, which depends not upon the equality of the angles DBC, EBD, FBE, but upon the strength**[1] of the ever increasing angle DBC, EBC, FBC, which strength approximates the sine (so called by the geometers). The ascent's being changed into descent thereby gradually, by a continuous diminution, is more probable than if the planet were said suddenly to turn its prow in the other direction—which we indeed said in Ch. 39 clearly conflicts with observational results. And since the measure of this reciprocation points the finger at a natural mode, its cause will also be natural; that is, it will be some natural—or better, corporeal—faculty, and not a planetary mind.

Also, in Ch. 39, for the best reasons, one of our suppositions was that a planet cannot make a transition from place to place by the bare effort of its inherent forces unless these be assisted or directed by an extrinsic force. This being the case, we must consider whether we should also carry over this reciprocation in part to the solar power itself. In our exertions to this end, we shall be obliged once again to take up our oars which were introduced in Ch. 39.[2]

270 Let there be a circular river CDE, FGH, and in it a sailor who revolves his oar once in twice the periodic time of the planet, by an inherent and perfectly uniform force. Thus at C let the line of the oar be at right angles to the line from

Terms:
What the circumferential distance is, and what the diametral.

*** The principle of this reciprocation is proved to be natural.

**This is the genuine measure of this reciprocation, supported by reason; in other words, it is the reason why the versed sine of the eccentric anomaly is the measure of this reciprocation.

Natural examples of reciprocations of this kind.

Oars.

---

[1] The phrase, "supported by reason" (in the marginal note) is an attempt at rendering the Greek word "ἀπολόγητος" which appears in Kepler's text. However, there seems to be no such word in classical Greek. The word "ἀπολογητικός" means "suitable for defense, apologetic:" perhaps this is what Kepler intended.

[2] The oars are to be found in Ch. 38, not 39. See p. 300, above.

the sun, and let it direct now the bow and now the stern forward at alternate returns. At *F*, however, let the line of the oar be part of the line from the sun, and at other positions let it have an intermediate inclination. Now the stream, flowing down upon the oar at *DE*, will push the ship[3] down towards *A*, while from *C* it will push very little, since the ship is also but slightly inclined. The same is true at *F*, because at this moment the stream strikes the oar directly. At *D* and *E*, however, it pushes down more strongly, because here the oar is greatly disposed to such an approach by its inclination. The opposite happens in the ascending semicircle. For the river, coming beneath the oar at *G* and *H*, drives it away from the sun.

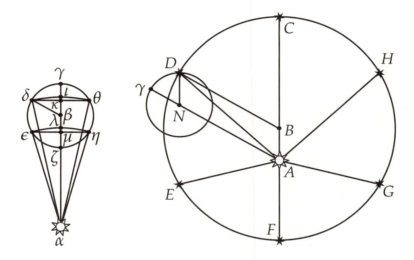

At the same time it will also happen, other things being equal, that the impulse is less at *C* than at *F*, since our river is weak at *C* and strong at *F*. And this is is also in accordance with our wishes, since our reciprocation was following equal spaces on the eccentric, and the planet spends longer in the upper ones than in the lower.

This example shows only the possibility of this arrangement. In itself it is rather inadequate, since the rotations of the oar and the river are accomplished, not in the same time, but a double time. Furthermore, to those looking at them from earth, the faces of the planets should appear to change, while the face of the moon, although it participates with the planets in that motion which we are discussing, does not change over the course of a month. Instead, it always is turned towards the earth, whence its eccentricity is reckoned. In addition, while the force of a river is material (for its water acts by its weight and material impetus), the force of the sun is immaterial. Therefore, the comparison with the planets ought to be different: they will need no oar, no physical instrument, for catching hold of the force of some weighty thing (for that motive *species* of the sun has no weight). Nor do we deem it fitting that the stars have a corporeal oar, seeing that we hold them to be round.

Defect of the example.

---

[3]Note that the "ferryman" and the "skiff" of Chapter 38 (p. 300), which suggested ties to Charon and the classical underworld, have now become simply "sailor" and "ship". The change in words primarily represents transformation of a concrete example to a more general and abstract model. Yet the language is also marvelously suggestive of the way Kepler's new universe is breaking out of the restrictions of the old "cave world" and at the same time abandoning the old gods that made it live.

Example of the magnet.

But from this very refutation, there comes another example, which will perhaps be more suitable. The river and the oar are of the same quality. The river is an immaterial *species* of magnetic power in the sun. So why not have the oar too borrow something from the magnet? What if all the bodies of the planets are enormous round magnets? Of the earth (one of the planets, for Copernicus) there is no doubt. William Gilbert has proved it.

William Gilbert's magnetical philosophy.

Some magnetic arrangement in the body of the planet appears to be the cause of this reciprocation.

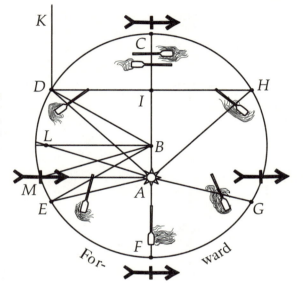

271

But to describe this power more plainly, the planet's globe has two poles, of which one will seek out the sun, and the other will flee the sun. So let us imagine an axis of this sort as a magnetic strip, and let its point seek the sun. But despite its sun-seeking magnetic nature, let it remain ever parallel to itself in the translational motion of the globe, except to the extent that over the ages it transfers the polar inclination from one of the fixed stars to another, thus causing the progressive motion of the aphelion. I nevertheless admit the possibility that a mind may be needed for both of these, of such a nature as to be adequately instructed by the animate faculty for performing this motion. For this is a motion, not of the entire body from place to place (which motion, in Ch. 39 above, was rightly denied to a motive cause inherent in the planets),[4] but of the parts about the center of the whole, as if at rest.

Example of the earth.

Here again, in the globe of the earth there is an example of this directional property of the axis, from Copernicus. For as long as the axis of the earth, in the annual circulation of its center, remains almost perfectly equidistant from itself in all its positions, summer and winter are brought about.* On the other hand, insofar as very long ages cause it to incline, the fixed stars are thought to move forward, and the equinoxes to retrogress.

* The precession of the equinoxes is like the progression of the aphelia.

Why, then, should we have doubts about attributing to all the planets, in order to save the phenomena of eccentricity, something that is observed to be in one of them (that is, the earth) because of the phenomena of the precession of the equinoxes and the sun's annual cycle of rising and falling?

Copernicus, for example, was deceived here when he thought that he needed a special principle to cause the earth to reciprocate annually from north to south and back so as to produce summer and winter, and to bring about the equality of the tropical and sidereal years (to the extent that they are equal) by constructing it in a manner commensurate with its revolution. For all those effects are obtained by having the earth's axis, about which the

---

[4]This is the fifth of the axioms introduced by Kepler at the beginning of Ch. 39.

diurnal motion is made, retain a single, constant direction: there is no need for extrinsic causes, except to account only for the extremely slow precession of the equinoxes. And so here, too, there is on no account any need for movers for the planet, which would carry its body about the sun in a parallel position, and at the same time perform the reciprocation. For the one will naturally depend upon the other. The only thing remaining to be considered is the extremely slow progression of the aphelia.

To continue: when the strip is at C and F, there is no reason why the planet should approach or recede, since it holds its ends at equal distances from the sun, and would undoubtedly turn its point towards the sun if it were allowed to do so by the force that holds its axis straight and parallel. When the planet moves away from C, the point approaches the sun perceptibly, and the tail end recedes. Therefore, the globe begins perceptibly to navigate towards the sun. After F, the tail end perceptibly approaches, and the head end recedes from the sun. Therefore, by a natural aversion, the whole globe perceptibly flees the sun. And when it is opposite A, where the length of the axis is pointed directly at the sun, its approach in the former situation, or its flight in the latter, is strongest. Furthermore, our earlier presuppositions derived from the observations postulated this. For there, of the parts of the reciprocation $\gamma\kappa$, $\kappa\mu$, $\mu\zeta$ which correspond to equal arcs on the eccentric, the parts at the middle, such as $\kappa\mu$, were longest, and those near γ and ζ were short.

The reason why the reciprocation is swiftest in the middle.

But it is also consistent that the observations would have $\gamma\kappa$, $\mu\zeta$ equal, although their arcs $\gamma\delta$, $\epsilon\zeta$, or better, CD, EF on the eccentric, though equal, are traversed in unequal times, longer for CD, so that the part of the reciprocation $\gamma\kappa$ is traversed in a longer time than $\mu\zeta$ which is equal to it. For similarly, magnets approach one another more slowly when they are at a greater interval, and more swiftly and more quickly at a shorter interval.

The reason why the reciprocation is slower at the top, and swifter at the bottom.

In fact, we can transfer that force which keeps the magnetic axis in a parallel position, and does not allow it to remain pointed towards the sun, from the attention of a mind, to which we had entrusted it a little earlier, to the functions of nature. It appears to be an objection to this, that nature always acts in one and the same way, while this retentive force appears to make its exertions differently at different times. This is seen, for example, in the tendency of the axis to incline towards the sun, for the impeding of which the retentive force is ordained, which tendency is evanescent at the middle elongations but most strongly evident at aphelion and perihelion. Nevertheless, what is there to prevent this force of retention's being in many places stronger than the tendency to incline towards the sun, so that the force is either not at all or but little wearied by such a weak adversary? Let us again take an example from the magnet. In it are manifestly mingled two powers, one of directing it towards the pole, and the other of seeking iron. Thus if a strip or nautical needle be directed towards the pole, while some iron approach from the side, the needle quickly declines from the pole and inclines towards the iron, thus indulging somewhat in its intimacy with the iron, but in such a way that it gives most of it to the pole. Indeed, Gilbert thinks this

272

The planet's axis of power is kept in a parallel position by natural force.

With, nevertheless, an exception.

A magnetic example.

The reason why a magnet declines somewhat from the pole.

to be the reason why a strip declines from the pole towards the continents of greatest magnitude, the cause of this declination thus lying in the tracts of land, being greater and having a more vigorous power in the vicinity to the extent that they are higher on the right or left.

The cause of the motion of the aphelia.

Therefore, we can ascribe the same tasks and a uniform action to both natural faculties, and by the interplay of the two we can show a cause for the translation of the aphelia which will be neither obscure, nor, by Hercules, imaginary. For suppose that this force of directing the axis towards the sun does detract somewhat from the retentive power, commensurate with the ratio of the two. Accordingly, in the aphelial semicircle, as at *C*, the point will gradually incline towards *H* (that is, backward), and the tail end will turn away from the sun, gradually overcoming the retentive force. Thus the aphelion will become retrograde. But in the perihelial semicircle, as at *F*, the same point will incline towards *G* (that is, forward), again overcoming the contrary retentive force. Thus the aphelion will then be made to move forward, and to be fast. But because *AF* is shorter than *AC*, and the sun is closer to *F* than to *C*, the force tending to turn the magnetic axis towards the sun is therefore stronger at *F* than at *C*. Thus more will be detracted from the retentive [force] at *F* than at *C*. So the perihelial forward inclination not only compensates for the aphelial backward inclination, but even overcomes it. And so the reason

Why the aphelia do not retrogress.

is clear why the apsides progress, and do not retrogress. Thus the aphelion we have found will have that value only at an equated anomaly of 90° and 270° when the axis of power is directed straight at the sun, which is its correct place. And the motion of the aphelion will be spiral, as will become clear below in Chapter 68 for the motion of precession of the equinoxes also, which exists through another cause. So the direction of the magnetic axis in its parallel position, or the force which is its custodian, will not respect one or another of the fixed stars, but only the position of its body, as it is at any particular time. And, to think the matter through simply: because this direction is more like rest than motion, it is more appropriately sought in the material, and in the disposition of the body, than in some mind.

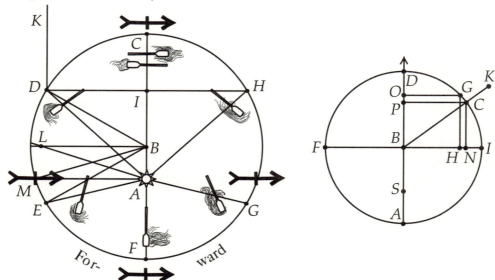

It is in accord with reason that any magnet correctly arranged would perform reciprocations of this sort.

But come: let us track more closely this similarity of the planetary reciprocation to the motion of a magnet, and that by a most beautiful geometrical demonstration, so that it may be clear that magnets have such a motion as that which we perceive in the planet. *Let DFA be either a round magnet or the body of Mars, DA the line along which the magnetic power is oriented, D the pole that seeks the sun, A the pole that flees the sun.* You will note, first, that in this theory it is all the same whether we consider the entire globe of the magnetic body, or one single physical line of its power, parallel to *DA*.

For this magnetic power is corporeal, and divisible with the body, as the Englishman Gilbert, B. Porta, and others, have proved,[5] surely because its globe consists of an infinite number of physical lines, as it were, parallel to *DA*, whose power is extended in a straight line and in one direction in the world. Therefore, the judgements made about individual parts with respect to the quality of their motion will be the same as those concerning them all joined together, and vice versa. *So let the central axis DA be proposed for theorizing, in place of the whole body and all its filaments. Let DA be bisected at B, and let FBI be drawn perpendicular to DA. Thus, when the planet is so positioned that BI points toward the center of the sun, there will be no [tendency to] approach. For the angles DBI, ABI are equal, and thus have equal strength, the former for approaching, and the latter for receding. So this is like an equipoise in mechanics.[6] Under these conditions, the center of Mars B is on the apsides, at aphelion, say, at its greatest distance from the sun. Now let some arc IC be taken, measuring the angle of equated anomaly, and let BC be drawn and extended to K. And let the planet be so situated that BC points towards the sun, which is understood to be indicated by K.* The first thing to be sought is the measure of the strength of the planet's approach. *Now the approach occurs because the seeking pole D is inclined towards the sun K at the angle DBK, while the fleeing pole A is turned away at the angle ABK. And since the strength of this angle is natural, it will follow the same ratio as the balance. But when a line CP is drawn from C perpendicular to DA, between DP and PA there will be the ratio of the balance. For if a pair of scales is suspended from the balance support KB, and the arms come to rest at the angle DBK, the weight of the arm BD will be to the weight of the arm BA as DP is to PA, just as, if the balance arms were suspended from CP at P, and the weight of the arm BA were applied to PD while the weight of BD were applied to PA, then DA would be at right angles to the hanging balance support CP. See my* Optics,[7] *and do not be easily swayed by insufficiently careful experimentation. Therefore, as DP is to PA, so is the strength of angle ABC*

Thou, O magnet, showest sailors the hidden track: what wonder that the Wanderers follow thy pleasure?

What is the measure of the speed of reciprocation at any point.

This reciprocation has the ratio of the balance; hence its name [lat. *libratio*].

274

---

[5]The *Magia naturalis* of the Neapolitan Giovanni Battista della Porta was distributed in a very large number of editions in various languages in the second half of the sixteenth century. Although the first edition, divided into four books (Naples, 1558) hardly mentions magnetic phenomena, in the comprehensive edition of 20 books, which first appeared in 1589 in Naples, the entire Book 7 is devoted to magnetism. With respect to the present passage, it is Chapter 5 of that book that is chiefly of interest.

William Gilbert, whose work has already been mentioned above in Chapter 34 (see the footnote on p. 289), treated the phenomena consequent upon the division of a magnet in book I Ch. 5 (P. F. Mottelay trans., pp. 28–30).

[6]In his *Optics* (Ch. 1 Prop. 20, English trans. pp. 27–34), Kepler criticized earlier accounts of the balance, and presented his own theory. In that book, he used his law of the lever as a way of understanding refraction.

[7]*Optics* Ch. 1 Prop. 20, English trans. pp. 27–34. Kepler's warning about "insufficiently careful experimentation" may reflect the inability of this theory of the balance to describe the phenomena fully (although under certain circumstances it can give correct results).

to the strength of angle *DBC. Thus DP here measures the force of fleeing, and PA the seeking force. From PA subtract a magnitude equal to DP, and let this be AS. Therefore, SP is the measure of the seeking power alone, with the impediment of fleeing [power] subtracted, and it will be in the same proportion in which AD measures the single greatest force. But where the half DB measures the greatest force, the half of PS, which is PB, or the sine CN of the equated anomaly CBI, measures the net force of the planet's approach towards the sun at this location.* So the sine of the equated anomaly is the measure of the strength of the planet's approach towards the sun in this place. And this is the measure of the increments of power.

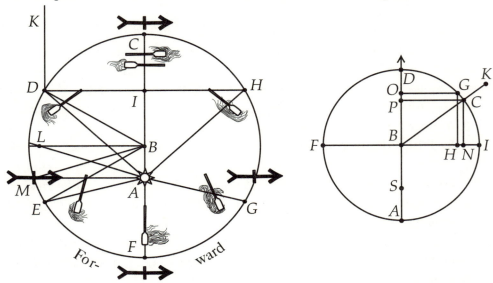

The measure of the distance of the reciprocation traversed by these continuous increments of power is quite another thing. For the observations show that if the eccentric anomaly *GI* corresponds to the equated anomaly *IC*, the versed sine *IH* of the arc *GI* is the measure of the reciprocation accomplished. If this can also be deduced from the previously indicated measure of the speed *CN*, then we shall have reconciled experience with the demonstration involving the balance. *Since the sine of any arc is the measure of the strength of that angle, the sum of the sines will be an approximate measure of the sum of the strengths or impressions over all the equal parts of the circle. And the completion of the entire reciprocation is the effect of all of these in common. Furthermore, letting IC and IG, though they are unequal elsewhere, be equal here to avoid confusion, the sum of the sines of the arc IG is to the sum of the sines over the quadrant, approximately as the versed sine IH of that arc IG is to the versed sine IB of the quadrant.* Approximately, I say. For at the beginning, when both the versed sine and its increments are small, it is less by half than the sum of the sines. For: *Let the quadrant be taken as 90 units. The sum of the 90 sines is 5,789,431.*[8] *In this instance I have added them all in order.*[9] *The sum of the sines at an arc of 1°, that is, the first sine, is 1745. And the former sum is to the latter as 100,000 is to 30. On the other hand, the versed sine*

What is the measure of the space traversed in the reciprocation up to a given moment.

What is the ratio of the versed sine of some arc to the sum of the sines of all preceding degrees.

---

[8]Evaluated by modern methods, this sum is 5,779,433. Most likely, somewhere in the course of computation, an 8 was substituted for the second 7.

[9]This may mean that, instead of taking half the sum of the sines through 180°, which he already had computed in Chapter 43 using the Cardan/Bürgi rule (p. 334), he actually added up the sines through 90° here.

of the quadrant is 100,000, *and the versed sine of* 1 *degree is* 15, *which is half of* 30.

The reader should not be at all deterred by this geometrical *faux pas*[10] and fallacious principle. For before this becomes a perceptible portion of the reciprocation, the effects of the two procedures differ imperceptibly. *For the sum of* 15 *sines, which is* 208,166, *gives* 3594[11] *[as a fourth proportional]. And the versed sine of* 15° *gives* 3407/100,000, *which is only a little less than the other. Likewise, the sum of* 30 *sines, which is* 792,598, *shows, by the rule of proportions, a part of the reciprocation which is* 13,691 *out of* 100,000. *And the versed sine of* 30° *shows* 13,397. *Also, the sum of* 60 *sines, which is* 2,908,017, *shows a little more than* 50,000, *while the versed sine of* 60° *is* 50,000.[12]

> The ratio is approximately constant within the limits of sense perception.

It has been demonstrated that if any magnet be set out as we have supposed the bodies of the planets to be set out in the heavens with respect to the sun, a reciprocation of the magnetic body will result that is such as is measured by the versed sine, as regards the space traversed. And indeed, the observations testify that the planet's body reciprocates according to the measure of the versed sine of the eccentric anomaly. It is therefore perfectly consistent that the bodies of the planets be magnetic, and so disposed to the sun as we have described.

> Application of the magnetic reciprocation, just demonstrated, to the observed reciprocation of the planet.

I must now show that it was not a great mistake to have taken the arcs *IC* and *IG* as equal. *When I say that the arc IC on the body of the planet is the measure of the equated anomaly, I am speaking properly, and CN is then the genuine measure of the strength possessed by the planet when it has the sun on the line BK.* However, when I say that *IG* is the measure of the eccentric anomaly which corresponds to the [equated] anomaly *IC*, I am speaking improperly, incorrectly using the circle of the planet's body to represent the eccentric. But on the eccentric's descending semicircle, since a greater arc of eccentric anomaly corresponds to a smaller arc of equated anomaly (namely, *IG* to *IC*), we will be adding up considerably more sines on *IG* than on *IC*, and rightly so. For since the sine measures the strength, and the strength acts in proportion to the time and to the closeness to the sun (magnets being stronger when closer)—that is, to put it briefly, in proportion to the arc *IG*—just as many sines should be set up on *IC* as are found on *IG*.

> 275
>
> The ratio of the versed sines of the eccentric anomalies is the same as that of the sums of the sines of the equated anomalies corresponding to those eccentric anomalies, quite precisely.
>
> To the extent that the planet is slower on any arc, the parts of the equated anomaly are to be made smaller, so that the sum of their sines may be a true measure of the power sent forth over that equated anomaly.

Our only error is this, that we take those many sines to be longer than they should be, as *GH* is longer than *CN*.

But first of all this excess is in itself very small and imperceptible. For at the beginning of the quadrant the arcs *IC* and *IG* hardly differ, and the sines are small, and at the end of the quadrant, where the eccentric equation *CG* is greatest, the sines hardly differ.

And then this error is in accord with our wishes. For the sums of the sines always come out a little greater than the versed sines; and here we are keen to accommodate and reconcile the libratory and magnetic ratios to those com-

> The defect in the ratio which we have established between the versed sine and the sum of the sines is compensated by our contrary error when in using the eccentric anomaly instead of the equated anomaly we take sines which are too long.

---

[10] ἀγεωμετρήτῳ. As in other places, a French phrase that is common in English has been utilized to render Kepler's Greek (despite some slight inaccuracy).

[11] This value, as well as the corresponding figure at 30°, are somewhat low owing to the error in the sum of the sines to 90°.

[12] For a careful discussion of this perplexing argument, see Bruce Stephenson, *Kepler's Physical Astronomy*, 113–116.

mended by experience. Therefore, this present error of ours, of accumulating long sines instead of short ones, is avoided if we use the versed sines instead of the sums of the sines themselves, for the sums of the sines are not exactly equal to the versed sines, but exceed them because of the effect of the reciprocation.

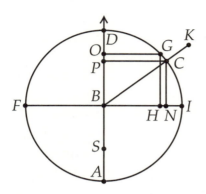

Therefore, by the best reasoning at our disposal, we have brought the calculation within the limits of observable error.[13] Let us conclude that the body of the planet, like a magnet, approaches and recedes according to the law of the lever along an imaginary diameter of the epicycle tending towards the sun, and that the body's diameter of power, its true diameter DA, tends towards the middle elongations, so that for this time BD tends toward 29° Taurus and BA towards 29° Scorpio. For the aphelion is at 29° Leo.

The magnetic force inhering in the bodies of the planets is excited and brought into action by a similar force of the solar body.

Thus this reciprocational approach is performed without the action of mind, by a magnetic force which, though it inheres in the planet and is independent, nevertheless depends for its definition upon the extrinsic body of the sun. For the force is defined as seeking the sun or fleeing it. And while the force between magnets tending to bring them together ought to be mutual, I have denied, in Chapter 39 above, that the sun has the planets' attracting force: it was instead understood to be purely attractive only, as is clear from the argument presented there.[14] The planets' force, on the other hand, is supposed to be simultaneously attractive on one side and repulsive on the other. Alternatively, one might suppose that the sun, like unmagnetized iron, is only sought after, and does not in turn seek other things. For in the above passage, its filaments were circular, while those of the planets are here supposed to be straight.

Difficulty and imperfection of this magnetic example.

276

I am satisfied if this magnetic example demonstrates the general possibility of the proposed mechanism. Concerning its details, however, I have doubts. For when the earth is in question, it is certain that its axis, whose constant and parallel direction brings about the year's seasons at the cardinal points, is not well suited to bringing about this reciprocation or this aphelion. The sun's apogee, or earth's aphelion, today closely coincides with the solstitial points, and not with the equinoctial, which would fit our theory; nor will it have remained at a constant distance from the cardinal points. And if this axis is unsuitable, it seems that there is none suitable in the earth's entire body, since there is no part of it which rests in the same position while the whole body of the globe revolves in a ceaseless daily whirl about that axis.

On the mental basis of this reciprocation. I am afraid to say "rational" for fear that it would be understood to denote a discursive faculty.

So indeed, there may be absolutely no material, magnetic faculty that can accomplish the tasks entrusted to the planets individually, since there may be a lack of means, that is, no suitable diameter of the body which remains equidistant to itself as the body is moved around. For this lack has just been

---

[13]*Intra sensus propinquitatem*; literally, "within the nearness of sense."

[14]See the argument near the end of Ch. 39, p. 307.

made apparent in one of the planets, namely, the globe of the earth. Therefore, let a mind be summoned, which, as was said in Chapter 39, arrives at a knowledge of the distances it traverses by contemplating the growth of the sun's diameter.[15] Let this mind govern a faculty, either animate or natural, that keeps its globe in a parallel position in a manner allowing it to be suitably impelled by the solar power and to reciprocate with respect to the sun. (For a mere mind, unassisted by a faculty of a lower order, could not by itself do anything in a body.[16]) At the same time care should be taken that the periodic time of the reciprocation not be made exactly equal to the periodic return of the planet, so that the apsides will move. The plausibility of these things is argued in Chapter 39 above.[17]

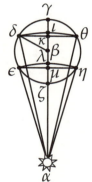

Now that we have obtained from the observations the laws and quantitative characteristics of this reciprocation by which the sun's apparent diameter is varied, matters of which we were still ignorant in Ch. 39, it remains for us to see whether those laws may be such that the planets may plausibly know them. The laws of the reciprocation were that the versed sine of the eccentric anomaly is the measure of the part of the reciprocation completed.

To begin, therefore, I say that admitting as given the observational evidence, namely, that after equal arcs of the eccentric are traversed, the planet is found at $\gamma, \kappa, \mu, \zeta$ rather than at $\gamma, \iota, \lambda, \zeta$, the increment of the sun's diameter presents a legitimate measure of the versed sine of the equated anomaly,[18] no less so than we know the versed sines of the eccentric anomaly to be a measure of the reciprocation.

The increases of the sun's diameter are proportional to the versed sines of the equated anomaly.

Now, as was said in Ch. 39,[19] the planet's mind (if it has such an adjunct) perceives the spaces it traverses in the reciprocation in no other way than by the evidence provided by the increase of the sun's diameter. It will therefore be fitting that it know the versed sine of the equated anomaly, in order that by approaching it might increase the sun's diameter to its prescribed size.

The proof is as follows. *Let the planet be at $\gamma, \kappa, \mu, \zeta$ after traversing equal arcs*

---

[15]See Ch. 39, p. 305.

[16]This passage is a clear reference to the three faculties thought at the time to govern animal physiology. As Kepler is describing them, they are the natural faculty, controlling nutrition and growth; the animate faculty, controlling motion; and the faculty of mind, controlling purposive action.

[17]See p. 306.

[18]

$$\text{The versed sine of the} \left\{ \begin{array}{c} \text{eccentric} \\ \\ \text{equated} \end{array} \right\} \text{anomaly measures the} \left\{ \begin{array}{l} \text{planet's reciprocation} \\ \\ \text{Increase of the sun's} \\ \text{diameter, as seen by a} \\ \text{spectator supposed to} \\ \text{be on the planet, and} \\ \textit{vice versa.} \end{array} \right.$$

—Kepler's footnote.

[19]See p. 305.

of the imperfect eccentric CD, DE, EF, and let the points D and H be joined, the line
intersecting the diameter CF at I. Therefore, since the straight lines δκθ, εμη cut the
epicycle into arcs similar to those on the eccentric, by construction, as CF is to CI, so
will γζ be to γκ, one section being a measure of the other.

277

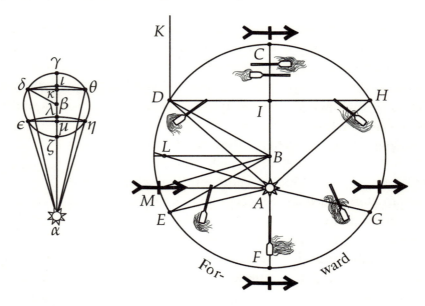

These things being so, I say it will also follow that the diameters of the
sun at α, as observed from γ, κ, μ, ζ, will be augmented by the same measure, namely, the measure by which the versed sine of the equated anomaly
increases. It would be inconvenient to prove this solidly here. It will, however, easily be understood as holding solidly if we prove it for both ends and
the middle. *At C the equated anomaly is nothing, and the versed sine is nothing, and*
the sun, observed from γ, appears at its minimum, so that the amount of its increase
is again nothing. At F the equated anomaly is 180°. The versed sine is equal to the
whole diameter, 200,000. And the sun, observed from ζ, appears at its maximum, so
that it shall have acquired all of its increase.

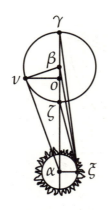

Now for an equated anomaly of 90°, from A let AM be set
up perpendicular to CF, and let MB be joined. Also, from α
let a line be drawn tangent to the epicycle at ν, and the tangent
point ν be joined with the center β. Now since ανβ is right, by
Euclid III. 18, and MAB is right by construction, and βν, BA
are equal by construction, as well as βα, BM, therefore, the
triangles are equal and congruent. So νβα, ABM are equal.
From ν let νo be drawn perpendicular to γζ. Therefore, since
νoβ is right, it is equal to MAB, and νβo will be equal to
MBA. So the triangles are similar, and as νβ is to βo, so is
MB to BA, and vice versa. And since νβ, βγ, and βζ are
equal, and also MB, BC, and BF, as νβ, βo together, or γo, is
to oζ, so are MB, BA together, or CA, to AF. Therefore, since CA is the versed sine of
the eccentric anomaly CBM, and is supposed to be the measure of the corresponding
part of the reciprocation, γo will be that part. Therefore, at this eccentric anomaly
CBM, or equated anomaly CAM of 90°, the planet will be at o.

But the versed sine of the equated anomaly of 90°, the angle CAM, is half the total
diameter, or 100,000. I say also that the apparent magnitude of the diameter of the

*sun at A, α, as seen from o, will be a mean between the magnitudes as seen from γ
and ζ, so that it shall have acquired half of its increase when the planet is at o below
β.*

*For let the diameter of the sun's body be αξ, and the apparent angles formed by
joining ξ with ζ, o, γ, be ξζα, ξoα, ξγα. And because AF, ζα are equal, as well as
AC, αγ, and as CA is to AF, so is γo to oζ, therefore, as γα is to αζ so is γo to
oζ. But γξ differs imperceptibly from γα, and ζξ from ζα. Therefore, as γξ is to ζξ,
so is γo to oζ, within the limits of perception. So in the triangle γξζ, the angle ξ is
divided by the line ξo so that the base γζ is divided in the same ratio as the sides γξ,
ζξ. Therefore, by the converse of Euclid VI. 3, the angle γξζ is divided into two equal
parts by the line ξo, and γξo is half of γξζ, the total increase of the sun's diameter.
Q. E. D.* It is therefore certain at both ends and the middle that in this way,
if the diameter of the reciprocation is divided by the planet in proportion to
the versed sines of the eccentric anomaly, the sun's diameter would increase in   278
proportion to the versed sines of the equated anomaly.

To make it more evident, this is clear in part from the following. *Let the
straight line BL be drawn from B perpendicular to CF, and about center A, with
radius equal to BC, let an arc be drawn intersecting BL at L, and let AL be joined.
Since the eccentric anomaly CBL is 90°, the versed sine will be CB, 100,000, half of
the whole diameter, and consequently the reciprocation will be γβ, half of the whole
γζ. Also, the distance will be βα. But AL is equal to it by construction. Thus the
planet will be at L. And because AL is equal to BC or BM, and BA is a common
side, and LBA is right, as well as MAB, therefore, the triangles BMA, ALB are
congruent. So BL is equal to AM. But AM is equal to αν, as above, and therefore
BL is equal to it also. But αν, which lies opposite the right angle αον, is longer than
αo, which subtends the acute angle ανo. Therefore, BL is also longer than αo, and
AL is longer than BL. Thus AL is much longer than αo. Therefore, the sun appears
smaller at the distance AL than at the distance αo. But the distance αo was just now
seen to be the mean between the maximum and the minimum. Thus at distance AL
the sun appears less than the mean.* So at L, even though half the semicircle of
the eccentric has been traversed, less than half of the increase has been added
to the sun's diameter. This is, of course, because the equated anomaly *LAC* is
less than the half, 90°. And this was the problem that had tied us in knots in
Ch. 39, as was said in the preceding chapter (Ch. 56). For if the planet's orbit
had been a perfect circle, the increase of the sun's diameter would have been
a measure of the increases of the versed sines of the eccentric anomaly, whose
observation is more foreign to the planet's mind than is the observation of the
equated anomaly, as we shall shortly hear. You can see from this contrast just
how conveniently this measure is attributed to the planet, and how plausibly.

We might suppose that the measure of the reciprocation (the versed sine
of the eccentric anomaly, as the observations show) is to be grasped directly
by the mind. But then the planet's mind would be deprived of the assistance
of the variable solar diameter, because it does not adjust itself to the versed
sines of this anomaly of the eccentric. For the planet's path is not a circle. And
the planet's mind would have to intuit the parts of the reciprocation, or the
distances to be traversed in them, without any indicators. This we long ago
rejected as absurd. It would also have to intuit the eccentric anomaly, which is
the angle between two straight lines projected from the center of the eccentric,

*The planet cannot obtain
knowledge of the
eccentric anomaly.*

one through the aphelial point and the other through the center of the planetary globe. In the diagram, it is *DBC* (or if the line *DK* be projected from *D* parallel to *BC*, *KDB* is then the supplement of the same eccentric anomaly). Therefore, if the mind perceives the angle *KDB*, it must perceive the triad of points *K*, *D*, *B*. Concerning the point *D* there is no problem, since it is the center of its globe. I am not much concerned about *K*, because, owing to the infinite distance of the fixed stars, *BC* and *DK* ultimately coincide at the same location among the fixed stars, and the fixed stars are real bodies. Therefore there is no absurdity in holding that the planet's mind uses some hidden sense to keep in view that fixed star which provides lodging for the aphelion at any particular time. Of *B* only is it denied that any sensation of it belongs to the planet's mind, because *B* is not clothed in any body.

And besides, in the natural method proposed just above, there was no need for this idea.

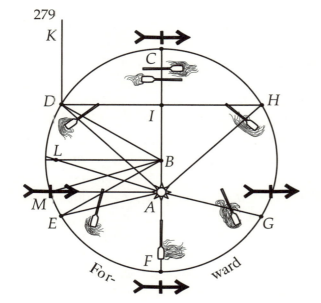

279

Furthermore, when one removes the cause for keeping watch on *B*, the effect is also removed. But *B* ought to be watched if the circle *CD* is to be traversed. However, the planets' orbits are not perfectly circular, as was proven from the observations in Chapter 42. Therefore, the planets do not take aim at *B*. And thus this putative center *B* is actually secondary to the path *CD*. But if it were watched by the planet, it would be prior to the path.

For these reasons, therefore, I deny that the versed sine of the eccentric anomaly provides the planet with a measure of its reciprocation, not because this is not such a measure, but because even if it is, it is not discerned by the planet's mind.

But if we suppose it to be the increasing and decreasing of the sun's diameter that serves as the means or aid by which the planet arrives at the correct distances (imperceptible in themselves) in its reciprocations, and then for the variation of this diameter of the sun, from the demonstration just completed, we posit a rule or measure, to be perceived by the planet's mind, [namely,] the equated anomaly of the eccentric, *DAC*, or rather, *KDA*, in the diagram, we now stand closer to the truth. For both measures are perceptible: for the reciprocation, the increasing and decreasing magnitude of the sun's diameter, and for the measure, or angle, three points clothed in bodies. For at *A* there is the sun itself, at *D* the planet, and at *K* the fixed star that indicates the aphelia.

The planet can obtain knowledge of the equated anomaly.

Perhaps it ought therefore to be said (as indeed we also just embraced above in Ch. 39, when we supposed that the forces of Nature were insufficient to administer the celestial motions) that there is attributed to the planet an ability to sense the light of the fixed stars and the sun, the confluence of whose radiations at the center of the planetary body gives an estimate of this angle of equated anomaly.[20]

There is but one difficulty to clear up. For what reason is it not the angle it-self that is made to be the measure of the planetary operation (that is, to make the sun's diameter increase by approaching the sun), but the versed sine in place of the angle?*[21] And by what means might the planet perceive the sine of the equated anomaly? Does it proceed in the way humans do, by geometri-cal reasoning? Nevertheless, hitherto no faculty of administering the celestial motions belongs to the planet's mind that could not have been acquired by a divine inspiration imparted at the very beginning of the world and lasting even to this day, without any reasoning whatever.

Therefore, what was said just above should be repeated, namely, that the sine of the equated anomaly is the index of the strength of the angles KDA: on this point, see Aristotle's *Mechanics*,[22] and what was said above in this chapter. For when the two balance arms are disposed at an obtuse angle, they are more easily directed than when they are at a right angle, the ease of direction being proportional to the sines. And, on the other hand, when the two arms are connected at an acute angle, they are more easily made to move together into a single line, head-to-head, than if they were connected at a right angle. Refer again to the demonstration contained in what was just presented.

Thus, in one way, if it makes sense for the planet to have a sense of the strength of the angles, there will be no absurdity in our saying (using our hu-man conception) that the sines of the angles are known to it. But why would it take note of the natural strength of the angles? (We are evidently returning to natural principles). As before, let there be certain regions of the planetary body in which there is a magnetic force of direction along a line tending towards the sun. However, contrary to the previous case, let it be an attribute, not of the nature of the body, but of an animate faculty of the sort that governs the body of the planet from within, that as it is swept along by the sun, it keeps that magnetic axis always directed at the same fixed stars, except to the extent that it turns the axis a little over the ages. The result will be a battle between the animate faculty and the magnetic faculty, and the animate will win. It is no different from what we had said in Ch. 34, that the bodies of the planets nat-urally seek rest, but are moved by the extrinsic force of the sun.

Alternatively, here is a more apt example. The weight of the human arm naturally tends towards the center of the earth. However, in a flag bearer the animate faculty takes over and makes the weight extend over his head and wave in a circulatory motion. Here the animate faculty overcomes the natural weight, and would do so forever if the body of the flag-bearer together with all its faculties had not been created mortal.

Marginal notes:

The planet's mind, if it is indeed intent upon the angle of the equated anomaly, does not estimate its magnitude, but its sine.

* Just as a little earlier the sine of the eccentric anomaly (or of the corresponding equated anomaly) was the index of the strength of the reciprocation, while the versed sine of the eccentric anomaly was the index of the amount of the reciprocation traversed, so here the sine of the equated anomaly is the index of the speed by which the sun's diameter increases, while the versed sine of the equated anomaly is the index of the amount of increase occasioned by all the antecedent [degrees of] speed.

A way in which the planet could acquire knowledge of the versed sine.

280

---

[20]See p. 305.

[21]The resemblance of Kepler's thinking here (in the marginal note) to certain of Galileo's ideas is intriguing. See, for example, *Two New Sciences*, Third Day, "On Naturally Accelerated Motion," Proposition I Theorem I, in Galileo, *Opere*, Vol. VIII pp. 208–9.

[22]Aristotle, *Mechanica*, Chapter 2, esp. 850a 2–29. This is not regarded today as by Aristotle himself, but as a product of his school.

On the basis of these presuppositions, the planet's mind will be able to intuit and perceive the strength of the angle from the wrestling match between the animate faculty, which is designed to keep the magnetic axis in line, and the magnetic power of directing it towards the sun.

This arrangement seems also to be confirmed by the example of the moon, which is incontestably more strongly propelled when it is on the diametral line of the sun and the earth, perhaps because of this strength of the angles.

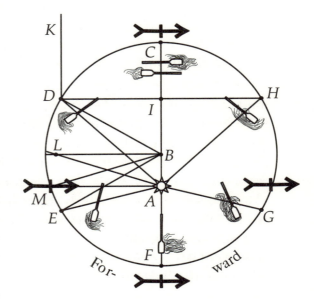

The character of the celestial motions, if mind concurs in them.

The final conclusion, then, will be this. A planet situated at aphelion makes no endeavor in the direction of the sun, but is carried along in proportion to the distance *AC*. The angle *KDA* results from its forward motion. In accord with the proportion of strength of this angle, the planet causes the sun's diameter to increase by approaching the sun. In its approach, it diminishes the distance, making it *AD*. Since the distance is diminished, the forward motion is increased. Therefore, the angle *KDA* is changed more rapidly. Therefore, the planet causes the sun's diameter to increase more rapidly (other things being equal). Thus is established a perpetual circulation which does not occur by leaps such as we have supposed in our thinking and calculation, ignoring imperceptible errors, but is quite continuous.

A comparison of the mental principle with the magnetic.

What I have said so far holds conditionally, if the reciprocation supported by the observations cannot be performed by some magnetic power, implanted in the bodies of the planets, and if it has become absolutely necessary for us to have recourse to a mind. Otherwise, if a comparison between the natural motion and the mental one is in order, the former stands on its own, requiring nothing external, while the latter, the mental motion, appears to give evidence of the magnetic one, and to require its assistance, no matter how you equip it with an animate faculty of moving the body. For in the first place, mind by itself can do nothing to a body. It is therefore necessary to provide for the mind an adjunct faculty that performs its functions in making the planet's body reciprocate. This faculty will be either animate or natural and magnetic. It cannot be animate, for an animate faculty cannot transport its body from place to place (as this reciprocation requires) without the power of another assisting body. Therefore, it will be a magnetic, that is, natural, faculty of sympathy between the bodies of the planet and the sun. Thus the mind calls upon nature and the magnets for assistance.

281   Second, at the halfway point in its pattern, which is the equated anomaly, it has traversed a greater part *γο* of its reciprocation above, and a smaller part *οζ* below, while completing half of its task, which consists of increasing or decreasing the sun's diameter. Nor do *γο, οζ* correspond to parts of the time.

For more time is consumed on γo than its excess over $o\zeta$ required. Nor do the parts increase continuously from ζ to γ, the ones about γ, κ being smaller, as well as those about μ, ζ. The operations of mind, however, are accustomed to being constant.

There was consequently a need for us to equip the mind with an animate faculty, as well as a magnetic one, and to contrive a battle between the two which would remind the mind of its duties, of which it could not have been reminded by the equality of either the times or the spaces traversed. So again we have asked nature to assist the mind.

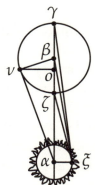

On the other hand, all these modifications really appertain to the workings of the sun's extrinsic magnetic power, and of the magnetic [power] joined to it, which inheres in the planet, as was explained above. If, therefore, the magnetic powers can do the job on their own, what need have they of the directing function of mind?

Although we have remained uncertain about the magnetic force inherent in the planetary bodies, through our consideration of the earth's axis, which is different from the sun's line of apsides, this difficulty is common to both explanations. For even when we supposed a mind, we were compelled to admit the kind of axis that we wanted in the earth, through whose mediation the mind could apprehend the strength of the angle, or its versed sine. On the contrary, probability strongly urges us to ascribe this reciprocation of the planets, which without doubt is in accord with the laws of nature, entirely to nature, whatever may be the means by which it occupies the planet's bodies.

Moreover, I do not know whether I have given sufficient proof to the philosophical reader of this perceptual cognition of the sun and the fixed stars, which I myself so easily accept, and bestow upon the planet's mind.

Furthermore, in those very modes of operation which we have prescribed to the mind, the soundest of all those which were deemed possible appear to involve some geometrical uncertainty. I am not sure whether this might not be repudiated by God Himself, as to this point He has always been seen to proceed by the path of demonstration. For if a planet, insofar as it may have approached the sun partly by its inherent force, comes into one and another degree of extrinsic power from the sun (as it does indeed come); and if the different degrees also reciprocally intensify the planet's own force of approaching while they increase the angle, which is posited as the standard of measure[23] of the approach, or of the increase of the sun's diameter; then the planet's own striving finally becomes in part its own measure, and simultaneously prior and posterior in the intensification of the planet['s force]. For in its parts it is unequal, and for this reason it requires a measure. Thus, the search for the forces tempering both powers will be concluded by a kind of iterative method[24] rather than deductively, so that they may complete their cycles in

---

[23] *Regula* (with initial capital).

[24] *quasi per regulam falsi.* This procedure, also called "false position," involves assuming a value

the same time, and in the same revolution of the body.

282
A cause for the progression of the aphelia that follows from the supposition of a mind.
On occultation, once again.

Someone might, however, want to think he had found the cause of the progression of the aphelia in the ungeometrical[25] nature of this measure. But in Ch. 35 we left it undecided whether this category of motion might not exist through another cause, namely, occultation.[26] That is, just as a plate of iron intercepts the force of a magnet on a strip of iron, the bodies of the planets might also mutually intercept the magnetic powers proper to them, by which they incline towards the sun. And so that this might not happen to the solar power—so that, I mean, the solar power, common to all, could not be intercepted for one planet by the interposition of another—we have drawn a distinction between the essence of the solar body and that of the planets' bodies. So, since we have not drawn a distinction between the bodies of the planets themselves, this seems still to be a possible cause. It could, of course, not have been arranged thus unless the exact magnetic disposition of the planet's body, by which the reciprocation is administered, were known.

Occultation does not transpose the aphelia, neither through natural means;

But to give an example of reasoning: let the planet have a magnetic disposition of the sort which, though we had introduced it somewhat earlier, we later denied the earth to possess. In this disposition, impediment through occultation does not have any place. For because it was the effect of the magnetic power to tend towards the sun and to recede from the sun, meanwhile keeping the fibers of its magnetic seat in line, if another planet, coming between the sun and the planet, impedes this travel towards the sun, or recession from it, while not impeding the common motion from the sun, the planet will approach or recede less than it should, and thus the size of the circuit will be altered along with the periodic time, over the ages, and will again be corrected by contrary eclipses. However, the aphelion will not change position through this occultation. So the cause for the motion of the aphelia previously proposed by us still reigns alone, without peer or rival.

Nor through the supposition of a mind.

Further, if a mind should preside over the reciprocation in the manner described, occultation will still do no harm. For as was said, the mind would use the angle of equated anomaly as its measure for increasing the sun's diameter; and, while deprived of the perception of it for a slight amount of time (that is, while the sun is hidden), it would be possible (the gods willing) to compensate for what the mind would have missed, upon the sun's re-emerging and restoring the equated anomaly into view. For mind (where there is one) is master of the animate faculty, and uses it differently and unequally according to circumstances. So why should it not use the animate faculty in an unusual manner here too, in removing the discrepancy between the measure (the equated anomaly) and the measured quantity (the sun's diameter) which had insinuated itself through the means of the sun's eclipse? What about other slow motions of this kind, such as the precession of the equinoxes arising from the earth's axis being directed at one or another of the fixed stars, and not at

---

for the solution, which is known to be incorrect, in order to test it against the given conditions and approach a more correct value by successive approximation.

[25] ἀγεωμετρήτῳ.

[26] Here and in the following three paragraphs, Kepler uses the Greek word ἀντιφράξις where the translation reads "occultation."

the sun? For here, the removal of the sun's light can have no effect, since its presence has no effect either.

We would like to avoid the inconvenient effects of magnetic occultations upon the reciprocations proper to the planets, just as we did in Ch. 35 for the common revolving effect of the sun. It should therefore be said that the bodies of the planets can indeed be similar in their magnetic dispositions, but either (1) they are so far from one another that the planets' orbs of power would not overlap, or (2) the power coming from the sun is so strong (that which activates the planets' proper powers no less than that which makes them revolve in their orbs) that it could not be impeded at all by the interposition of a small, weak body, so that it would pass on through, just as light passes through a globe of water, or (3) the bodies of the planets are so meager that they would have no effect, nor is the sun ever substantially blocked by any planet for any of the other planets moved by the sun, as the sun is never substantially blocked from the earth by the moon. For although the whole sun can be covered for the moon for several hours, the moon performs its reciprocation with respect to the earth, not the sun, and it can never be deprived of its perception of the earth since there is no body between the earth and the moon.

What the natural scientist could say to deny [the effects of] occultation.

283

Nevertheless, it might appear plausible to someone that the transposition of the apogees is instantaneous, and occurs through the cause of the sun's being eclipsed. Let him say, if he please, that to prevent the reciprocation's undergoing a sudden leap of speed when it is interrupted by an eclipse, during which the planet is moved by the sun to another angle and another degree of its strength, this angular leap is compensated by the planet itself, by having its axis incline towards the sun at the same angle after the eclipse as it was at the beginning of the eclipse. For thus a transposition of the aphelia will be obtained, but one occurring by leaps, and remaining in the same sidereal position for many years, until there happens to be another occultation of the planet.

Under what conditions, a mind being given, can the motion of the aphelia be ascribed to occultation?

On the other hand, the prior cause of the transposition of the aphelia, arising from the reciprocation's aberration from its sidereal circuit, produced by the ungeometrical[27] interconnection of its components, favors the uniform transposition of the apogees.

Another cause of the motion of the aphelia on the supposition of a mind.

Finally, if neither of these causes obtains, let the mind, furnished with an animate faculty which presides over the constant direction of the magnetic axis, have the additional task of inclining the axis over the ages. But if none of these causes stands, nor even the general idea of a mind, let us be satisfied with nature, which, as she has allowed everything else to be disentangled, has also shown a splendid occasion for the motion of the aphelia.

A third.

---

[27] ἀγεωμέτρητον.

CHAPTER 58

In what manner, while the reciprocation discovered and demonstrated in Chapter 56 holds good, an error may nevertheless be introduced in a wrongheaded application of the reciprocation, whereby the path of the planet is made puff-cheeked.

> With an apple Galatea seeks me, the lusty wench:
> She flees to the willows, but hopes I'll see her first.[1]

It is perfectly right that I borrow Virgil's voice to sing this about Nature. For the closer the approach to her, the more petulant her games become, and the more she again and again sneaks out of the seeker's grasp just when he is about to seize her through more twists. Nevertheless, she does not cease to invite me to seize her, as though delighting in my mistakes.

Throughout this entire work, my aim has been to find a physical hypothesis that not only will produce distances in agreement with those observed, but also, and at the same time, sound equations, which hitherto we have been driven to borrow from the vicarious hypothesis of Chapter 16. So, while trying to use a false method to do the same thing through this hypothesis, which is itself perfectly correct, I began once again to fear for the whole undertaking. *On the line of apsides, about centers A and B, let the equal circles GD, HK be described. And let AB be the eccentricity of the circle GD. Also, let the eccentric anomaly, or its number of degrees, be the arc GD or HK, by the equivalence established in Chapter 2.[2] Next, about center K, with radius KD which shall be equal to AB, let the epicycle LDF[3] be described, which will intersect the circle GD at D, through the equivalence established in Ch. 2.[4] Let AK be drawn, and extended to intersect the epicycle at L, so that the arc LD is similar to the eccentric anomaly GD or HK. And let BD be joined. Now, from D let perpendiculars be drawn to GA, LA, and let these be DC, DE.* Therefore, by what has previously been demonstrated in Ch. 56, AE will indubitably be the correct distance at this eccentric anomaly. The question remains how much time was taken to arrive at it. Now the versed sine of its arc, GC, which, after multiplication, becomes LE, when subtracted from GA, yielded the correct distance AE. These indications persuaded me that the other end of AE should be sought, not on the line DC (which was actually perfectly correct), but at the point I of the line DB, such that if I were to draw the arc EIF about center A with radius AE, it would intersect DB at I. Thus, according to this persuasion, AI would be the correct distance, both in position

284

---

[1] Virgil, *Eclogues*, 3. 64.

[2] Here all the Latin editions have "III"; however, the equivalence Kepler refers to is established in Ch. 2.

[3] Should be "LD", since point F is not on the epicycle but on line CD.

[4] "III" in all editions.

and length, and IAG would be the true equated anomaly. But it is manifest that the arc EIF would intersect the line DC at a higher place, namely, at F. Thus the angles IAG and FAG differ by the quantity IAF.

I therefore erred in taking the line AI instead of AF.[5] I first discovered the error empirically. For when I explored the quantity of the area DAG, either using all the distances or using the small area DAB, and then fitted the angle IAG, rather than FAG, to this area DAG, now converted into time, in the upper part of the semicircle I found $5\frac{1}{2}'$ more, and in the lower half $4'$ less, than the vicarious hypothesis gave with sufficient certainty. And so, since the equations disagreed with the truth, I began once more to accuse these perfectly correct distances AE, and the planet's reciprocation LE, of the crime for which my false method, which took I in place of F, was to be blamed. What need is there for many words? The very

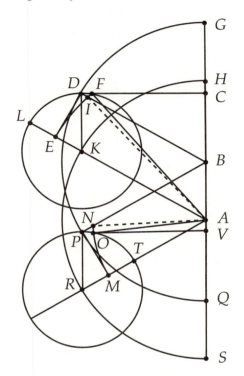

truth, and the nature of things, though repudiated and ordered into exile, sneaked in again through the back door, and was received by me under an unwonted guise. That is, I rejected the reciprocation on the diameter LE, and began by recalling the ellipses, quite convinced that I was thus following an hypothesis far, far different from the reciprocation, although they coincide exactly, as will be demonstrated in the following chapter. The only difference was that the errors I committed in method earlier were corrected by this procedure, and that F was used instead of I, as it should have been.

My line of reasoning was like that presented in Ch. 49, 50, and 56. The circle of Ch. 43 errs in excess, while the ellipse of Chapter 45 errs in defect. And the excess of the former and the defect of the latter are equal. But the only figure occupying the middle between a circle and an ellipse is another $\quad$ 285 ellipse. Therefore, the ellipse is the path of the planet, and the lunule cut off from the semicircle has half the breadth of the previous one, namely, 429.

Moreover, if the planet's path were an ellipse, it was clear enough that I could not be taken in place of F, for if this is done, the planet's path is made to be puff-cheeked.[6] *For let the angles QBP, SAR in the lower part be equal to GBD, HAK, and about center R let the epicycle PT again be described, equal to the previous*

---

[5]Max Caspar's analysis of this hypothesis (cited in *KGW* 3 p. 480) shows that it reaches a maximum error of $7'$ at a mean anomaly of $45°$.

[6]*Buccosam.* Kepler chose this name because he believed that the resulting path bulged outwards towards aphelion, presumably more so than the "face-shaped" path of Ch. 45. The name has misled some commentators into describing the path as bulging out at *perihelion*. However that may be, at the eccentricity of Mars's orbit, the asymmetries of the two curves are too small to be of any significance: what is in question in the present chapter is how to match the correct distances with their appropriate times.

one, and from P, the intersection of the epicycle with the eccentric, let perpendiculars PV, PM, be dropped to BQ, AR [respectively], and let PB be joined. And about center A, with radius AM, let the arc MN be described, intersecting PV at O, and PB at N. So, by analogy with the above, just as we are taking I in place of F, let us now take N for O, and let us consider that just as AN is the correct distance in length, it is also correct in position. Now the points I, N, and the like do indeed make the planet's path puff-cheeked. For the arcs GD and QP are equal. And BD, BP, projected from a common center, intersect the lunule cut off. But DI and PN, the breadths of the lunule extended towards the center, are unequal. And DI is smaller, and PN greater. For since ED and MP are equal, and EDI, MPN are right, while EI is a greater circle, since its radius AE is greater, and MN is a smaller circle, since its radius AM is smaller, therefore, PN will definitely be greater, and DI smaller. Therefore, the lunule cut off is narrower above, at D, and broader below, at P. In the ellipse, in contrast, this lunule is of equal breadth at points equally removed from the apsides G and Q. So it is clear that the path is puff-cheeked, so it is not an ellipse. And since the ellipse gives the correct equations, this puff-cheeked path should by rights give incorrect ones.

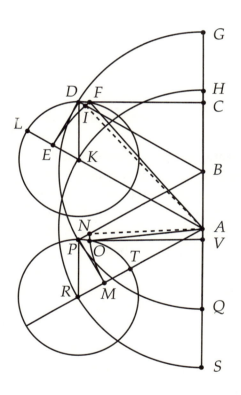

Nor was there any need to compute the equations anew from the ellipse. I knew they were going to perform their function without further prompting. I was only concerned about the distances, that if they were taken from the ellipse they might cause me trouble. But although this did happen, I had already prepared a refuge, namely, the uncertainty of 200 units in the distances. Consequently, I did not hesitate much here, either. The greatest scruple by far, however, was that despite my considering and searching about almost to the point of insanity, I could not discover why the planet, to which a reciprocation *LE* on the diameter *LK* was attributed with such probability, and by so perfect an agreement with the observed distances, would rather follow an elliptical path, as shown by the equations. O ridiculous me! To think that the reciprocation on the diameter could not be the way to the ellipse! So it came to me as no small revelation that the ellipse is consistent with the libration. This will be made clear in the following chapter, where it will be demonstrated at the same time, through the agreement of arguments from physical principles with the body of experience, mentioned in this chapter, that is contained in the observations and in the vicarious hypothesis, that no orbital figure is left for the planet other than a perfectly elliptical one.

CHAPTER 59

Demonstration that when Mars reciprocates on the diameter of an epicycle, its orbit becomes a perfect ellipse; and that the area of the circle measures the sum of the distances of points on the circumference of the ellipse.

PROTHEOREMS[1]

I

If an ellipse be inscribed within a circle, touching it at its vertices at opposite points, and a diameter be described through the center and the points of contact, and further, if perpendiculars be drawn to the diameter from other points on the circumference of the circle, all these lines will be cut in the same ratio by the circumference of the ellipse.[2]

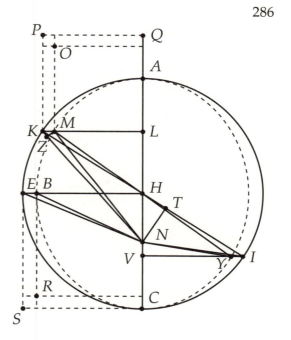

Using Book I page[3] 21 of the Conics *of Apollonius, Commandino[4] proves this in his commentary on Proposition 5 of Archimedes'* On Spheroids.

Let there be the circle AEC with el-lipse ABC inscribed in it, touching the circle at A and C, and let the diameter be

---

[1] *Protheoremata,* in Latin. The word was used by Martianus Capella (5th century CE), presumably as a direct transliteration from the Greek, where it meant "preliminary discussion." It is clear from the context that this is not what Kepler meant; however, it is not clear what he did mean by this word, or why he used it instead of calling these simply "*theoremata*". He also has used it in the *Optics* (e. g., Ch. 1, English trans. p. 33), where it denoted a subsidiary demonstration. It therefore seems best to call them "protheorems," rendering Kepler's Latin neologism with an English one.

[2] In the 1609 edition, the enunciations of the Protheorems were typeset exuberantly, using a much larger font than had appeared before, to emphasize the significance of these demonstrations. Experiments with similar typography for the translation had unsatisfactory results; however, Kepler's intention of highlighting these formal proofs is emulated by setting the enunciations in boldface, an option not available to Kepler.

[3] Although Kepler writes "page 21," the correct reference is to Proposition 21.

[4] F. Commandinus, *Commentaria in opera nonnulla Archimedis* (Venice, 1558), second part, folium 31^v Note that there are two series of folium numbers: the first series comprises only the Latin translation of Archimedes' works, while the second series contains Commandino's commentaries.

drawn through the points of contact A and C, passing through the center H. Then from the points K and E on the circumference let the perpendiculars KL, EH be dropped, cut by the circumference of the ellipse at M, B. BH will be to HE as ML is to LK, and so on for all other perpendiculars.

II

The area of an ellipse thus inscribed in a circle is to the area of the circle in the same ratio as the lines just mentioned.

For as BH is to HE, so is the area of the ellipse ABC to the area of the circle AEC. This is Proposition 5 of Archimedes' On Spheroids.

III

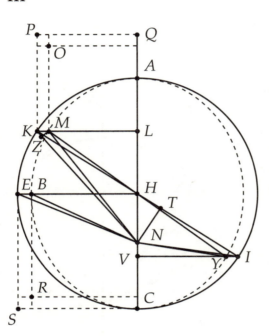

If from a given point on the diameter lines be drawn to the points on the perpendiculars where the circumferences of the circle and the ellipse intersect them, the areas cut off by these lines will again be as the segments of the perpendiculars.

Let N be the point on the diameter and KML the perpendicular, and let K and M be connected with N. I say that as ML is to LK, or (by Protheorem I) as the shorter semidiameter BH is to the longer HE, so is the area AMN to AKN. For the area AML is to the area AKL as ML is to LK, by Archimedes' assumptions in On Spheroids, *Prop. 5, which Commandino demonstrates under letters C and D in his commentary on this proposition. But the altitude NL of the right triangles NLM, NLK is the same, and the bases are LM, LK; and consequently MLN is to KLN as ML is to LK. Therefore, by composition, the whole area AMN is to the whole area AKN as ML is to LK. Q. E. D.*

287

IV

If the circle be divided into any number of equal arcs by perpendiculars such as these, the ellipse is divided into unequal arcs; and the arcs that are near the vertices adopt the greatest ratio [to the arcs of the circle], while those in the middle positions adopt the least ratio.[5]

For about the vertices the ratio of the arcs is close to the ratio of the perpendiculars cut off [by them], to which they closely approximate in length, although they are less.

---

[5]Kepler's use of the terms "greater ratio" and "smaller ratio" differs from Euclid's (*Elements* V Definition 7). Kepler's greater ratios are those which are farther from equality, regardless of which magnitude is greater. See, for example, *Stereometria doliorum vinariorum* (Linz, 1615), Proposition V, in *KGW* 9 pp. 82 ff.

About the middle positions they are nearly equal, but the elliptical arcs are smaller because they are less curved than the circular ones. This is self-evident.

V

The entire elliptical circumference is approximately the arithmetic mean between the circle on the greater diameter and the circle on the smaller diameter.

For it was proved in Chapter 48 above that that circumference[6] *is longer [than the ellipse] whose diameter is the mean proportional between the diameters of the ellipse, the area of which circle, by Archimedes,* On Spheroids *Prop. 7, being equal to the area of the ellipse. But, too, the arithmetic mean is longer than the mean proportional. Therefore, they are approximately equal.*

VI

The gnomons[7] of squares divided proportionally are to one another as the squares.

Let there be two squares, PL and SH. Let their sides KL, EH be divided proportionally at the points M, B. Let the gnomons KOQ and CRE be described. Therefore, because ML is to LK as BH is to HE, OL will also be to LP as RH is to HS. But the gnomons are the differences of the squares. Therefore also, as LP is to its gnomon, so is HS to its, and permuted, as PL is to HS, so is the gnomon KOQ to the gnomon CRE.

VII

If from the end of the shorter semidiameter on the circumference of an ellipse, a line equal to the longer semidiameter be extended, ending at the longer semidiameter, the distance between that point of intersection and the center is equal in square to the gnomon that the square of the longer semidiameter places about the square of the shorter semidiameter.[8]

From the end B of the shorter semidiameter HB, let the straight line BN be ex- 288
tended, equal to the longer semidiameter AH. I say that HN is equal in square to the gnomon ERC, that is, that it is the mean proportional between EB and the remainder of the circle's diameter. This was proved in Chapter 46 above. But it is demonstrated more easily and briefly in the pure case here. For the gnomon is the difference of the

---

[6]Reading "circumferentiam eam" for "circumferentia ea."

[7]According to Euclid (Book II, Definition 2), "In any parallelogrammic area let any one whatever of the parallelograms about its diameter with the two complements be called a **gnomon**." Hence, the gnomon is the figure remaining when in a parallelogram one draws lines through a point on a diagonal parallel to the sides and removes from the whole parallelogram one of the two small parallelograms through which the diagonal passes.

[8]Here Kepler proves that in the ellipse, the square of the eccentricity (in the astronomical sense) is equal to the difference of the squares of the two semiaxes. This use of the semimajor axis to determine the eccentricity is the nearest Kepler comes to introducing the focus of the ellipse. Although he himself had originated the term "focus" in his *Optics* (Frankfurt, 1604), Ch. 4 Sect. 4 English trans. p. 107 (*KGW* 2 p. 91), he does not use it in *Astronomia Nova* (except in the "Epigrams"—see footnote 8 to the "Epigrams", above), and he clearly did not acknowledge at this time that the eccentric point and the focus are the same.

squares BH and HE or HA, by the sixth of these protheorems. But the square on HN is also the difference of the squares BH and BN, that is, HE or AH, by Euclid I. 46.[9] *Therefore, the square on HN is equal to the gnomon ERC. Q. E. D.*

VIII

If a circle be divided into any number (or an infinity) of parts, and the points of division be connected with some point within the circumference of the circle other than the center, and also be connected with the center, the sum of the lines drawn from the center will be less than the sum of those from the other point.

Also, a pair of lines close to the line of apsides drawn to opposite points[10] from a point other than the center will be approximately equal to two drawn to opposite points from the center, while a pair so drawn at intermediate locations will be much greater than those drawn from the same center.

This was proved in Chapter 40.[11]   So[12] that excess does not increase uniformly with the number of lines, much less with the sines. For their differences vanish at the end, while the differences of the said excesses are greatest at the end. And since the area of the circle *KNA* increases uniformly, its part *KHA* increasing with the number of lines, by construction, and its part *KNH* with the sines of the arcs to which the lines are drawn, multiplied by *HN*, by Chapter 40, the area of the circle is therefore not adapted to the measure of the sum of the distances to its circumference.[13]

IX

If, on the other hand, instead of the lines from the point other than the center, those lines be taken which are bounded by perpendiculars drawn from that point to the lines which are drawn through the center—that is, if, in
289   **the terms of Ch. 39 and 57, the diametral distances are taken in place of the circumferential ones—then their sum equals the sum of those drawn from the center.**[14]

---

[9]The propositions meant is obviously the "Pythagorean theorem," I. 47, not I. 46 as stated.

[10]That is, opposite through the center of the eccentric.

[11]The demonstration is found on p. 313

[12]The following is in effect a corollary to the protheorem.

[13]An elucidation may be useful here. Kepler knew from Ch. 40 that the area *KNA* is proportional to the sum of the areas of the sector *KHA* and the triangle *KHN*, which is proportional to the sine of angle *KHA*. Could this area also be represented by lines from *N* to points on the equally divided arc *KA*? Kepler thought not, reasoning thus. The difference between *KH* (whose sum manifestly measures the area of the circle) and *KN*, if it were the required sine function, would increase maximally at the apsis and minimally at the quadrant. However, its rate of increase is actually the opposite. Therefore, the distances *KN* cannot represent the area.

Kepler's qualitative evaluations are correct: the area is a function of the sine of angle *KHA*, while the line *KN* is a function of its cosine. However, Kepler needed to compare changes in the area with changes in the *sum* of the distances *KN*. This comparison is not so easy, and it seems clear that Kepler never made it. Already aware that he would need to use the diametral distances instead of the circumferential ones, he apparently thought of this argument as a way to show more clearly why the circumferential ones could not be used.

[14]Kepler clearly intends this to apply only to the whole semicircle (as in the previous Protheorem),

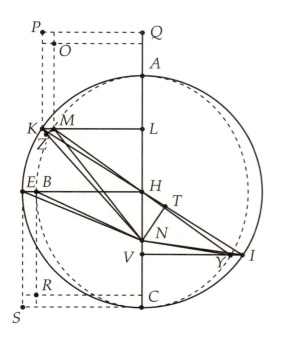

For let any point whatever on the circumference of the circle be chosen, K in the present instance, and from K let a straight line be drawn through H to the opposite part of the circumference I. Now from N let a perpendicular be dropped to KI, and let this be NT. Then KH, HI together, are equal to KT, TI together. And any sum of the pairs KH, HI, is equal to an equal sum of the pairs KT, TI.[15] And since, when AK is divided into any number of equal parts, the sum of the lines AN, KT to those parts increases partly with the number of lines HA, HK and partly with the sines multiplied by HN, the sum therefore increases uniformly with the area KNA, by the foregoing. Thus the area of the circle, and the parts KNA, are a measure of the sums of the diametral distances.

X

The ratio of distances from a point not at the center of an ellipse to equal arcs of the ellipse, no less than those on the circle in Protheorem 8, is contrary to the ratio of arcs of the circle and the ellipse to one another, explained in Protheorem 4. For the pair drawn from the point not at the center exceeds the pair drawn from the center in opposite directions, in the least ratio, and nothing at all, at the apsides; but at the middle elongations they exceed the latter in the greatest ratio.

This is clear from Chapter 40.[16] So, again, as in Protheorem 8, the area of the

---

for elsewhere the sum of the distances *AN, KT,* and those in between, cannot equal the sum of *AH, KH,* and those in between.

[15] At this point, Kepler has proved the theorem stated above (for the full circle only, although the proof could easily be revised so as to apply to the complete semicircle as well). However, he continues with an attempt to show that the sum of the lines *KT* is equivalent to the area *KNA*. His reasoning is as follows. The lines *KT* are made up of *KH* and *HT,* the former of which is the radius, and the latter of which can be defined in terms of the angle *KHA* and the eccentricity *HN.* Here he apparently made a hasty mistake: he thought that *HT* was equal to the product of *HN* and the sine of *KHA,* although it is actually the product of *HN* and the sine of the complement of *KHA,* in Kepler's terminology (this is of course the cosine, a word which in fact is a contraction of *complementi sinus,* "the sine of the complement."). His attention was no doubt distracted from the error by his excitement in recalling that he had already proved in Ch. 40 that the area *KNA* is equal to the sum of the sector *KHA* (that is, the sum of the distances) and the area *KNH* which is proportional to the product of *HN* and the sine of *KNA.* Thus, he thought the area would be equal to the sum of the distances *KT.*

To prove this extension of Protheorem IX, Kepler would have had to evaluate a sum that, in effect, is the integral of a function of the cosine of angle *AHK,* an operation that was beyond his powers or those of any mathematician of his time.

[16] See the demonstration on p. 313. Kepler implicitly extends the proof to equal elliptical arcs.

ellipse is not suited to measuring the sums of the distances of equal arcs of its elliptical circumference.[17]

XI

With these preliminaries completed, I shall now proceed to the demonstration.

If in an ellipse divided by perpendiculars dropped from equal arcs of the circle, as in Protheorem 4 above, the points of division of the circle and the ellipse be connected to the point that was found in Protheorem 7, I say that those that are drawn to the circumference of the circle are the circumferential [distances], while those that are drawn to the circumference of the ellipse are the diametral, which are established at an equal number of degrees from the apsides of the epicycle.

290   *From the point I, opposite K from the center H, let IV be dropped perpendicular to AC, intersecting the elliptical circumference at Y. And from the point N found in Protheorem 7 let the lines NK, NM, and also NI, NY be drawn to the points of intersection K, M, and also I, Y made by the two perpendiculars, respectively. Further, let the diagram of Ch. 39 and 57 be brought back, and let the semidiameter of the epicycle βγ be equal to the eccentricity HN, and the arc γδ beginning from the apsis αγ be similar to AK beginning from the apsis, and let αβ equal the semidiameter HA. I say that NK is the circumferential distance αδ (this was proven in Ch. 2) and NM is the diametral distance ακ.*

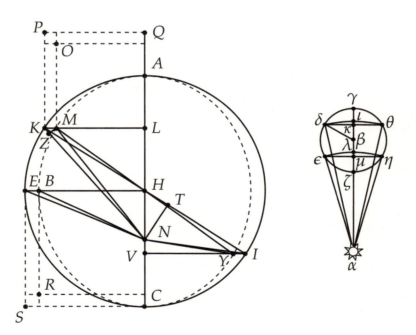

First, KN is equal in square to the sum of the squares on KL and LN. Likewise, MN in square is equal to the sum of the squares on ML and LN. Let LP be the square on LK, and LO the square on LM. Thus when the square on LN and the square on LM (that is, the square LO), common to both, are subtracted, there remains

---

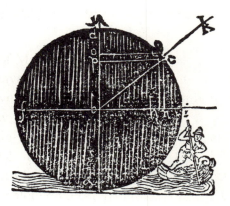

Kepler's woodcut of the planet's magnetic body (p. 414), showing the boatman and his oar pulling the planet along.

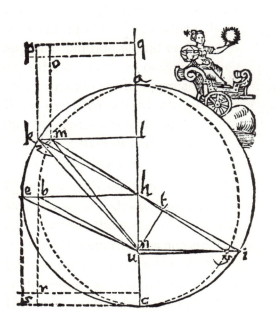

Kepler's woodcuts for the diagrams in Protheorem 11 (opposite). Note Urania in her triumphal chariot coming to crown Kepler with a laurel wreath for his conquest of the war god; also, the two angels, one with carpenter's tools and the other bearing what may be the blueprint for the universe.

the gnomon KOQ, by which the square on KN exceeds the square of, or on, MN.[18]
*Now as KL is to EH, so is KM to EB, by the first protheorem. Therefore also, as
KQ, the square on KL, is to EC, the square on EH, so is the gnomon KOQ to the
gnomon ERC, by Protheorem 6. And further, as KL, the sine of the arc AK, is to EH
or AH, the whole sine, here in the eccentric circle, so is the perpendicular δκ in the
epicycle (from the point δ of the arc γδ, which is similar to AK, to the diameter of
the apsides βγ) to the semidimeter of the epicycle βγ. Therefore also, as the gnomon
KOQ is to the gnomon ERC, so is the square on δκ to the square on βγ. But HN is
equal to βγ. And the square on HN is equal to the gnomon ERC, by Protheorem 7.
Therefore the square on βγ is also equal to the gnomon ERC, and consequently, the
square on δκ, the perpendicular from the point on the epicycle just mentioned, will
equal the gnomon KOQ. But the square on that perpendicular δκ is the excess of the
square on the circumferential distance δα over the square on the diametral distance
κα. Therefore also, the gnomon KOQ, equal to it, is the excess of the square on δα
over the square on κα. But KN is equal to δα. Therefore [the square on] KN exceeds
[the square on] κα by the gnomon KOQ. But it also exceeds the square on MN by the
same gnomon. Therefore, the diametral distances MN and κα are equal. Q. E. D. It
also will be demonstrated likewise concerning NY, that it is equal to αμ, where ζη is
similar to CI. And so for all.*

XII

Again, it is also clear from the same that

**The area of the circle, both as a whole and in its individual parts, is the
genuine measure of the sum of the lines by which the arcs of the elliptical
planetary path are distant from the sun's center.**

*For by Protheorem 9, if the area of the whole circle is set equal to all the diametral
distances of all the arcs of the division chosen, the parts of the area, as KNA, bounded
at the point N from which the eccentricity originates, are made equal to those diametral
distances that belong to the arc KA enclosing that area.*

*But by Protheorem 11, preceding, the diametral distances KT, TI, that is, κα, μα,
by Chapter 40, are the same as the distances MN, NY of the points M, Y of the ellipse.*

*Therefore, as the area of the circle is to the sum of the distances of the ellipse, so
is the part of the area of the circle KNA bounded at the sun's center N, whence the
eccentricity is measured, to the sum of the distances on the ellipse belonging to the
elliptical arc AM having the same number of degrees as the arc of the circle enclosing
the area AK.*[19]

291

---

[18]That is, the quantity $(LN^2 + LO)$ is subtracted from both sides of each equation. Thus the first
equation becomes:

$$KN^2 - LN^2 - LO = \text{gnomon } KOQ;$$

and the second:

$$MN^2 - LN^2 - LO = 0;$$

hence,

$$KN^2 - MN^2 = \text{gnomon } KOQ.$$

[19]Kepler's manner of stating this theorem shows that he had still not accepted what we now know
as "Kepler's Second Law" as anything more than a convenient way of approximating the true
law, stated in Chapter 32, that the increments of time over equal small arcs are proportional to

XIII

However, the following doubt arises: if the area *AKN* is equivalent to all the distances of as many points on the elliptical arc *AM* from *N* as we have taken to be present on *AK*, what, then, would that elliptical arc be; that is, where would it end? For it seems that it should not end at the perpendicular line *KL*. The reason for this is that in this way, by Protheorem 4, unequal elliptical arcs correspond to equal arcs on the circle, and thus the arcs are less about the vertices *A, C,* and greater about *B*. However, it appears necessary to take equal arcs of the elliptical orbit, should we wish to estimate and compare the times of the planet to traverse them. To be specific: because it is certain that the end

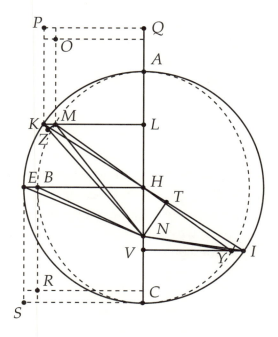

of this arc should be at the distance *MN* from *N*, therefore, as in Chapter 58, an arc *MZ* drawn about center *N* with radius *NM* somewhere indicates a point bounding this arc of the ellipse, and it appears that that point is going to be not *M* but *Z*, at which the arc intersects the line *KH*, making that arc of the orbit *AZ*.

The reply is made that the arc of the ellipse whose time increments the area *AKN* measures should by all means be divided into unequal parts, with those near the apsides being smaller.

For let it be that the actual path ABC of the planet be divided into equal arcs. Since the planet takes a longer time on arc A than on arc C proportionally as NA is longer than NC, while NA and NC taken together equal the longer diameter of the ellipse, and HB is the shorter semidiameter of the ellipse, the planet's amount of time on the arc at B and the opposite arc together will therefore be less than on the equal arcs at A and C together.[20] Therefore, to make the amount of time at A and C shorter, and at B and its opposite longer, thus making the amount of time on any two opposite arcs taken together the same, the arcs at A and C should be made smaller, and at B and

---

the distances of those arcs from the sun. Since the total elapsed time, or the mean anomaly, is the sum of those distances, if Kepler could show that the area is an exact measure of the sum of the distances, he could then use the area instead of the sum. So the next three theorems represent an effort to show beyond a doubt that the area is such a measure if the distances are taken as described. Kepler actually used the area of the (imaginary) circle instead of that of the ellipse, which shows how far the "area law" is removed from the physical truth as Kepler saw it.

[20]Later, in the *Epitome of Copernican Astronomy* (p. 669), Kepler remarks on this line: "The single small word *erit* ["will be"] introduced a great deal of obscurity. If you change it to *computaretur* ["could be computed"], everything will be clearer. I should, however, say that this was the more obscurely expressed, and made the more laborious, in that the distances were not considered as triangles there, but as numbers and lines."

its opposite, longer. But this is accomplished by the perpendiculars KML, as is clear from the objection itself.

But by this solution we only discover with certainty that the arcs about *A*, *C* should be somewhat shorter. Whether the particular arcs determined by the perpendiculars *KML* are exactly the required arcs, is not yet established. But it will now be made clear, in the following manner.

XIV

If someone were to divide an ellipse *AMC* into any number of equal arcs, assigning to each individually its distance from *N*, while taking the areas *AMN*, *ABN*, *ABCNA* in place of the sum of the distances on *AM*, *AB*, *ABC*, by Protheorem 10 he would bring about the same error that occured in Ch. 40 above when we tried to do on a perfect circle what we are here supposing to be tried on an ellipse. That is, two lines *MN*, *NY* from two points *M*, *Y* opposite one another through center *H*, are taken as equivalent to the shorter line *MHY*.

Suppose, however, that that same person were to divide an ellipse *AMC* into the same number of unequal arcs, contrary to Protheorem 10, according to the following law: the circle *AKC* being first divided into equal arcs, perpendiculars *KL* would then be drawn to *AC* from the ends of the individual arcs, cutting the ellipse *AM* into arcs also; and the elliptical area would be taken for the distances of these arcs from *N*. In that case, a remedy would be provided for the error which has been committed: a most perfect compensation.

I shall prove this for the beginnings of the quadrants, *A* and *C*; for the ends, *B*; and the motion in between.

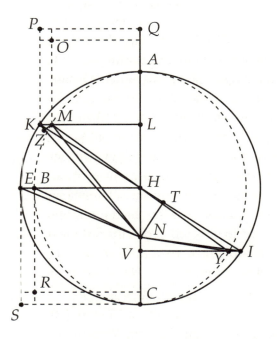

At the beginnings of the quadrants A, C, if the two lines NA, NC be taken for the line AHC, there is no error. At the end, however, if for BN (that is, for EH) I take BH, the consequent error or defect is a maximum, by Protheorem 10, the amount being BE. And by Protheorem 7 of this chapter, as HE is to EB, so is the required length to the error committed at this position. Now, suppose that the sum total of all the distances receives a measure erring in defect, namely, the area of the ellipse. Then when the defect is distributed among the individual distances by the force of our operation or computation, it will happen that the distances NA, NC are taken too small with respect to this measure of all of them. We are thus deceived in thinking that all the lines err equally in defect, since NA and NC are not in fact in error. They do indeed add up to the correct sum, but when the sum is in turn distributed, NA and NC will not

have received their correct value, because certain lines about B will have defrauded the whole.

Let us now see how we can remedy this error in the same proportion.

By Protheorem 4 of this chapter, the least arcs AK, AM about the apsides A or C are in the same ratio as KL to LM, that is, EH to HB. This was the ratio by which the lines about B formerly erred in defect. And at B, in turn, the least arcs of the circle and the ellipse, KE and MB, say, are equal; just as, before, the straight lines AN, NC together were equal to the line AHC. Therefore, as in the previous consideration of the straight lines, so will it be here in the consideration of the arcs: when the mean and uniform measure of the arcs is conceived, the arc at the apsides A or C will be short with respect to it, and the arc at the middle intervals B will be long. Thus where the distances are too short with respect to their erroneous sum, in the faulty area of the proposed ellipse, the arcs will be small with respect to their mean value, as at *A, C*; and where the distances are too long, the arcs are too long, as at *B*. And so to the extent that we accumulate too small an interval of time in our calculation, owing to the rather short distances at the apsides, there are that many more distances on that arc, it being cut into small parts each of which has its own assigned distance. And inversely, to the extent that about the middle intervals *B* more time is accumulated by the individual distances than is fitting in our calculation, when we carried over to the innocent apsides *A, C*, the part of the defect at this location, the calculation has collected correspondingly fewer distances, they having been obtained from the large parts of the arc by begging. At that place (*A* and *C*), what the individual distances cannot do owing to their brevity in the calculation, they accomplish by their being closely spaced, with the result that they accumulate the correct time intervals. And here, the error arising in the calculation occasioned by their excessive length is again removed by their being more widely and loosely spaced.

Of the beginning and end, I have said that both the arcs of the circle at *A* and *C* and the correct distances differ at the beginning from those that the area of the ellipse accumulates, in the same ratio that *EH* has to *HB*; and that the arcs at *BE* and the distances at *A* and *C* end up differing by the same ratio, namely, the ratio of equality.

We must now state the same for the progress in between.

The lines NA, NC, from small beginnings, by swift increments exceed the lines AHC by a perceptible amount. On the other hand, where the excess is greatest, as that of BN over HB, the increments are slowed down perceptibly. They are greatest in the middle, near an eccentric anomaly of 45°.

This is to some extent shown by the angle and secants of the equation. For BN differs from BH by about the same amount as the secant of the angle of the optical equation differs from the whole sine, while the opposite angles of the equations assist each other in the same ratio. Now the increments of the secants of the optical equation at about 45° are near a maximum, and are slow at the beginning and end of the quadrant. Concerning these, see the end of Ch. 43.

Furthermore, the increments of the elliptical arcs marked off by the perpendiculars KL progress in the same ratio. For at the beginnings, A and C, the arc AK, always beginning from A, is to its increment as LK is to KM. But the whole arc itself is small, and so the increment is small as well. At the end, near B, the ratio of AE to

293

AB is reduced nearly to equality, even though the arc AB is large, since it is near the quadrant. Therefore, the increment is again small. So it is at the middle, about 45°, that the increment of the arcs is most evident.

It is thus clear that in the intermediate progress, too, the ratios are equal, so far as minute consideration can be carried.

Although the demonstration is most certain, it is likewise *gauche* and ungeometrical,[21] at least in that part pertaining to the progress of the intermediate increases. As elsewhere, I would like to have this small part carried out geometrically and with *finesse*,[22] so that even an Apollonius would be satisfied. Meanwhile, until someone else discovers and provides us with this small part, we should be content with what we have.

XV

294

But let us complete the proof.

The arc of the ellipse whose time is measured by the area AKN, **should end on** LK, **so it would be** AM.

For hitherto we have been proceeding on the fiction that if anyone had so much leisure as to want to compute the area of the ellipse, it would turn out that in using the area of the ellipse AMN in place of the same number of distances of AM as there are equal arcs on AK, he would not miss the mark. Let this serve us as the previously demonstrated major premise of the proposition.

I shall now add the minor premise, derived from Protheorem 3 of this chapter. Here it was shown that the area AKN is to the area AMN as the area AKC is to the area AMC. The conclusion therefore is, since the ratio of equimultiples is the same, that the area of the circle AKN also measures the sum of the diametral or elliptical distances (such as KT, TI) on AM, there being as many as there are parts in AK. Whence it is clear that I correctly assigned more closely spaced distances to the parts of the ellipse about A, C: that is, the same number of distances as there were intersections made by the perpendiculars KL coming from equal arcs of AK.

So that no one may doubt the truth of this, confused by the subtlety and perplexity of the argument, this truth previously came to be known through experience, in the following manner. At the individual degrees of eccentric anomaly, I set up the diametral lines KT, TI in place of the distances from N. I also added each in order to the sum of the previous ones. When all were collected, the sum was $36,000,000$, as is fitting. Next, when the individual sums were compared with the whole, following the rule of proportions, the sum $36,000,000$ would be to 360 degrees (the nominal value of the whole periodic time) as the individual sums would be to the increments of time they signify. This produced exactly the same results, down to the last second, as would have come out had I multiplied half the eccentricity by the sine of the eccentric anomaly, [added the area of the sector contained by the eccentric anomaly], and compared [the sum] to the area of the circle, which would be given the same value of 360 degrees (the nominal value of the periodic time).[23]

---

[21] ἄτεχνος et ἀγεωμέτρητος.

[22] ἐντέχνως.

[23] In the diagram, the eccentric anomaly is KHA, whose sine is KL. The eccentricity is HN, and

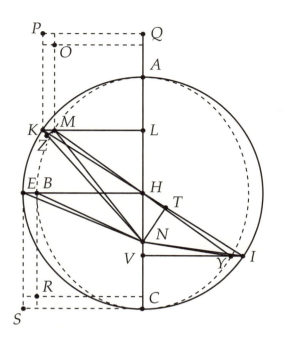

Then, when I was of the opinion that the correct distance *NM* should be applied to the line *KH*, becoming *ZN*, and had thus investigated the equated anomaly *ZNA*, attributing it to the mean anomaly *AKN*, the equations disagreed obviously with my vicarious hypothesis of Chapter 16. At about 45°, the difference between the equated anomaly and the true value found through experience in the observations was a defect of about $5\frac{1}{2}$ minutes, and near 135° about 4 minutes. But when *NM*[24] was so applied as to end on *KL*, then when the equated anomaly *MNA* was applied to the mean anomaly *AKN*, it agreed exactly with the vicarious hypothesis, that is, with the observations. And when the fact was established, I was afterwards driven, once I had settled on the principles, to seek the cause of the matter which I have revealed to the reader in this Chapter as skilfully and lucidly as possible. And unless the physical causes that I had originally taken in the place of principles had been good ones, they would never have been able to withstand an investigation of such exactitude.

If anyone thinks that the obscurity of this presentation arises from the perplexity of my mind, I shall myself only thus far acknowledge to him my guilt, that I was unwilling to leave anything untested, no matter how utterly obscure, and not strictly necessary to the practice of astrology, which many deem the sole end of this celestial philosophy. But as for the subject matter, I urge any such person to read the *Conics* of Apollonius. He will see that there are some matters which no mind, however gifted, can present in such a way as to be understood in a cursory reading. There is need of meditation, and a close thinking through of what is said.

295

---

when its half is multiplied by *KL*, the product is the area *KHN*. When this is added to the area of the segment *KHA*, the sum is the area *AKN*, which represents the mean anomaly, which is the time. Kepler discovered empirically that the ratio of this area to the area of the circle is the same as the ratio of the distances on *AM* (at points determined by the perpendiculars) to the sum of all distances on the circle. In this passage he failed to mention the adding of the sector to the area of the triangle, clearly an oversight in a syntactically involved sentence. The omission has been remedied by the bracketed words.

[24]Reading "NM" instead of "AM".

CHAPTER 60

A method, using this physical—that is, authentic and perfectly true—hypothesis, of constructing the two parts of the equation and the authentic distances, the simultaneous construction of both of which was hitherto impossible using the vicarious hypothesis. An argument using a false hypothesis.

In Chapters 56, 58, and 59, the planet was assumed to approach the sun and recede from it along a diameter directed towards the sun, thus making an elliptical orbit, and further, it was assumed to spend time at each individual point in proportion to the distance of the point from the sun.[1] Thus we happen upon a most convenient short cut through the preceding Chapter 59, for evaluating the sum of any number of time intervals all at once. For it was shown that when a line is drawn from a circle perpendicular to the longer diameter of an ellipse inscribed in that circle (in the previous diagram, let KL be dropped to AC), so as to intersect the ellipse at M, and supposing that the sun is at N, the sum of all the distances of points on the arc AM from the sun N is contained in the area AKN.

Given an eccentric anomaly, to find the mean anomaly corresponding to it,

For an arc AM of the ellipse being supposed, which is defined by the arc of the circle AK, the area AHK is given, which is the sector of the arc AK, by which arc that sector is also measured, in units of which the whole area of the circle is 360°.

Or the physical part of the equation.

And because the arc AK is given, its sine KL is also given. But as KL is to the whole sine EH, so is the area HKN to the area HEN, as was proved in Ch. 40.[2] Also, since the eccentricity HN is given, half of it multiplied by HE will describe the area HEN. This value is found at once at the beginning, so that it may be known what this small area amounts to, when the whole area of the circle has the value of 360° of time.

And so, once the area HEN is known, it is very easy to find the area HKN by the rule of proportions. For as EH is to KL, so is NEH to the area NKH, or its value in de-

296 *grees, minutes, and seconds; and this, added to the value of KHA, establishes a value for KNA, which is the measure of the time which the planet takes on AM.[3] This, then, is one of the parts of the equation, the one I call "physical"[a], namely, the*

Terms.
a) The physical part of the equation.

_____

[1] This is the first succinct statement of what we now know as "Kepler's First and Second Laws." It is, however, notable that the "Second Law" is stated as a proportionality of time to distance, not to area. This is entirely in accord with Kepler's physical principles, enunciated in Chapter 32, and indicates that the area law was at this point a convenient means of computation and not a physically accurate principle.

[2] See p. 312.

[3] Kepler's sketch of his procedure may require some explication.

$$\frac{KL}{KH} = \sin \widehat{AK} = \frac{NKH}{NEH}; \text{ or } NKH = NEH \sin \widehat{AK}.$$

area *HKN*. Yet I so arrange the tables that there is no need to mention the equation, nor is there a separate column showing the optical[b] part of the equation, that is, the angle *NKH*. The [meanings of the] terms, "mean anomaly," "eccentric anomaly," and "equated anomaly" will be more personal to me. The mean anomaly[c] is the time, arbitrarily designated, and its measure, the area *AKN*. The eccentric anomaly[d] is the planet's path from apogee, that is, the arc of the ellipse *AM*, and the arc *AK* which defines it. The equated anomaly[e] is the apparent magnitude of the arc *AK* as if viewed from *N*, that is, the angle *ANK*.

b) The optical part.

c) The mean anomaly.

d) The eccentric anomaly.

e) The equated anomaly.

Now the angle of equated anomaly is found as follows. *The arc AK being given, the sine of its complement LH is given.   And as the whole is to LH, so is the whole eccentricity to the part to be added*[4] *to 100,000 (subtracted, below 90°) to give the correct distance of Mars from the sun, namely, NM.*[5] *So in triangle MLN, the angle at L is right, and MN is given, and LN is also given. For it is made up of LH, the sine of the complement of AK, the distance from apogee, or the eccentric anomaly; and the eccentricity HN. Below 90°, in*

Given the eccentric anomaly, to find the equated anomaly.

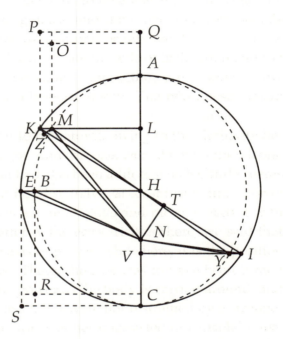

place of the sum LH, HN their difference should be taken, and in place of the complement of the eccentric anomaly, its excess. Therefore, the angle of equated anomaly LNM will not be hidden.   Here anyone who wishes can easily figure out what has to be changed in the other semicircle.

On the other hand, given the eccentricity and the equated anomaly, the eccentric anomaly is given, a little more laboriously, whether we proceed demonstratively or by analysis.[6]

Given the equated anomaly, to find the eccentric anomaly, and thus the mean as well.

It can be investigated demonstratively by this method: namely, by measuring the angle under which is viewed the planet's incursion *KM* made from any point *K* on the circle as if seen from the center of the sun. This method depends upon several protheorems.

Preparation for this.

---

[4]That is, *HT*.

[5]Since *NM* = *KT*.

[6]In the task of computing the eccentric anomaly when the equated anomaly is given, Kepler proceeds quite haphazardly. The simplest way would proceed analytically; however, Kepler provides this analytic solution only in the second place. In the first place, he presents the unsystematic but original geometrical method of the following five theorems.

I

The small lines of the planet's incursion towards the diameter of the apsides increase in proportion to the sines of the eccentric anomaly.

For as EH is to KL, so is EB to KM. This was established in Chapter 59, and demonstrated in the Conics.

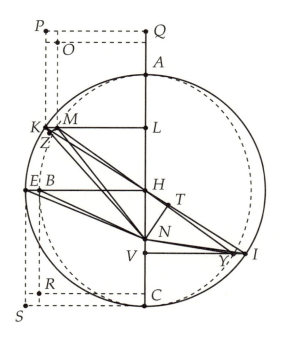

II

The ends of one of the small lines being connected to the center, and it being supposed that the small line remains the same in quantity at all points of the eccentric, the tangent of the angle at the center decreases approximately in proportion to the sines of the complement of the eccentric anomaly.

297

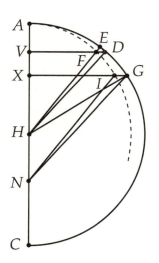

Let DF be the small line, part of DV, the sine of the eccentric anomaly AD. Let the ends D, F be connected to H, and let HF be extended. And let the straight line ED be tangent to the circle at D, intersecting HF at E. Therefore, since DVH is right, VDH will be the complement of VHD, the eccentric anomaly. And since EDH is also right, HED will be less than a right angle by the quantity EHD. This is of hardly any significance, since where it is greatest it does not exceed 8 minutes. And for the same reason, VFH, that is, EFD, is greater than the complement FDH of the eccentric anomaly, but by the quantity FHD, which is clearly of no significance. And since FED is somewhat more acute than a right angle, the arc circumscribed about FED will be somewhat longer than a semicircle. Therefore,

ED is to DF as the sine of an angle which somewhat exceeds the complement of the eccentric anomaly, is to the sine which is slightly — really, hardly at all — smaller than the whole sine. Now if FD retains this length throughout the whole quadrant, ED is made approximately proportional to the sines of the complement of the eccentric anomaly. For if FD remains the same in length, and the end D is at A, the angle FDH is right, and thus FHD is a maximum, and then DFH is at its most acute, and consequently the arc above FD is at its longest. From that point, as FD moves down from A, the arc FED decreases and the angle FED increases, until at degree 90 FD

becomes part of the line DH. Thus HF belongs to HD, and ED vanishes, and there,
by analogy, the arc above FD is equal to the semicircle, and is at its least.

III

The ends of the small line of the planet's incursion towards the diameter of
the apsides being connected, however long the line happens to be at any eccentric anomaly, the tangents of the angles at the center (and thus the angles
themselves as well, when they are at their smallest), increase approximately in
the ratio compounded of the ratio of the sines and the ratio of the sines of the
complements of the eccentric anomaly; that is, in proportion to the rectangles
on the quadrant formed by multiplying the sines of the angles by the sines of
their complements. Thus, the greatest rectangle at 45 degrees is to the greatest
angle at the same eccentric anomaly of 45° as the remaining rectangles are to
the angles of the remaining eccentric anomalies.

Term.
What is the rectangle of the quadrant?

For at these angles, such as EHD, two factors are compounded: the length of the
incursion, varying from nothing to a maximum, and its apparent magnitude, from
nothing to a maximum. But, by I, the incursions increase in proportion to the sines,
and by II, the tangents of the angles of apparent magnitude of these incursions, as
if viewed from the center of the eccentric, decrease in proportion to the sines of the
complement. By the former cause it happens that the angle is nothing at A when the
sine is zero, and by the latter cause the angle is zero at an eccentric anomaly of 90,
when the sine of the complement is zero; and further, at both places the rectangle has
vanished entirely. But at an anomaly of about 45°, FD has now turned out greater
than half [its maximum], because the sine, 70,711, is greater than 50,000, half the
whole sine. And its angle EHD is greater than half by still more, because the sine
of the complement is also greater than half, namely, 70,711 also. Consequently, the
rectangle of the quadrant is the greatest of all, and at the same time is a square, equal
to half the square on the radius, namely, 5,000,000,000.

298

IV

The angle of the planet's incursion from the circumference of the circle towards
the diameter of the apsides is the same at the eccentric anomaly, about the center of the eccentric, and at the circular equated anomaly, of the same number
of degrees, about the center of the sun.

Let the equated anomaly ANG be constructed equal to the eccentric anomaly
AHD at the circumference of the circle G; that is, let NG be drawn parallel to HD.
And from G let GX be drawn perpendicular to AC, and on it let GI be the correct
incursion of the planet. And let I and N be joined. Now XG is to GI as VD is to DF,
by I, while XG is to GN as VD is to DH, because of the similarity of the triangles.
Therefore, IG is to GN as FD is to DH. Also, FDH and IGN are equal. So FHD
and ING are also equal. And H is the center of the eccentric, while N is the center of
the sun. Therefore, the angle, et cetera. Q.E.D.

Term.
This anomaly is called the "circular equated anomaly" because it is not really the equated anomaly; rather, it is what the equated anomaly would be if the planet's orbit were a circle.

V

The authentic and truest measure of the angle by which the fictitious equated
anomaly, which depends upon the circle, differs from the true equated anomaly,

which ends on the ellipse, is the rectangle contained by the sine of the ficiti-tous equated anomaly and the sine of the complement of the true equated anomaly.

In the same diagram, when the sine of the angle AHD was multiplied by the sine of the angle VFH, the authentic measure of the angle FHD was going to result, by III. But by IV, the sine of the equal angles VHD and XNG is the same, and also, the sine of VFH, XIN is the same. Therefore, when the sine of the angle XNG, the fictitious equated anomaly, is multiplied by the sine of the angle XIN, the complement of XNI, the true equated anomaly, there results the authentic measure of the angle FHD, that is, by IV, of the angle ING, which is the difference between XNG and XNI.

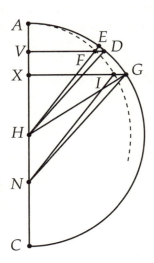

Corollary

Because the difference *ING* is small, and is never greater than 8 minutes, the difference between the rectangles of the sines of *XIN* and *XGN* is going to be still smaller in effect.

299
Given the equated anomaly, to find the corresponding eccentric anomaly,

From this, the following procedure will arise. *The angle of the true equated anomaly being given, let its sine be multiplied by the sine of its complement. Let double the product, with the last five digits dropped, be multiplied by the maximum angle of incursion, at an anomaly of 45°. The product will be the angle of incursion at the given anomaly. This, added to the true equated anomaly XNI, gives the fictitious, XNG. By this angle, and the known sides NH, HG, the eccentric anomaly AHG is*

* And the mean anomaly.   *found, and the value of the triangle HGN*, as before.*

Moreover, it is not difficult to find the maximum angle at anomaly 45°. *Let VHD be 45°. Therefore, as the whole sine is to 70,711, so is the maximum incursion, or maximum breadth of the lunule, of 429 (or, more correctly, 432) to FD, 305. And since at 45° HV, VD are now equal, subtract FD, 305, from VD, 70,711. The remainder, VF, is 70,406. This, with HV, gives the angle VHF, 44° 52' 34'', which differs from 45° 0' 0'' by only 7' 26''. And this is the maximum of the angle ING.*

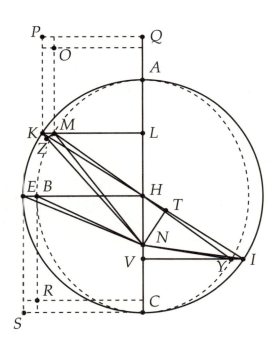

The following is another method using analysis, whose fundamentals are these. *In the diagram of Ch. 59, given the angle MNL, the ratio of the lines MN, NL is given, and I know that MN and LN are composed of parts in a known permuted proportion. For MN contains the (known) whole sine, and LN contains the known eccentricity HN. The*

remainder of MN has the same ratio to the remainder of LN, which is LH, as the eccentricity HN has to the whole sine.[7] If you prefer, you may also refer to the diagram in Chapter 58. Therefore, let MN be $100,000 + 1x$,[8] LN from the angle MNL, 30°, be $(8,660,300,000 + 86,603x)/100,000$, and NH be 9265 or $926,500,000/100,000$, so that HL would be $(7,733,800,000 + 86,603x)/100,000$. But as HN, 9265, is to 1x, so is 100,000 to LH. Therefore, in the second instance, HL is $(100,000/9265)x$, that is, $(1,079,320/100,000)x$. Previously, it was $(7,733,800,000 + 86,603x)/100,000$. With the denominators removed, as well as whatever can be subtracted equally from both sides, what remains is that $992,717x$ is equal to the number $7,733,800,000$. Therefore, the single root is 7790. And MN is 107,790. And because as HN is to this root, so is the whole to LH, LH will therefore be 84,084, which is the sine of KE, 57° 14′, the complement of the eccentric anomaly AK, 32° 46′. Now that this is found, the area AKN, the measure of the time or the mean anomaly, is found as shortly above. These things are clearest in the diagram of Ch. 58. Let GQ be the eccentric, AB the eccentricity, GD or LD the eccentric anomaly, FAC the equated anomaly, FA or EA the distance. So as AK is to AB, so is BC to KE. And at the equated anomaly CAO, as AR is to AB, so is BV to RM.[9] So EK or RM is supposed to be one root. The rest is as above.

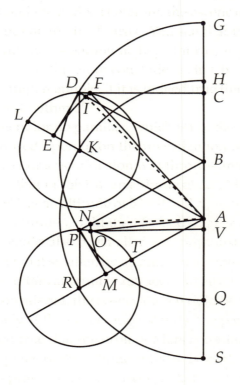

---

[7]Since $MN = KT = KH + HT$, the remainder of MN is HT. The remainder of LN, as Kepler says, is LH. And because the triangles KHL and NHT are similar, $HT : HN :: LH : KH$, or the remainder of MN is to the remainder of LN as the eccentricity HN is to the radius (or "whole sine") KH.

[8]Letting the unknown, x, be the magnitude of HT.

[9]As Kepler says, it is clearer here. In modern terms,

$$r : e :: r\cos\beta : e\cos\beta,$$

where r is the radius, e the eccentricity, and β the eccentric anomaly.

Given the mean anomaly,
to find the eccentric
anomaly and thus the
equated anomaly.

But given the mean anomaly, there is no geometrical method of proceeding to the equated, that is, to the eccentric anomaly. For the mean anomaly is composed of two parts of the area, a sector and a triangle. And while the former is numbered by the arc of the eccentric, the latter is numbered by the

300   sine of that arc multiplied by the value of the maximum triangle, omitting the last digits. And the ratios between the arcs and their sines are infinite in number. So, the sum of the two being set out, we cannot say how great the arc is, and how great its sine, corresponding to this sum, unless we were previously to investigate the area resulting from a given arc; that is, unless you were to have constructed tables and to have worked from them subsequently.

This is my opinion. And insofar as it is seen to lack geometrical beauty, I exhort the geometers all the more to solve me this problem: Given the area of a part of a semicircle and a point on the diameter, to find the arc and the angle at that point, the sides of which angle, and which arc, enclose the given area. Or, to cut the area of a semicircle in a given ratio from any given point on the diameter.[10]

It is enough for me to believe that I could not solve this *a priori,* owing to the heterogeneity[11] of the arc and the sine. Anyone who shows me my error and points the way will be for me the great Apollonius.

---

[10]Here the "Keplerian Problem" is formulated with perfect clarity.

In the Kepler manuscripts at the Russian Academy of Sciences in St. Petersburg (vol. XIV p. 422, quoted in *KGW* 3 pp. 482–3, and at greater length in *Kepleri Opera*, ed. Frisch, vol. 3 pp. 503–505) there appears a fragment of a letter (in the form of a copy) to an unknown correspondent, in which Kepler remarks upon the computation of the equated anomaly from the mean anomaly, as follows:

"Now something from my Astronomy. Ch. 59 and 60 are thick with errors in the letters, not all of which are corrected in your copy. For me the anomalies are three: the mean, given by the time, which I number by the area AKN; the eccentric, which, improperly, is the area of the circle AKH, or the arc AK, or the angle AHK, or properly, the arc of the ellipse AM; and the equated, which is the angle MNA. Given the mean anomaly AKN, I have no way of finding the eccentric anomaly AK other than by trial and error. For I suppose an arc AK, and multiply its sine KL by the value of the maximum area EHN, which is $19,110$ seconds (this is provided by multiplying EH by half the eccentricity HN, and comparing it to the area of the circle, which I arbitrarily suppose to be 360°). Thus I obtain the area KHN in seconds, which I add to the supposed arc AK, or the area AHK (since the measure of the two is now the same), to give AKN. If this is equal to the given mean anomaly, then I supposed AK well. When tables are constructed, as I have indeed done, there is no further difficulty here, for it is extracted immediately. But I am now concerned with a method of computing some single equation. Through the given arc AK and LH, to which I add the eccentricity HN, there results LN. KH being drawn, and its perpendicular NT. KT will be the measure of the distance of the planet M from the sun N. For KT and MN are equal. I did in fact say this in the commentaries, but did not explain it in this diagram in Ch. 59 and 60. And now MN, NL are known, and L is right, and therefore, the equated anomaly MNL is given."

The procedure outlined here is still the best way to compute the equated anomaly from the mean.

[11]ἑτερογένειαν.

Part V

OF THE COMMENTARIES

ON
THE MOTIONS OF THE STAR

MARS

ON

THE LATITUDE

This number is computed at the position of the sheel. However, since the position given is 37 miles farther north, the latitude one minute east, and the computed latitude about 4° 27', we do not get to the observed position north.

The correction mentioned in the preceding footnote would make this angle 1° 8'.

At 47°, with the correction mentioned above.

We should have been 1° 4' 27" or 1° 2' 55" with the corrected inclination.

Chapter 61

An examination of the position of the nodes.

The ratio of the orbs of Mars and the earth, the eccentricity of each, and the shape of their paths, have all been found with great certainty in the preceding chapters. Therefore, we can now easily accomplish here what we sought out in an approximate way in Chapters 11, 12, 13, and 14.

Let us begin with the nodes. On 1593 December 10, at $7^h 0^m$ in the evening, Mars was observed at 4° 44′ Aries, with latitude 0° 1′ 15″ south, parallax not accounted for.[1] Its altitude being $35\frac{1}{2}$°, it was not subject to refraction. After the 687 days of one complete revolution of Mars, on 1595 October 28 at $11^h 30^m$ pm Mars was found at an altitude of 51° in 18° 35′ Taurus, with latitude $4\frac{1}{2}′$ south, parallax not accounted for.[2] And again, 687 days previously, on 1592 January 23, at 10^h pm, it again had a southern latitude of 2′, with an altitude of 25°.[3] And finally, subtracting another 687 days, so that we come to 1590 March 7, Mars was observed on March 4 at 7^h, at an altitude of 14°, to have a latitude of 3′ 20″ south.[4] This would have appeared larger, except that Mars was low enough to be refracted, and appeared too high. For the refraction at this altitude is $3\frac{1}{2}′$, of which about 2′ is accounted to the latitude; thus, the apparent latitude would be 5′ south. But since we are anticipating by three days the date corresponding to the others, three minutes are removed from the inclination by the approach to the node of $1\frac{1}{2}$° made in this space of time. When this is converted into latitude, however, the effect is somewhat less, so that the latitude remaining to Mars on the 7th would be $2\frac{1}{2}′$, and perhaps a little less, if the refraction were less. For its quantity is not perfectly constant.

Let the latitude in 1590 be 1 minute; 1592, $1\frac{1}{2}′$; in 1593, $2\frac{1}{2}′$; in 1595 at 11^h, $4\frac{1}{2}′$, as we might allow an error from one source or another of one minute either way. These latitudes will indicate to us an inclination of $1\frac{1}{2}′$, which requires a distance from the nodes of about 40′. This is only for the sake of consensus.

We will nevertheless accomplish our aim more accurately using the year 1595. For while on October 28 at 12^h the latitude was $4\frac{1}{2}′$ south, six days later, on the following November 3, at the same time, the latitude was 19′ 45″ north.[5] Therefore, over 6 days the latitude was changed by 24′. So it changed 4′ per

---

[1] *TBOO* 12 p. 296.

[2] *TBOO* 12 p. 452. This is part of a long series of observations of the 1595 opposition, starting on *TBOO* 12 p. 446 and ending around p. 455. The observations at the time Kepler gives appear on p. 452, but the numbers that Kepler gives are not taken directly from the observation book. On p. 453, Brahe gives a parallax-corrected longitude of 18° 29′ 20″ Taurus, latitude 0′ 25″ N., at 11:05 pm.

[3] *TBOO* 12 p. 219 (raw observation only).

[4] *TBOO* 12 p. 44 (raw observations only).

[5] *TBOO* 12 p. 455.

day. And since on October 28 at 12^h its eccentric position was $16°\ 8\frac{1}{3}'$ Taurus,[6] and the remaining latitude was $4\frac{1}{2}'$: let this be traversed in one day and one eighth, after which time 37′ are added to Mars's position. Therefore, the node will be at $16°\ 45\frac{2}{5}'$ Taurus, at the beginning of November of 1595.

303   About the other node, there was not such a crowd of observations. Therefore, the year 1589 alone will uphold the trustworthiness of this operation. For since on 1589 May 6 Mars had $6\frac{2}{3}'$ of northern latitude,[7] it traversed this in $2\frac{1}{3}$ days, according to the proportion of the latitudinal motion of the preceding days, [arriving at the node on] May 8 at 20^h, at which time its eccentric position is found to be $16°\ 42'$ Scorpio. In 1595, this would be $16°\ 47'$ Scorpio, the position of the ascending node, while previously we found the ascending node to be at $16°\ 45\frac{2}{5}'$ Taurus. Therefore, at the end of 1595, the nodes are at $16°\ 46\frac{1}{3}'$ Taurus and Scorpio.

---

[6]This appears inconsistent with the position of $18°\ 35'$ Taurus, at nearly the same time, given above.

[7]*TBOO* 11 p. 334 (only measured distances from stars given).

Chapter 62

An examination of the inclination of the planes.

On 1593 August 25 at $17^h 27^m$, Mars was observed at opposition to the sun at 12° 16′ Pisces.[1] On the 23rd its latitude was 6° 7′ 30″. On the 24th it was 6° 5′ 30″. On the 29th it was 5° 52′ 15″. Therefore, in 5 days the latitude decreased by 13′ 15″, while during one day before opposition, by 2′. Therefore, according to this proportion, if the latitude on the day and hour of opposition is taken to be 6° 2′ 30″, there will not be half a minute's error.

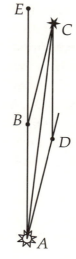

These latitudes were observed when Mars was at an altitude of 22°, which is now thought to be enough to free the fixed stars from refraction. Now since the equated anomaly was 166° 36′, the distance between Mars and the sun was 138, 556, and between the earth and the sun, 100, 666. *Hence, in the diagram of Chapter 13, if A is the sun, B the earth, C Mars, and AB is* 100, 666, *AC* 138, 556, *and EBC* 6° 2′ 30″, *the declination BAC of the orbit from the ecliptic at this point is shown to be* 1° 39′ 22″. And since the node is at 16° 43′ Taurus,[2] I subtract from this 12° 16′ Pisces. There remains an arc of 64° 27′. And as the sine of this is to the inclination here of 1° 39′ 22″, so is the whole sine to 1° 50′ 10″, the inclination of the southern limit.[3]

But since the position is rather far from the limit, in order to cut off any opportunity for suspicion, let us consult observations at positions other than acronychal, where Mars is near the limit. In undertaking this, I shall also present a universally applicable demonstration of the ratio between the inclination and the observed latitude. On 1593 July 21 at 14^h (in astronomical terms),[4] the planet was observed at 17° 45¾′ Pisces, with latitude 5° 46¼′ south.[5] At this hour the eccentric position of Mars is found to be 20° 1½′ Aquarius, while the sun's position was 8° 26′ Leo.

---

[1] This is Observation 7 from the table at the end of Chapter 15. No observation was entered on that day, but the three latitudes that follow are on *TBOO* 12 p. 290.

[2] According to the position given at the end of the preceding chapter, together with the motion of the node given in Chapter 17, this should be 16° 44¾′ Taurus; the effect upon the computation, however, is negligible.

[3] The translator finds Mars's distance to be somewhat less (138, 547). The resulting inclination is 1° 39′ 11″, and the inclination at the limit would accordingly be 1° 49′ 56″.

[4] That is, at 2 am on July 22.

[5] *TBOO* 12 p. 203.

*In the present diagram, let EA be at 8°
26' Leo, and KA at 20° 1½' Aquarius. EAK,
the true angle of relative motion, will be 11°
35½'. Also, let EK be at 17° 45¾' Pisces.
I say that the sine of AEK is to the sine of
EAK as the sine of the inclination of K is to
the sine of its observed latitude. For let the
inclination of K be understood as a straight
line dropped perpendicularly from the body of
the planet to the ecliptic. So, as the distance
EK is to the distance AK, so will the sine of
the apparent angle of the line K as seen from*

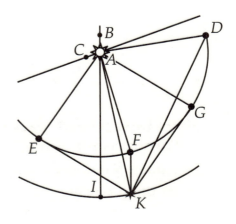

*A be to the sine of its apparent angle as seen from E. But as the sine of EAK is to the
sine of AEK, so is the distance EK to the distance AK. Therefore, as the sine of EAK
is to the sine of AEK, so is the sine of the apparent angle of the line K as seen from A
to the sine of its apparent angle as seen from E.*

304   *The minor premise is known from trigonometry, and specifically, from Book 3
Number 14 of Lansberg's trigonometry.*[6] *The major premise requires proof. There-
fore, let there be the straight line VO, from two points of which, P and M, let two
perpendicular and equal lines PQ and ML be set up.
And let the ends Q and L be joined with a point on the
line VO, and let this be O. Now, about center O, with
radius OL, let an arc be described intersecting QO at
N, and from N let a perpendicular NR be dropped to
VO. Therefore, as PQ is to QO, so is RN to NO. But
ML is equal to PQ. Therefore, as ML is to QO, so is
RN to LO. Now ML is the sine of the angle LOM, un-
der which the magnitude PQ or LM is observed from
nearby, when LO, which is the shorter distance of the
end L, is the whole sine. But QO is the longer distance
of the magnitude ML, or of the end of PQ, namely, Q.
And RN is the sine of the angle NOR, under which
LM is observed, or the more remote line PQ, where
NO, or LO, is again the whole sine. Therefore, as the*

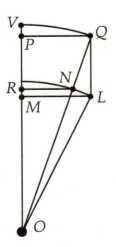

*sine of the apparent angle from nearby is to the longer distance, so is the sine of the
apparent angle from afar to the shorter distance. And, permuted and converted, as the
shorter distance is to the longer, so is the sine of the apparent magnitude from afar to
the sine of the apparent magnitude from nearby.* And in the present investigation,
and universally as well, as the distance of Mars from the earth is to its distance
from the sun, so is the sine of the latitude to the sine of the inclination of the
planes. And, in turn, as the distance from the sun is to the distance from the
earth, so is the inclination to the latitude. *Q.E.D.*

*Since these things are certain, and since the line designated by K appeared to be 5°
46¼' from E, multiplying the sine of this by the sine of EAK, and then dividing by the
sine of AEK, results in the sine 3188, whose arc is 1° 49' 37". And this is the amount
of the inclination of the point K as it would appear from A.* And since Mars is at

---

[6]Philip Lansberg, *Triangulorum geometriae libri IV* (Leiden 1591).

$20°\ 1\frac{1}{2}'$ Aquarius, and the node is at $16°\ 43'$ Taurus,[7] and thus the elongation of Mars from the node is $86°\ 42'$, therefore, as the sine of this elongation is to the whole sine, so is the sine of $1°\ 49'\ 37''$ to the sine of the maximum inclination, 3200.[8] Therefore, as before, this again gives $1°\ 50'\ 2''$ south.

For the northern inclination, at midnight following 1585 January 31, at an altitude of $53°$, the latitude of Mars, now decreasing, was $4°\ 31'$ north.[9] But the true opposition was 16 hours 46 minutes previously, at $21°\ 36\frac{1}{6}'$ Leo.[10] Accordingly, the latitude then was $4°\ 31'\ 10''$. And since the [full-circle] complement of Mars's equated anomaly was $7°\ 6'\ 23''$,[11] its distance from the sun was 166,334, and the sun's distance from the earth, 98,724. *So, again in the diagram of Chapter 13, if AC is 166,334, AB 98,724, and EBC 4° 31' 10'', BCA comes out to be 2° 40' 50''. This, subtracted from EBC, leaves BAC, 1° 50' 20''.* But because we are $5°$ from the limit, the inclination of the limit will be about $25''$ greater, namely, $1°\ 50'\ 45''$. Before, the southern inclination was $1°\ 50'\ 8''$. The difference of 37 seconds is clearly of no significance. The average of the two is $1°\ 50'\ 25''$, the perfectly correct inclination, the same amount found in Ch. 13 above with various methods and operations, to which I again draw your attention here.

Now, if I compute the latitudes of Mars at opposition to the sun using this inclination of the limits, I find the following.[12]

305

	Year	Distance of Mars	Distance of sun	Inclination			Apparent latitude			Our table in ch. 15				
1	1580	152,976	98,223[13]	0°	37'	42''	1°	$45\frac{1}{2}'$	N.	1°	40'			
2	1582	162,255	98,233	1°	36'	6''	4°	$3\frac{1}{3}'$	N.	4°	6'	or	4°	3'
3	1585	166,335	98,724	1°	50'	3''	4°	$30\frac{1}{2}'$	N.	4°	$31\frac{1}{6}'$			
4	1587	164,635	99,641	1°	25'	42''	3°	37'	N.	3°	37'	or	3°	41'
5	1589	157,045	100,860	0°	23'	20''	1°	$5\frac{1}{3}'$	N.	1°	$7\frac{1}{3}'$	or 1°	$12\frac{3}{4}'$	
6	1591	144,744	101,777	1°	11'	9''	3°	$59\frac{1}{6}'$	S.	4°	$1\frac{1}{2}'$	or	3°	56'
7	1593	138,556	100,666	1°	39'	40''	6°	$3\frac{3}{4}'$	S.	6°	$2\frac{1}{2}'$	or	5°	58'
8	1595	148,817[14]	98,756	0°	1'	39''	0°	$5\frac{1}{5}'$	N.	0°	8'	approximately		
9	1597	159,200	98,203	1°	19'	17''	3°	20'[15]	N.	3°	33'			
10	1600	165,406	98,478	1°	49'	24''	4°	$30\frac{1}{4}'$	N.	4°	31'			
11	1602	166,004	99,205[16]	1°	39'	35''	4°	$7\frac{2}{5}'$	N.	4°	8'	or	4°	10'
12	1604	160,705	100,359	0°	52'	9''	2°	$18\frac{3}{5}'$	N.	2°	$21\frac{1}{2}'$	or	2°	26'

---

[7] This position is about two minutes low—an insignificant error, as was remarked above.

[8] Recalculation shows the sine to be 3193, and the corresponding angle $1°\ 49'\ 26''$

[9] *TBOO* 10 p. 388.

[10] This is Observation 3 from the table at the end of Chapter 15.

[11] This is the exact figure obtained using the aphelial position for the beginning of 1584. Kepler apparently used the wrong row in the table of Ch. 17. The correct anomaly is $7°\ 7'\ 32''$. This has no significant effect upon the distance, however.

[12] Where recomputation reveals a substantial difference, the translator's figure is given in a footnote.

[13] 98,415

[14] 148,321

[15] $3°\ 26'$

[16] 99,268

In the first, an observation on the day was lacking, as you have seen in Ch. 15. In the second there was an uncertainty of three minutes in the observation, since they occasionally used 34° 7′ as the altitude of the pole, which was 34° $5\frac{1}{2}'$.[17] The third has served as our foundation. The fourth agrees to a hair, if you neglect parallax, through which the observed latitude is wrongly corrected so as to be 3° 41′, as you have seen in Ch. 15. In the fifth, we are wanting 2 minutes. It is surely the observation that is too high, on account of refraction, since Mars was no higher than $22\frac{1}{2}°$, as you know from Ch. 15. In the sixth, you may note a slight defect of about two minutes. But the quantity of the refraction is not so reliable: what if it was two minutes higher? The seventh, again, served us as a foundation. The eighth without doubt had an erroneous declination, for at that time (8^h) Mars was not at the meridian. And the armillary spheres, by which the declination is measured elsewhere than at meridian, err more easily than the quadrants. Furthermore, a comparison with nearby dates, as in Ch. 15, shows that the latitude was 0° 5′ N., the same as we have computed. The ninth observation is not worthy of trust.[18] However, the accurately examined calculation for December 10 closely agrees with the Fabrician latitude [19] of 3° 23′, for it gives 3° $21\frac{2}{3}'$ N. The tenth comes close to the calculation. The eleventh corresponds to a hair when refraction is excluded. The twelfth is barely two minutes greater than the calculation. I believe this is because there is that much uncertainty in my instruments. For in my quadrant of six cubits, two minutes are not easily discerned. We therefore have the acronychal latitudes determined accurately enough throughout the entire circumference of the circle, using this inclination of 1° 50′ 30″. An examination of the remaining latitudes at observations at positions other than acronychal, of which there are many closely spaced examples in this book, I leave to more diligent scholars.

---

[17] Kepler inadvertently gives the zenith distances of the pole rather than the polar altitudes here.

[18] Note, however, that Kepler's computed latitude is in error, and should have been 3° 26′.

[19] That is, the latitude as determined by the Dutch astronomer David Fabricius, with whom Kepler frequently corresponded while composing *Astronomia Nova*. See James R. Voelkel, *The Composition of Kepler's* Astronomia Nova (Princeton University Press, 2001), 170–210.

CHAPTER 63

Physical hypothesis of the latitude.

It was said in Chapter 57 that if the diameter of the body or globe of Mars is supposed to possess a magnetic force, and to be directed towards the middle elongations, and also to remain parallel to itself in that disposition throughout its entire circuit, the physical hypothesis of the eccentricity is complete.

This supposition is all the more probable in that now the reason for the latitude too is explained using a closely related theory: that there be supposed some diameter of latitude in the body or globe of Mars that is directed towards the sidereal position of the limits, and remains parallel to itself in this disposition throughout its entire circuit. The ratio of this power to the former is that which [the power] of direction towards the pole in our magnets has to the force of attracting iron.

That is, the former seeks the sun or flees it, while the latter, rather than seeking by sailing towards, or fleeing, those sidereal positions beneath which the limits of the latitude are reached, is only directed towards those positions, as a magnet towards the pole (for likewise, a magnet does not sail towards the polar region even if it floats freely).[1]

In fact, the excursion of the planet from the plane of the ecliptic to either side follows the direction towards which this axis of inclination, on the side that leads in the motion of its body, is pointed. Let *CBAD* be the ecliptic, *A, C* the nodes; *B, D* the limits. And let the axis of latitudes in the body of the planet be *GNH, EAF, LOM, ICK*. Now since we are supposing this axis to remain equidistant from itself throughout its circuit, it will happen that as the body moves from the ascending node *C* to the northern limit *B*, the axis *IK* of the body, which initially, at the node *C*, was tangent as it were to the imaginary circle of its circulation through *CNAO*, later intersects it at right

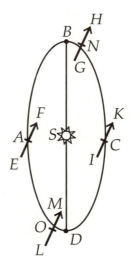

angles at the limits *N, O*, being directed towards the center of the world *S*, that is, towards the sun. But also, owing to a certain amount of declination from the royal road[2] *CBA*, this axis had thus far been enticing the body of the planet

---

to leave that path in the direction of *N*, towards which it had turned the preceding part *K*. Now, at the limits, although it has indeed remained inclined to the plane of the ecliptic *CBS* (for we have said that it remains equidistant from itself in all positions, and so once it is inclined to the plane of the ecliptic it will always be inclined), it nevertheless does not continue to decline from the royal road itself, that is, from the circumference of the plane of the ecliptic *CBAD*, once it is placed at *GH*. For it does not incline forward towards *A*, nor back towards *C*, but only to the side or towards the pole, places to which its path is not directed. So when the planet is moved forward beyond *B*, the other part of the axis, *G*, which inclines towards the south, is now in front, and it thus leads the planet from its greatest northern inclination *N* through the descending node *A* to the greatest southern inclination *O*.

307

This axis of inclination is somewhat like an oar, in that just as boatmen use oars to move from one bank to the other, the planet brings this about through the inclination of the axis, moving it from north to south and back, while the river, that is, the immaterial *species* of the sun, proceeds along the direct path *CBAD*.

As for geometrical dimensions, there is no need for verbosity. A straight line moved on a rectilinear course while remaining parallel to itself creates a plane through its motion. This axis is itself a straight line, and it is moved in the direction it points (this pointing, moreover, presupposes a straight course). It therefore describes a plane. And if this plane be extended, it intersects the sphere of the fixed stars in the shape of a great circle, *FEGH* in the diagram of Chapter 13, because it intersects the plane of the ecliptic *DC* at the center of the world or the sun, *A*. To further convince yourself of this, you should consider that the points of intersection or nodes are in opposite positions about the center of the sun *A*, as you see in the diagram. This is shown by experience: see Ch. 62. And so since there is a plane in which the orbit of Mars is moved about, its inclination to the plane of the ecliptic will follow a pattern. That is, when two equal circles are described, one, *DC*, in the plane of the ecliptic, and the other, *FE*, in the plane of Mars's orbit, about a common center *A* of the sun (i. e., on one and the same sphere of the fixed stars, concentric with the sun), the sine of the arc *BD* between the intersection of the circles and some point on Mars's circle, *D* say, is to the whole sine as the sine of the inclination *DF* of the point *F* is to the sine of *CE*, the greatest inclination, at the limit *E*. Furthermore, it was proved in Ch. 13 above, using an ingenious treatment of the observations, that the declinations of all the points of the circuit from the plane of the ecliptic are arranged by the same measure. So no instance can be urged against our hypothesis.

---

fixed stars, Kepler introduced a "mean ecliptic" under the name "royal road," in order to avoid giving the plane of the earth's orbit a privileged status. There, the royal road is the equatorial region of the sun, which he assumed to rotate.

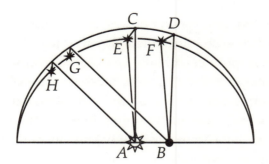

But there are still two difficult questions to be answered. One concerns how this declination of the axis originates, and the other concerns the axis itself. First, is the inclination of the axis natural or rational, the work of the body's nature or of an angel? And second, are the axis of inclination and the magnetic axis that seeks the sun the same? And if they are different, how do they exist in the same globular planetary body? The two questions are interrelated.

I might almost have believed it to be natural, owing to the similarity between the natural power in the magnet and this one, were it not for the successive transposition of the nodes that had also been added, which clearly seems the work of a reasoning faculty, if not discursive, at least instinctual. For to maintain its equidistant position is less marvelous and more in accord with nature than previously in the matter of the eccentricity. There, we said that it is the sun that is sought by the axis of power, while here, it is the position beneath the far distant fixed stars. There, the axis was to have been turned about by the force of this magnetic power as the body is carried around, and would not remain equidistant to itself, if it were not restrained by a stronger directional force, or by an animate force, either unassisted or capable of reasoning in some manner. Here, the axis follows these equidistant positions by the force of our directional power itself, with no need of an animate power or of reasoning. Someone might, however, consider it the act of a mind, that the diameter that effects the latitude points directly towards the center of the sun when the planet is located at the limits, thus making a great circle of the planet's orbit, and causing the nodes to be at opposite positions with respect to the sun.   308

To this argument, in Ch. 39 above, I affirmed that the planet moves with respect to the sun. However, it is not just any kind of relationship with the sun that argues for the assistance of reason. It is of course true that he who first ordained the heavenly motions so directed this axis as to point at the sun when at the appointed position, and did so deliberately and with perfect rationality. But this relationship with the sun can now be maintained without a mind, by the constancy of the magnetic faculty alone. For it is more like rest than motion, and hence is material, not mental.

Therefore, it is only the variation of this inclination which we call the translation of the nodes over the ages that still makes a case for a motive force that is more than natural, or physical, as are magnetic powers.

Nevertheless, I would prefer to think that the two must be conjoined, rather than to suppose that the rational faculty acts alone. Let the magnetic faculty be subordinate; let the rational be in charge, ruling over it, just as we said before in Chapter 57 concerning the power of seeking the sun.

Once this question is settled, there follows the other. If this directive power arises from magnetic, physical, natural [powers], its substrate will be a body. Could it therefore happen that the same diameter that seeks the sun or flees from it also governs the planet's deviation from the ecliptic, by being inclined to it? If the nodes were connected to the apsides and the limits to the middle elongations then the diameter would be the same in all respects, administering both the eccentricity and the latitude.

For it was said in Ch. 57 that the diameter that causes the eccentricity is directed towards the middle elongations, while it was just now said that the diameter that causes the latitude is directed towards the limits. Therefore, if the limits were connected to the middle elongations, both diameters would have the same direction, and, their positions thus being in agreement, nothing would prevent their being identical. However, the nodes, or intersections with the true ecliptic, do not coincide with the apsides. For Mars, the northern limit is 12 degrees before the aphelion, for Jupiter, the northern limit and the aphelion coincide exactly, for Saturn, the node follows the aphelion by 24 degrees, and for the moon, owing to its short orbit, everything becomes interchanged with everything else. For now the node is at apogee, now at the middle elongations, now at perihelion [*sic*]. So since these two powers differ in time and position, it follows that they are not identical.

There is, however, nothing to impede their residing in one and the same planetary body as a whole, except the motion or rotation of the globe. Thus if the planets are moved like the moon, which does not rotate, but always shows us the same face, nothing prevents our saying that the two are interwoven, as the weft is interwoven with the warp. For since the entire body of the planet would then stay in the same sidereal position as it is carried about the sun, any of its rectilinear parts, among which are numbered those two diameters, will stay in the same sidereal position. If, on the other hand, it is the earth's globe that is in question, which has a daily rotation in addition to its annual revolution, we are left in great doubt, no less than before, in Ch. 57. For if the body rotates, only one single diameter of power, that which is parallel to the axis of its rotational motion, remains constant and equidistant from itself. So if you were to say that in addition to that diameter, there is interwoven with it another altogether different one, a power of another sort, which causes the latitude, it will observe the same direction as the axis of rotation, since it circumscribes a cone about that axis, successively traversing each of its parts, and since it inclines, now to the right, now to the left side, it finally leads the body towards the middle position, whither the axis of rotation points.

Therefore, if the globe rotates, the subject of this declinational power is either not a body but something spiritual, or is not the same body. If it is something spiritual, how does it look to certain regions of the world, which are corporeal? And how does it impart this kind of motion (declination from the royal road) to the body? Is it perhaps that the body is more easily inclined, and departs from the royal road more easily (meanwhile receiving the cause of its translational motion extrinsically, from the sun), than it is carried from place to place by the force of its own proper mover? If, on the other hand, we prefer a corporeal subject, some mechanism will be brought into being for us,

Terms.
The diameter that causes the eccentricity is one thing, the diameter of the reciprocation another. The former is something real, while the latter is imaginary, devised in order to give an image of its effect. The one is the same everywhere, being directed perpendicular to the line of apsides, or towards the sidereal position of the middle elongations; the other, as was said in Ch. 39, is always directed towards the body of the sun.

309

like those spherical oil lamps which, though thrown and spun around, do not spill any oil. For within is enclosed a little flask which, being drawn down by a weight in its belly,[3] and held there, does not follow the convoluted motion of the surrounding sphere.

Is there then also some interior globe within this globe of the earth, to which the diurnal motion of the earth's exterior does not penetrate, but which is held in place by a very strong inclination towards certain sidereal positions, so as not to follow the revolving exterior of the body? For as we shall see in Chapter 68, this question pertains to the earth as well. We shall also see there whether, if some mean ecliptic be proposed for the six planets, that which we were requiring a little earlier can be accomplished, namely, that the nodes of each of the planets correspond to the apsides.

Or is it rather to be believed that there are some possible modes of celestial motion which, though physical like the magnetic [powers], cannot be comprehended by anyone on earth owing to the lack of examples? For if we had lacked the example of the magnet (which was indeed unknown at one time), we would have been ignorant of most of the causes of the celestial motions.

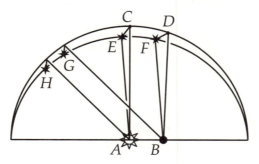

Those who believe in solid orbs can easily set everything right, following what was said in Ch. 13. For they will attribute to the plane of Mars's eccentric *FE* an inclination to the plane of the ecliptic *DC* that does not librate, but is fixed and constant, above the diameter *BA* of intersection of these planes, drawn through the center of the world (the center of the sun, for Brahe), and they will say that over the ages it rotates about the center *A* beneath the ecliptic *DC*.

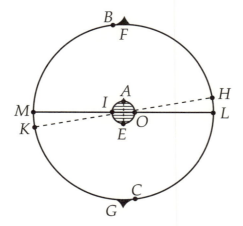

And since the poles (*F, G* and *B, C* in the present diagram) of two great   310

---

[3]Reading *ventriculoso* instead of *ventricoso*.

circles (*ML* and *KH*) are distant by an amount equal to their maximum declination *MK*, *LH*, the poles of Mars *B*, *C* will therefore describe small circles about the poles of the ecliptic *F*, *G*, with radius *FB*, *GC*, of 1° 50′ 25″. These people will also say that the poles of the Martian sphere *B*, *C* revolve forward with a motion quantitatively the same as that which was expressed above in Ch. 17, and which will be corrected below in ch 69.

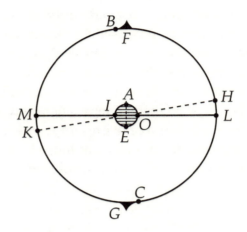

CHAPTER 64

Examination of the parallax of Mars through the latitudes.

In Chapter 61, the two nodes were found to be at positions exactly opposed, a marvelous agreement and one which excludes all parallax.

Let it be the case that Mars's parallax is at least 1′ and 2′ [respectively] when at opposition to the sun (and nearer to the earth than the sun) in 1595 and 1589, and that on the former date Mars was about 38° from the zenith, and on the latter about 66°. Accordingly, in 1589, when it was thought to be at the node, it would still have been nearly 2′ to the north. Therefore, it would still have been one degree before the node. So the node would be, not 16° 46′ Scorpio, but 17° 46′ Scorpio. In 1595, on the other hand, it would have 1′ of parallax. Therefore, on the day on which it was thought to be at the ascending node, it would now actually have had a latitude of 1′, and it would thus now have been about 30′ beyond the node. Therefore, the ascending node would be, not at 16° 46′ Taurus, but at 16° 16′ Taurus. You see that the descending node is at $17\frac{3}{4}°$ Scorpio, and the ascending at $16\frac{1}{4}°$ Taurus, if you make use even of the least parallax. Let us therefore conclude, as in Ch. 11, that Mars's diurnal parallax is entirely imperceptible, if it is indeed true that the two observations of latitude are correct within 2′.

Another argument for no parallax, which is not dissimilar, will arise for us out of Ch. 62, premised upon upon the investigation of the inclination of the planes, perfectly true unless refraction will throw something off.

Let it be the case that in 1593, at an altitude of 22°, Mars had a parallax of at least 2′, while in 1585, at an altitude of 53°, it had a parallax of one minute. The observed southern latitude would therefore be smaller then the northern, and so the inclination would also be smaller. But now, just above, without parallax, it appears somewhat smaller, by an amount attributable to a small error in observation or to a certain amount of refraction at an altitude of 23°. Therefore, when parallax is considered, the observation would be charged with a greater error, and conversely, if the observation stands, the parallax is entirely eliminated, if it is indeed true that the orbit of Mars is contained in a perfect plane that intersects the plane of the ecliptic at the very center of the sun.

But the same is proved much more certainly from the latitudes observed at other acronychal positions, especially those which the circumstances of observation or refraction did not render dubious. As I began to say in Ch. 15, this matter has so far been impossible to settle.[1] For in 1587, when Mars was 55 degrees from the zenith, if it had had a parallax of 4′ its latitude of 3° 37′ would have been increased to 3° 41′. But in Chapter 62, it was found to be no greater than 3° 37′.[2] And in 1589, when the nonagesimal was 64° from the zenith, if

311

---

[1]See Ch. 15, Observation IV, p. 175. Observation from *TBOO* 11 184 (under 5 March).

[2]See line 9 of the table on p. 457.

Mars's parallax had been $5\frac{1}{2}'$ (judging from the sun's horizontal parallax of 3'), then the northern latitude, instead of the observed 1° 7', would have been 1° $12\frac{1}{2}'$ freed of parallax.[3] But we have computed no more than 1° $5\frac{1}{3}'$, although a slight error of 2' could have occured in the observation, such as if Mars at an altitude of 22°, still subject to refraction, had appeared 2' higher (to the north) than was correct, as was said in both Ch. 62 and Ch. 15.[4] And in 1602, when with a parallax correction the observed latitude was found to be 4° 10', and without the correction, 4° $7\frac{1}{2}'$, we computed 4° $7\frac{2}{5}'$, very precisely.[5] Similarly in 1604 we did not agree perfectly with the observed quantity of northern latitude.[6] Therefore, we shall complain that it is much less when it is increased through the removal of parallax.

By these three procedures, we have overcome our uncertainty about Mars's parallax. However, we have not completely proved that it is utterly imperceptible, since the matter of refraction eludes us, and besides, the observations do not descend to within 2 or 3 minutes. So if anyone wishes to attribute to Mars a maximum latitudinal parallax of 2 or $2\frac{1}{2}$ minutes, these Brahean observations do not significantly disagree with him. For the inclination, too, will be accommodated to this view, becoming 1° 51' 0''.

---

[3]See line 5 of the table on p. 457.

[4]See line 5 of the table in Ch. 62, p. 457, and observation V in Ch. 15, p. 176.

[5]See line 11 of the table in Ch. 62, p. 457.

[6]See line 12 of the table in Ch. 62, p. 457.

Chapter 65

Investigation of the maximum latitude in both regions: in conjunction as well as opposition with the sun.

Once the inclination is established, it is easy to define the maximum latitude, and this can be done in two ways. For one can find the maximum for all time, or how great it could be in our time. Today the two hardly differ, since the limits are the midpoints between the apsides of Mars and of the sun or earth, and they are no more than 54 degrees from one another, and the eccentricity of the sun or earth is not great. Nonetheless, let it be the case (as it once was) that the apsides of Mars and the sun coincide, along with the limits of Mars's latitudes. And let the ecliptic maintain its sidereal position. Now since, in the diagram of Ch. 13, Mars's greatest distance AC is $166,465$, the sun's least distance AB is $98,200$, and BAC is $1° 50\frac{1}{2}'$, the maximum northern latitude at opposition to the sun computed from these data is $4° 29' 10''$. At conjunction with the sun, when the sun's distance from the earth is $101,800$, this is decreased to $1° 8'$ 34''. But the southern latitude, from Mars's distance of $138,234$, and the sun's of $101,800$, is computed to be $6° 58' 24''$, a little less than $7°$. At conjunction with the sun, when the sun's distance is $98,200$, this is decreased to $1° 4' 36''$. If, however, one considers the contrary case, in which the sun's apogee coincides with Mars's perihelion, the maximum northern latitude at opposition comes out to be $4° 44' 12''$, and at conjunction $1° 9' 32''$, while the southern latitude at opposition is $6° 20' 50''$, and at conjunction $1° 3' 32''$.

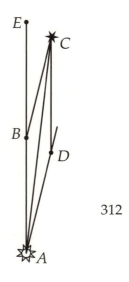

And this is how it would be if the apsides and the limits were to coincide at some time, but whether this is going to happen before the whole fabric comes to ruin is uncertain. It is certain that Ptolemy attributed equal motions to the apsides and nodes, and if this were so, that conjunction would never happen. But even though today they appear to undergo different motions, the observations of the ancients are not sufficiently reliable, and the difference of these motions even in modern astronomy is not sufficiently great, for us to conclude how many myriads of years apart these conjunctions of the apsides and the limits occur.

Therefore, let us return to our era, that which extends between us and Ptolemy. And here, one who is looking for geometrically precise determinations is presented with a manifold obstacle to computation.[1]

First, the apsides of the sun and Mars are not in conjunction, and second, the planets' orbits are not perfect circles. So even if we project a new line of

---

[1] ἀμηχανία.

apsides through the centers of the circles of Mars and the earth (through *B, C* in the diagram of Chapter 52), it will still be possible for the nearest approach of the celestial bodies to occur elsewhere than on this line.

Finally, even if the position of the nearest approach were established, the position of the northern and southern limit is different. For example, the limit is at 16° 50′ Leo. But the straight line *BC* projected through the centers of the circles is directed towards $24\frac{1}{2}$ Leo and Aquarius, approximately; in the same direction, that is, in which Brahe put his line of apsides *HF*, to which our line *BC* runs parallel, both eccentricities being bisected, *AF* at *C*, and *AH* at *B*.

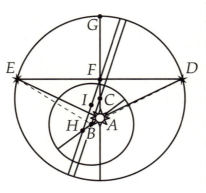

And I was now about to choose the mean between 17° Leo and 25° Leo, namely, 21° Leo, but the year 1585 gave me pause, since in that year the latitude observed at 21° 36′ Leo was clearly not a maximum. For while the opposition was on the night following January 30, the latitude observed on the 24th, preceding the opposition, was 4° 31′, still increasing, while on January 31, 16 hours past opposition, the observed latitude was again 4° 31′. It is therefore evident that on the 24th, if the opposition had occurred at that eccentric position, a latitude greater than 4° 31′ would have been observed, for two reasons: first, because the celestial body was nearer the earth than when it was not at its acronychal position, and second, because Mars was farther from apogee, and was lower.

313   Therefore, let the maximum latitude occur about 19° Leo and Aquarius, where Mars was on January 24. And since the supplement of the equated anomaly was 10°, the distance of Mars will be 166, 200, and of the sun, 98, 670. And so the maximum northern latitude will be about $4° 31\frac{3}{4}′$.[2] At conjunction with the sun, since its distance is 101, 280, this appears as 1° 8′ 30″.[3]

For the maximum southern latitude, Mars's equated anomaly of 170° shows us a distance of about 138, 420, and the sun at 19° Leo has a distance of 101, 280. Hence it is concluded that the maximum southern latitude will be about 6° 52′ 20″,[4] and at conjunction it appears as 2° 4′ 20″.[5]

---

[2]This number is consistent with a position at the limit. However, since the position given is 2° from the limit, the angle is about one minute less, and the computed latitude about 4° 29′, which does not fit the observations so neatly.

[3]The correction mentioned in the preceding footnote would make this angle 1° 8′.

[4]6° 47′, with the correction mentioned above.

[5]This should have been 1° 4′ 20″, or 1° 3′ 55″ with the corrected inclination.

CHAPTER 66

The maximum excursions in latitude do not always occur at opposition to the sun.

Concerning the maximum latitude that can occur in any particular period of Mars, however, it is a much more complicated business to define its exact positions geometrically, and also involves this great paradox, which I found emphasized among the observations of 1593 in Tycho Brahe's hand, in the following words:[1]

"It is worthy of consideration that on about the tenth day of August Mars had its maximum southern latitude, and that it decreased afterwards, so that at opposition on the 24th it was about one fourth of a degree nearer to the ecliptic. However, the Canons[2] do not show this at 18 Aquarius, even when the position of maximum latitude is corrected, no matter how that maximum latitude is derived there. The cause of this needs to be looked into carefully."

When I later had come to him in Bohemia, and frequently inquired about how the latitudes are arranged, he answered that the nodes are at opposite positions, and the line of intersection passes through the point of the sun's mean position, or through the center of its epicycle (for which see Ch. 67 below), and enumerated many other things. Reminded by this mention, he said, regarding the present matter, "this is remarkable, that the latitudes reach their maximum before or after opposition to the sun." Mention was also made of this above in Chapter 15.

The cause of this occurrence is in fact contained in the true hypothesis of the latitude established in this fifth part; however, you would have almost as much trouble finding the boundaries of the maximum latitudes geometrically, as Apollonius of Perga had in finding the boundaries of the stations.

314

For in the business of the stations, a certain condition can be described through which the position of the stations may be known (and that condition is this, that the line of vision of Mars, the earth being in motion, remains parallel to itself). But the position of the stations cannot be demonstrated *a priori* from this condition without multiple calculations, owing to the confluence of many causes. And matters stand just the same with the maximum latitude for any given occurrence. For the latitude is greatest when the distance of Mars from the earth is increasing or decreasing in the same ratio in which the lines of Mars's inclinations increase or decrease. And the latitude is increasing when the ratio of the distance decreases more than the ratio of the lines of inclination, or when the former is decreasing while the latter, on the contrary, is increasing. And, in turn, the latitude is decreasing either when the

On the station points.

On the points of the maximum latitudes.

---

[1] *TBOO* 12 p. 291. Kepler's transcription differs in a few insignificant particulars from Dreyer's transcription.)

[2] This may refer to planetary models, or to the tables constructed from them.

distance of Mars from the earth increases more than the lines of inclination, each in its own proportion, or when the distance is increasing while the lines are decreasing.

These conditions are satisfied indiscriminately, now at opposition, now before, now after, depending on whether the opposition falls at the limit, or before, or after the limit.

That these results follow from the hypothesis of this work, my ephemerides prove.[3] In 1604, about Feb. 25 or March 6, the northern latitude was a maximum, while opposition followed by an entire month. On Sept. 27 or October 7, in turn, the southern latitude was a maximum, while Mars was between its quintile and sextile aspects to the sun.[4] Again, at the end of 1605 the northern latitude was maximum, while the sun was moving from quintile to quadrature with Mars. And, in turn, at the end of July of 1606, the southern latitude was maximum when the sun was trine with Mars.[5] But in 1607, the maximum northern latitude occurred a little after the conjunction of Mars with the sun.

The reason why these things would appear remarkable in ancient astronomy is chiefly that Ptolemy and his imitators had fabricated the extremely intricate motions of inclination, deviation and reflection.[6] For since Ptolemy clung to his invention of the epicycle, as soon as he saw that when the planet was at opposition to the sun (and was thus visible) the epicycle went out to one side, he immediately indulged in conjecture, asserting that at conjunction with the sun, when it is not visible, the epicycle goes out in the other direction, and generally, that at conjunction the epicycle does the opposite of what he observed it to do at opposition. This is done in order that there be some compensation and equality of return and coherence with the sun. However, this is not discovering the true by observing, but fabricating the observations by a falsely conceived fancy. Nevertheless, it should be condoned in him, since he had few observations. On this subject, see Ch. 14 also.[7]

But come, let us see whether our calculation gives the observed latitude on August 10. For we are sure of July 21 and August 25 of that year, since the calculation yields the observations upon which it is based.[8]

315   So, on August 10 at 13^h 45^m, Mars's eccentric position on the ecliptic is computed to be 2° 41′ 18″ Pisces, the sun was 27° 37′ 49″ Leo, the angle at the sun 5° 3′ 29″, the angle at the earth 18° 25′, and from the calculation Mars was at 16° 3′ Pisces, while it was observed at 16° 7′ Pisces.[9] And since 2° 40′

---

[3]Kepler published no ephemeris prior to 1617 (no. 52 in *Bibliographia Kepleriana*. The present statement shows that Kepler had been preparing ephemerides for his own use prior to the publication of *Astronomia Nova* in 1609.

[4]Quintile is 72°, and sextile is 60°.

[5]That is, 120° from Mars.

[6]These devices are constructed in Book XIII of the *Almagest*.

[7]See p. 172.

[8]These observations are reported at the beginning of Chapter 62, and formed the basis of the revised hypothesis of the latitudes presented in that chapter.

[9]*TBOO* 12 p. 285.

48″ Pisces, the position on the orbit, is distant from 16° 43′ Taurus[10] by 74° 2′, the inclination will therefore be 1° 46′ 10″. From this and the two angles mentioned, using the method given in Ch. 62, the observed latitude is found to be 6° 21′ 14″, still two minutes more than the observation has. But lest the angle's small magnitude trip us up, let us use the true distances of Mars from the earth and the sun (as the method given above requires), or in their place, the true angles. In the diagram of Chapter 20 you see that CB, BA differ from CL, LA. And our method did not say that the sine of the angle LAB is to the sine of the angle LCB as CB is to BA, but as CL is to LA. Let the ecliptic position be 2° 41′ 18″ Pisces, Mars standing beneath the point λ, and κ be the position opposite the sun, 27° 37′ 49″ Aquarius. Therefore, $\kappa\beta$ is 5° 3′ 29″, and $\beta\lambda$ is 1° 46′ 10″. From this and the right angle $\lambda\beta\kappa$, $\kappa\lambda$ or CAL is given as 5° 21′ 36″, to which corresponds the true distance of Mars L from the sun A.[11] So in triangle CAL, from the sides CA, $101,077$, and AL, $138,261$, and from the angle just found, let LCA be sought, which is found to be 160° 33′. Its supplement is 19° 27′, to which corresponds the true distance of Mars L from the earth C.

So now, using these angles of the operation, I find the apparent latitude LCB to be 6° 19′ 10″, very nearly the same as the observed value.[12]

Thus the hypothesis established in this work shows this very thing whose cause Brahe advised was diligently to be sought, and which ancient astronomy, for all its apparatus, cannot show. And, I would add, it shows this in all its simplicity, in that the plane of the eccentric is given a constant inclination or obliquity, and this is variously increased or diminished, not in reality, but optically only, insofar as our sighting approaches it or recedes from it, or (for Brahe and Ptolemy) it approaches or recedes from our sighting.

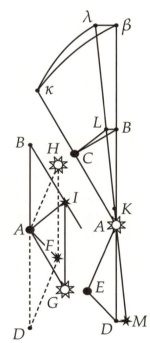

---

[10] Again, as in Chapter 62, Kepler takes a position for the node that differs from the position determined in Chapter 61. It should have been at 16° 45⅓′ Taurus, after correction for the motion of the node. The presence of these aberrant numbers in Chapters 62 and 66 suggest that the node was originally about 2′ farther back, and that Kepler later revised Chapter 61 without changing the data in other chapters.

[11] Here and in the final sentence of this paragraph, Kepler appears to have gotten the sun and the earth mixed up: surely the distance corresponding to angle CAL is CL, the distance of Mars L from the earth C, while the distance corresponding to angle LCA is LA, the distance of Mars L from the sun A.

[12] Compare this with the latitudes for August 25 (the date of the opposition), 6° 2′ 30″ S., and July 21, 5° 46′ 15″ S.

CHAPTER 67

From the positions of the nodes and the inclination of the planes of Mars and the ecliptic, it is demonstrated that the eccentricity of Mars takes its origin, not from the point of the sun's mean position (or, for Brahe, the center of the sun's epicycle), but from the very center of the sun.

316   The end is a reply to the beginning. In Chapter 6, I argued on physical grounds that when solid orbs are denied, the eccentricities of the planets cannot take their origin from any point other than the very center of the sun. I postponed part of the geometrical proof of this, based upon the observations, to Chapters 22, 23, and 52, in which places I think I have satisfied even the sharpest-eyed critic. The other part I shall now expound. This is done first through the positions of the nodes. It was proven in Chapter 61 that when Mars's eccentricity is constructed from the very center of the sun, or, what is the same, using acronychal observations taken when the planet is at opposition to the sun's apparent position, the nodes fall at positions that are very precisely opposite in relation to the sun's center; that is, that the diameter of the apsides and the diameter of the intersection of the planes of the ecliptic and of Mars coincide, or intersect one another, at the same center from which the eccentricity is computed, namely, at the center of the sun. The question now is, if we use the sun's mean motion instead of its apparent motion, will the nodes as a result still be at opposite positions about the point whence the eccentricity is computed? Not at all. Consider again the Copernican diagram in Chapter 6.[1] *In it let κδ now be the line of the limits, at $16\frac{3}{4}°$ Leo and Aquarius (not, as in Ch. 6, the line of apsides at 29° Leo). Therefore, the line [NN'] drawn through κ perpendicular to κδ will be the diameter of the nodes. But if we use the mean sun instead of the apparent sun, then we are given β instead of κ as the point from which the eccentricity is reckoned. So from β let βς be drawn perpendicular to κδ [and extended to PP']. This will fall at positions exactly opposite about β, but will not fall at the positions of the nodes, because the former perpendicular [NN'] through κ, falling at*

317   *the positions of the nodes, is above βς by the distance κς.* It is desirable to enquire into the magnitude of the angles at the circumference of the eccentric when the point κ is connected to the points of intersection of the line βς with the circumference of the eccentric. *Since, by supposition, ςκ is at 16° 45' Leo, and βκ is at about 5° 45' Cancer, the angle βκς will be 41°; and since βςκ is right, κβς will*

---

[1]To save the expense of having extra diagrams engraved, Kepler reused the same figure when he could. This sometimes resulted in diagrams that did not quite fit the case under discussion. In this chapter, Kepler would have wanted the line of nodes, through κ perpendicular to κδ, and an extension of the line βς. These lines have been added to the present diagram: they are the dashed lines NN' and PP', respectively. References to these lines are editorial, and have been enclosed in square brackets.

472

be 49°. And since κβ is the sun's eccentricity, 3600, where the orb of the earth or the sun is 100,000, therefore, as the whole sine of the angle ς is to βκ, 3600, so is the sine of angle β to κς, 2717. And in the same units (where the semidiameter of the earth's orb is 100,000), the semidiameter of Mars's orb, from Ch. 54, is 152,350. Therefore, where the semidiameter of Mars's orb is 100,000, κς will be 1790, showing an angle of 1° 1′ 33″ in the [table of] sines.

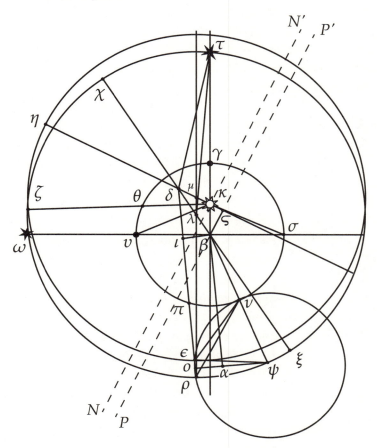

The ascending node [P′] should therefore have been moved backward, and the descending node [P] moved forward, by that number of degrees, minutes, and seconds, if I had been mistaken in taking the sun's center κ instead of the Ptolemaic, Copernican, and Brahean point β. But in turn, where the observations are referred to the mean sun, and thus the point β is taken, if this is done in error, and κ should have been chosen instead, the ascending node [N′] found from β should be in a place farther forward, and the descending node [N] farther back, so as to shorten the northern semicircle [N′ηN] by an arc of 2° 3′ 6″.

Let us see whether it happens in this way. In the observations of Chapter 12, considered approximately, on 1595 October 28 Mars was considered to have been at the node.[2] From the Brahean equations, which rely on the point β, its eccentric position was found to be 16° 48′ Taurus. And on the morning of 1589 May 9 we supposed Mars to have been at the other, descending, node.[3] Using the same Brahean equations, we computed Mars's eccentric po-

---

[2]See Ch. 12, ascending node, Obs. 4, p. 162.

[3]See Ch. 12, descending node, Obs. 2, p. 163.

sition to be 15° 44$\frac{1}{2}'$ Scorpio at that time. So what I said should happen, does happen: there are 1 degree and 3$\frac{1}{2}$ minutes less in the northern semicircle. If the observations are treated more accurately, as in Ch. 61, Mars arrives at the ascending node one day and 15 hours late. Therefore, about 50 minutes are added to the eccentric position, so that the planet falls at 17° 38' Taurus, in its eccentric motion. Accordingly, the abbreviation of the upper semicircle is 1° 53$\frac{1}{2}'$, virtually equal to the computed value of 2° 3'.

Therefore, the point κ is entirely confirmed, and β is rejected. For why will the diameter of the intersection of the planes not intersect the diameter of the apsides in the center from which the eccentricity originates, as above? What would be the cause of such a thing?

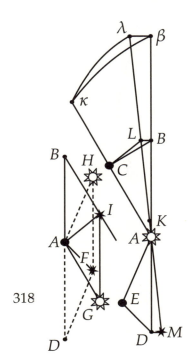

The same is also demonstrated through the incli-nation of the planes demonstrated in Ch. 62, using the diagram of Chapter 20. *The inclination, that is, the angle LAB under which the digression of the northern limit appears when seen from the sun A, was there found to be 1° 50' 45". But the angle MAD, under which the southern limit's digression from the ecliptic appears when viewed from the sun A, was found to be nearly equal to it, namely, 1° 50' 8". So the angles at A, above and below, are equal, and the line drawn from A to the ecliptic posi-tions of the limits B, D, is one line (since it is in the one plane of the ecliptic). It was therefore concluded from this that the other line, drawn from A to the limits themselves L, M, is also one line; and further, that what is enclosed within the orbit of Mars is a single plane. Furthermore, if the common intersection of the planes were not at κ in the former diagram (which is A in the present one), but at $\beta\varsigma$ (that is, below A in the present diagram), when the limits L, M are connected with some point on the line BD below A, the angle under which LB appears from that point would be smaller, and the angle under which MD appears would be larger, by about two minutes.*

It is true that if we are allowed the liberty of making the parallax as great as we please, the arguments of this chapter are easily weakened. But it is a well-documented certainty that it is impossible to allow a parallax great enough to fully enervate this demonstration.

Also, since the point of this chapter was demonstrated most soundly in Ch. 52, I could take another tack, and instead of demonstrating this point by denying parallax, I could deny parallax, as in Ch. 64, by maintaining this point, which has its own demonstration in Ch. 52.

It does not matter which way you do it. For both points have other demon-strations. The present way occurred to me first, and suited my purpose of showing how everything is in agreement.

CHAPTER 68

Whether the inclinations of the planes of Mars and the ecliptic are the same in our time and in Ptolemy's. Also, on the latitudes of the ecliptic and the nonuniform circuit of the nodes.

It was said in Chapter 14 that in any one period of Mars whatever, the obliquity or inclination of Mars's plane to the plane of the ecliptic remains fixed. There is, however, some doubt whether this obliquity is the same, and fixed, for all ages. The reason for the doubt is this.

In the first volume of the *Progymnasmata*, p. 233,[1] Brahe demonstrated that the latitudes of the fixed stars are different today than at the time of Ptolemy, the difference being this: that in the region of the summer solstice, the latitudes of the northern stars increased and those of the southern stars decreased; and, in turn, in the region of the winter solstice, the latitudes of the northern stars decreased and those of the southern stars increased. As one goes from these boundaries towards the equinox points, the alteration of the latitudes diminishes, until near the equinox points there is none at all. This observation of our time we shall accommodate to our principles laid down in Ch. 63, thus:

On the altered latitude of the fixed stars.

It is established that the sphere of the fixed stars is raised above the planets by an immense interval, and it is accordingly not affected by those motions that are in the planets. Copernicus puts the matter very simply: the fixed stars are not subject to any motion from place to place, and thus are truly fixed forever in the same places.

319

What is the ecliptic?

The ecliptic, in turn, is the great circle in the sphere of the fixed stars beneath which, for us on earth, the sun ever appears, and which it is seen to traverse annually. And whether this motion belongs to the sun or the earth, in either case it belongs to one of the planets. Therefore, the fixed stars do not themselves contain the cause of the ecliptic: it only results from the annual motion of the earth or of the sun about the center of the world.

Thus, since the ecliptic is found to have changed its position with respect to the fixed stars, it is not the fixed stars that have moved away from the ecliptic, but the latter that has moved away from the fixed stars.

The ecliptic is transported to other fixed stars.

The reason for this translation is shown beyond doubt by our principles of Chapter 63, if, indeed, they are sound. *Since the sun, through its most rapid rotation in its space which, for Copernicus, is the center of the world, sets the planets*

The reason for the changed ecliptic.

---

[1] Brahe's investigation begins on p. 233, but extends to p. 247. See *TBOO* 2, pp. 233–247.

in motion through an emitted species, *this rotation will have determinate poles. In the last diagram of Chapter 63, let the body of the sun be IO, and the poles of rotation be A, E, above which stand the points F, G on the sphere of the fixed stars. The great circle IO of the rotating solar body will thus be arranged beneath some great circle of the fixed stars: let this be ML. This is doubtless one and the same circle beneath the fixed stars, the poles F and G remaining constant, and the dignity of its body declaring that it first instils motion into the others. Nevertheless, the planets are found to move on various circles that are inclined to one another, owing to the natural principles explained in Chapter 63. Therefore, beyond doubt, the various circles of all the planets depend upon this "royal circle" ML, described by the rotation of the solar body about its axis AE, and each of them will keep its inclination to this circle constant in quantity, though having a translational motion, since we know by experience that the nodes are transposed.*

There exists a mean ecliptic.

Since the ecliptic, too, is one of the planetary circles, either the sun's or the earth's, it is consistent for it, too, to have some inclination to the royal circle ML, described among the fixed stars by the great circle IO of the solar body. For what would be the reason why the other planets would decline from one another, while the ecliptic alone, standing above the solar or the terrestrial path, coincides exactly with this royal circle ML?

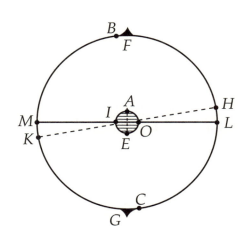

Let this therefore be granted: that the ecliptic, properly so called, is inclined to the royal solar circle. Let it be represented to us by the circle KH drawn among the fixed stars, and let its poles be BC. Under these conditions, we easily discover the occasion of the alteration of the fixed stars' latitudes: as the name suggests, these are computed from the true ecliptic and not from that royal solar circle, hitherto unknown. For the intersections or common nodes of the ecliptic, truly and properly so called (as a result of eclipses' occurring only beneath that line along which the sun proceeds), with the circle ML, which we might call the "mean ecliptic", will be carried along no less than the nodes of the other planets. Nevertheless, the maximum obliquity MK or LH, which is measured by the distance of the poles FB, GC, remains fixed and constant, as in the rest of the planets. That is, if about centers F, G, with constant radii FB, GC, small circles be described upon which we suppose the poles of the ecliptic B, C, to revolve, then the circle KH as well will depart from its original position on the sphere of the fixed stars FMG, and over the ages will make the southern limit come to be near the same fixed stars where the northern limit once was. Over a shorter period, however, it will be as follows. Since the limits K, H have not moved far from their fixed stars, their latitudes will be changed by some imperceptible quantity. However, since the nodes have progressed by the same amount from their fixed stars, the latitudes of their fixed stars will be altered more evidently. This is because at the end of the quadrant, near the limit, the sines of the inclinations increase by imperceptible increments, while at the beginning, near the nodes, these increments are quite perceptible.

From the supposition of a changeable ecliptic, all the alterations observed in the changes in the latitudes of the fixed stars follow.

320

Hence, because no change in the latitudes of the fixed stars is perceived near the equinoxes, while it is noticeable enough near the solstices, we correctly conclude that the limits of the ecliptic's latitudes are near the equinoxes, and the nodes are near the solstices. Therefore, the points K, H will be near the equinoxes. We likewise conclude this: that since the northern part of the true ecliptic flees from the north, in that the northern latitudes are increasing in Gemini and Cancer, the ecliptic's northern limit is therefore either in Libra, if the nodes progress, or in Aries, if (as is more probable) they retrogress. For the moon's nodes also retrogress, traversing the zodiac in 19 years, while the apogee progresses, traversing it in $8\frac{1}{2}$ years.

Where the true ecliptic and the mean one intersect one another.

Now the sun's apogee, or the earth's perihelion, is at $5\frac{1}{2}^{\circ}$ Cancer, and thus by Chapter 57, the diameter of power, causing the eccentricity, points at the sun when the earth is at $5\frac{1}{2}^{\circ}$ Aries. But also, by Chapter 63, the diameter of power that causes the latitude points at the sun when the earth is at the limit, which is in Aries by the present Chapter 68. Therefore, by the same Chapter 63, both powers can be effected by the same diameter of the earth's body. Hence one may argue plausibly that this invisible circle or mean ecliptic and the true one known to us coincide at $5\frac{1}{2}^{\circ}$ Cancer and Capricorn.

If the aphelia of all the planets were arranged on a single great circle, we could say that this is what we are seeking. For then it could be true of all planets, as it is here in the earth's circuits, that the nodes coincide with the apsides, and thus both variations—that of the eccentricity (in height) and that of the the obliquity (in latitude)—are effected by the same diameter of power. This would free us from the great difficulties with which we were left in Chapter 63.

A probable examination of the mean ecliptic.

And in fact the apogees of the sun, Mars, Jupiter, and Saturn fit approximately. For the aphelia of all three superior planets are in the same semicircle, and at the same time in the same northern direction. Therefore, the southern limit of the true ecliptic would be in Libra, and the northern in Aries, which agrees with the above.

A full consideration of this question must, however, be deferred until the motions of all the planets are examined with reference to the true ecliptic, the one known to us.

Further confirmation of this opinion of a hidden royal circle, projected from the sun among the fixed stars, is provided by the obliquity of the ecliptic that is in common use, which is computed from the equator, but which we might more correctly call the equator's latitude from the ecliptic. Now the equator is the great circle of the earth's body that is intermediate between the poles of the earth's daily rotation on its axis. And the same name of "equator" or "equinoctial" is given to that region of the sphere of the fixed stars that stands above the terrestrial equator in any era. The same name of "poles" is given to the points of the fixed stars that stand above the earth's poles in any era. Hence this axis, and this great circle, are inclined to the ecliptic differently in different eras. For to the extent that the northern latitude of the fixed stars in Cancer, and the southern latitude in Capricorn, is greater today, the equator's latitude from the ecliptic is smaller than it was once, since this obliquity is greatest in Cancer and Capricorn. It was once 23° $51\frac{1}{2}'$, while today it is 23° $31\frac{1}{2}'$, the difference of 20' being the change in latitude of the fixed stars.

Another argument for the mean ecliptic.

321

What is the equator or equinoctial?

It is, however, reasonable to suppose that the circle of the equator with its axis and poles would forever decline from the poles of this ecliptic *HK* by an equal and fixed distance, if the true ecliptic were the world's primary circle. But the ecliptic has changed, and the inclination of this axis to the ecliptic (and with it the inclination of the equator, to which this axis belongs) has been altered, so that to the extent that the ecliptic has receded from the fixed stars in Cancer, it has approached the equator. Therefore, the equator appears to maintain a constant inclination to some other circle. So a great cause, and a great dignity, ought to belong to this hidden circle. And thus from all these plausible arguments there arises a royal circle *LOM*, middle among the circles of the planets, to which all the planets, and Mars with them, maintain a constant inclination.

The example of the moon should not trouble us, whose inclination to the ecliptic, but not to any other great circle, is a constant 5°, both in the past and today, even though the ecliptic has been moved. For there is an enormous difference between the moon and the other planets. The orbs of the others encircle the center of the world. The moon's orb alone (roughly speaking) is outside the center and is transported from place to place. The others in common circle the sun, while the moon circles the earth. The eccentricities of the others and the whole theory of longitude and latitude originate from the sun, while those of the moon originate from the moving earth. The sun sweeps the others around in a circle, while the earth so moves the moon. What wonder, then, if the moon keeps the limits of its latitude constant with respect to the changeable ecliptic *HK*, beneath which lies the terrestrial circle, while the other planets do so with respect to some other invariable circle, such as *LOIM*? So the moon should not prevent us from giving credence to this theory.

It is therefore granted that Mars's orbit is inclined at a constant angle to some circle that maintains its position beneath the same fixed stars, such as *LOIM*. It follows that this same orbit of Mars has different inclinations to the ecliptic *HK* in different ages, since in certain of its parts it leaves the fixed stars it originally lay beneath and moves on to others. This only follows, however, if we grant that the nodes of Mars and the nodes of the earth, that is, the intersections of these orbits with the invisible circle *LOIM*, are not always carried over the same intervals in the heavens, some being faster than others. An authentic example of this was just given. For since the equator maintains a constant inclination to this invisible circle *LOIM*, while the ecliptic is meanwhile moved, the declination of the equator from the ecliptic is consequently perceived to be changeable.

Let A be the pole of the mean ecliptic, or the point upon which the straight line falls that is drawn from the center of the sun through the pole of the solar body. About center A with radius AB of 23° 42′ (or thereabouts) let a smaller circle be described, and let B, C be the positions of the north pole of the world, or the points upon which falls a line from the center of the earth's body through the pole of the daily rotation on the same body, B at the time of Ptolemy, and C at our time. If the nodes of the ecliptic also retrogress, the northern limit must be placed near the fixed stars in the region of Aries and Pisces. For the northern latitude of the fixed stars in Gemini and Cancer has increased, as was previously said. Let the midpoint D between B and C be taken, marking the position of the pole of the equator at an intermediate time, and let AD be joined. Thus the circle AD extended will pass through the solstice of the

Whether the inclinations of the planes of Mars and the ecliptic vary.

322

intermediate time. From A at right angles to AD let AE be drawn, which, being extended, will pass through the vernal equinox of the intermediate time. Therefore, close to the line AE there would be the pole of the circle beneath which the orbit and circuit of the earth was once arranged. And because the northern limit is in Aries, let EA be extended in the direction of A, and let the point I be taken on the extension below A. Thus the Ptolemaic pole of the ecliptic would be at I. About center A with radius AI let a small circle be described, on which let another point O be taken, nearer to C than I is to B. And let O be the present pole of the ecliptic, 23° 31½′ from C, while I, the Ptolemaic pole of the ecliptic, is 23° 51½′ from B. This will be the theory of the ecliptic's altered obliquity and of the altered latitude of the fixed stars, except that the size of the small circle OI is not known to us. For the ecliptic's 20′ alteration of obliquity can be produced in various ways.

And because O is today's pole of the ecliptic, and OC points towards the beginning of Cancer, let CP be the eighth part of the circle, and P the middle of Leo, where the northern limit of Mars is today. Let PO be extended beyond O, and let GI be drawn through I nearly parallel to PO, but slanting somewhat forward in longitude (for the sidereal position of Mars's limit was once a little farther forward than today), and let it be extended beyond I. And about A let a small

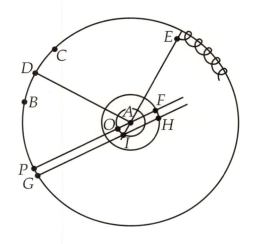

circle be described, intersecting PO at F and GI at H. Let the size of the circle be sufficient to make OF greater than IH. And let the pole of the circle beneath which the circuit of Mars is arranged, be placed at F today, and at H in the past. Today's obliquity OF, or the inclination of the plane of Mars to the ecliptic, will be greater, and the Ptolemaic obliquity HI will be smaller. Nevertheless the pole of Mars's orbit H, F would have moved from H to F keeping at a constant distance AH, AF from A.

And since the pole of the Martian orbit has traversed a fairly large arc from H to F, whether forward or backward, but at the same time the pole of the ecliptic has gone from I to O about the same point A, the pole of Mars would appear to be nearly at rest, since IH and OF are nearly parallel.

A great inequality in the motion of the nodes must indeed follow if it is true that the poles of the individual planets circle some common pole in different times.

For there is an anomaly originating from this source in the precession of the equinoxes, whose treatment is very much like the present ones.

I have stated what is in agreement with the principles established in this work, and by what hypotheses this can be brought about, so that the inclinations of the planes can be made different in different ages. Let us now examine the observations of Ptolemy. For since the northern latitude of Mars is with Cor Leonis, a northern star, while the southern latitude is with the southern stars of Capricorn, it is reasonable that the same happened to Mars's

---

The earth's pole does not proceed exactly along the small circle BCE, but in the spiral depicted at E, with one loop per year, a similar spiral being described by the opposite pole. Each depends upon the other, and from this interconnection arises the progression of the equinoxes and the solstices. But the magnitude of each of these spirals is about the same as that of Copernicus's *orbis magnus*, or of the orb of the sun, for the others. That is, the ratio of this spiral to the surface of the sphere of the fixed stars is imperceptible. It can thus be taken as the simple line BCE. However, for the correct imagining of this motion, it should be noted that the axis of the terrestrial equator, extended both ways to the fixed stars, describes a cylinder each year whose magnitude is the same as that of one of these spirals, and which has the sun's body in its middle. Further, over the ages the same axis of the earth describes two cones with equal vertices at the sun, except that they are confused by the earth's circuit about the sun, so that the vertex of each cone is contained within the other, owing to the coincidence of all the cylinders; while the base is BCE. Thus the cone is compounded of many cylinders.

What is the reason why the translation of Mars's nodes is so slow?

323
On the inequality of the precession of the equinoxes.

maximum latitudes as happened to those stars: they both increased. For the stars' latitudes did increase, the northern ones around the summer solstice, and the southern ones around the winter solstice. Ptolemy therefore said that Mars's maximum observed northern latitude was 4° 20′, while today it is 4° 32′.[2] This confirms our opinion, since he shows a maximum latitude that is 12′ smaller than today's, while the nodes stayed at approximately the same distance from aphelion as they are today. On the other hand, he makes the southern latitude about 7°, while today it could also be that much, namely, $6° 52\frac{1}{3}′$. We are therefore left undecided by his observations. For concerning those 12′ in the northern latitude, it should be noted that the smallest graduations on his instruments were 10 minutes, and that he usually supposed an error amounting to one of these parts. Also, the difference between the Greek symbols for 20′ and 40′ is very small and slippery, often neglected by translators.[3] Nevertheless, the Arabic says 20′ here.

There is nothing besides this in Ptolemy that can lead us to a judgement of how these matters stood in antiquity. For the observation examined in the following Chapter 69 is shown to be in error. Therefore, as long as we are wanting suitable observations from antiquity, circumstances compel us to leave this discussion of the motion of the nodes, along with many other matters, to posterity, if, indeed, it should please God to vouchsafe the human race a length of time in this world sufficient to work through such remaining questions thoroughly.

---

[2]Ptolemy, *Almagest*, Book 13 Chapters 3 and 5.

[3]The Greek numeral for 20 is κ′, while that for 40 is μ′.

CHAPTER 69

A consideration of three Ptolemaic observations, and the correction of the mean motion and of the motion of the aphelion and nodes.[1]

From all antiquity, there have survived no more than five written observations of the star Mars, as well as one of extreme antiquity noted by Aristotle, who saw Mars occulted by the dark part of the half moon. However, neither the year nor the time of day were given. Nevertheless, I have discovered, using a very lengthy process of induction extending over the 50 years from Aristotle's fifteenth year to the end of his life, that this could not have happened on any other day than the evening of April 4 in the 357th year before the commonly accepted epoch of Christ, when the 21-year-old Aristotle was, as we know from Diogenes Laertius, a student of Eudoxus.[2] The second observation, obtained from the Chaldeans, was preserved for us by Ptolemy. This was made on the morning of January 18, 272 BC, when Mars occulted the northern star in the head of Scorpius.[3] Here again, no particular time was given. The other four were by Ptolemy himself, using an astrolabe to measure Mars's distance from fixed stars. However, he reports only the zodiacal position at the exact moment of Mars's opposition to the sun's mean motion.[4]

Upon these few observations, arguments of the greatest moment are to be founded; or, if this is not possible, astronomy must remain incomplete. First, through the four Ptolemaic observations, the epoch of the mean motions, related to the fixed stars, belonging to Ptolemy's time, is to be found, and by a comparison of this with the modern data the mean motion itself is to be determined. Next, it seems possible to use the Chaldean observation to inquire whether the solar eccentricity was really once greater than it is today. And then, using both this observation and the Aristotelian one, if the time were known, one could hazard a guess at Mars's latitude at those times.

But by God immortal, what is this path upon which we shall be tread-

---

[1] This chapter shows signs of having been written in great haste. The number of careless errors is extraordinary, even for Kepler. Yet the Keplerian virtuosity in handling data is also evident, and the main conclusions stand despite the mistakes.

[2] In the Kepler manuscripts at the Russian Academy of Sciences at St. Petersburg (vol. XIV fol. 295 v) there is an elaboration of the present passage, which has recently been published as part of Kepler's "Mars Notebook" (*KGW* 20.2, 497–499). From the dates given in Kepler's "induction" mentioned above (*KGW* 20.2 p. 498, a birth year for Aristotle of 375 BCE may be deduced, although the year presently accepted is 384 BCE. Further, Kepler gives Aristotle's age as 22, and the year of the occultation as Nabonasser 393, or 357 BCE.

According to a more recent account, the date was 4 May. See the note on p. 205 of Aristotle, *On the Heavens*, trs. W. K. C. Guthrie (Harvard University Press, Cambridge 1960).

[3] β Scorpii. The observation is in the *Almagest* Book 10 Chapter 9.

[4] The three observations considered in the present chapter are in the *Almagest*, Book 10 Chapter 7. The fourth is in Book 10 Chapter 8.

ing? For we have hardly anything from Ptolemy that we could not with good reason call into question prior to its becoming of use to us in arriving at the requisite degree of accuracy.

I

How astronomers may investigate the beginning of the zodiac or of the ecliptic.

First, at the times set out, he records the mean position of the sun by a calculation that depends upon observation of the equinoxes and solstices. The sun reveals the beginning of Aries, not by pointing a finger at the place, but by a blind conjecture of the time. For we call the beginning of Aries that point that the sun occupies when the day is observed to equal the nights. What if Ptolemy had been in error as to the time? We are not wanting conjectures.

The Ptolemaic observation of the equinox is suspect.

First, he does not give his method of observing it. My hope is that he observed the meridian altitudes, for by induction from these the moment of the sun's entering the northern hemisphere is obtained without error. But what if he had observed it using Alexandrian armillaries, where refraction could have put him off? He himself clearly suggests that he did so, when he says that the equinox was measured twice on the same day using these armillaries. He ascribes this to instrumental error; I, however, suspect that the error arose from refraction.[5]

Difficulty in accepting the day of the equinox given by Ptolemy.

Let it stand, however: let him have observed it through meridian altitudes. There is another suspicion that unwelcomely but most forcefully insinuates itself: the moments of the equinoxes recorded by Ptolemy do not agree within a day and a half when compared with the previous observations of Hipparchus and the later ones of Albategnius[6] and Brahe, all of which conspire in a single equality. The Ptolemaic equinoxes alone deviate. This fact has given rise to many extremely complicated opinions on the heavens, and gave birth to the motions of trepidation and libration, all of which are brought to ruin once one sees that the observations after Ptolemy always agree to the point of equality with the most ancient ones of Hipparchus.

On behalf of the Ptolemaic observation of the day of the equinox.

325

Ptolemy nevertheless supports himself by comparing the vernal equinoxes with the autumnal. For if, as a result of instrumental error, he had pronounced the true equinox to be on the following day, when it had occurred on the previous one, the autumnal equinox would have been pronounced to be on the previous day when it belonged on the following one. If two days were thus subtracted from the length of the summer, a great alteration in the sun's eccentricity would have resulted. Nevertheless, following his observations, he left this at the same quantity found by Hipparchus. So no alternative is left us but to trust Ptolemy in believing him to have observed correctly the time at which the sun stood at the beginning of Aries.

---

[5]Kepler considers this matter in the *Optics* (Frankfurt, 1604) p. 146, or p. 157 of the English translation.

[6]al-Battānī (c. 858–929), an Arab astronomer well known to Mediaeval and Renaissance Europeans, who was particularly distinguished for his work on the solar theory. He discovered the motion of the sun's apogee.

II

Once a beginning is made, and the obliquity of the ecliptic is found through observation, it is a trivial matter to use the sun's declinations on each day to report its elongation from the point occupied by the sun at the time stated to be the moment of the equinox, whatever it might be or in whichever sphere it might be established. For various authors thought up a variety of spheres for this purpose: after the eighth and ninth spheres established by Ptolemy, others have set up a tenth, and most recent authors an eleventh and a twelfth, all through the merest of speculations, against which overabundance of mechanisms[7] Brahe vehemently inveighed. He never, however, told me what he intended to substitute in their place, nor did he leave it in writing anywhere. Copernicus, on the other hand, acted ingeniously and wittily (in the common opinion), and wisely (in mine), in removing his eyes from the heavens and seeking that point in the globe of the earth, above which, in the sphere of the fixed stars, there stands a point, determinate for any age you please, as was said in Ch. 68. However, this is not the place to discuss this matter further.

How observations are used to learn the sun's zodiacal position, even when the zodiac's sidereal position is unknown.

The theory of the eighth sphere is difficult to understand in the ancient account, and inconsistent;

Easy in Copernicus.

III

There follows a demonstration of the equation, which depends upon the sun's observed entry into the beginnings of the cardinal signs. For when the equation is subtracted from or added to the sun's apparent position, the sun's mean motion is established with respect to the point which the sun is observed to occupy at the time of the equinox. Here too, as to the quantity of the equation, there is greater uncertainty than before over the equinox or the beginning of the zodiac. For today that equation appears to be 20′ less than the quantity which Hipparchus appears to have demonstrated for himself, and which Ptolemy retained. Nor is there sufficiently good reason why we should say that the ratio of the orbs is different today from what it once was. For affirmations of the greatest moment require the most solid evidence, and this we are lacking. And those observations cannot be that accurate, especially concerning the entry into Cancer and Capricorn. If we substitute the modern equations for Ptolemy's, we shall not change his observations as much as Ptolemy himself says he can discern in observing, nor as much as the Ptolemaic observations can be vitiated by the matter of refraction. For we can with certainty name the day of the Ptolemaic equinox observation, though meanwhile remaining uncertain about the time of day. And here, the partnership of the vernal and autumnal equinoxes is no such defense against the small error we are considering here as it was before against a large one.

Were the equations of the sun once greater?

The time of the Ptolemaic equinox is uncertain.

That the equations of Ptolemy's day were really equal to ours, the constancy of the modern ones argues. For those found today by Brahe, and those found several centuries ago by Albategnius and Arzachel,[8] are nearly the same.

---

[7]Overabundance of mechanisms: πολυπραγμοσύνην.

[8]Abū Ishāq Ibrāhīm al-Zarqālī, the earliest noteworthy astronomer of Spain (fl. ca. 1080). His

There is therefore suspicion that the equation of the sun used by Ptolemy is in error, since it is deduced from erroneous apparent positions of the sun.

326   Consequently, Ptolemy did not relate Mars to either the sun's mean or its apparent position, without the chance of error.

Nevertheless, there is this consolation, that we need to make use of the sun's apparent position, the comprehensive argument for which was given above.

With many thanks for the Ptolemaic observations, the modern equations of the sun are retained.

We can, however, proceed on a twofold path: either we believe Ptolemy on the equinoxes, or, using the modern equations, we apply a correction to the Ptolemaic equinoxes, making the vernal equinox three hours later than that noted by Ptolemy, and the autumnal earlier by the same amount, the result being an error of 8′ in the sun's declination at each place. Ptolemy's instruments were doubtless not calibrated with divisions any finer than 10′. And Hipparchus assigns an uncertainty of one such division. For this reason, the times, too, that the sun takes to traverse the quadrants of the zodiac, were not expressed more precisely than to within a quarter of a day. So much for the true length of summer and winter.

IV

The Ptolemaic position of the sun's apogee has an uncertainty of many degrees.

But what shall we now say of the sun's entry into Cancer and Capricorn, upon which the apogee and the setting out of the equations depends? And of how easily a quarter of a day can diminish the vernal quadrant of the zodiac, and increase the autumnal? For the sun's entry into Cancer is quite imperceptible. Nor can I be persuaded that Hipparchus and Ptolemy examined the very moment of this entry, ignoring the intermediate points. I find it more credible that throughout the summer they were carefully noting the sun's declinations, and were matching up the equal ones on both sides of the solstice, taking as the true entry of the sun into Cancer the time intermediate between the moments of equal declinations. In this way, if a comparison had been made of positions near to the solstice, there would indeed be little error, but there still could have occurred as much as a quarter of a day, during which time 15′ of the sun's motion elapse. Therefore, even if the equinoxes were perfectly certain, it is nonetheless possible that there be an excess or defect of a quarter of a degree in the sun's position in the two segments about the solstice, and that the apogee should then fall eight degrees farther forward or farther back. So much for the sun's motion.

V

Now, as for the observations of Mars itself, even if Ptolemy in fact managed to line the astrolabe up accurately on the fixed stars, there is still clearly no more certainty about Mars's zodiacal position (just as in the previous consideration of the sun's position) than there is about the positions of our fixed stars. If

---

name has often been associated with the Toledan Tables, although he was but one of many contributors to that work.

For Albategnius (al-Battānī) see the note earlier in the present chapter.

Ptolemy committed a error in assigning a fixed star its degree of elongation from the point of the equinox, the same error will be committed in proclaiming the position of Mars. Furthermore, not even the elongation of the fixed stars from the sun (and thus from the point of Aries from which the sun's elongation is known through its declination) is free from suspicion of error. For consider both the manner in which it is found and the argument of error. In the year 2 of Antonine Ptolemy sought it through the half-illuminated moon.[9] Using the astrolabe, he found the moon's elongation from the sun, and that of Cor Leonis from the moon. Therefore, when the sun's elongation from the point of the equinox is given, the elongation of the fixed star from the same point is also given*. Now, in measuring the elongation of the moon from the sun, an error of half a degree appears to have been committed. For the measurement was made at sunset. But the sun when setting appears through refraction higher than it should by about half a degree. Therefore, the moon's elongation appears less than it should, and so also that of Cor Leonis from the sun, and likewise from the equinoctial. Thus it appears that half a degree must be added to the positions of the fixed stars at the time of Ptolemy.

Therefore, when Ptolemy considered Mars (when observed in relation to the fixed stars) to be at opposition to the sun's mean position, it would now really have been half a degree beyond this opposition. So when these four observed positions of Mars are presented by Ptolemy: 21° 0′ Gemini, 28° 50′ Leo, 2° 34′ Sagittarius, 1° 36′ Sagittarius; we should take these: 21° 30′ Gemini, 29° 20′ Leo, 3° 4′ Sagittarius, 2° 6′ Sagittarius. Now Ptolemy actually defended himself against such presumption, affirming that he frequently had sought out this one thing, namely, the elongation of the fixed stars from the moon, of the moon from the sun, and hence the distance of the fixed stars from the sun and from the equinox point, and had always found it to be the same. So although he produced only one observation in order to demonstrate the method, it is nonetheless credible that he consulted several observations, at both the rising and the setting of the sun or moon, finally choosing the one which he saw to be intermediate among many operations that produced various positions.

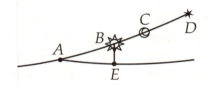

Now this argument over 30′ seems to be irrelevant to Mars's mean motion, if indeed on these four occasions Mars, as it was observed with respect to the fixed stars, can be referred to them, without consideration of the equinox point, whose distance is uncertain. This is the method I used in Ch. 17 above to investigate the position of the aphelion in Ptolemy's times. We are nevertheless hindered in this respect, that the observed positions of Mars are to be referred to a point opposite the sun's apparent position. This task can never proceed correctly unless the distances of both Mars and the sun from the common equinox point is previously known, since the arc of the true elongation of Mars from the sun cannot be deduced otherwise than through these com-

The Ptolemaic zodiacal positions of the fixed stars are not without suspicion of an error of about 20 minutes.

327

* *A* is the hidden point of the equinox. *B* is the sun, *C* the moon, *D* the fixed star, all visible; *BE* is the sun's declination. *AB* is obtained most easily by observations of *BE* at noon. *BC* is known through instruments during the day, and *CD* through instruments and by night. *AB*, *BC*, and *CD* being combined, the elongation *AD* of the fixed star from *A* is at length obtained, *A* being a previously invisible point which now is revealed once it is related to the fixed star *D*. Afterwards, the planets are related to the fixed stars through observation, and thus their elongation from the beginning of the zodiac *A* is known.

In defense of the Ptolemaic longitudes of the fixed stars.

The extent to which the uncertainty in the positions of the fixed stars affects the observations of Mars.

---

ponents (so to speak).

If, at the moment taken as the true opposition of the bodies, the planet should appear to be 30′ beyond the sun's true positions, the planet still has some involvement with the second inequality, and is not yet ready for an enquiry into the first inequality. And at apogee, these 30′ in the equation of the orb occupy a large arc on the eccentric, to which corresponds an even larger portion of the time or the mean motion. At perigee the opposite takes place. For this equation of the center occupies a small arc on the eccentric, to which corresponds an even smaller portion of the mean motion. Therefore, anyone who says that on these four occasions Mars was observed 30′ farther along on the zodiac, says in effect that at the equinoctial point Mars's mean motion was many minutes farther back at apogee, and a few minutes back at perigee. And since the arc on the eccentric is smaller than this 30′ arc resulting from faulty observation, neither Mars's eccentric position nor even its sidereal position are as far forward as Mars itself appears to have moved with respect to the fixed stars, the difference being that quantity by which the arc on the eccentric differs from the 30′ arc in the observation. And since this arc is large at aphelion, differing but little from the 30′ arc in the observation, and the oppo-

328   site at perigee, it therefore finally will follow that at aphelion a small amount, and at perihelion more, must be subtracted from Mars's mean motion with respect to the fixed stars, if we accept that the fixed stars are 30′ farther forward on the zodiac. Thus not only is the mean motion made smaller (although by a much smaller quantity than the 30′ resulting from faulty observation), but also the arrangement of the three acronychal observations used by Ptolemy is disturbed, whence must arise another aphelion and another eccentricity. However, this will cause us no trouble later. For we may neglect it, even if the observations introduce something large, as long as there is no suspicion of error in the fixed stars, since it is certain that they do not have the same precision as the Brahean ones. So we shall use the form of equation found through the Brahean observations, as if they remain the same throughout the ages.

Since we have encountered three forks in the road, one concerning the sun's eccentricity, another concerning the position of the sun's apogee, and the third concerning the zodiacal positions of the fixed stars and of Mars, there are therefore eight ways of establishing the mean motions and the aphelion at the moments of observation, even if, ignoring the zodiac, we compute only with reference to the fixed stars.

Let the first investigation retain all the Ptolemaic data concerning the sun and the fixed stars.

Reduction of the Ptolemaic observations to apparent opposition to the sun.

Since the positions of the sun's mean motion were 21° 0′ Sagittarius, 28° 50′ Aquarius, and 2° 34′ Gemini, and the sun's apogee was 5° 30′ Gemini, the apparent positions of the sun were 21° 40′ Sagittarius, 1° 13′ Pisces, and 2° 41′ Gemini, all three beyond opposition.[10] The true opposition therefore preceded

---

[10]For these computations, Kepler used Ptolemy's value for the sun's eccentricity. In the *Almagest* (Book 3 Chapter 4), Ptolemy finds this to be 2^p 29′ 30″ where the radius of the eccentric is 60^p. This amounts to 4153 where the radius is 100,000. Compare this with the Tychonic value of

them. And since the diurnal motion at 21° Gemini (Cancer, today) is about 23', and that of the sun, 61', and the sum is 1° 24', those 41' therefore require 8 hours,[11] at which time Mars was visible at 21° 8' Gemini, opposite the sun's apparent position. Likewise, at 29° Leo (Virgo, today) Mars's diurnal motion is usually taken to be 24', the sun's diurnal motion 59', and the sum 1° 23'. Therefore, a difference of 2° 23' requires 1 day 17 hours 21 minutes, at which time Mars was visible at 29° 31' Leo. Finally, at 3° Gemini (Cancer, today) Mars's diurnal motion is 23', the sun's 57', the sum 1° 20', by which it is shown that for 7' there are required $2^h 6^m$, at which time Mars was visible at 2° 36' Sagittarius.

The corrected times are therefore these:[12] Positions:

Hadrian 15	Tybi 26	$5^h\ 0^m$	21°	8'	Gemini
Hadrian 19	Pharmouthi 4[13]	$15^h\ 39^m$	29	31	Leo
Antonine 2	Epiphi 12	$7^h\ 54^m$	2	36	Sagittarius[14]
Interval of $\Big\{$ 4, 68 days		$10^h\ 39^m$	68	23	
Egyptian years $\Big\{$ 4, 97 days		$16^h\ 16^m$	93	5	

To the first interval there corresponds a mean sidereal motion of 80° 57' 14" beyond the complete cycles, and to the second, 96° 16' 24". But in the former instance the apparent motion of Mars was 68° 21' 20" beyond the complete cycles, with precession over the interval subtracted, in the amount that it was at that time.[15] In the latter instance, the apparent motion was 93° 2' 20".

Now let the hypothesis previously under investigation, established on the

---

about 3600, which Kepler accepted (in Chapter 23).

[11] The time corresponding to these figures is more nearly 11 hours. Although this change has no immediate significant effect upon the positions, Kepler repeats this error (which implies a combined diurnal motion of 2°) many times in this chapter.

[12] The translator's recomputation shows slightly different figures, partly as a result of Kepler's error in the diurnal motion. Although the changes are slight, they affect later computations, and so are given here:

Hadrian	15	Tybi	26	$2^h\ 0^m$	21°	10'	Gemini	
Hadrian	19	Pharmouthi	4	$16^h\ 14^m$	29	31	Leo	
Antonine	2	Epiphi	12	$7^h\ 54^m$	2	36	Sagittarius	
Eg. yrs.	4	days	68	$14^h\ 5^m$	68	21		
Eg. yrs.	4	days	97	$15^h\ 40^m$	93	5		
		Mean motion for the first interval:			81°	5'	20"	
		for the second:			96	19	17"	

[13] Incorrectly stated as "VI" in the *KGW* text; the 1609 edition has "IV," which is correct.

[14] Tybi, Pharmouthi, and Epiphi are the fifth, eighth, and eleventh months of the 365 day Egyptian year. See G. J Toomer's introduction to his translation of Ptolemy's *Almagest* (Duckworth, 1984) pp. 9–14 for an explanation of Ptolemaic chronology and calendars.
  The Ptolemaic observations (from *Almagest* X.7, trs. Toomer p. 484) are:

Hadrian	15	Tybi	26/7	1 hour after midnight	21°		Gemini
Hadrian	19	Pharmouthi	6/7	3 hours before midnight	28	50'	Leo
Antonine	2	Epiphi	12/13	2 hours before midnight	2	34	Sagit.

[15] Ptolemy caused later astronomers endless difficulty by adopting an annual precession of 36", while data from all other ages yield a precession of about 51". Kepler followed accepted procedure in accepting Ptolemy's word that the precession really was that small then.
  Applying the Ptolemaic precession to the intervals of 4 years $2\frac{1}{4}$ months and 4 years $3\frac{1}{4}$ months gives 2' 30" and 2' 34", respectively. From the numbers Kepler gives, it is clear that he intended to use 2' 40" for both intervals, but erred in subtraction when correcting the first angle. It should have been 68° 20' 20", or 68° 18' 20" using the corrected diurnal motion.

basis of the most recent observations, be brought in, and let the question be raised, at what anomalistic position do the apparent motions on the eccentric correspond to the mean motions, as I have just now given them? After a few

329   trials, this is found: If for the last time the aphelion of Mars is placed at 0° 41′ Leo, and for the remaining times somewhat before that, owing to the precession of the equinoxes, while at the first time the mean anomaly is 46° 37′, at the second, 34° 21′, and the third, 130° $37\frac{1}{2}'$; and thus the elongation from the equinox at the middle time was 5^s 4° 59′ 20″, then by the modern hypothesis of the equations the star Mars is placed at 21° 7′ Gemini for the first time, 29° [31′][16] Leo for the second, and 2° $37\frac{1}{2}'$ Sagittarius for the third, fortuitously precise. For the foundations are not such as to allow one to hope for such precision. Had Ptolemy made note of more oppositions of his day, we would doubtless be experiencing greater difficulty. For with three solar oppositions it is handled easily.[17] Compare this aphelion with Ch. 17.

Second, the Ptolemaic equation and apogee being retained, let 30 minutes be added to the fixed stars.

The result will be slightly different. For since Mars is half a degree beyond opposition to the sun, the corrected opposition will follow. The sums of the diurnal motions were 1° 24′, 1° 23′, and 1° 20′. Therefore, for the extra 30′ the corresponding times come out almost exactly the same, to be added to all three, namely, about 8 hours 40 minutes. To this there corresponds $8\frac{1}{2}'$ of Mars's apparent motion, which is to be subtracted from these 30′. The remaining $21\frac{1}{2}'$ are to be added to the planet's positions, placing it at 21° $29\frac{1}{2}'$ Gemini, 29° $52\frac{1}{2}'$ Leo, and 2° $57\frac{1}{2}'$ Sagittarius. The intervals both of time and of zodiacal positions will remain almost exactly the same. Thus the distribution of the mean anomaly among these observations, which was just now found, will also be the same. Only the aphelion will be transposed by the same number of minutes, so that on the last date it is at 1° $2\frac{1}{2}'$ Leo. It therefore has to be moved $8\frac{1}{2}'$ back among the fixed stars. And the mean motion from the equinox will be increased by the above mentioned $21\frac{1}{2}'$, but it will be 8 hours 40 minutes

---

[16]The number of minutes was omitted here. However, in case 3 below, Kepler states that the second longitude was unchanged and gives it as 29° 31′ Leo, a figure that agrees closely with the aphelion and mean anomaly given here.

[17]Since observations 1 and 2 are on opposite sides of the aphelion and at comparable distances from it, they are not much affected by changes in the position of the aphelion. Accordingly, if the angle between them does not fit the theory, there is not much that can be done. Fortunately, the angle fits fairly well, as Kepler remarks.

Observations 2 and 3, on the other hand, are on the same side of the aphelion, and the predicted heliocentric positions can be changed appreciably by moving the aphelion. So the way to adjust the aphelion is to try different positions, computing only positions 2 and 3, until the observed positions, run *back* through the theory, give mean anomalies that are at the correct interval from each other. That this was indeed Kepler's procedure is suggested by the translator's test of the method, in which the computed mean anomalies are 96° 15′ 36″ apart, almost exactly Kepler's interval. It should also be noted that (as Kepler shows in Chapter 60) it is much easier to find the mean anomalies from the longitudes than the other way around.

The same procedure applied to observations 1 and 2 give mean longitudes that are too near each other by 9′. Kepler apparently compensated for this by increasing all the mean anomalies, especially that of observation 1, thus keeping 2 and 3 at a constant interval while increasing the interval between 1 and 2.

longer. And to these hours there correspond 11′ 24″ of mean motion. Therefore, at the proposed time, the mean motion from the equinox will be only 10′ greater than before. But the positions of the fixed stars are 30′ farther removed from the equinox. Therefore, Mars's mean motion with respect to the fixed stars has proceeded 20′ less than before.

Third, the sun's apogee being transposed by 11 or 12 degrees,[18] while the longitude and equation of the fixed stars remains the same.

Then on the first date the sun will be 20′ back in position;[19] on the middle date practically nothing will be changed; and on the last date it will be 21′ farther along in position, owing to the altered equations of the sun. Therefore, the first opposition will be 4 hours later,[20] and Mars will be the same number of minutes farther back in position, while the last opposition will occur $4\frac{1}{3}$ hours earlier,[21] with Mars the same number of minutes farther along in position.

The results:[22]		H.	M.		Positions:	
Tybi	26	9	0	21°	4′	Gemini
Pharmouthi	4[23]	15	39	29	31	Leo
Epiphi	12	3	37	2	40	Sagittarius
Egyptian	{ 4, 68 days	6	39	68°	27′	
intervals	{ 4, 97 days	12	0	93°	9′	

The first interval of time becomes smaller. So also the corresponding mean   330
motion is 5′ 15″ smaller, so as to be 80° 53′. The second interval of time again becomes smaller. Therefore, the mean motion corresponding to it is 5′ 40″ smaller, namely, 96° 10′ 48″. So, since to the two mean motions, both smaller, there corresponds a greater apparent motion than before, and on the supposition that the mean anomaly is the same as before for both, the apparent motion is greater by about 9′, it therefore appears that Mars has to move down from aphelion. However, the first interval is unchanged unless a great descent is made, while the second [is corrected] by a descent of about 36′.[24] Therefore,

---

[18] At the end of Section IV of this chapter, Kepler mentions changing the apogee by 8°, a change that would better agree with the 20′ change in the sun's position mentioned in the next sentence.

[19] Here, the three observations are on the same side of the line of apsides (the sun's, this time). The first observation is near perigee, the second near the quadrant, and the third near apogee, all in the ascending semicircle. In this configuration, the equations of the second are not much affected by a change in the apsides, but those of the first and third are altered, in opposite directions. Specifically, when the apogee is moved forward, position 1 is moved back and position 3 is moved forward. This is the situation Kepler describes; hence, he has moved the apogee forward.

[20] Here again is Kepler's odd implicit overestimation of the combined diurnal motion: 20′ in 4 hours would amount to 2° per day. The correct time would be $5\frac{3}{4}$ hours.

[21] $6\frac{1}{3}$ hours.

[22] The effect of the above changes is to increase all the differences, so that the positions are all several minutes farther apart and the times or mean anomalies are several minutes of arc closer together. However, the earlier error in the diurnal motion decreased the first longitude, so the position Kepler gives, 21° 4′ Gemini, is correct.

[23] Given incorrectly as "VI" in both the 1609 edition and *KGW*.

[24] This is because the first two observations are on opposite sides of aphelion, while the second and third are on the same side, so positioned as to allow their interval to be changed easily. See

if we were to indulge in our enquiry, and not take the modern hypothesis as given, we would arrive at a completely different hypothesis with a new eccentricity.[25] And if, on the other hand, these three observations of Ptolemy were perfectly certain, this would constitute the basis for an argument that he established the sun's apogee correctly.

But when 36′ are subtracted from Mars's aphelion, placing it at 0° 3′ Leo[26] for the last time, and when its mean motion is so adjusted that for the middle time the [mean] anomaly is 34° 58$\frac{1}{2}$′ with longitude from the equinoctial of 5^s 5° 0′ 50″, the following observations result:[27]

First	21°	7′	Gemini	should	21°	4′	Gemini	3+	
Second	29	28	Leo	have	29	31	Leo	3–	difference
Third	2	37	Sagittarius	been	2	40	Sagittarius	3–	

Again, an accurate enough approach. For we cannot hope that the observations were of such certainty. So, whether the sun's apogee is known correctly or not, the distance of the mean motion from the equinoctial is certain within 1$\frac{1}{2}$′.

Fourth, the same things will be changed in the computed positions of the second case, and in establishing the mean longitude, that is, by transposing the apogee and the fixed stars.[28]

Fifth, the sun's apogee and the Ptolemaic longitude of the fixed stars remaining the same, the modern solar eccentricity is used.

Thus, while the first and last positions of the sun remain almost exactly the same, the sun's apparent position will be changed in the middle observation by 20′. For the former fall near the sun's apsides, where the equation is small, while the latter is near the middle longitude, where the equation caused by the eccentricity is maximum. And since in Aquarius the equation is additive, when 20′ are taken away from the equation, the sun will be moved back through the same number of minutes, and will be, not in 29° 31′ Aquarius, but in 29° 11′ Aquarius. The correct and truest opposition therefore follows by 4

---

footnote 17 above.

[25]The first two observations are so situated as to make it practically impossible to adjust the predictions to the times by moving the aphelion. This would accordingly have to be accomplished by changing the eccentricity, a course of action upon which Kepler was understandably reluctant to embark.

[26]Subtracting 36′ from 41′ leaves 5′, a figure also in accord with the position obtained by subtracting the mean anomaly of 34° 58$\frac{1}{2}$′ from the mean longitude of 5^s 5° 0′ 50″.

[27]The corrections mentioned above tend to decrease the intervals between the predicted positions (first column), and increase the intervals between those obtained from the observations (second column). The resulting discrepancy, however, is about what Kepler thought, except that the third difference should be $-7\frac{1}{2}$′. This could, however, be reduced by further adjustments in the aphelion and mean longitude, so Kepler is in principle correct in his claim that the theory can be made to fit the data.

[28]This has been expressed somewhat confusedly, as the translation suggests. Kepler's meaning appears to be that, when both the apogee and the position of the fixed stars are changed, the resulting positions and times are as in case 2. This will be so provided that moving the apogee does not significantly affect the longitudes.

hours.[29] The planet will then be at 29° 27′ Leo.[30] The earlier time interval and its mean motion is increased, and the apparent motion is decreased, while the later time interval is decreased, and the apparent motion is increased. So once again, more evidently than before, the application of this correction calls upon us to change the hypothesis, unless we had sworn allegiance to the words and numbers of the hypothesis of this era by our best judgement. For to move the planet forward a smaller distance in a greater time near apogee, and a greater distance in a smaller time near perigee, nothing can suffice but to increase the eccentricity. If everything were kept the same, as in the first case, the results for the first and third times would indeed again be the same as then, namely, 21° 7′ Gemini and 2° 37$\frac{1}{2}$′ Sagittarius. But in the middle position, it would come out to be 29° 36$\frac{1}{2}$′ Leo, where it ought to have been 29° 27′ Leo, a difference of 9$\frac{1}{2}$′.[31] To eliminate this, the aphelion ought to remain in about the same place, but the mean motion should omit 3$\frac{1}{2}$′. Then the results will be:[32]

331

For the first	21°	4′	Gem.	should	21°	8′	Gem.		−4
Second	29°	33$\frac{1}{2}$	Leo	have	29°	31	Leo	Difference	+2$\frac{1}{2}$
Third	2°	38$\frac{1}{2}$′	Sag.	been	2°	36	Sag.		+2$\frac{1}{2}$

Sixth, the same change of the second case will occur, if we at once change both the sun's eccentricity and the longitude of the fixed stars.

Seventh, if on the other hand we at once change both the eccentricity and apogee of the sun, combining the third and fifth cases, the fundamentals will be these.[33]

---

[29] Here again is Kepler's erroneous diurnal motion of 2°. The correct time is 6 hours.

[30] When the diurnal motion is corrected, this becomes 29° 25′ Leo.

[31] Because of the error in the diurnal motion, the computed longitude is 2′ greater, the longitude of true opposition 2′ less (as was noted above), and the difference is therefore 13$\frac{1}{2}$′, much worse than Kepler thought. There is no way to accommodate the theory to so great a difference, as the corrected figures below show.

[32] The numbers in the first column are very curious indeed. Although the first longitude is in accord with Kepler's assumptions, the second and third are actually greater than those in case 1. This would be the result of an increase in equated anomalies (since both are on the descending semicircle). However, a decrease in the mean longitude at these positions would result in a decrease in the mean anomalies. It is something of a mystery how Kepler managed to decrease the mean anomalies while increasing the equated anomalies!

The longitudes should have been 29° 35$\frac{1}{2}$′ Leo and 2° 34′ Sagittarius. The figures in the second column are also in error. Because of the error in the diurnal motion in case 1, the first longitude should be 21° 10′ Gemini. For the second longitude, Kepler gives the original position from case 1, apparently forgetting that he had changed the position of true opposition to 29° 27′ Leo (29° 25′ corrected). The differences are thus −6, +9$\frac{1}{2}$, and −2: obviously no adjustment of aphelion or mean longitude could accommodate such differences.

[33] Because of Kepler's errors in the diurnal motion and in the change resulting from moving the sun's apogee, nearly all the times and positions in this table are wrong. The correct figures for the first and third dates are obtained from the corrected table under case 3, and the correct figures for the second date are from case 5, corrected. The result is:

	H.	M.		Positions	
Tybi	26	9	0	21°	4'	Gemini
Pharmouthi[34]	4	19	39	29	27	Leo
Epiphi	12	3	37	2	40	Sagittarius
Inter-vals	D. 68 H. 10 M. 39		68°	23'	
	D. 97 H. 8 M. 0		93	13	

The first interval remains the same as in the first case, while the last is much altered. And because more of the path is traversed in a smaller time, it must descend farther down towards perigee. To 8 hours of mean motion correspond 10' 30", to which add the extra 8 minutes of travel.[35] The total is thus $18\frac{1}{2}'$, which we shall make up if we move the aphelion back by 1° 12', putting it at 29° 29' Cancer for the last time, with a mean anomaly of 131° 45'.[36] Therefore, [the position of] its mean motion is 11° 4' Sagittarius,[37] which, in the first case, was 11° $18\frac{1}{2}'$ Sagittarius. From this we compute:[38]

For the first time:	21°	$3\frac{1}{2}'$	Gemini		should	21	4	Gemini
Second	29°	$26\frac{1}{2}'$	Leo		have	29	29	Leo
Third	2°	41'	Sagittarius		been	2	40	Sagittarius

Finally, with alterations in all three data that we have taken from Ptolemy, the result will be the combined effect of the seventh and second cases.

It is therefore apparent that the epoch of the mean motion with respect to the equinox and the fixed stars is not much changed by an alteration in the sun's

Tybi	26	7^h	54^m	21°	4'	Gemini
Pharmouthi	4	22^h	14^m	29	25	Leo
Epiphi	11	23^h	54^m	2	42	Sagittarius
Inter-vals	D. 68 14^h 20^m			68°	21'	
	D. 97 3^h 20^m			93	17	

The mean motion corresponding to the second interval is 96° 3' 8".

[34] In this table, the row heads "Pharmouthi" and "Epiphi" are switched in the 1609 edition, an error corrected in *KGW*. However, while the 1609 edition correctly states the day of Pharmouthi as "IV," *KGW* incorrectly has "VI."

[35] The correct figures are: $12\frac{1}{3}$ hours of mean motion, to which 16' 10" correspond, extra travel being 12'. The total is thus 28' 10".

[36] These figures are corrected as follows. The first requisite is to find two mean anomalies at an interval of 96° 3' 8" which yield equated anomalies at an interval of 93° 17'. After several trials, the following are found:

Mean anomalies:	36°	11'		132°	14'	8"
Equated anomalies:	30°	27'	43"	123°	44'	43"

If we now place the aphelion at 28° 56' 0" Cancer and 28° 58' 40" Cancer, respectively, the required longitudes are obtained, approximately (the slight discrepancy results from neglecting motion of the aphelion in the first computation). Thus the aphelion is moved back 1° 42' 20", not 1° 12' as in Kepler's computations.

[37] Kepler's figures would require this to be 11° 14'. The translator's revision results in nearly the same value, 11° $12\frac{3}{4}'$.

[38] Again, almost everything changes. The table should be:

For the first time:	21°	1'	Gemini		should	21°	4'	Gemini	−3'
Second	29	24	Leo		have	29	25	Leo	−1
Third	2	43	Sagittarius		been	2	42	Sagittarius	+1

Note that the second longitude on the right has inadvertently been increased by 2'.

eccentricity, or in the apogee, or in both at once, but that it is only changed when the positions of the fixed stars are altered. For the third case adds 1′ 30″, the fifth subtracts 3′ 30″, and the seventh subtracts 4′ 30″. The second case alone subtracts 10′[39] from the mean motion measured from the equinoctial, and 20′ measured from the fixed stars.[40]

As a result of this, two epochs of motion at the time of Ptolemy are estab-   332
lished.

But what if we make something suitable by combining the second and fifth cases, by which we can hold simply to the Ptolemaic longitude of the fixed stars, eliminating any need for us to suspect that there might be two epochs of Mars's mean motion? For Ptolemy explicitly affirms that in his observation he found the moon's distance from the sun to be 92° 8′, the same amount he computed from his hypothesis of the moon's motions. Ptolemy would have spoken truly: he would have been skillful enough in observing, and would have plainly seen this distance on his instrument to be the same as that prescribed by his hypothesis of the moon's motions, which was not in error near the quadratures. From this, I argue thus. If the sun had been at 3° 5′ Pisces,[41] where Ptolemy placed it using his eccentricity, the moon could not have appeared 92° 8′ from it, the measure that is just and computed from the hypothesis, for the reason that the setting sun reaches the eye by refraction, and appears higher than it actually is (and thus 30′ farther to the east). But because the arc from the moon to the sun was observed to be 92° 8′, and because of refraction, this was in actual fact 92° 38′, the sun was therefore not at 3° 3′ Pisces, but at 2° 33′ Pisces. And this is in agreement with the fifth case, where we said that Ptolemy's maximum additive equation (which occurs at 5° Pisces) becomes 20′ smaller when today's eccentricity is used, thus putting the sun at 2° 43′ Pisces instead of 3° 3′ Pisces. And so, on the supposition that refraction is universal throughout all places and times, as is discussed in the *Optics*,[42] and supposing this observation to stand, we arrive at the conclusion that the sun's eccentricity is less than that reckoned by Ptolemy.

It should not trouble you that I spoke of a refraction of 30′, while this diminution is but 20′. For if you consider well, since 30° Taurus had culminated, 1° Pisces was then setting at Alexandria, and the sun, being at 3°

How it happens that through two errors that cancel one another the Ptolemaic elongation of the fixed stars from the beginning of Aries remains the same.

---

[39] In the second case Kepler *added* 10′ to the mean motion measured from the equinox, and subtracted 20′ from the mean motion measured with respect to the fixed stars. Here again one can see the haste with which this chapter was composed.

[40] With the data corrected, the third and fifth cases could not produce any satisfactory theory. However, Kepler's evaluation (with corrections) holds for the remaining cases.

[41] The longitude given by Ptolemy is 3° 3′ Pisces, as Kepler has it below.

[42] Kepler's conclusion in the *Optics* is carefully nuanced. He writes,

> And so, to finish up this section, let this be certain, that in different places and weather the refractions are different, and that on extraordinary occasions they are extraordinary. For example, if the altitude of the place be high, the refraction will be none; if there is an exceptional releasing of vapors, the refraction will be prodigious. If on the other hand, the places and weather keep themselves moderate, the refractions will be approximately the same. (English trans., p. 150)

The constancy and variability of atmospheric refraction is discussed in *Optics* Ch. 4 Sect. 8–9, English trans. pp. 146–155.

Pisces, consequently had an altitude of two degrees, or perhaps even more, and therefore the refraction was less than 30′; nor was all the refraction simply longitudinal. Thus these two causes[43] were almost exactly the same in quantity, and cancelled one another.

However, anyone who knows anything of the Ptolemaic reckoning of the fixed stars will not consider this difference of ten minutes worth mentioning.

Ptolemy's values for the positions of the fixed stars are not scrupulously accurate.

For example, Ptolemy comes up with an interval of 54° 10′ between Cor Leonis and Spica Virginis, although in the heavens themselves it is not more than 53° 59′.

Establishment of the mean motion.

Let us therefore follow whither our inclinations and our arguments lead: as in the first case, in the second year of Antonine, on the 12th day of Epiphi, at the 8th hour, at Alexandria in Egypt, let Mars's mean motion from the equinoc-

Epoch of the mean motion at the time of Ptolemy.

tial be 11° 18′ 30″ Sagittarius. This time corresponds to the common year of Christ 139 May 27. The difference of meridians between Hven and Alexandria is nearly two hours, from the most recent geograpical tables. Therefore, at Hven in the year of Christ 139 May 27 at 6^h the mean motion was 8^s 11° 18′ 30″. But in that year, Cor Leonis had a longitude of 2° 30′ Leo, that is, 4^s 2° 30′ 0″. Therefore Mars's mean motion was 4^s 8° 48′ 30″ from Cor Leonis. But on

333   1599[44] May 27 at 6^h Mars's mean motion was 0^s 0° 47′ 30″ from the equinoctial, while the distance of Cor Leonis from that point, as demonstrated by Brahe, was 4^s 24° 15′ 45″. Therefore Mars was 7^s 6° 31′ 45″ from Cor Leonis.

On	139	May	27	at	6^h :	4^s	8°	48′	30″
	1599	May	27	at	6^h :	7	6	31	45
An interval of	1460	Julian Years		The Prutenics	2	27	43	15
	1461	Egyptian Years		give	2	28	5	56
				Difference			22	41

For each year nearly one second must be subtracted. Therefore, at noon on 1 January in the first year of Christ, at Hven, it is elongated in its mean motion by 5^s 8° 52′ 45″ from Cor Leonis.[45]

And so much for Mars's mean motion with respect to the fixed stars.

The motion of the aphelion will come out a little different from what it was above in Ch. 17. For in the year of Christ 139 May 27 it was at 0° 41′ Leo, while Cor Leonis was at 2° 30′ Leo. It therefore preceded the latter by 1° 49′. But

---

[43] That is, refraction and the altered eccentricity.

[44] This year was clearly chosen advisedly: it is 1460 Julian years from Ptolemy's date, and also exactly 1461 Egyptian (that is, 365 day) years from that date, because 1460 times $365\frac{1}{4}$ is equal to 1461 times 365.

[45] This should have been 9^s 8° 52′ 45″, as the following computations show:

First, Kepler's value for Mars's mean diurnal motion is required. In 1460 Julian years Mars completes about 776.25 sidereal cycles. By Kepler's figures here (which are consistent with the mean longitudes and dates elsewhere in this work), the total mean motion is thus 87° 43′ 15″ beyond 776 full cycles in 533,265 days. So the mean diurnal sidereal motion is 0.52403162°.

Next, the total mean motion must be counted from a known position to the epoch. On 139 May 27 at 6 pm the elongation from Cor Leonis was 4^s 8° 48′ 30″. From this date back to 1 CE January 1 at noon is 50,550.25 days, to which corresponds a mean motion of 209° 55′ 45″ beyond full cycles. Since we are counting back in time, this must be subtracted from the elongation in 139; or its full circle complement, 150° 4′ 15″, must be added. The sum is 278° 52′ 45″, or 9^s 8° 52′ 45″, exactly as given by Kepler except for the number of signs.

Incidentally, Kepler's value for Mars's sidereal period, computed from the above mean sidereal motion, is 686 days 23^h 33^m 17^s.

today, in 1599 May 27, it is at 28° 58′ 50″ Leo, while Cor Leonis is at 24° 15′ 45″ Leo.

Therefore, today's aphelion follows by . .	4°	43′	5″
While for Ptolemy it preceded by	1°	49′	0″
a progress of	6	32	5

over an interval of 1460 Julian years, which makes an annual motion of a little greater than 16″. Therefore, the root of the Christian era, at noon on January 1, has this aphelion 2° 27′ before Cor Leonis.

On the mean motion of the sun with respect to the fixed stars, treated in passing, for future reference.

In the year of Christ 139 Pharmouthi 9, which is February 23, at sunset at 5^h 30^m, 3^h 30^m at Hven, the apparent position of the sun was computed as 3° 3′ Pisces; therefore, the mean position was 0° 43′ Pisces. But the longitude of Cor Leonis was found to be 2° 30′ Leo. Therefore, the mean motion of the sun preceded Cor Leonis by 5^s 1° 47′ 0″. But on 1599 February 23 at 3^h 30^m at Hven the mean motion of the sun was at 12° 47′ 41″ Pisces, and Cor Leonis 24° 15′ 30″ Leo. So the mean motion of the sun preceded Cor Leonis by 5^s 11° 27′ 49″.

Over 1460 Egyptian years,[46] 9° 40′ 49″ are removed.[47]

Our conclusion is 2′ 42″ less than that from the Prutenics over the same number of years, and the epoch will be 5^s 7° 14′ 36″ from Cor Leonis at the root of the Christian era on January 1.[48]

Similarly, the progression of the sun's apogee is found to be 8° 23′, and at the root of the Christian era it was 1^s 27° 48′ 0″ before Cor Leonis.[49]

---

[46]This must be 1461 Egyptian years: see footnote 44.

[47]From this, one can compute a Keplerian value for the sidereal year. The total number of cycles (1460) is multiplied by 360, and 9° 40′ 49″ are subtracted, to find the total angular motion in degrees. This is then divided by the number of days (533,265), the quotient being the mean diurnal motion, 0.9856081°. The inverse of this, multiplied by the 360° in the full circle, gives the number of days in a sidereal year as 365.2567. The modern value is 365.2564.

[48]This is computed as follows:

The time interval from Ptolemy's date back to 1 January of the year 1 is $50,456^d$ $15\frac{1}{2}^h$. Using the mean diurnal sidereal motion computed from Kepler's data in the preceding footnote, and subtracting full cycles, one finds the remaining angle to be about 50° 28′ 45″. This must be subtracted from the position in Ptolemy's observation, so that the sun's mean position is now 6^s 22° 15′ 45″ before Cor Leonis, or 5^s 7° 44′ 15″ beyond it. It would thus appear that the figure in the text is incorrect: very likely the minutes were inadvertently changed from 44 to 14. This was not a printer's error, however, since Kepler uses the same elongation in Chapter 70: see footnote 7 to that chapter.

[49]The computation would have been something like this:

Date	Position of Apogee	Position of Cor Leonis			Elongation		
1599	$95\frac{1}{2}°$	144°	15′	45″	−48°	45′	45″
139	$65\frac{1}{2}°$	122°	30′	0″	−57°	0′	0″
1461 Egyptian years					+8°	14′	15″.

Motion over the additional 139 years and several months would be about 46′ 50″, resulting in an elongation of −57° 46′ 50″ at the beginning of the Christian era. This agrees well with Kepler's elongation; however, it is not clear how he arrived at the figure of 8° 23′ for the progression.

CHAPTER 70

Consideration of the remaining two Ptolemaic observations, in order to investigate the latitude and ratio of the orbs at the time of Ptolemy.

334   It is true, as I have more than once remarked, that Ptolemy had at his disposal many more observations than were presented in his Opus. This may be seen in the presentation of the method for investigating the ratio of the orbs, where he uses a single observation that is within three days of the opposition. For it was said in Chapter 53 that observations that are so close result in a very large error if they are off by even one minute. Nevertheless, let us follow his footprints, and upon the hypothesis just established, erected upon the foundation of the first case, let us compute this fourth position[1] as well.

Epiphi	12	8^{h2}		— Anomaly	130° 37′ 30″	
	15	9^h				
days	3	1^h		Mean motion	1° 35′ 39″	
Equated [Anomaly]	123°	43′	34″	[Mean] Anomaly	132 13 9	
Aphelion	120	41	0			
Eccentric position	4° 24′ 34″ Sag.			Distance	143,660	

The sun's true position on the 12th was 2° 36′ Gemini.[3] Add the motion of the three days and one hour, near apogee, which is 2° 53′ 40″ from modern experience. This makes it 5° 29′ 40″ Gemini, and let the present distance of the apogee be used, 101,800. Therefore, the point opposite the sun and Mars's eccentric position differ by 1° 5′ 6″. This arc appears to be 3° 43′ 14″, so that Mars would appear at 1° 46′ 26″ Sagittarius.

If, however, we use the Ptolemaic eccentricity of the sun, the sun's motion over the three days will be 1′ smaller, and the sun will be at 5° 28′ 40″ Gemini. The difference is thus 1° 4′ 6″. The apparent magnitude of this arc, using the Ptolemaic distance of the sun and earth, 102,100, will be 3° 45′ 45″,[4] so the planet will fall at 1° 43′ Sagittarius. But Ptolemy said that it was seen at 1° 36′ Sagittarius. We have therefore come out 7′ to 10′ beyond the correct figure. But the least division of the Ptolemaic instrument, which can always be considered as the uncertainty, has the value 10′.

---

[1] This is the observation from the *Almagest* Book 10 Chapter 8, of 2 Antonine, 15/16 Epiphi, 3 hours before midnight (139 May 30 at 9 pm), when the sun's mean position was 5° 27′ Gemini and Mars was observed at 1° 36′ Sagittarius.

[2] This is the time of the true opposition, from "case 1" of Chapter 69.

[3] This is the position of true opposition, from Chapter 69.

[4] The apparent arc resulting from this distance is 3° 41′ 16″. It is not clear how Kepler obtained his figure, which is consistent with a distance of 102,900. However, a complete recomputation from Ptolemy's data yields an arc of 3° 43′ 5″, which would place the planet at 1° 46′ 18″ Sagittarius, a change that does not affect Kepler's conclusions.

You should also note that if we have erred by two minutes in the eccentric position, we shall now err by seven minutes in the observed position. For let Mars be moved back to 4° 22′ Sagittarius on the eccentric: it will now appear at 1° 36′ Sagittarius.

Above, on Epiphi 12, there was also an excess of $1\frac{1}{2}'$. So these results are in agreement.

And because at such proximity to opposition a difference in the eccentricity has little effect, let us also consult the more ancient observation.[5] Between the morning of January 18 during the year 272 before Christ,[6] and noon on January 1 of the year 1 of Christ, there are 272 Egyptian years, 51 days, and several hours. For since at Alexandria, the sun at 25° Capricorn rises at 7^h, and the morning observation of Mars was made one hour earlier, as dawn was breaking, it was therefore made at the sixth hour, which is the fourth hour at Hven, from which time there are eight hours until noon. From this time interval, by the principles laid down above, the sun's mean motion is found to have gone 5ˢ 25° 32′ 50″ beyond Cor Leonis,[7] with an anomaly of 234° 54′ 34″. The corresponding equation from Ptolemy is 2° 0′ 30″, and from Brahe, 1° 42′ 54″, additive; and for the former, the sun's distance from the earth is 98,790, and for the latter, 98,976. But Mars's mean motion was then 2ˢ 6° 7′ 12″ beyond Cor Leonis. Also, since the aphelion is 3° 40′ 20″ before Cor, Mars's anomaly will be 69° 47′ 32″, the equated anomaly 60° 15′ 27″, and the distance 158,320.

An investigation of the ratio of the orbs through the more ancient observation.

335

Here we shall follow a twofold path to the end of the calculation. The first is through the Ptolemaic eccentricity and equation. Then the sun's elongation from Cor Leonis is 5ˢ 27° 33′ 20″,[8] differing from Mars's eccentric longitude of 1ˢ 26° 35′ 7″ by 4ˢ 0° 58′ 13″. This arc length, and the distances of the earth and Mars from the sun, show an apparent elongation from the sun of 82° 43′ 46″. Thus the apparent elongation of Mars from Cor Leonis is 3ˢ 4° 49′ 34″.[9]

And the second path is through the Brahean eccentricity and equations, if they are assumed to have been the same then as well. The sun's apparent position will be 17′ 36″ farther back, or 5ˢ 27° 15′ 44″.[10] Thus the angle of commutation is 4ˢ 0° 40′ 37″. Through this, together with our value for the sun's distance from earth, as if it too were the same then, Mars's apparent

---

[5]That is, the Chaldean observations from the *Almagest* Book 10 Chapter 9, of 476 Nabonassar, on the morning of Athyr 21 (18 January 272 BCE, historical style) when Mars was observed to occult β Scorpii (which, according to Ptolemy, was at 6° 20′ Scorpio).

[6]This is according to the historical style; in the astronomical reckoning it would be −271.

[7]This shows two things: first, that Kepler used the elongation given in Chapter 69, 5ˢ 7° 14′ 36″, and thus that the incorrect number of minutes given there (the correct number being 44′) was not a printer's error; and second, that he used a diurnal motion here that differs slightly from the one resulting from his data in Chapter 69 (see footnote 46 of that chapter). These data result in a figure of 0.985607°, corresponding to a sidereal year of 365.2572 days. The elongation should have been about 30′ greater.

[8]This is again about 30′ too small, owing to the error in Chapter 69, and subsequent figures should be adjusted accordingly.

[9]Recomputation using corrected data shows this to be about 3ˢ 4° 56′.

[10]Again, too small: this elongation should have been about 30′ greater.

elongation from Cor Leonis is shown as 3^s $4°$ $51'$ $28''$.[11] The difference between the two calculations is very small and of no significance. Is it then true that

Mars was observed as if placed upon or fitted to the northern star in the brow of Scorpius,[12]

as the description of the observation proclaims? Let us see. For Ptolemy, Cor Leonis is at $2°$ $30'$ Leo, and the Northern Bright Star in the Brow of Scorpius is at $6°$ $20'$ Scorpio, at an elongation of 3^s $3°$ $50'$ $0''$. For Brahe, Cor Leonis is at $24°$ $17'$ Leo. The brow of Scorpius is at $27°$ $36'$ Scorpio. The elongation is 3^s $3°$ $20'$ $0''$.[13] But Mars's elongation was just computed to be 3^s $4°$ $51'$ $28''$. The difference is a degree and a half.

That Ptolemy, although he pretended to have tested it with another observation, here seems to have demonstrated an erroneous ratio of the orbs using an erroneous observation.

Since he had confidence in this observation, it being the most ancient of those upon which he could have depended, Ptolemy doubtless established that ratio of the orbs which we have hitherto discovered in his numbers, and which this observation appeared to require. For in the mean motion computed for this time, he differs from me by no more than 20 minutes. The remaining discrepancy therefore comes from the ratio of the orbs. Now because he pretends to investigate this ratio using an observation three days from opposition, he made different things seem to be deduced from different phenomena. And so since the ancient observation was to be reserved for investigating the mean motions, he substituted the other one for finding the ratio of the orbs, which had already been found using the ancient one. As was just said, it is absurd to test the ratio of the orbs using an observation as close to opposition as the one by which Ptolemy pretends to have demonstrated this ratio.

336   Let no one therefore wonder at our differing by a degree and a half from the observation that Ptolemy summoned from antiquity: let him rather examine Ptolemy's ratio of the orbs, so different from those proven by present day observations, and consider that in order to keep this observation, Ptolemy corrupted the ratio of his orbs.

Ptolemy did not correctly understand the words describing the observation.

As for the observation itself, of which this is the verbal description: ἑῷος ὁ τοῦ Ἄρεως ἐδόκει προστεθεικέναι τῷ βορείῳ μετώπῳ τοῦ σκορπίου· (In the morning, the [star] of Mars appeared to have just come upon the northern brow of Scorpius),[14] I believe that an error was committed by Ptolemy, who understood the first star of Scorpius, while the observer nodded towards the fifth.[15] This is proven from the words themselves. For the brow of Scorpius has six bright stars. Of these, there are three prominent ones, of third magnitude, or

---

[11] Recomputation using corrected data shows this to be about 3^s $4°$ $58'$. It is therefore true, as Kepler says, that the difference between the two methods of computation is insignificant. However, the elongations are greater than Kepler's by an amount not entirely negligible.

[12] β Scorpii.

[13] This is what appears in the text, although the difference between the longitudes is obviously 1' less.

[14] *Almagest* 10.9. Kepler's Greek has a slightly different word order from Heiberg's text, and where Kepler has προστεθεικέναι, Heiberg has ἐπιπροσθετεικέναι. Heiberg also notes προστεθεικέναι as appearing in three of the manuscripts.

[15] The numbering represents the order in which the stars appear in both Ptolemy's and Tycho's tables. The fifth star is ν Scorpii.

better, second. The remaining three are of fourth, or, by my estimate, third magnitude, and one is higher than the three bright ones, and farther north. Now if the observer called the "Bright Star in the Brow" (which Brahe correctly pronounces to be of second magnitude, and which Ptolemy understood to be the intended star) the "Northern Brow", did he not speak ambiguously in saying simply "northern" rather than "brightest of the northern", since the star was not the northernmost? Thus I, much more prudently, will take it to be the northernmost, the fifth in number, that was described by the observer.

Furthermore, my computed longitude of Mars agrees with this, and not with the Bright Star of the Brow; and this while the hypothesis remains valid which the modern Brahean observations have generated. For Brahe places the northernmost star at 29° $3\frac{1}{2}'$ Scorpio. Subtract Cor Leonis at 24° 17' Leo. The difference will be its elongation from Cor, 94° $46\frac{1}{2}'$. But our calculation puts Mars at an elongation of 94° $49\frac{1}{2}'$ or 94° $51\frac{1}{2}'$ from Cor Leonis.[16] The difference is 3 or 5 minutes, not greater.

I do not deny that the latitude presents a difficulty for me, in that I interpret the words, ἐδόκει προστεθεικέναι, as if he said, "It appeared to approach so near that the two stars could be taken as if they were one, that they appeared to touch one another." The Arabic, however, translates it [with a word signifying] "to have covered up," as if the Greek had read, ἐπιπροστεθεικέναι. Accordingly, in the *Optics*, p. 304, I used the word, 'superimposed'.[17] The best word in German is "drangesetzt." From this I reasoned as follows. Whether Mars ran beneath it centrally, or grazed its northern or southern margin, it could not have been removed from the star latitudinally by any great distance. For indeed, the latitudes are less uncertain than the longitudes, since the manner of their variation is simpler and more consistent, as is proven in this book. We now know that the node retrogresses with respect to the fixed stars, by 4° 15' during one "year of the Dog,"[18] as was proven in Ch. 17. For Ptolemy, the northern limit was considered to precede Cor Leonis by $3\frac{1}{2}$ degrees. For us, over the intervening 410 years, it had retrogressed one degree, so that at the time of observation it would be $2\frac{1}{2}°$ before Cor Leonis. Therefore, the node is $87\frac{1}{2}°$ past Cor Leonis. But Mars is 56° 35' past Cor Leonis. Therefore, it is 31° from the node, making the inclination $57\frac{1}{2}'$, which, by the parallax of the orb, results in a true latitude of 1° 7'.[19] But now it is clear from Brahe that the latitude of the Bright Star of the Brow is 1° 5', while that of the Northernmost

Whether Mars's latitude would have allowed it to cover up the star?

337

---

[16] The recomputed elongations were 94° 56' and 94° 58'. The difference is thus somewhat greater than Kepler claims, although his argument retains its plausibility.

[17] In the passage cited, Kepler used "placed beneath" (*suppositum*) rather than "superimposed" (*superpositum*). See *Optics*, English trans. p. 315: where Kepler quotes Ptolemy's Greek, the translation is given in quotation marks; hence, the apparent difference in meanings.

[18] Since the Egyptians reckoned their year as 365 days, the seasons are displaced in their time reckoning. They found that after 1461 such years, that is, after 1460 tropical years, the first appearance of Sirius, which was most important to them, fell once again at the same season. Hence, this length of time was called an *annus cynicus* (year of the dog). Exactly one such dog-star period lies between the times of writing of the *Almagest* and of *Astronomia Nova*, a coincidence that proved useful to Kepler in the previous chapter.

[19] That is, Mars's inclination from the ecliptic has been optically lengthened by the earth's proximity to Mars. The method for computing this "parallax of the orb" is to be found in Chapter 62.

Star of the Brow is 1° 42'. So the latitude appeared to refute me concerning the Bright Star of the Brow, leading me to believe that this star was occulted by Mars, and not the other.

But this collusion of numbers is fortuitous. For in the latitude of the Northernmost Star of the Brow Brahe and Ptolemy are in agreement, the former pronouncing it to be 1° 46', and the latter, 1° 42'. In the latitude of the Bright one they differ. Ptolemy has 1° 20'; Brahe, 1° 5'. But the former numerical equality results from an error, and the latter difference is really more like an agreement. For the latitudes of the northern stars in Scorpius, Sagittarius, Capricorn, and Aquarius are smaller today than they once were by about 16' 20", and those of the southern stars are greater by the same amount, since the ecliptic has been transposed and the declinations of the degrees of the ecliptic have been altered by the same amount, as Brahe proved and as we have said in Ch. 68. Thus if it is true—and it is very true—that the latitude of the Bright Star in the Brow of Scorpius is 1° 5' today, at the time of Ptolemy and Hipparchus it was no less than 1° 20', probably greater. And so Mars has a smaller northern latitude than either of the stars mentioned, and passed beneath both. For it is certain that even if we are too high by a whole degree in [the position of] the node, the latitude in the calculation was wrong by no more than three minutes. Also, it has now been shown in Ch. 64 to be entirely uncertain whether the northern latitude for Mars was also once greater in the southern signs. Therefore, my clever interpretation of the word "προστεθεικέναι was in vain. It can only be explained as denoting the stars' being placed side by side in the same longitude; and on this ground, the one that I favor is just as good a candidate as the Bright, its greater latitude notwithstanding.

The words of the observation have their common signification. Consider whether the meaning could be this: that since in the northern part of the brow there are three stars in the form of a triangle, Mars was sighted in the middle of them, and was thus "placed upon the northern brow" of Scorpius, it having simply been made one of that number of stars that are in the northern part of the brow of Scorpius.

This interpretation is furthered by the observer's having said "northern brow" rather than "northern star of the brow", since he is denoting, not one single star, but an entire part of the constellation.

So these two ancient observations are of no use to us in estimating either the latitude or the ratio of the orbs at that time. Therefore, since there are no observations to the contrary to impede us, while the extreme likelihood of our position confirms us in it, let us conclude that the ratio of the orbs is also the same today as it was once, while the maximum latitudes today are somewhat altered.

Appendix A
Observations Used or Cited

TABLE 1: OBSERVATIONS BY ORDER OF APPEARANCE IN *ASTRONOMIA NOVA*

Chapter	Date	Time	Position(s)	Source
Ch. 8	1580 Nov. 17	9h 40m	6°46′10″ Gem.	Brahe; unknown
Ch. 8	1582 Dec. 28	12h 16m	16°46′10″ Can.	Brahe; unknown: cf. *TBOO* 10 196–203
Ch. 8	1585 Jan. 31	19h 35m	21°10′26″ Leo	Brahe; unknown
Ch. 8	1587 Mar. 7	17h 22m	25°10′20″ Vir.	Brahe; unknown
Ch. 8	1589 Apr. 15	13h 34m	3°58′10″ Sco.	Brahe; unknown
Ch. 8	1591 Jun. 8	16h 25m	26°32′0″ Sag.	Brahe; unknown
Ch. 8	1593 Aug. 24	2h 13m	12°43′45″ Pis.	Brahe; unknown
Ch. 8	1595 Oct. 29	21h 22m	17°56′15″ Tau.	Brahe; unknown
Ch. 8	1597 Dec. 13	13h 35m	2°28′0″ Can.	Brahe; unknown
Ch. 8	1600 Jan. 19	9h 40m	8°18′0″ Leo	Brahe; unknown
Ch. 10	1580 Nov. 12	10h 50m	8°36′50″ Gem.	*TBOO* 10 84
Ch. 10	1582 Dec. 28	11h 30m	16°47′ Can.	*TBOO* 10 176–7
Ch. 10	1585 Jan. 31	12h 0m	21°18′11″ Leo	*TBOO* 10 388
Ch. 10	1587 Mar. 7	19h 10m	25°10′20″ Vir.	*TBOO* 11 184 (under 5 March)
Ch. 10	1589 Apr. 15	12h 5m	3°57′11″ Sco.	*TBOO* 11 333
Ch. 10	1591 Jun. 6	12h 20m	27°15′ Sag.	*TBOO* 12 142
Ch. 10	1593 Aug. 24	10h 30m	12°38′ Pis.	*TBOO* 12 288
Ch. 10	1595 Oct. 30	8h 20m	17°48′ Tau.	*TBOO* 12 453
Ch. 10	1597 Dec. 10	8h 30m	3°45′30″ Can.	*TBOO* 13 112
Ch. 10	1600 Jan. 13	11h 50m	10°48′36″ Leo	*TBOO* 13 221
Ch. 11	1582 Nov. 23	all	constant	*TBOO* 10 174
Ch. 11	1582 Dec. 26	8h 28m	17°38′ Can.	*TBOO* 10 176; cf. 10 196
Ch. 11	1582 Dec. 27	19h 15m	17°28′20″ Can.	*TBOO* 10 177
Ch. 11	1582 Dec. 29	7h 47m	29°38′30″ from southern foot of Erichthonius	*TBOO* 10 177
Ch. 11	1582 Dec. 30	8h 8m	29°13′30″ from southern foot of Erichthonius	*TBOO* 10 178
Ch. 11	1583 Jan. 16	7h 30m	23°29′ from bright star in foot of Erichthonius	*TBOO* 10 242–3

Continued on next page

Chapter	Date	Time	Position(s)	Source
Ch. 11	1583 Jan. 16	10h 30m	23°27′ from bright star in foot of Erichthonius	*TBOO* 10 243
Ch. 11	1583 Jan. 17	10h 36m	23°12′30″ from bright star in foot of Erichthonius	*TBOO* 10 244
Ch. 11	1583 Jan. 17	5h 20m	23°16′ from foot of Erichthonius	*TBOO* 10 243
Ch. 11	1583 Jan. 17	15h 0m	23°9′ from foot of Erichthonius	*TBOO* 10 244
Ch. 11	1583 Jan. 18	5h 5m	23°1′30″ from foot of Erichthonius	*TBOO* 10 244
Ch. 11	1583 Jan. 18	16h 45m	44°27′20″ from Regulus	*TBOO* 10 245
Ch. 11	1583 Jan. 18	16h 52m	7°59′ from 7 Gem.	*TBOO* 10 244
Ch. 11	1583 Jan. 18	7h 34m	7°51′ from 7 Gem.	Not found
Ch. 11	1583 Jan. 18	8h 52m	44°22′ from Regulus	*TBOO* 10 244
Ch. 11	1583 Jan. 19	7h 3m	44°32′30″ from Regulus	*TBOO* 10 245
Ch. 11	1604 Feb. 17	Corvus culminating	26°56′ Lib.	Kepler's obs.
Ch. 11	1604 Feb. 19	Mars rising, alt. 11°	29°18′20″ from Arcturus	Kepler's obs.
Ch. 11	1604 Feb. 22	ζ Leonis culminating	29°15′ or 19′ from Arcturus	Kepler's obs.
Ch. 11	1604 Feb. 29	half hour before Cor Hydrae culm.	29°10′ from Arcturus	Kepler's obs.
Ch. 11	1604 Feb. 29	half hour before Cor Hydrae culm.	9°26′+ from Spica	Kepler's obs.
Ch. 12	1590 Mar. 4	7h 10m	24°22′56″ Ari.	*TBOO* 12 44
Ch. 12	1592 Jan. 23	10h 15m	11°34′30″ Ari.	*TBOO* 12 219
Ch. 12	1593 Dec. 10	eve.	at the node, lat 0°	*TBOO* 12 296
Ch. 12	1595 Oct. 27	12h 20m	lat. 2′20″ N.	*TBOO* 12 450
Ch. 12	1595 Oct. 28	eve.	lat. 0′45″ S.	*TBOO* 12 451–2
Ch. 12	1595 Jan. 4	19h 10m	13°36′40″ Sag.	*TBOO* 12 441
Ch. 12	1589 Apr. 15	night	lat. 1°7′ N.	*TBOO* 11 333
Ch. 12	1589 May 6	11h 15m	lat. 6′40″ N.	*TBOO* 11 334
Ch. 13	1588 Nov. 10	18h 30m	25°30′ Vir.	*TBOO* 11 275
Ch. 13	1588 Dec. 5	18h	9°19′24″ Lib.	*TBOO* 11 276
Ch. 13	1586 Oct. 22	18h	0°7′ Vir.	*TBOO* 11 64
Ch. 13	1586 Nov. 2	16h 40m	5°52′ Vir.	*TBOO* 11 64
Ch. 13	1586 Dec. 1	19h 30m	20°4′30″ Vir.	*TBOO* 11 66
Ch. 13	1583 Apr. 22	9h 45m	1°17′ Leo	*TBOO* 10 250
Ch. 13	1596 Mar. 9	8h	15°49′ Gem.	*TBOO* 10 47
Ch. 13	1589 Sep. 15	7h 15m	16°47′20″ Sag.	*TBOO* 11 335
Ch. 13	1589 Nov. 1	6h 20m	20°59′15″ Cap.	*TBOO* 11 335
Ch. 13	1584 Nov. 12	13h 30m	23°14′ Leo	*TBOO* 10 321
Ch. 13	1585 Apr. 26	9h 42m	21°26′ Leo	*TBOO* 10 399
Ch. 13	1591 Oct. 16	7h 30m	1°27′20″ Aqu.	*TBOO* 12 146

Continued on next page

Chapter	Date	Time	Position(s)	Source
Ch. 13	1591 Oct. 10		lat. 2°18′40″ N.	*TBOO* 12 146
Ch. 13	1591 Oct. 2		lat. 2°38′30″ N.	*TBOO* 12 146
Ch. 13	1593 Aug. 24		lat. 6°3′ N.	A. N. Ch. 8 obs. 7
Ch. 15	1580 Nov. 12	10h 50m	8°37′ Gem.	*TBOO* 10 84
Ch. 15	1582 Dec. 28	11h 30m	16°47′ Can.	*TBOO* 10 176–7
Ch. 15	1585 Jan. 31	12h 0m	21°18′11″ Leo	*TBOO* 10 388
Ch. 15	1587 Mar. 4	13h 16m	26°26′17″ Vir.	*TBOO* 11 184
Ch. 15	1589 Apr. 15	12h 5m	3°58′20″ Sco.	*TBOO* 11 333
Ch. 15	1591 Jun. 6	12h 20m	27°15′ Sag.	*TBOO* 12 142
Ch. 15	1593 Aug. 24	10h 30m	12°38′ Pis.	*TBOO* 12 288
Ch. 15	1595 Oct. 30	8h 20m	17°48′ Tau.	*TBOO* 12 453
Ch. 15	1597 Dec. 10	8h 30m	3°45′30″ Can.	*TBOO* 13 112
Ch. 15	1600 Jan. 13	11h 50m	10°48′36″ Leo	*TBOO* 13 221
Ch. 15	1602 Feb. 18	10h 30m	13°19′30″ Vir.	Kepler's observation
Ch. 15	1602 Feb. 15	17h 0m	14°21′18″ Vir.	Fabricius's observation
Ch. 15	1602 Feb. 23	12h 0m	11°19′20″ Vir.	Fabricius's observation
Ch. 15	1604 Mar. 29	9h 43m	18°21′30″ Lib.	Kepler's observation
Ch. 19	1585 Jan. 31	12h 0m	21°18′11″ Leo	*TBOO* 10 388
Ch. 19	1593 Aug. 24	10h 30m	12°38′ Pis.	*TBOO* 12 288
Ch. 19	1582 Dec. 28	11h 30m	16°47′ Can.	*TBOO* 10 176–7
Ch. 20	1600 Mar. 5	12h	29°12′30″ Can.	*TBOO* 13 224 (under March 15)
Ch. 20	1593 Jul. 30	13h 45m	17°39′30″ Pis.	*TBOO* 12 284
Ch. 22	1585 May 18	10h 30m	0°50′45″ Vir.	*TBOO* 10 402
Ch. 22	1591 Jan. 22	19h	22°33′ Sco.	*TBOO* 12 138
Ch. 24	1590 Mar. 4	7h 10m	24°22′ Ari.	*TBOO* 12 44
Ch. 24	1592 Jan. 23	7h 15m	11°34′30″ Ari.	*TBOO* 12 219
Ch. 24	1593 Dec. 10	7h 20m	4°45′ Ari.	*TBOO* 12 296
Ch. 24	1595 Oct. 27	12h 20m	18°52′15″ Tau.	*TBOO* 12 450
Ch. 26	1590 Mar. 4	7h 10m	24°22′56″ Ari.	*TBOO* 12 44
Ch. 26	1592 Jan. 23	7h 20m	11°32′44″ Ari.	*TBOO* 12 219
Ch. 26	1593 Dec. 7	8h 0m	3°6′50″ Ari.	*TBOO* 12 296
Ch. 26	1595 Oct. 25	8h 10m	19°39′25″ Tau.	*TBOO* 12 447
Ch. 27	1585 May 7	11h 26m	25°55′ Leo	*TBOO* 10 401
Ch. 27	1585 May 12	10h 8m	28°3′30″ Leo	*TBOO* 10 401
Ch. 27	1587 Mar. 27	9h 40m	18°21′45″ Vir.	*TBOO* 11 187
Ch. 27	1587 Apr. 1	9h 30m	17°11′ Vir.	*TBOO* 11 188
Ch. 27	1589 Feb. 11	17h 13m	8°48′ Sco.	*TBOO* 11 330
Ch. 27	1590 Dec. 27	19h 8m	8°6′ Sco.	*TBOO* 12 45
Ch. 27	1591 Jan. 4	18h 50m	12°44′24″ Sco.	*TBOO* 12 138
Ch. 28	1583 Apr. 22	9h 40m	1°17′ Leo	*TBOO* 10 250
Ch. 28	1585 Mar. 9	9h 10m	11°49′6″ Leo	*TBOO* 10 395
Ch. 28	1585 Mar. 11	5h 0m	11°45′30″ Leo	*TBOO* 10 395
Ch. 28	1585 Mar. 12	5h 0m	11°45′45″ Leo	*TBOO* 10 396
Ch. 28	1587 Jan. 25	17h 0m	4°41′45″ Lib.	*TBOO* 11 180

Continued on next page

Chapter	Date	Time	Position(s)	Source
Ch. 28	1587 Jan. 27	17h 0m	4°41'0" Lib.	*TBOO* 11 180
Ch. 28	1588 Dec. 4	18h 30m	9°23'0" Lib.	*TBOO* 11 276
Ch. 28	1588 Dec. 14	18h 12m	14°35'40" Lib.	*TBOO* 11 276
Ch. 28	1590 Oct. 30	18h 15m	2°57'20" Lib.	*TBOO* 12 44
Ch. 42	1585 Feb. 17	10h 0m	15°12'30" Leo	*TBOO* 10 392
Ch. 42	1586 Dec. 26	16h 0m	29°42'40" Vir.	*TBOO* 11 66
Ch. 42	1586 Dec. 31	19h 8m	1°4'36" Lib.	*TBOO* 11 177
Ch. 42	1588 Nov. 9	18h 30m	25°31'0" Vir.	*TBOO* 11 275
Ch. 42	1588 Dec. 4	18h 0m	9°19'24" Lib.	*TBOO* 11 276
Ch. 51	1589 May 6	11h 20m	27°7'20" Lib.	*TBOO* 11 334
Ch. 51	1594 Dec. 27	19h 15m	8°43'18" Sag.	*TBOO* 12 348
Ch. 51	1595 Jan. 4		lat. 9'20"	*TBOO* 12 441
Ch. 51	1595 Dec. 17	7h 6m	11°31'27" Tau.	*TBOO* 12 458
Ch. 51	1597 Oct. 29	17h	12°16' Can.	Kepler's observation; cf. *TBOO* 13 11
Ch. 51	1589 Apr. 5	11h 33m	7°31'10" Sco.	*TBOO* 11 332
Ch. 51	1591 Feb. 18	17h 30m	7°34'20" Sag.	*TBOO* 12 139
Ch. 51	1582 Nov. 11	18h 45m	26°35'30" Can.	*TBOO* 10 174
Ch. 51	1596 Mar. 9	7h 40m	15°49'12" Gem.	*TBOO* 13 47
Ch. 51	1584 Nov. 12	13h 26m	23°14'5" Leo	*TBOO* 10 321
Ch. 51	1589 Feb. 11	17h 13m	8°48' Sco.	*TBOO* 11 330
Ch. 51	1585 Jan. 24	9h 0m	24°9'30" Leo	*TBOO* 10 387
Ch. 51	1586 Dec. 15	18h 30m	26°6'24" Vir.	*TBOO* 11 66
Ch. 51	1588 Nov. 5	18h 50m	23°16' Vir.	*TBOO* 11 275
Ch. 51	1591 May 13	13h 40m	2°24'30" Cap.	*TBOO* 12 140
Ch. 51	1589 Dec. 3	5h 39m	15°25'33" Aqu.	*TBOO* 11 335
Ch. 51	1591 Oct. 16	6h 28m	1°27'18" Aqu.	*TBOO* 12 146
Ch. 51	1593 Sep. 8	10h 38m	8°53'51" Pis.	*TBOO* 12 291
Ch. 51	1595 Jul. 21	14h 40m	4°11'10" Tau.	*TBOO* 12 441
Ch. 51	1593 Jun. 29	13h 30m	13°37'22" Pis.	*TBOO* 12 282
Ch. 53	1582 Nov. 23	4h 0m	26°38'30" Can.	*TBOO* 10 175
Ch. 53	1582 Dec. 26	8h 30m	17°40'30" Can.	*TBOO* 10 176
Ch. 53	1582 Dec. 30	8h 10m	16°0'30" Can.	*TBOO* 10 178
Ch. 53	1583 Jan. 26	6h 15m	8°20'30" Can.	*TBOO* 10 247 et alibi
Ch. 53	1584 Dec. 21	14h 0m	1°13'30" Vir.	*TBOO* 10 332
Ch. 53	1585 Jan. 25	9h 0m	24°7'30" Leo	*TBOO* 10 387
Ch. 53	1585 Feb. 4	6h 40m	19°47'30" Leo	*TBOO* 10 389
Ch. 53	1585 Mar. 12	10h 30m	11°46'0" Leo	*TBOO* 10 396
Ch. 53	1587 Jan. 25	17h 0m	4°42'0" Lib.	*TBOO* 11 180
Ch. 53	1587 Mar. 4	13h 24m	26°25'40" Vir.	*TBOO* 11 182
Ch. 53	1587 Mar. 10	11h 30m	24°5'15" Vir.	*TBOO* 11 186
Ch. 53	1587 Apr. 21	9h 30m	15°48'20" Vir.	*TBOO* 11 188
Ch. 53	1589 Mar. 8	16h 24m	12°16'50" Sco.	*TBOO* 11 330
Ch. 53	1589 Apr. 13	11h 15m	4°43'20" Sco.	*TBOO* 11 333
Ch. 53	1589 Apr. 15	12h 5m	3°58'20" Sco.	*TBOO* 11 333

Continued on next page

Chapter	Date	Time	Position(s)	Source
Ch. 53	1589 May 6	11h 20m	27°7'20" Lib.	*TBOO* 11 334
Ch. 53	1591 May 13	14h 0m	2°20'0" Cap.	*TBOO* 12 140
Ch. 53	1591 Jun. 6	12h 20m	27°15'0" Sag.	*TBOO* 12 142
Ch. 53	1591 Jun. 10	11h 50m	25°2'36" Sag.	*TBOO* 12 143
Ch. 53	1591 Jun. 28	10h 24m	21°10'0" Sag.	*TBOO* 12 144
Ch. 53	1593 Jul. 21	14h 0m	17°45'45" Pis.	*TBOO* 12 283
Ch. 53	1593 Aug. 22	12h 20m	13°10'15" Pis.	*TBOO* 12 287
Ch. 53	1593 Aug. 29	10h 20m	11°14'0" Pis.	*TBOO* 12 290
Ch. 53	1593 Oct. 3	8h 0m	7°50'10" Pis.	*TBOO* 12 293
Ch. 53	1595 Sep. 17	16h 45m	26°7'12" Tau.	*TBOO* 12 444
Ch. 53	1595 Oct. 27	12h 20m	18°51'15" Tau.	*TBOO* 12 450
Ch. 53	1595 Nov. 3	12h 0m	16°18'30" Tau.	*TBOO* 12 455
Ch. 53	1595 Dec. 18	8h 0m	11°40'0" Tau.	*TBOO* 12 460
Ch. 61	1593 Dec. 10	19h 0m	4°44'0" Tau.	*TBOO* 12 296
Ch. 61	1595 Oct. 28	23h 30m	18°35' Tau.	*TBOO* 12 452
Ch. 61	1592 Jan. 23	22h	2'0" S. Lat.	*TBOO* 12 219
Ch. 61	1590 Mar. 4	7h	3'20" S. Lat.	*TBOO* 12 44
Ch. 61	1595 Nov. 3	23h 30m	19'45" N. Lat.	*TBOO* 12 455
Ch. 61	1589 May 6		6'40" N. Lat.	*TBOO* 11 334
Ch. 62	1593 Aug. 25	17h 27m	12°16' Pis.	*TBOO* 12 290
Ch. 62	1593 Aug. 23		6°7$\frac{1}{2}$' [S.] Lat.	*TBOO* 12 290
Ch. 62	1593 Aug. 24		6°5$\frac{1}{2}$' [S.] Lat.	*TBOO* 12 290
Ch. 62	1593 Aug. 29		5°52$\frac{1}{4}$' [S.] Lat.	*TBOO* 12 290
Ch. 62	1593 Jul. 22	2h 0m	17°45'45" Pis.	*TBOO* 12 283
Ch. 66	1593 Aug. 11	1h 45m	16°7' Pis.	*TBOO* 12 285

TABLE 2: OBSERVATIONS BY DATE

Date	Time	Chapter	Position(s)	Source
1580 Nov. 12	10h 50m	Ch. 10	8°36′50″ Gem.	*TBOO* 10 84
1580 Nov. 12	10h 50m	Ch. 15	8°37′ Gem.	*TBOO* 10 84
1580 Nov. 17	9h 40m	Ch. 8	6°46′10″ Gem.	Brahe; unknown
1582 Nov. 11	18h 45m	Ch. 51	26°35′30″ Can.	*TBOO* 10 174
1582 Nov. 23	all	Ch. 11	constant	*TBOO* 10 174
1582 Nov. 23	4h 0m	Ch. 53	26°38′30″ Can.	*TBOO* 10 175
1582 Dec. 26	8h 28m	Ch. 11	17°38′ Can.	*TBOO* 10 176; cf. 10 196
1582 Dec. 26	8h 30m	Ch. 53	17°40′30″ Can.	*TBOO* 10 176
1582 Dec. 27	19h 15m	Ch. 11	17°28′20″ Can.	*TBOO* 10 177
1582 Dec. 28	11h 30m	Ch. 10	16°47′ Can.	*TBOO* 10 176–7
1582 Dec. 28	11h 30m	Ch. 15	16°47′ Can.	*TBOO* 10 176–7
1582 Dec. 28	11h 30m	Ch. 19	16°47′ Can.	*TBOO* 10 176–7
1582 Dec. 28	12h 16m	Ch. 8	16°46′10″ Can.	Brahe; unknown: cf. *TBOO* 10 196–203
1582 Dec. 29	7h 47m	Ch. 11	29°38′30″ from southern foot of Erichthonius	*TBOO* 10 177
1582 Dec. 30	8h 8m	Ch. 11	29°13′30″ from southern foot of Erichthonius	*TBOO* 10 178
1582 Dec. 30	8h 10m	Ch. 53	16°0′30″ Can.	*TBOO* 10 178
1583 Jan. 16	7h 30m	Ch. 11	23°29′ from bright star in foot of Erichthonius	*TBOO* 10 242–3
1583 Jan. 16	10h 30m	Ch. 11	23°27′ from bright star in foot of Erichthonius	*TBOO* 10 243
1583 Jan. 17	10h 36m	Ch. 11	23°12′30″ from bright star in foot of Erichthonius	*TBOO* 10 244
1583 Jan. 17	5h 20m	Ch. 11	23°16′ from foot of Erichthonius	*TBOO* 10 243
1583 Jan. 17	15h 0m	Ch. 11	23°9′ from foot of Erichthonius	*TBOO* 10 244
1583 Jan. 18	5h 5m	Ch. 11	23°1′30″ from foot of Erichthonius	Unknown
1583 Jan. 18	7h 34m	Ch. 11	7°51′ from 7 Gem.	Not found
1583 Jan. 18	16h 52m	Ch. 11	7°59′ from 7 Gem.	*TBOO* 10 244
1583 Jan. 18	8h 52m	Ch. 11	44°22′ from Regulus	*TBOO* 10 244
1583 Jan. 18	16h 45m	Ch. 11	44°27′20″ from Regulus	*TBOO* 10 245
1583 Jan. 19	7h 3m	Ch. 11	44°32′30″ from Regulus	*TBOO* 10 245
1583 Jan. 26	6h 15m	Ch. 53	8°20′30″ Can.	*TBOO* 10 247 et alibi
1583 Apr. 22	9h 40m	Ch. 28	1°17′ Leo	*TBOO* 10 250
1583 Apr. 22	9h 45m	Ch. 13	1°17′ Leo	*TBOO* 10 250
1584 Nov. 12	13h 26m	Ch. 51	23°14′5″ Leo	*TBOO* 10 321
1584 Nov. 12	13h 30m	Ch. 13	23°14′ Leo	*TBOO* 10 321
1584 Dec. 21	14h 0m	Ch. 53	1°13′30″ Vir.	*TBOO* 10 332
1585 Jan. 24	9h 0m	Ch. 51	24°9′30″ Leo	*TBOO* 10 387

Continued on next page

Date	Time	Chapter	Position(s)	Source
1585 Jan. 25	9h 0m	Ch. 53	24°7′30″ Leo	*TBOO* 10 387
1585 Jan. 31	19h 35m	Ch. 8	21°10′26″ Leo	Brahe; unknown
1585 Jan. 31	12h 0m	Ch. 10	21°18′11″ Leo	*TBOO* 10 388
1585 Jan. 31	12h 0m	Ch. 15	21°18′11″ Leo	*TBOO* 10 388
1585 Jan. 31	12h 0m	Ch. 19	21°18′11″ Leo	*TBOO* 10 388
1585 Feb. 4	6h 40m	Ch. 53	19°47′30″ Leo	*TBOO* 10 389
1585 Feb. 17	10h 0m	Ch. 42	15°12′30″ Leo	*TBOO* 10 392
1585 Mar. 9	9h 10m	Ch. 28	11°49′6″ Leo	*TBOO* 10 395
1585 Mar. 11	5h 0m	Ch. 28	11°45′30″ Leo	*TBOO* 10 395
1585 Mar. 12	5h 0m	Ch. 28	11°45′45″ Leo	*TBOO* 10 396
1585 Mar. 12	10h 30m	Ch. 53	11°46′0″ Leo	*TBOO* 10 396
1585 Apr. 26	9h 42m	Ch. 13	21°26′ Leo	*TBOO* 10 399
1585 May 7	11h 26m	Ch. 27	25°55′ Leo	*TBOO* 10 401
1585 May 12	10h 8m	Ch. 27	28°3′30″ Leo	*TBOO* 10 401
1585 May 18	10h 30m	Ch. 22	0°50′45″ Vir.	*TBOO* 10 402
1586 Oct. 22	18h	Ch. 13	0°7′ Vir.	*TBOO* 11 64
1586 Nov. 2	16h 40m	Ch. 13	5°52′ Vir.	*TBOO* 11 64
1586 Dec. 1	19h 30m	Ch. 13	20°4′30″ Vir.	*TBOO* 11 66
1586 Dec. 15	18h 30m	Ch. 51	26°6′24″ Vir.	*TBOO* 11 66
1586 Dec. 26	16h 0m	Ch. 42	29°42′40″ Vir.	*TBOO* 11 66
1586 Dec. 31	19h 8m	Ch. 42	1°4′36″ Lib.	*TBOO* 11 177
1587 Jan. 25	17h 0m	Ch. 28	4°41′45″ Lib.	*TBOO* 11 180
1587 Jan. 25	17h 0m	Ch. 53	4°42′0″ Lib.	*TBOO* 11 180
1587 Jan. 27	17h 0m	Ch. 28	4°41′0″ Lib.	*TBOO* 11 180
1587 Mar. 4	13h 16m	Ch. 15	26°26′17″ Vir.	*TBOO* 11 184
1587 Mar. 4	13h 24m	Ch. 53	26°25′40″ Vir.	*TBOO* 11 182
1587 Mar. 7	17h 22m	Ch. 8	25°10′20″ Vir.	Brahe; unknown
1587 Mar. 7	19h 10m	Ch. 10	25°10′20″ Vir.	*TBOO* 11 184 (under 5 March)
1587 Mar. 10	11h 30m	Ch. 53	24°5′15″ Vir.	*TBOO* 11 186
1587 Mar. 27	9h 40m	Ch. 27	18°21′45″ Vir.	*TBOO* 11 187
1587 Apr. 1	9h 30m	Ch. 27	17°11′ Vir.	*TBOO* 11 188
1587 Apr. 21	9h 30m	Ch. 53	15°48′20″ Vir.	*TBOO* 11 188
1588 Nov. 5	18h 50m	Ch. 51	23°16′ Vir.	*TBOO* 11 275
1588 Nov. 9	18h 30m	Ch. 42	25°31′0″ Vir.	*TBOO* 11 275
1588 Nov. 10	18h 30m	Ch. 13	25°30′ Vir.	*TBOO* 11 275
1588 Dec. 4	18h 0m	Ch. 42	9°19′24″ Lib.	*TBOO* 11 276
1588 Dec. 4	18h 30m	Ch. 28	9°23′0″ Lib.	*TBOO* 11 276
1588 Dec. 5	18h	Ch. 13	9°19′24″ Lib.	*TBOO* 11 276
1588 Dec. 14	18h 12m	Ch. 28	14°35′40″ Lib.	*TBOO* 11 276
1589 Feb. 11	17h 13m	Ch. 27	8°48′ Sco.	*TBOO* 11 330
1589 Feb. 11	17h 13m	Ch. 51	8°48′ Sco.	*TBOO* 11 330
1589 Mar. 8	16h 24m	Ch. 53	12°16′50″ Sco.	*TBOO* 11 330
1589 Apr. 5	11h 33m	Ch. 51	7°31′10″ Sco.	*TBOO* 11 332
1589 Apr. 13	11h 15m	Ch. 53	4°43′20″ Sco.	*TBOO* 11 333

Continued on next page

Date	Time	Chapter	Position(s)	Source
1589 Apr. 15	night	Ch. 12	lat. 1°7′ N.	*TBOO* 11 333
1589 Apr. 15	12h 5m	Ch. 10	3°57′11″ Sco.	*TBOO* 11 333
1589 Apr. 15	12h 5m	Ch. 15	3°58′20″ Sco.	*TBOO* 11 333
1589 Apr. 15	12h 5m	Ch. 53	3°58′20″ Sco.	*TBOO* 11 333
1589 Apr. 15	13h 34m	Ch. 8	3°58′10″ Sco.	Brahe; unknown
1589 May 6		Ch. 61	6′40″ N. Lat.	*TBOO* 11 334
1589 May 6	11h 15m	Ch. 12	lat. 6′40″ N.	*TBOO* 11 334
1589 May 6	11h 20m	Ch. 51	27°7′20″ Lib.	*TBOO* 11 334
1589 May 6	11h 20m	Ch. 53	27°7′20″ Lib.	*TBOO* 11 334
1589 Sep. 15	7h 15m	Ch. 13	16°47′20″ Sag.	*TBOO* 11 335
1589 Nov. 1	6h 20m	Ch. 13	20°59′15″ Cap.	*TBOO* 11 335
1589 Dec. 3	5h 39m	Ch. 51	15°25′33″ Aqu.	*TBOO* 11 335
1590 Mar. 4	7h	Ch. 61	3′20″ S. Lat.	*TBOO* 12 44
1590 Mar. 4	7h 10m	Ch. 12	24°22′56″ Ari.	*TBOO* 12 44
1590 Mar. 4	7h 10m	Ch. 24	24°22′ Ari.	*TBOO* 12 44
1590 Mar. 4	7h 10m	Ch. 26	24°22′56″ Ari.	*TBOO* 12 44
1590 Oct. 30	18h 15m	Ch. 28	2°57′20″ Lib.	*TBOO* 12 44
1590 Dec. 27	19h 8m	Ch. 27	8°6′ Sco.	*TBOO* 12 45
1591 Jan. 4	18h 50m	Ch. 27	12°44′24″ Sco.	*TBOO* 12 138
1591 Jan. 22	19h	Ch. 22	22°33′ Sco.	*TBOO* 12 138
1591 Feb. 18	17h 30m	Ch. 51	7°34′20″ Sag.	*TBOO* 12 139
1591 May 13	13h 40m	Ch. 51	2°24′30″ Cap.	*TBOO* 12 140
1591 May 13	14h 0m	Ch. 53	2°20′0″ Cap.	*TBOO* 12 140
1591 Jun. 6	12h 20m	Ch. 10	27°15′ Sag.	*TBOO* 12 142
1591 Jun. 6	12h 20m	Ch. 15	27°15′ Sag.	*TBOO* 12 142
1591 Jun. 6	12h 20m	Ch. 53	27°15′0″ Sag.	*TBOO* 12 142
1591 Jun. 8	16h 25m	Ch. 8	26°32′0″ Sag.	Brahe; unknown
1591 Jun. 10	11h 50m	Ch. 53	25°2′36″ Sag.	*TBOO* 12 143
1591 Jun. 28	10h 24m	Ch. 53	21°10′0″ Sag.	*TBOO* 12 144
1591 Oct. 2		Ch. 13	lat. 2°38′30″ N.	*TBOO* 12 146
1591 Oct. 10		Ch. 13	lat. 2°18′40″ N.	*TBOO* 12 146
1591 Oct. 16	6h 28m	Ch. 51	1°27′18″ Aqu.	*TBOO* 12 146
1591 Oct. 16	7h 30m	Ch. 13	1°27′20″ Aqu.	*TBOO* 12 146
1592 Jan. 23	7h 15m	Ch. 24	11°34′30″ Ari.	*TBOO* 12 219
1592 Jan. 23	7h 20m	Ch. 26	11°32′44″ Ari.	*TBOO* 12 219
1592 Jan. 23	10h 15m	Ch. 12	11°34′30″ Ari.	*TBOO* 12 219
1592 Jan. 23	22h	Ch. 61	2′0″ S. Lat.	*TBOO* 12 219
1593 Jun. 29	13h 30m	Ch. 51	13°37′22″ Pis.	*TBOO* 12 282
1593 Jul. 21	14h 0m	Ch. 53	17°45′45″ Pis.	*TBOO* 12 283
1593 Jul. 22	2h 0m	Ch. 62	17°45′45″ Pis.	*TBOO* 12 283
1593 Jul. 30	13h 45m	Ch. 20	17°39′30″ Pis.	*TBOO* 12 284
1593 Aug. 11	1h 45m	Ch. 66	16°7′ Pis.	*TBOO* 12 285
1593 Aug. 22	12h 20m	Ch. 53	13°10′15″ Pis.	*TBOO* 12 287
1593 Aug. 23		Ch. 62	6°7$\frac{1}{2}$′ [S.] Lat.	*TBOO* 12 290

Continued on next page

Date	Time	Chapter	Position(s)	Source
1593 Aug. 24		Ch. 13	lat. 6°3′ N.	A. N. Ch. 8 obs. 7
1593 Aug. 24		Ch. 62	6°5½′ [S.] Lat.	*TBOO* 12 290
1593 Aug. 24	2h 13m	Ch. 8	12°43′45″ Pis.	Brahe; unknown
1593 Aug. 24	10h 30m	Ch. 10	12°38′ Pis.	*TBOO* 12 288
1593 Aug. 24	10h 30m	Ch. 15	12°38′ Pis.	*TBOO* 12 288
1593 Aug. 24	10h 30m	Ch. 19	12°38′ Pis.	*TBOO* 12 288
1593 Aug. 25	17h 27m	Ch. 62	12°16′ Pis.	*TBOO* 12 290
1593 Aug. 29		Ch. 62	5°52¼′ [S.] Lat.	*TBOO* 12 290
1593 Aug. 29	10h 20m	Ch. 53	11°14′0″ Pis.	*TBOO* 12 290
1593 Sep. 8	10h 38m	Ch. 51	8°53′51″ Pis.	*TBOO* 12 291
1593 Oct. 3	8h 0m	Ch. 53	7°50′10″ Pis.	*TBOO* 12 293
1593 Dec. 7	8h 0m	Ch. 26	3°6′50″ Ari.	*TBOO* 12 296
1593 Dec. 10	eve.	Ch. 12	at the node, lat 0°	*TBOO* 12 296
1593 Dec. 10	7h 20m	Ch. 24	4°45′ Ari.	*TBOO* 12 296
1593 Dec. 10	19h 0m	Ch. 61	4°44′0″ Tau.	*TBOO* 12 296
1594 Dec. 27	19h 15m	Ch. 51	8°43′18″ Sag.	*TBOO* 12 348
1595 Jan. 4	19h 10m	Ch. 12	13°36′40″ Sag.	*TBOO* 12 441
1595 Jan. 4		Ch. 51	lat. 9′20″	*TBOO* 12 441
1595 Jul. 21	14h 40m	Ch. 51	4°11′10″ Tau.	*TBOO* 12 441
1595 Sep. 17	16h 45m	Ch. 53	26°7′12″ Tau.	*TBOO* 12 444
1595 Oct. 25	8h 10m	Ch. 26	19°39′25″ Tau.	*TBOO* 12 447
1595 Oct. 27	12h 20m	Ch. 12	lat. 2′20″ N.	*TBOO* 12 450
1595 Oct. 27	12h 20m	Ch. 24	18°52′15″ Tau.	*TBOO* 12 450
1595 Oct. 27	12h 20m	Ch. 53	18°51′15″ Tau.	*TBOO* 12 450
1595 Oct. 28	eve.	Ch. 12	lat. 0′45″ S.	*TBOO* 12 451–2
1595 Oct. 28	23h 30m	Ch. 61	18°35′ Tau.	*TBOO* 12 452
1595 Oct. 29	21h 22m	Ch. 8	17°56′15″ Tau.	Brahe; unknown
1595 Oct. 30	8h 20m	Ch. 10	17°48′ Tau.	*TBOO* 12 453
1595 Oct. 30	8h 20m	Ch. 15	17°48′ Tau.	*TBOO* 12 453
1595 Nov. 3	12h 0m	Ch. 53	16°18′30″ Tau.	*TBOO* 12 455
1595 Nov. 3	23h 30m	Ch. 61	19′45″ N. Lat.	*TBOO* 12 455
1595 Dec. 17	7h 6m	Ch. 51	11°31′27″ Tau.	*TBOO* 12 458
1595 Dec. 18	8h 0m	Ch. 53	11°40′0″ Tau.	*TBOO* 12 460
1596 Mar. 9	7h 40m	Ch. 51	15°49′12″ Gem.	*TBOO* 13 47
1596 Mar. 9	8h	Ch. 13	15°49′ Gem.	*TBOO* 10 47
1597 Oct. 29	17h	Ch. 51	12°16′ Can.	Kepler's observation; cf. *TBOO* 13 11
1597 Dec. 10	8h 30m	Ch. 10	3°45′30″ Can.	*TBOO* 13 112
1597 Dec. 10	8h 30m	Ch. 15	3°45′30″ Can.	*TBOO* 13 112
1597 Dec. 13	13h 35m	Ch. 8	2°28′0″ Can.	Brahe; unknown
1600 Jan. 13	11h 50m	Ch. 10	10°48′36″ Leo	*TBOO* 13 221
1600 Jan. 13	11h 50m	Ch. 15	10°48′36″ Leo	*TBOO* 13 221
1600 Jan. 19	9h 40m	Ch. 8	8°18′0″ Leo	Brahe; unknown
1600 Mar. 5	12h	Ch. 20	29°12′30″ Can.	*TBOO* 13 224 (under March 15)

Continued on next page

Date	Time	Chapter	Position(s)	Source
1602 Feb. 15	17h 0m	Ch. 15	14°21′18″ Vir.	Fabricius's observation
1602 Feb. 18	10h 30m	Ch. 15	13°19′30″ Vir.	Kepler's observation
1602 Feb. 23	12h 0m	Ch. 15	11°19′20″ Vir.	Fabricius's observation
1604 Feb. 17	Corvus culminating	Ch. 11	26°56′ Lib.	Kepler's obs.
1604 Feb. 19	Mars rising, alt. 11°	Ch. 11	29°18′20″ from Arcturus	Kepler's obs.
1604 Feb. 22	ζ Leonis culminating	Ch. 11	29°15′ or 19′ from Arcturus	Kepler's obs.
1604 Feb. 29	half hour before Cor Hydrae culm.	Ch. 11	29°10′ from Arcturus	Kepler's obs.
1604 Feb. 29	half hour before Cor Hydrae culm.	Ch. 11	9°26′+ from Spica	Kepler's obs.
1604 Mar. 29	9h 43m	Ch. 15	18°21′30″ Lib.	Kepler's observation

Appendix B
On the Table of Oppositions
in Chapter 15

by Yaakov Zik

TABLE 1:
INITIAL POSITIONS OF MARS IN CHAPTER 15 COMPUTED WITH GUIDE 9
USING JPL DE430.

General notes:

1. The time is measured from midnight; Gregorian dates are obtained by adding 10 days to Kepler's dates. The dates are given in old style.

2. Geographical coordinates: Uraniborg 12 41 46 E, 55 54 30 N [50m]; Wandesburg (Wandsbek-Hamburg) 10 05 53 E, 53 35 10 N [15m]; Prague 14 26 E, 53 35 10 N [250m].

3. The longitudes are given in geocentric ecliptic coordinates.

Obs. no. / fn.	Date (old style)	Time (LMT)	Long. Mars	Lat. Mars	Long. Sun
I (fn. 3)	12 Nov. 1580	22:50	8° 37′ 37″ Gem.	1° 33′ N	0° 49′ 30″ Sag.
II (fn. 6)	28 Dec. 1582	23:30	16° 48′ 18″ Can.	4° 5′ 35″ N	17° 12′ 4″ Cap.
III (fn. 7)	31 Jan. 1585	23:59:59	21° 18′ 47″ Leo	4° 32′ 2″ N	22° 15′ 18″ Aqr.
IV (fn. 8)	5 Mar. 1587	01:16	26° 27′ 7″ Vir.	3° 39′ 40″ N	23° 51′ 58″ Pis.
V (fn. 13)	16 Apr. 1589	00:05	3° 59′ 49″ Sco.	1° 1′ 30″ N	5° 31′ 23″ Tau.
VI (fn. 16)	7 Jun. 1591	00:20	27° 13′ 41″ Sag.	3° 55′ 19″ S	24° 58′ 59″ Gem.
VII (fn. 18)	24 Aug. 1593	22:30	12° 37′ 48″ Psc.	6° 7′ 34″ S	11° 9′ 4″ Vir.
VIII (fn. 19)	30 Oct. 1595	20:20	17° 48′ 32″ Tau.	0° 4′ 1″ N	16° 55′ 55″ Sco.
IX (fn. 20)[1]	10 Dec. 1597	20:30	3° 44′ 20″ Can.	3° 23′ 28″ N	29° 5′ 24″ Sag.
X (fn. 21)[2]	13 Jan. 1600	23:40	10° 40′ 16″ Leo	4° 29′ 10″ N	3° 21′ 36″ Aqr.
XI (fn. 29)[3]	18 Feb. 1602	22:30	13° 19′ 41″ Vir.	4° 9′ 50″ N	10° 9′ 32″ Pis.
XII (fn. 37)[4]	29 Mar. 1604	21:43[5]	18° 22′ 34″ Lib.	2° 17′ 13″ N	19° 13′ 26″ Ari.

---

[1] measured at Wandesburg.

[2] Measured at Prague. The time is adjusted to Uraniborg time.

[3] Measured at Prague.

[4] Measured at Prague.

[5] Determined by the culmination of the back of Leo (Delta Leonis).

TABLE 2:
TABLE OF OPPOSITIONS OF MARS, 1580–1604
COMPUTED WITH GUIDE 9 USING JPL DE430.

Compare with the table at the end of Chapter 15 (reproduced below). Longitudes are given in geocentric ecliptic coordinates; dates are in old style.

Opposition	DD.MM.YY	HH:MM:SS (LMT)	Mars long. ° : ′ : ″	Mars lat. ° : ′ : ″
I	18.11.1580	11:37:54	06:29:46 Gem.	01:47:28
II	28.12.1582	16:09:56	16:55:44 Can.	04:05:17
III	31.01.1585	08:32:25	21:34:26 Leo.	04:32:10
IV	06.03.1587	23:34:20	25:42:18 Vir.	03:37:34
V	14.04.1589	21:25:31	04:24:18 Sco.	01:04:16
VI	08.06.1591	16:19:19	26:43:12 Sag.	– 04:01:12
VII	26.08.1593	06:08:44	12:15:54 Pis.	– 06:04:30
VIII	31.10.1595	11:27:47	17:34:30 Tau.	00:06:40
IX	14.12.1597	02:13:03	02:27:04 Can.	03:28:37
X	19.01.1600	04:00:19	08:36:50 Leo.	04:31:55
XI	22.02.1602	06:16:54	12:25:08 Vir.	04:08:20
XII	29.03.1604	07:57:00	18:35:38 Lib.	02:18:18

Kepler's table, for comparison

	Date, Old Style					Longitude				Latitude			Mean Long.			
	Year	D.	Month	H	M	D	M	S	Sign	D	M		S	D	M	S
I	1580	18	Nov.	1	31	6	28	35	Gemini	1	40	N.	1	25	49	31
II	1582	28	Dec.	3	58	16	55	30	Cancer	4	6	N.	3	9	24	55
III	1585	30	Jan.	19	14	21	36	10	Leo	4	$32\frac{1}{6}$	N.	4	20	8	19
IV	1587	6	Mar.	7	23	25	43	0	Virgo	3	41	N.	6	0	47	40
V	1589	14	Apr.	6	23	4	23	0	Scorpio	1	$12\frac{3}{4}$	N.	7	14	18	26
VI	1591	8	Jun.	7	43	26	43	0	Sagitt.	4	0	S.	9	5	43	55
VII	1593	25	Aug.	17	27	12	16	0	Pisces	6	2	S.	11	9	55	4
IIX	1595	31	Oct.	0	39	17	31	40	Taurus	0	8	N.	1	7	14	9
IX	1597	13	Dec.	15	44	2	28	0	Cancer	3	33	N.	2	23	11	56
X	1600	18	Jan.	14	2	8	38	0	Leo	4	$30\frac{5}{6}$	N.	4	4	35	50
XI	1602	20	Feb.	14	13	12	27	0	Virgo	4	10	N.	5	14	59	37
XII	1604	28	Mar.	16	23	18	37	10	Libra	2	26	N.	6	27	0	12

Kepler's Observation of Opposition XII (1604)

Kepler corrected the initial time for his computations of the opposition that occurred in 1604 on the basis of observations he made in Prague. Figure 1 below shows the sky map of Prague on the night between March 29 and 30, 1604 at $9^h 43^m$, old style. Mars was placed on the line from Arcturus to Spica. The star at the back of Leo, that is, Delta Leonis (Zosma), was at the meridian and its right ascension 163° 13'.[6] The Sun's ecliptic longitude at noon time was 18° 56' 24″ Aries, and its right ascension 17° 27' 55″.[7] Kepler's computation for the difference of ascensions is 163° 13' − 17° 27' 55″ = 145° 45' (should be 145° 51' 16.2″), which he resolved into a time difference of 9 hours 43 minutes.

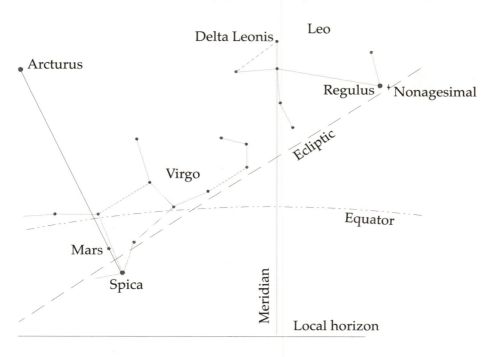

Figure 1

It is instructive to follow the method by which Kepler modified and extrapolated the observational data to define the positions of Mars and Sun at the time of oppositions. The rising point at which the ecliptic intersects the Eastern horizon was set at 232° 30' (Scorpio 22° 30'). The longitude of the nonagesimal (marked by + in Figure 1) was 142° 30',[8] the horizontal altitude 51°, and its distance from the zenith 39°.[9] Kepler assumed the distance of

---

[6]The transit time of Delta Leonis, according to modern computation was 21:42:06 and the right ascension 163° 14' 20.4″.

[7]Modern values are longitude 18° 49' 41″ Aries, and right ascensions 17° 21' 47.88″.

[8]The nonagesimal (medium coeli) is a virtual point at the ninetieth degree of the ecliptic reckoned from its point of ascension on the eastern horizon (i.e., the rising point). Located at the highest point of the part of the ecliptic which is above the horizon, the nonagesimal does not represent a point immediately overhead, but a point at which that meridian intersects with the ecliptic. The elevation of the nonagesimal above the horizon denotes its altitude, and its distance from the equinoctial point at Aries defines its longitude.

[9]According to modern computations the rising point was at 233° 39' 47″, the longitude of the nonagesimal 143° 39' 47″, the altitude 50° 55' 48″, and its distance from the zenith 39° 4' 12″.

Mars from Earth to be a little greater than half of the distance of the Sun from Earth. Kepler computed the diurnal latitudinal parallax of Mars to be 3′ 28″, from which he determined the latitude of Mars without parallax to be 2° 25′. Since Mars's distance from the nonagesimal was 56° and its ecliptic latitude 2° 25′, Kepler computed the value of the diurnal longitudinal parallax (i.e., the difference between Mars's distance from the nonagesimal and its ecliptic longitude) to be 3′ 32″.[10] Kepler's computations for the latitudinal and longitudinal parallax resulted in positions for Mars's ecliptic longitude of 198° 21′ 30″ (18° 21′ 30″ Libra), and its latitude 2° 21′ 30″. The longitude of the true Sun was at 19° 20′ 8′ Aries (Figure 2A). The deviation of elongation between Mars and the Sun was 58′ 38″. Kepler determined the diurnal motions of the Sun and Mars to be 58′ 38″ and 22′ 36″ respectively, and the sum of the diurnal motions was 1° 21′ 14″. With the sum of the diurnal motions of Mars and the Sun (1° 21′ 14″), Kepler computed the time it takes for Mars and the Sun to move 58′ 38″, that is, about 17 hours 20 minutes.

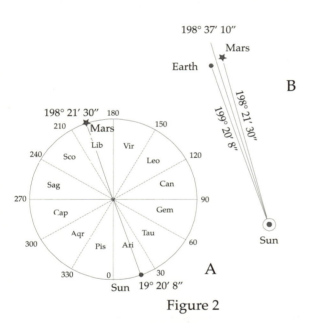

Figure 2

As shown in Figure 2B, the Sun is assumed to be at the center of the planetary system. The longitudes of Mars and Earth's in reference to the center of the Sun are 198° 21′ 30″ and 199° 20′ 8″ respectively. The deviation of elongation between Mars and the Earth remain 58′ 38″. To find the position of Mars at opposition, Kepler computed the angular distance that Mars and Earth—now substituting the place of the Sun—moved during 17 hours 20 minutes; Mars moved eastward about 16′ 20″ and the Sun westward about 42′ 18″. Accordingly, Kepler determined the longitude of Mars at opposition to be 198° 37′ 50″ from which he subtracted about 39″ in order to correct Mars's orbit; he got 198° 37′ 10″ (18° 37′ 10″ Libra).

The Sun moved westward and its longitude decreased from the time of observation to its position opposite to Mars. Therefore, the time of opposition is 17 hours 20 minutes before March 29, at 21:43, the time when the observation was made. Kepler determined the time of opposition on March 28, 4^h 23^m AM, old style.

---

[10]Kepler's observations yield topocentric horizontal coordinates delineated in azimuthal and altitudinal or zenith distance. The transformation of the measurements to geocentric, ecliptic longitude and altitude, or equatorial right ascension and declination coordinates is obtained by spherical trigonometry.

Appendix C
Overview of Keplerian constructions

Although there are numerous diagrams of different orbital constructions throughout *Astronomia Nova,* Kepler never shows us how these diagrams relate to each other, and in particular we never see how the epicyclic diagram (almost always shown in isolation) relates to the various eccentric circles, ovals, and ellipses. The adjacent diagram provides that overview.

The eccentric circle of the earlier chapters is CQ, which Kepler shows is equivalent to a concentric circle C_1N bearing an epicycle QU with center Z. In this construction the planet is at Q. Its distance from the sun A is marked off for comparison on AU, the extended diametral line of the epicycle, by the arc QK_c. Note that the angle UZQ remains equal to angle CBQ, the eccentric anomaly.

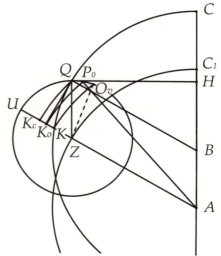

Diagram of Keplerian orbital constructions adapted from A. E. L. Davis, *A Mathematical Elucidation,* p. 205

In the oval hypothesis of Chapter 45, the planet moves *uniformly* around the epicycle as the epicycle is carried *nonuniformly* around the eccentric. Because the arc CQ is farther from the sun than other places on the eccentric, the epicycle moves more slowly there, and hence the angle CBQ is less than angle UZO_v, which is the angle traversed by the planet on the epicycle in its uniform motion. Thus the planet is at point O_v, and its path is an oval falling inside the eccentric. The planet's distance at this place is marked off as before by the arc O_vK.

Subsequent testing of this oval showed that it was too narrow and elongated: all the points O_v are too close to the sun, A. Kepler's solution involved having the planet move as if reciprocating in and out along the diameter of the epicycle as the point Q moves around the epicycle while QZ remains parallel to AC. To construct this, he drew a line QK_0 from Q perpendicular to AU, and then with center A drew an arc K_0P_0 intersecting QH, the perpendicular to AC, at P_0. This results in the point P_0 traversing an ellipse, as Kepler proves in Chapter 59. This ellipse provides both the correct distances AK_0 and the correct angular position CAP_0.

Bibliography

Primary Sources

Brahe, Tycho, *Tychonis Brahe Dani Opera Omnia,* ed. J. L. E. Dreyer, 15 vols. (Nielsen & Lydiche (Axel Simmelkiaer), 1913–1929). Brahe's observation books are transcribed in vols. 10–13. Cited as *TBOO* .

Kepler, Johannes, *Johannes Kepler Gesammelte Werke,* ed. Walther von Dyck, Max Caspar, Franz Hammer, Volker Bialas, et al., 22 vols in 25 (C. H. Beck, 1938–). *Astronomia Nova* is in vol. 3, *Astronomiae Pars Optica* in Vol. 2, and Kepler's working papers on Mars are in vol. 20.2. Cited as *KGW.*

Kepler, Johannes, *Ioannis Kepleri Astronomi Opera Omnia,* ed. Ch. Frisch, 8 vols. (Heyder & Zimmer, 1858–1871). Many of these still-useful volumes are available free online. *Astronomia Nova* is in Vol. 3.

Kepler, Johannes, *Optics,* trans. W. H. Donahue (Green Lion Press, 2000) Frequently referenced in *Astronomia Nova,* especially for its Parallactic Table.

Kepler, Johannes, *Astronomia Nova* (1609). Facsimile reprint (Culture et Civilisation, Brussels, 1968)

Biographies

Caspar, Max, *Kepler,* trans. C. Doris Hellman. Second edition (Dover, 1993). The standard biography, written by the editor of *Astronomia Nova.* The Dover edition has an introduction and annotations by Owen Gingerich—a very valuable addition, since Caspar did not see fit to reveal his sources.

Voelkel, James R., *Johannes Kepler and the New Astronomy* (Oxford, 1999). A fine short biography by the author of *The Composition of Kepler's Astronomia Nova* (below).

Thoren, Victor E., *The Lord of Uraniborg: A Biography of Tycho Brahe* (Cambridge University Press, 1990). The authoritative biography.

Christianson, John Robert, *On Tycho's Island* (Cambridge University Press, 2000). Much information on minor players in the Brahe-Kepler drama.

About *Astronomia Nova*

Davis, A. E. L. , *A Mathematical Elucidation of the Bases of Kepler's Laws,* Ph.D. Dissertation, Imperial College of Science and Technology, 1981. Available for purchase from ProQuest (ID: 8822352) or purchase or free download from Ethos (E-theses Online Service). A very detailed study of the geometry of the various orbits tried by Kepler.

Stephenson, Bruce, *Kepler's Physical Astronomy* (Princeton University Press, 1994) The best guide to *Astronomia Nova* and other works of Kepler.

Voelkel, James R., *The Composition of Kepler's Astronomia Nova* (Princeton University Press, 2001). An account of how *Astronomia Nova* was shaped by legal constraints and by the demands of persuasion.

OTHER WORKS CITED

Bibliographia Kepleriana, ed. Max Caspar and Martha List (C. H. Beck, 1968). Complete list of Kepler's known works.

Van Brummelen, Glen, *The Mathematics of the Heavens and the Earth: The Early History of Trigonometry* (Princeton University Press, 2009).

Index

518

About the Author

Johannes Kepler was born in 1571 in Weil-der-Stadt, a small town in southern Germany. He was educated in the duchy of Württemberg, first in the Latin school and then in the seminaries at Adelberg and Maulbronn. At the age of sixteen, he won a scholarship to the University of Tübingen, where, while preparing to become a Lutheran minister, he studied astronomy under Michael Maestlin, who introduced Kepler to the heliocentric system of Copernicus. Kepler became an ardent Copernican, although he had no intention of becoming an astronomer.

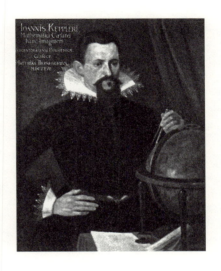

Upon graduation, he was asked by his patron, the Duke of Württemberg, to teach mathematics at a provincial school in Graz, Austria. Here he developed his unique cosmology involving nested Platonic solids that determined the planetary orbits. This system was published in 1596 as *Mysterium Cosmographicum*, a book which brought Kepler to the notice of the brilliant observational astronomer Tycho Brahe. Brahe invited Kepler to join his research team, which Kepler was eventually driven to do owing to Brahe's removal to Prague and anti-Lutheran purges in Graz.

In Prague, Kepler was assigned to work on Mars, an event in which Kepler saw the hand of Providence (see Chapter 7 of *Astronomia Nova*). Although relations with Brahe, and later, with Brahe's heirs, were difficult, Kepler was appointed Brahe's successor as Imperial Mathematician, a position he held for the rest of his life. While in Prague, he did much of his finest work, reforming the science of optics with his *Astronomiae Pars Optica* of 1604, and starting a revolution in astronomy with *Astronomia Nova* in 1609.

Over the subsequent decades, despite extremely difficult circumstances (for example, the events of the Thirty Years' War of 1618–1648), Kepler made numerous contributions to both astronomy and mathematics. His life's work culminated in the publication of the massive *Tabulae Rudolphinae* of 1627, the accuracy of which established his difficult theories as the standard of excellence. He died in Regensburg in 1630, while on a journey attempting to collect his salary from the imperial treasury.

About the Translator

William H. Donahue has been studying Kepler's works, especially *Astronomia Nova*, for fifty years. His interest in Kepler was first sparked as a student at St. John's College in Annapolis, where selections from Kepler's *Epitome of Copernican Astronomy* were part of the college's celebrated great books program. Hearing the claim that Kepler's greatest work was the hitherto untranslated *Astronomia Nova*, he wrote his senior essay on it, reading it in Latin. Later, while on the faculty of the Santa Fe campus of St. John's, he translated selections for an *Astronomia Nova* preceptorial which he led.

Leaving St. John's to help found a small innovative school, Donahue continued to work on the translation while teaching mathematics and science and developing the school's extensive outdoor program. In 1981, a grant from the National Science Foundation allowed him to take time off to complete the translation, which was eventually published by Cambridge University Press in 1992. That edition has long been out of print, and demand has been strong for a new edition, which is met at last by the present book.

Donahue received further support from the National Endowment for the Humanities for a guided study of *Astronomia Nova* and for the first English translation of Kepler's *Optics*. These books were later published by Green Lion Press, which Donahue and his wife Dana Densmore founded in 1995.

In 2005, Donahue returned to the faculty of St. John's College, Santa Fe, where he has served as Director of Laboratories, supervising the college's unique science curriculum based on study of original sources in science, from the ancient Greeks to modern quantum physics and ecology.